THE
HUMAN
BODY
BOOK

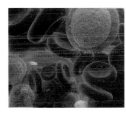

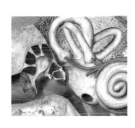

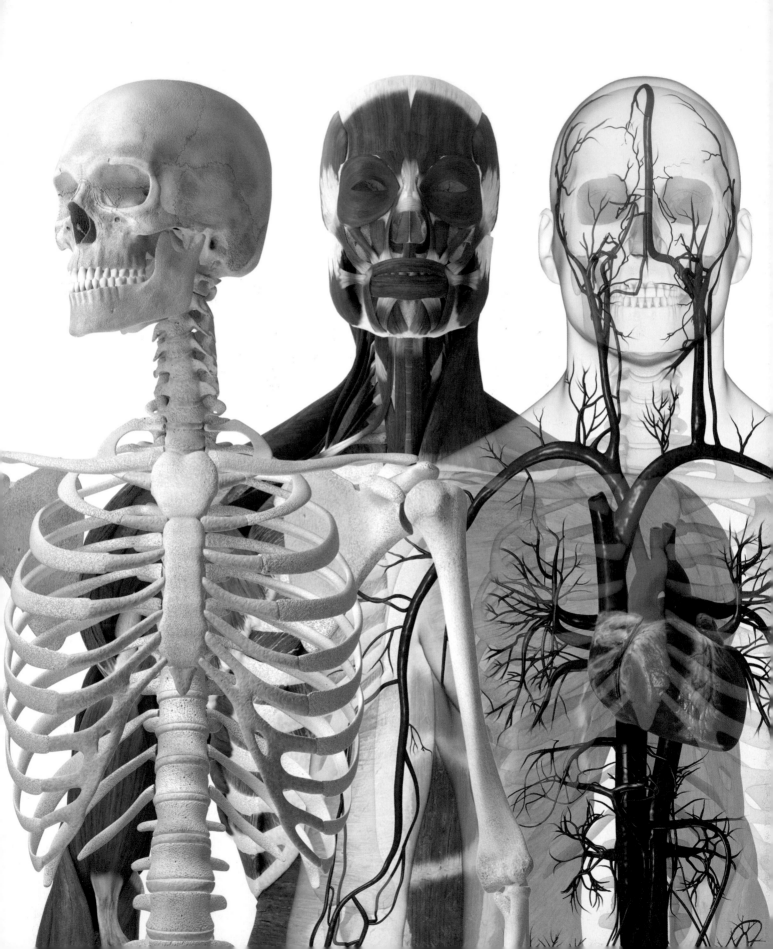

THE HUMAN BODY BOOK

STEVE PARKER

FOREWORD BY
ROBERT WINSTON

CONTENTS

DK | Penguin Random House

PROJECT EDITOR Rob Houston
PROJECT ART EDITOR Maxine Lea

EDITORS Ruth O'Rourke, Rebecca Warren,
Mary Allen, Sean O'Connor, Kim Bryan,
Tarda Davidson-Aitkins, Jane de Burgh,
Salima Hirani, Miezan van Zyl

EDITORIAL ASSISTANTS Tamlyn Calitz,
Manisha Thakkar

DESIGNERS Matt Schofield, Kenny Grant,
Francis Wong, Anna Plucinska

MANAGING EDITOR Sarah Larter
MANAGING ART EDITOR Phil Ormerod
PUBLISHING MANAGER Liz Wheeler
REFERENCE PUBLISHER Jonathan Metcalf
ART DIRECTOR Bryn Walls

PICTURE RESEARCHER Louise Thomas
JACKET DESIGNER Lee Ellwood
DTP DESIGNER Laragh Kedwell
PRODUCTION CONTROLLER Tony Phipps
INDEXER Hilary Bird
PROOFREADER Andrea Bagg

CONTRIBUTORS Mary Allen, Andrea Bagg,
Jill Hamilton, Katie John, Janet Fricker,
Jane de Burgh, Claire Cross

MEDICAL CONSULTANTS Dr. Sue Davidson,
Dr. Penny Preston, Dr. Ian Guinan, Dr. Aviva Schein

ILLUSTRATORS
CREATIVE DIRECTOR Rajeev Doshi
3-D ARTISTS Arran Lewis, Olaf Louwinger,
Gavin Whelan, Monica Taddei

Medi-Mation
Medical & Scientific Visualization

ADDITIONAL ILLUSTRATORS Peter Bull Art Studio,
Kevin Jones Associates, Adam Howard

REVISED EDITIONS

PROJECT EDITOR Martyn Page
PROJECT ART EDITOR Anna Hall
SENIOR EDITORS Peter Frances, Rob Houston

EDITORS Claire Gell, Priyanjali Narain
US SENIOR EDITOR Rebecca Warren
US EDITORS Megan Douglass, Jill Hamilton

ASSISTANT ART EDITOR Garima Agarwal
ART EDITOR Shailee Khurana

PICTURE RESEARCHER Akash Jain
PICTURE RESEARCH MANAGER Taiyaba Khatoon

JACKETS EDITOR Emma Dawson
JACKET DESIGNER Surabhi Wadhwa
JACKET MANAGER Sophia MTT

PRE-PRODUCTION PRODUCERS Andy Hilliard,
Adam Stoneham, Bimlesh Tiwary
SENIOR PRODUCERS Meskerem Berhane, Alice Sykes
PRE-PRODUCTION MANAGER Balwant Singh

MANAGING ART EDITORS Sudakshina Basu,
Michelle Baxter, Michael Duffy
MANAGING EDITORS Angeles Gavira Guerrero,
Rohan Sinha

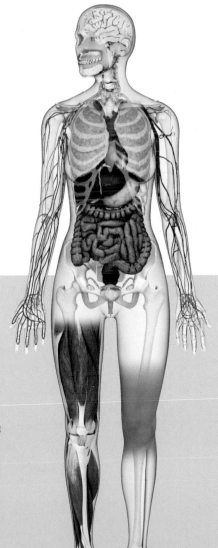

ART DIRECTORS Phil Ormerod, Karen Self
ASSOCIATE PUBLISHING DIRECTOR Liz Wheeler
PUBLISHING DIRECTOR Jonathan Metcalf

MEDICAL CONSULTANTS Dr. Kristina Routh,
Professor Susan Standring, Dr. Caroline Wigley

The Human Body Book provides information on a wide range
of medical topics, and every effort has been made to ensure
that the information in this book is accurate.
The book is not a substitute for medical advice, however,
and you are advised always to consult a doctor or other
health professional on personal health matters.

Some of the text in this book has been adapted from
The Human Body by Dr. Tony Smith,
first published in 1995.

This American edition, 2019
First American edition, 2007
Published in the United States by DK Publishing
345 Hudson Street, New York, New York 10014

A catalog record for this book
is available from the Library of Congress

ISBN 978-1-4654-8029-3

DK books are available at special discounts when purchased
in bulk for sales promotions, premiums, fund-raising, or
educational use. For details, contact:
DK Publishing Special Markets, 345 Hudson Street, New
York, New York 10014 or SpecialSales@dk.com

Color reproduction G.R.B. Editrice s.r.l.
in London, UK

Printed and bound in China

A WORLD OF IDEAS:
SEE ALL THERE IS TO KNOW

www.dk.com

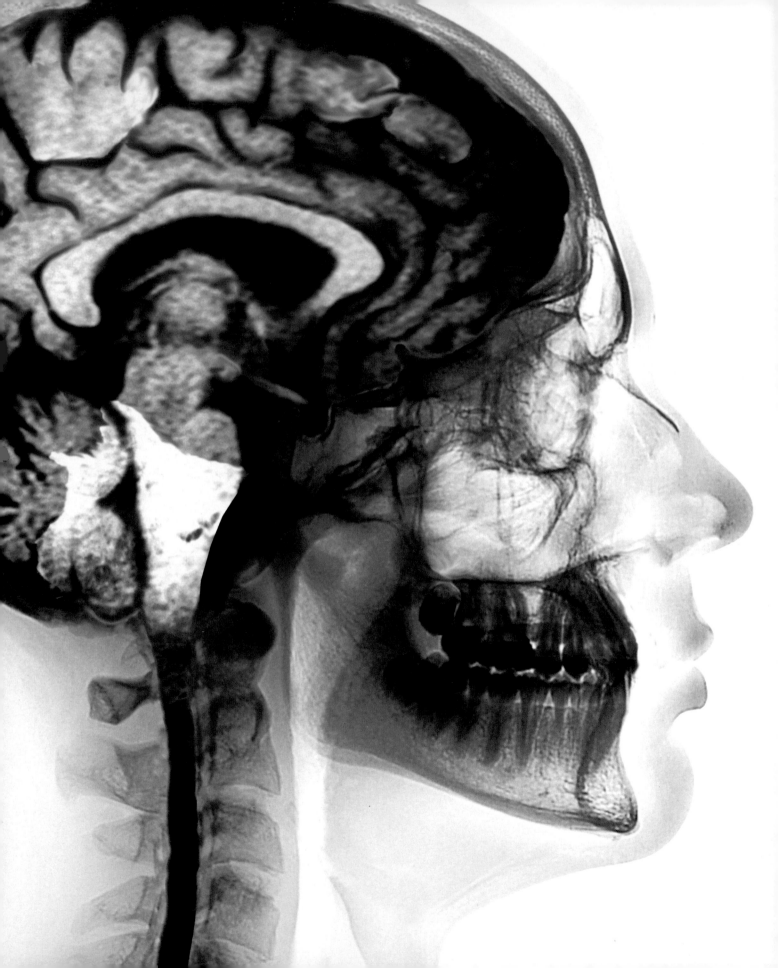

FOREWORD

This amazing book shows the detailed structure inside the human body as never seen before. It is only possible to produce these pictures because of major advances in technology. Although we have used dissection for several hundred years, new techniques help us reveal what lies under our skin in meticulous detail. Being able to see ourselves in this way was first made possible by computed tomography (CT), which is a series of X-rays that slice through the body, photographing it in sections. These images can then be combined using advanced computing, making it possible to construct accurate, elegant, three-dimensional images. More recently, tomographic techniques have been used with magnetic resonance scanning, which carries no risks. If your body is placed inside a massive magnet strong enough to rip the wristwatch from your arm, all the molecules in the tissues are harmlessly lined up like a needle in a compass. When radio waves are then directed at the magnetized tissues, different tissue structures vibrate in different ways. These vibrations can be detected and, after computation, it is again possible to produce a three-dimensional image. Consequently, we can now produce images of human anatomy with great accuracy. Of course, some pictures in this book are drawings of what is seen down a microscope. The combination of microscopic anatomy and three-dimensional images is highly instructive, and this book allows people to more than glimpse at the unique wonders inside the body. It will appeal not only to adults and young people, interested in how their body works, but also to medical experts, for whom it has professional relevance. How much more thrilling learning anatomy would have been when I was a medical student 40 years ago if we had been able to see beautiful, accurate images like these.

PROFESSOR ROBERT WINSTON

THE HUMAN BODY IS THE MOST DEEPLY STUDIED AND
FREQUENTLY PORTRAYED OBJECT IN HISTORY. DESPITE
ITS FAMILIARITY, IT IS INSTINCTIVELY ABSORBING AND
ETERNALLY FASCINATING. THE PAGES OF THIS BOOK
REVEAL, IN AMAZING VISUAL DETAIL, AND IN BOTH HEALTH
AND SICKNESS, THE INTRICATE INNERMOST WORKINGS OF
THE BODY'S CELLS, TISSUES, ORGANS, AND SYSTEMS. BUT
FIRST, MUCH OF THE FASCINATION LIES IN THE WAY THESE
PARTS INTERACT AND INTEGRATE AS EACH ONE RELIES ON
THE OTHERS TO FUNCTION AND SURVIVE.

INTEGRATED
BODY

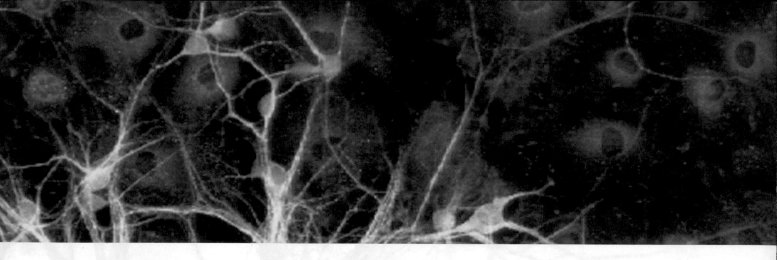

INTRODUCTION

The number of humans in the world has raced past seven billion (7,000,000,000). More than 250 babies are born every minute, while 150,000 people die daily, with the population increasing by almost 150 humans per minute. Each of these people lives, thinks, worries, and daydreams with, and within, that most complex and marvelous of possessions—a human body. An enduring feature of this body and its behavior is self-curiosity. We continually look inside ourselves, in enormous and ever-increasing detail, to comprehend the action within. This book aims to quench aspects of our curious nature by revealing every aspect of the human body.

LEVELS OF ORGANIZATION

To understand the inner structure and workings of the human body, this book takes the "living machine" approach, borrowed from sciences such as engineering. This views the body as a series of integrated systems. Each system carries out one major role or task. In the cardiovascular system, for example, the heart pumps blood through vessels to supply every body part with essential oxygen and nutrients. The systems are, in turn, composed of main parts known as organs. The stomach, intestines, and liver are organs of the digestive system. Moving through further levels in the hierarchy, the organs consist of tissues, and tissues are made up of cells.

Cells are often called the microscopic building blocks of the body. However, they are far from passive bricks in a wall—they are active and dynamic, they continually grow and specialize, function,

die, and replenish themselves, by the millions every second. The whole body contains about 100 million million cells, of at least 200 different kinds. Science is increasingly able to delve even deeper than cells, to the organelles within them, and onward and inward, to the ultimate components of ordinary matter—molecules and atoms.

ANATOMY

The study of the body's structure, and how its cells, tissues, and organs are assembled, is known as human anatomy. Its elements are often shown in isolation, using techniques such as cutaways, cross sections, and "exploded" views, which provide clarity and understanding. But in reality, the inside of the body is a crowded place. Tissues and organs push and press against one another. There is no free space, and no stillness either. Body parts shift continually in relation to each other, as we move about, breathe, pump blood, shift digestive matter, and eat. For example, swallowed food does not simply fall down inside the esophagus. The esophagus is normally pressed flat by internal chest pressure so that food must be forced down into the stomach by waves of muscular contraction.

PHYSIOLOGY

The anatomical drawing of a large factory or office would show the arrangement of rooms, location of machinery and furniture, and cables and pipes for electricity and water. It is a static snapshot of structure and layout. For a rounded understanding, we need to

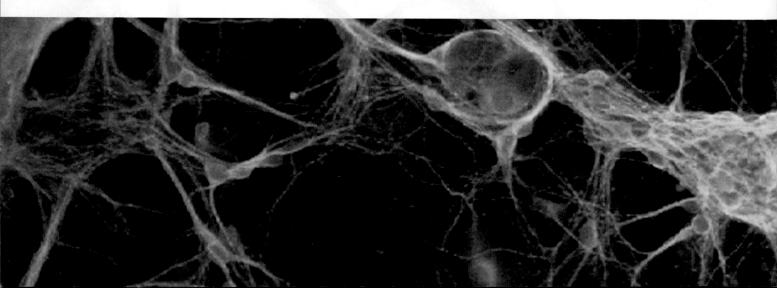

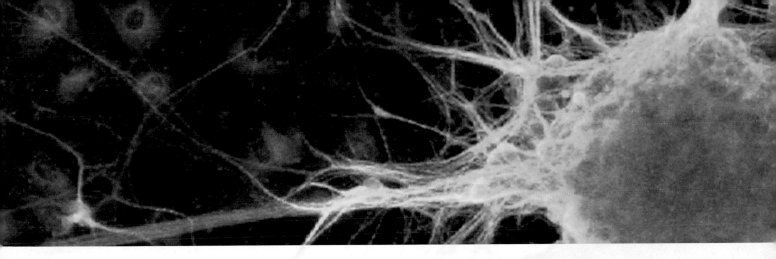

see the premises in action, with people, goods, and information on the move. Similarly, human anatomy is combined with its twin, physiology, which is the study of the body's workings and how it functions. Physiology focuses on the dynamic chemical minutiae at atomic, ionic, and molecular level. It investigates the workings of such processes as enzyme action, hormone stimulation, DNA synthesis, and how the body stores and uses energy from food. As researchers stare harder and closer, more biochemical pathways are unraveled, and more physiological secrets are unlocked. Much of this work is directed at preventing, treating, or alleviating disease, and allows us to appreciate the latest wonder treatment, or take a medication that makes us feel better.

HEALTH AND ILLNESS

Medical science amasses mountains of evidence every year for the best ways to stay healthy and avoid disease. At present, an individual's genetic inheritance, which is a matter of chance, is the given starting point for maintaining health and well-being. In coming years, treatments such as preimplantation genetic diagnosis (PGN), which is carried out as part of assisted reproductive techniques such as in-vitro fertilization (IVF), and gene therapy are able to remove or negate some of these chance elements. Many aspects of upbringing have a major impact on health. Factors such as diet—whether too rich, bringing with it the risk of obesity, or too poor, leading to malnutrition—particularly affect children, whose bodies are still developing. The body can be affected by many different types of disorder, such as infection by a virus or bacteria, injuries resulting from an accident or long-term repetitive activities, inherited faulty genes, or exposure to toxins in the environment.

ABOUT THIS BOOK

The pages that follow describe the structures and processes of the human body at all levels. First, the hierarchy of organization is described, from molecules such as DNA, to organelles and cells, to tissues and organs. Then the approach is basically functional, focusing on each major system in turn. Every section opens with an overview of its system, and subsequently explores its organs and tissues, to examine how they work and what they do.

At the end of each section common ailments relating to the system are explored. A variety of problems is discussed, including those caused by genetic variation, aging, infection, and injury.

The running order of the sections that follow moves from support and movement (bones and muscles), through control and coordination (nerves and hormones), to basic life support, protection, and nourishment (heart, lungs, skin, immunity, digestion, and waste disposal), and reproduction. The final section examines development, ageing, and inheritance.

COMMUNICATION NETWORK
This microscopic image of nerve cells (neurons) shows the thin strands (axons and dendrites) that connect the cell bodies. Neurons transmit electrical signals around the body, especially the brain and spinal cord; each one connects with hundreds of others to form a dense network.

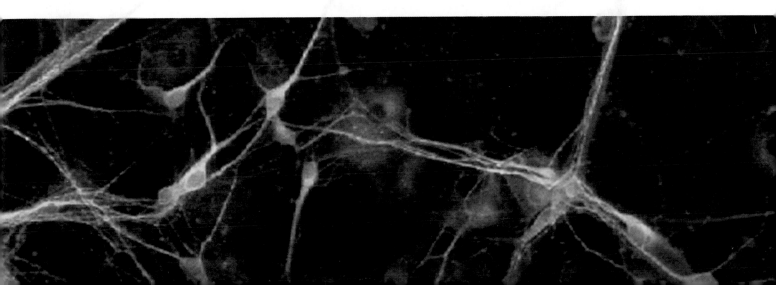

IMAGING THE BODY

IMAGING IS VITAL TO DIAGNOSE ILLNESS, UNRAVEL DISEASE PROCESSES, AND EVALUATE TREATMENTS. MODERN TECHNIQUES PROVIDE DETAILED INFORMATION WITH MINIMUM DISCOMFORT TO THE PATIENT AND HAVE LARGELY REPLACED SURGERY IN ESTABLISHING THE PRESENCE AND EXTENT OF DISEASE. MICROSCOPY HAS ALSO HELPED ADVANCE BIOLOGICAL RESEARCH.

The invention of the X-ray made the development of noninvasive medicine possible. Without the ability to see inside the body, many internal disorders could be found only after major surgery. Computerized imaging now helps doctors make early diagnoses, which in many cases greatly increase the likelihood of recovery. Computers process and enhance raw data to aid our visual ability, for example by coding and reinterpreting subtle shades of gray from an X-ray or scan into distinguishable colors. While enhanced images are valuable, sometimes direct observation is essential. Viewing techniques have also become less invasive with the development of instruments such as the endoscope (see opposite). This book makes extensive use of imagery from real bodies, as well as artistically rendered illustrations.

MICROSCOPY

Light microscopy (LM) uses magnifying lenses to focus light rays. In light microscopy, light passes through a thin section of material and enlarges it up to 2,000 times. Higher magnifications are achieved with beams of subatomic particles called electrons. In scanning electron microscopy (SEM) the beam runs across a specimen coated with gold film. Electrons bounce off the surface contours, creating a three-dimensional image.

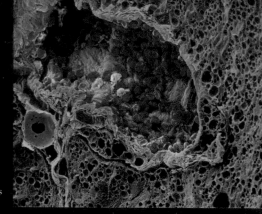

SEM OF TUMOR BLOOD SUPPLY
This freeze-fracture image, in which the specimen is frozen and then cracked open before being scanned, shows a blood vessel with blood cells growing into a melanoma (skin tumor).

TEM OF MITOCHONDRION
In transmission electron microscopy (TEM), enlargements of several million times are possible. This colored image shows a mitochondrion within a cell, magnified about 12,000 times.

LM OF TONGUE PAPILLAE
This light photomicrograph (LM) shows the tiny pimples, or papillae, on the tongue. Specimens for LM are usually stained with chemicals to color structures, such as cell nuclei.

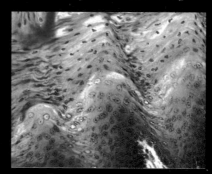

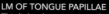

ANGIOGRAM
In this image, a contrast medium, which is here colored red, has been injected into the arteries of the shoulder, neck, and lower head. Bones show up white. This type of X-ray image is called an angiogram.

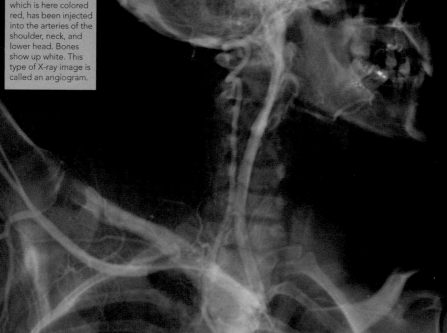

X-RAY

Like light rays, X-rays are electromagnetic energy, but of very short wavelength. When passed through the body to strike photographic film, they create shadow images (radiographs). Dense structures such as bone absorb more X-rays and show up white, while soft tissues, such as muscle, appear as shades of gray. To view hollow or fluid-filled structures clearly, these must first be filled with a substance that absorbs X-rays (a contrast medium). Fluoroscopy uses X-rays to gain real-time moving images of body parts, for instance to investigate swallowing.

X-RAY OF THE BREAST
A plain X-ray of the female breast (mammogram) is used as a routine screening test for breast cancer, which may show up as an unusually white area. This mammogram shows a healthy breast.

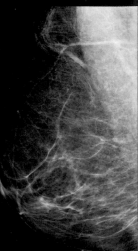

MRI AND CT SCANNING

Computerized tomography (CT) and magnetic resonance imaging (MRI) reveal detail about many tissue types. CT scans use weak levels of X-rays to produce an image. In CT, an X-ray scanner rotates around the patient as a computer records the levels of electromagnetic energy absorbed by tissues of different densities. A cross section is built from many layers of X-ray scans. In MRI, a person lies in a magentic chamber, which causes hydrogen atoms in the body to align. A pulse of radio waves is released, throwing the atoms out of alignment. As they realign, they emit radio signals, which are used to create an image.

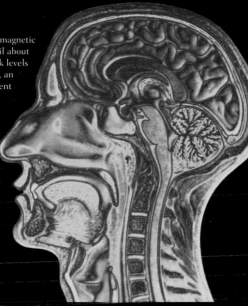

MRI SCAN OF HEAD
A colored MRI scan of the mid-line of the head in side view; visible structures including the brain and spinal cord, the nasal cavity, and the tongue.

CT SCAN OF THE HEART
A 3-D CT scan of the heart from the right side; showing the large aorta (main artery, center top) and some of the blood vessels of the lungs.

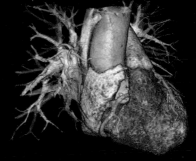

CT SCAN OF LUNGS
In a horizontal slice through the chest, the spongy tissues and airways of the healthy lungs (oranges and yellows) show up quite distinctly from their denser surroundings. The heart and major blood vessels between the lungs are blue, and the vertebrae (backbones), ribs, and sternum (breastbone) are dark blue.

NUCLEAR MEDICINE IMAGING

In nuclear medicine imaging, a radioactive substance (radionuclide) is injected into the body and is absorbed by the area to be imaged. As the substance decays it emits gamma rays, which a computer forms into an image. This type of imaging can help to diagnose many disorders including cancers, heart disease, digestive diseases, and neurological problems. Examples of nuclear medicine imaging scans are positron emission tomography (PET) and single-photon emission computed tomography (SPECT). These provide data about the function of a tissue rather than detailed anatomy. The PET or SPECT scan may be fused with a CT scan to give information about structure as well as function.

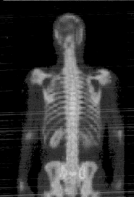

BONE SCAN
In this scan the radionuclide has concentrated in the spine, ribs, and pelvis, indicated by light blue, yellow, and orange areas. This method can reveal increased cell activity that could indicate cancer.

PET SCAN
These side views of the brain reveal its activity. The upper image was taken as the subject listened to spoken words; the auditory cortex is highlighted. The lower image shows the subject both listening to and repeating the words; a motor area of the brain becomes active to control the muscles of speech.

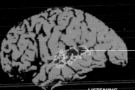

Auditory cortex Hearing region

LISTENING

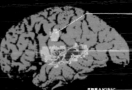

Motor control region

Auditory cortex Hearing region

SPEAKING

ULTRASOUND

Sound waves of very high frequency (too high-pitched for us to hear) are emitted by a device called a transducer as it is passed over the body part being examined. The sound waves echo back to the transducer according to the density of the tissues they encounter. A computer analyzes the reflections and creates an image. Ultrasound is used to monitor fetal development in the uterus. This technique is regarded as extremely safe because no radiation is used.

FETAL ULTRASOUND
A fetus of about six months, surrounded by amniotic fluid, is clearly visible in this image.

A modified form, echocardiography, shows the heart beating in real time.

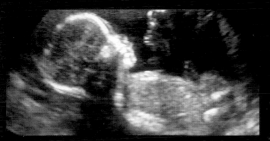

ENDOSCOPY

A variety of telescope-like endoscopes are inserted through natural orifices or incisions to produce images of the body's interior. Some types are rigid but most are flexible, utilizing fiberoptic technology, and can be bent and controlled as they are guided along. They carry their own light source and may be equipped with tubes to introduce or remove fluids or gases, blades for surgery, forceps to take samples (biopsy), and perhaps a laser to cauterize damaged tissue. Endoscopes have been developed to fit different body parts: a bronchosope for the airways, a gastroscope for the esophagus and stomach, a laparoscope for the abdomen, and a proctoscope for the lower bowel.

TRACHEA
A bronchoscope image of the interior trachea (windpipe) shows the hoops of cartilage that keep it from collapsing.

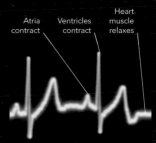

ELECTRICAL ACTIVITY

Sensor pads applied to the skin detect electrical signals coming from active muscles and nerves. The signals are coordinated, amplified, and displayed as a real-time trace, usually a spiky or wavy line. This technique includes electrocardiography (ECG) of the heart (see below) and electro-encephalography (EEG) of the brain's nerve activity.

Atria contract   Ventricles contract   Heart muscle relaxes

IMAGING THE HEAD AND NECK

TECHNIQUES FOR VISUALIZING THE HEAD AND NECK RANGE FROM ENDOSCOPY OF THE INSIDE OF HOLLOW STRUCTURES, SUCH AS THE LARYNX, TO COMPLEX, COMPUTER-AIDED TECHNIQUES FOR IMAGING STRUCTURES DEEP WITHIN THE BRAIN.

The head and neck region contains the brain, protected by the skull; the spinal cord and vertebrae; the eyes and ears; the nasopharynx (nasal cavity and throat) and larynx, which form the upper part of the respiratory system; and the teeth, tongue, and upper esophagus, at the start of the digestive system. Some of these structures can be viewed directly; for example, the larynx and nasopharynx can be viewed with an endoscope. For deeper or more detailed views, ordinary X-rays can produce images of the skull and vertebrae but do not generally produce clear images of soft tissues. For such tissues, CT and MRI scans give more detailed images. Techniques such as functional MRI (fMRI) and radionuclide scanning can also be used to reveal how tissues are functioning.

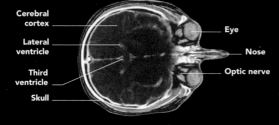

Cerebral cortex
Lateral ventricle
Third ventricle
Skull
Eye
Nose
Optic nerve

1 OBLIQUE TRANSVERSE MRI SCAN THROUGH THE UPPER HEAD

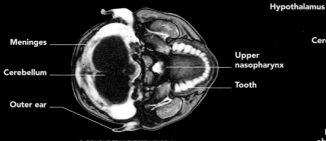

Meninges
Cerebellum
Outer ear
Upper nasopharynx
Tooth

2 OBLIQUE TRANSVERSE MRI SCAN THROUGH THE MIDDLE OF THE HEAD

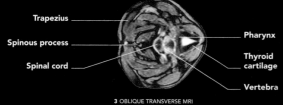

Trapezius
Spinous process
Spinal cord
Pharynx
Thyroid cartilage
Vertebra

3 OBLIQUE TRANSVERSE MRI SCAN THROUGH THE NECK

1, 2, 3 TRANSVERSE MRI SCANS THROUGH THE HEAD AND NECK
These transverse views show key structures at different levels: (1) the cortex and ventricles of the brain, and the eyeballs; (2) the cerebellum, upper nasopharynx, and teeth; and (3) the pharynx, spinal cord, and vertebra.

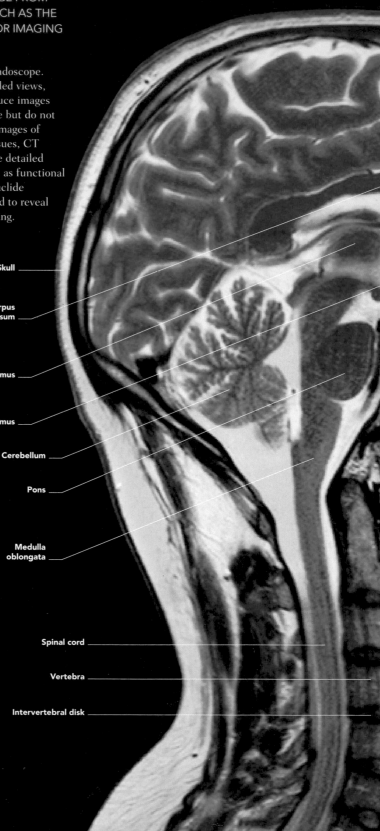

Skull
Corpus callosum
Thalamus
Hypothalamus
Cerebellum
Pons
Medulla oblongata
Spinal cord
Vertebra
Intervertebral disk

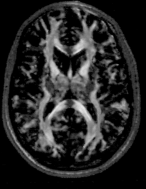

LEVELS OF SCANS

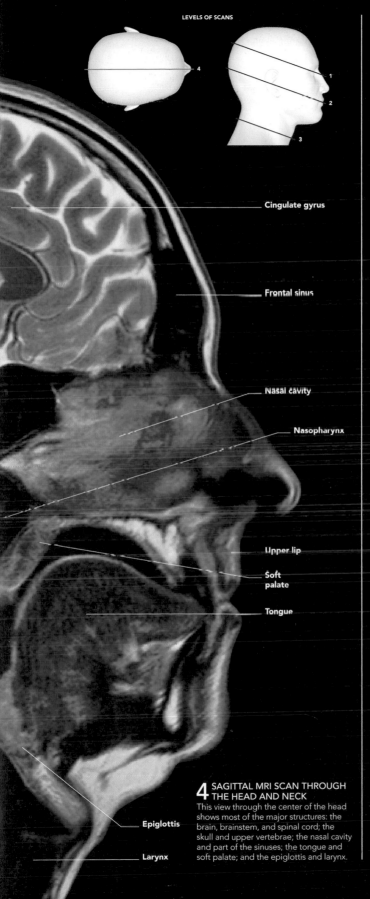

Cingulate gyrus

Frontal sinus

Nasal cavity

Nasopharynx

Upper lip

Soft palate

Tongue

Epiglottis

Larynx

4 SAGITTAL MRI SCAN THROUGH THE HEAD AND NECK

This view through the center of the head shows most of the major structures: the brain, brainstem, and spinal cord; the skull and upper vertebrae; the nasal cavity and part of the sinuses; the tongue and soft palate; and the epiglottis and larynx.

DTI SCAN OF THE BRAIN

Diffusion tensor imaging (DTI) is a form of MRI that can show details of tissue architecture. In this image, nerve fibers running from the front to the back of the brain are colored green, those running left to right are red, and those running top to bottom are violet.

NERVOUS SYSTEM

CT and MRI scanning enable the brain, brainstem, and spinal cord to be shown in detail. Most of these images show a 2-D "slice" through the tissues, but images may also be combined by computer to create a 3-D model of the brain. CT and MRI scans are commonly used in the diagnosis of disorders such as tumours or bleeding within the skull. Functional MRI (fMRI) can show blood flow through the brain, which indicates levels of neural activity in different parts of the brain. Radionuclide scans, such as PET and SPECT, can show levels of metabolic activity within brain tissues— for example, uptake of oxygen and glucose—that can reveal overactive areas, possibly indicating a tumor, or underactive regions, which may be a sign of Alzheimer's disease.

CARDIOVASCULAR SYSTEM

The carotid arteries, jugular veins, and other blood vessels in the head and neck can be revealed in detail by angiography. In contrast angiography, a chemical opaque to X-rays (the contrast medium) is injected into the blood vessels so that they will show up clearly on X-rays or on CT scans. Such scans can reveal blocked or narrowed areas, or abnormalities such as aneurysms (bulging, weak areas of artery wall). A noninvasive technique called Doppler ultrasound can show blood flow through the carotid arteries. All of these imaging techniques can help in assessing the risk of serious conditions such as stroke.

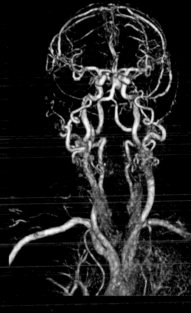

ANGIOGRAM OF THE HEAD AND NECK

This image shows the carotid arteries in the neck and the arteries in the brain (viewed from the front). A contrast medium has been used to highlight the arteries. Multiple 2-D "slices" were taken and combined to form this 3-D image.

RESPIRATORY SYSTEM

The upper respiratory system—from the nostrils through the nasal cavity and pharynx and down to the larynx—can be viewed directly by endoscopy. Direct views enable physicians to see the inside of the nasal cavity and structures such as the tonsils, adenoid, and vocal cords. They also make it possible to identify blocked or bleeding areas, and abnormalities such as polyps in the nose or nodules on the vocal cords. To view the nasal cavity and pharynx, a flexible endoscope may be passed into one nostril and then down into the throat. If only the pharynx or larynx needs to be viewed, the endoscope can be passed down through the mouth.

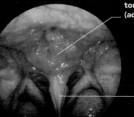

Pharyngeal tonsil (adenoid)

Nasal septum

ENDOSCOPE VIEW OF THE NASAL CAVITY

This view shows structures in the back of the nasal cavity, from where inhaled air passes into the nasopharynx, the uppermost part of the pharynx.

IMAGING THE THORAX

THE INTERNAL STRUCTURES OF THE THORAX ARE MOST COMMONLY IMAGED USING CONVENTIONAL X-RAYS AND CT OR MRI SCANS. OTHER TECHNIQUES MAY BE USED FOR SPECIFIC PURPOSES, SUCH AS CORONARY ANGIOGRAPHY TO IMAGE THE HEART'S BLOOD SUPPLY.

More commonly known as the chest, the thorax extends from the base of the neck to the diaphragm. It is protected by the rib cage, sternum (breastbone), and vertebrae of the thoracic spine. Most of the thoracic cavity is occupied by the lungs, heart, and major blood vessels, notably the aorta, venae cavae, and pulmonary arteries and veins. The esophagus, trachea, and bronchi are the other main structures in the thorax. Hollow structures, such as the trachea, bronchi, and esophagus, can be viewed by endoscopy. Other structures can be imaged by conventional X-rays, CT scanning, and MRI. Angiography can reveal blood vessels, and a type of ultrasound scanning known as echocardiography may be used to image the heart.

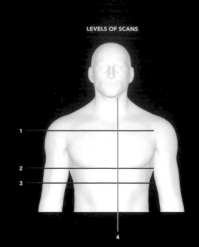

LEVELS OF SCANS

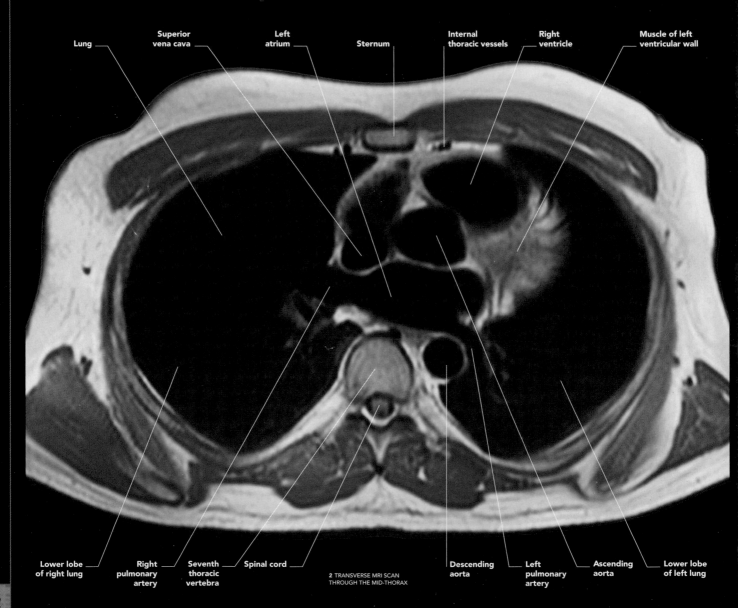

Lung

Superior vena cava

Left atrium

Sternum

Internal thoracic vessels

Right ventricle

Muscle of left ventricular wall

Lower lobe of right lung

Right pulmonary artery

Seventh thoracic vertebra

Spinal cord

2 TRANSVERSE MRI SCAN THROUGH THE MID-THORAX

Descending aorta

Left pulmonary artery

Ascending aorta

Lower lobe of left lung

SKELETAL SYSTEM

The ribs, spine, and sternum (breastbone), and associated bones such as the scapulae (shoulder blades) and clavicles (collarbones), can easily be seen by chest X-ray. This procedure is often used to assess injuries such as fractures, diseases such as bone cancer or osteoporosis, or abnormalities such as scoliosis. Finer details and more subtle signs of trauma (such as injury to soft tissue) can often be revealed with CT scans and MRI scans. These imaging procedures can be especially valuable in cases of injuries such as fractures to the ribs or vertebrae, where there is a risk that the fractured bones might have caused damage to the lungs, heart, or spinal cord.

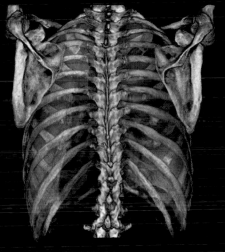

3-D CT SCAN OF THE THORAX
This colored scan shows the thorax from the rear. The spine is in the center; the ribs can be seen radiating from the thoracic spine; and the scapulae (shoulder blades) are visible at the top on each side. The blue area shows the position of the lungs.

RESPIRATORY SYSTEM

The lower part of the respiratory system, comprising the trachea, bronchi, bronchioles, and the lungs, can be imaged in various ways. Conventional X-rays are commonly used; the air-filled lungs appear dark, and the airways show as white. X-rays can show injuries such as a collapsed lung or signs of airway disease such as tuberculosis. A contrast X-ray technique called bronchography can be used to highlight the airways, although this technique has mostly been superseded by CT scanning and MRI, which can be used to give detailed 2-D or 3-D images of airway structures, lung tissue, and abnormalities such as tumors.

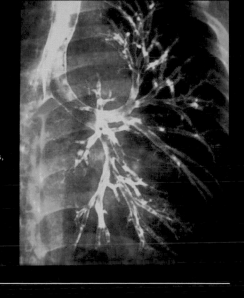

BRONCHOGRAM
This colored X-ray shows the trachea, and the left main bronchus and some of the bronchioles in a healthy left lung. A contrast medium has been used to coat the insides of the airways so that they show up more clearly.

CARDIOVASCULAR SYSTEM

Imaging can be used to assess the structure and function of the heart and main blood vessels. CT and MRI can show the chambers and valves inside the heart, and can reveal problems such as leaking heart valves. Angiography can be used to visualize the coronary and pulmonary arteries, allowing detection of any narrowed or blocked areas. Various nuclear medicine scanning techniques, including PET and SPECT (see p.13), may be used to assess blood flow through the heart's chambers and also within the

heart muscle. The pumping action of the heart and blood flow through the coronary vessels can also be viewed in real time by echocardiography, a form of ultrasound imaging.

CORONARY ANGIOGRAM
In a coronary angiogram, a contrast medium is passed into the arteries supplying the heart, and then an X-ray is taken. This image shows healthy coronary arteries; the arteries appear white as they encircle the heart.

1, 2, 3 TRANSVERSE MRI SCANS THROUGH THE THORAX
These transverse views through the thorax show: (1) the apex (top) of each lung, the first thoracic vertebra, and the spinal cord; (2) sections through the center of each lung and the chambers of the heart; and (3) the lower lobe of the left lung, the aorta, and the right lobe of the liver.

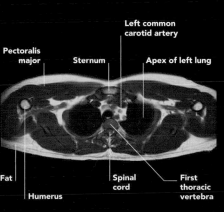

Pectoralis major · Sternum · Left common carotid artery · Apex of left lung

Fat · Humerus · Spinal cord · First thoracic vertebra

1 TRANSVERSE MRI SCAN THROUGH THE UPPER THORAX

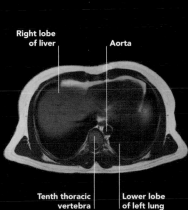

Right lobe of liver · Aorta

Tenth thoracic vertebra · Lower lobe of left lung

3 TRANSVERSE MRI SCAN THROUGH THE LOWER THORAX

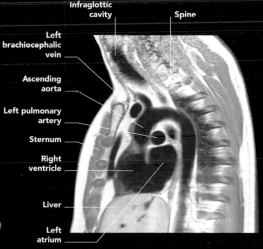

Infraglottic cavity · Spine · Left brachiocephalic vein · Ascending aorta · Left pulmonary artery · Sternum · Right ventricle · Liver · Left atrium

4 SAGITTAL MRI SCAN THROUGH THE THORAX
This scan shows two of the heart's chambers (right ventricle and left atrium); major blood vessels, such as the the aorta; the spine; and sternum. The liver (in the abdomen) can also be seen.

IMAGING THE ABDOMEN AND PELVIS

THE LOWER TORSO COMPRISES THE LUMBAR SPINE AND THE ABDOMINAL AND PELVIC CAVITIES, WHICH HOUSE THE URINARY AND REPRODUCTIVE SYSTEMS AND THE LOWER DIGESTIVE SYSTEM. THESE STRUCTURES CAN BE VIEWED DIRECTLY BY ENDOSCOPY, OR NONINVASIVELY BY VARIOUS IMAGING TECHNIQUES.

The abdominal cavity lies between the diaphragm and the iliac crest of the pelvic bones. The pelvic cavity is immediately underneath. It is basin-shaped and is enclosed by the two pelvic (innominate) bones at the front and sides, and the sacrum at the back. The abdominal and pelvic cavities contain the stomach, liver, kidneys, spleen, pancreas, small and large intestines, and internal reproductive organs, as well as the major blood vessels and nerves that supply the lower body. A layered membrane called the peritoneum, and visceral fat deep inside the abdomen, help protect these internal structures. Some of these structures can be viewed by endoscopy; they can also be imaged with plain and contrast X-rays, ultrasound scans, CT scans, and MRI scans.

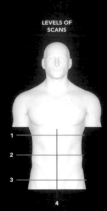

LEVELS OF SCANS

1
2
3
4

5

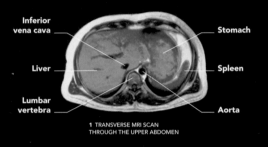

Inferior vena cava
Stomach
Liver
Spleen
Lumbar vertebra
Aorta

1 TRANSVERSE MRI SCAN THROUGH THE UPPER ABDOMEN

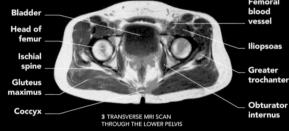

Bladder
Femoral blood vessel
Head of femur
Iliopsoas
Ischial spine
Greater trochanter
Gluteus maximus
Coccyx
Obturator internus

3 TRANSVERSE MRI SCAN THROUGH THE LOWER PELVIS

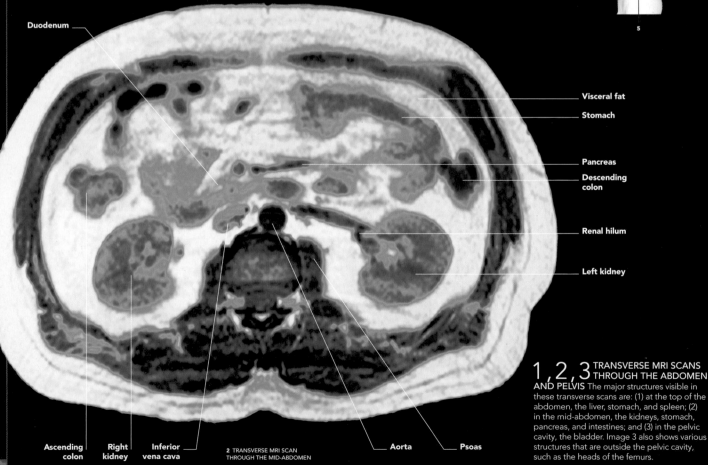

Duodenum

Visceral fat
Stomach
Pancreas
Descending colon
Renal hilum
Left kidney

Ascending colon
Right kidney
Inferior vena cava
Aorta
Psoas

2 TRANSVERSE MRI SCAN THROUGH THE MID-ABDOMEN

1,2,3 TRANSVERSE MRI SCANS THROUGH THE ABDOMEN AND PELVIS The major structures visible in these transverse scans are: (1) at the top of the abdomen, the liver, stomach, and spleen; (2) in the mid-abdomen, the kidneys, stomach, pancreas, and intestines; and (3) in the pelvic cavity, the bladder. Image 3 also shows various structures that are outside the pelvic cavity, such as the heads of the femurs.

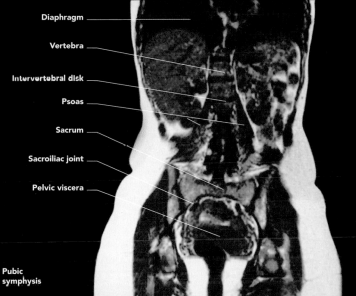

Diaphragm

Vertebra

Intervertebral disk

Psoas

Sacrum

Sacroiliac joint

Pelvic viscera

Pubic symphysis

5 CORONAL MRI SCAN THROUGH THE ABDOMEN AND PELVIS

Intervertebral disk

Lumbar vertebra

Sacrum

4,5 LONGITUDINAL MRI SCANS THROUGH THE ABDOMEN AND PELVIS

These scans of a female from the side (4) and the front (5) show how the soft tissues fit within the skeletal framework of the abdomen and pelvis. The sagittal view (4) reveals how shallow the abdominal cavity is in front of the lumbar spine.

4 SAGITTAL MRI SCAN THROUGH THE ABDOMEN AND PELVIS

URINARY SYSTEM

The urinary system—the kidneys, ureters, bladder, and urethra—can be visualized using intravenous urography (IVU), in which a contrast medium is injected into the bloodstream, travels to the kidneys, and enters the urine. A series of X-rays is taken as the contrast medium passes down the ureters and into the bladder; the structures containing the contrast medium show up as bright areas. This procedure reveals any blockages or abnormal areas. Another contrast X-ray procedure, cystography, involves introducing contrast medium into the bladder via the urethra; images may be taken as urine is passed (voiding cystourethrography). Ultrasound scans can be used to investigate the kidneys, bladder, and, in men, the prostate gland. Plain X-rays, CT scans, and MRI can also be used to show details of the urinary tract.

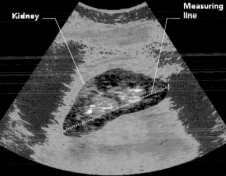

Kidney

Measuring line

ULTRASOUND SCAN OF A KIDNEY

This colored ultrasound scan shows a healthy adult kidney (the dark blue area in the center). The dotted line is being used to measure the kidney, which is about 4 in (10 cm) along its long axis—a normal size for an adult kidney.

DIGESTIVE SYSTEM

Abdominal ultrasound, CT scanning, or MRI may be used to image any of the digestive organs. The pancreas, bile ducts, and gallbladder may be examined using a technique called endoscopic retrograde cholangio-pancreatography (ERCP), which combines endoscopy and contrast X-ray imaging. The stomach, large intestines, and parts of the small intestine can be viewed directly with a flexible endoscope. They can also be examined using a barium contrast X-ray, in which the contrast medium is either swallowed or introduced via the rectum and X-rays are then taken.

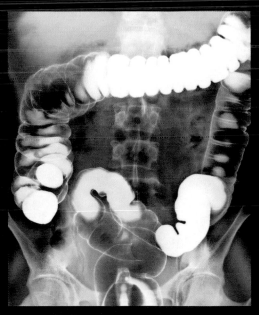

BARIUM X-RAY OF THE COLON

This colored contrast X-ray shows a healthy colon and rectum. A barium contrast medium, which appears white in this image, was used to make the colon and rectum stand out clearly.

REPRODUCTIVE SYSTEM

In males, ultrasound scans may be used to examine the testes, scrotum, penis, and some internal structures, and to investigate problems such as testicular tumors and hydrocele (collection of fluid in the scrotum). In females, the uterus, ovaries, and fallopian tubes are also commonly viewed by ultrasound scanning, including in pregnancy. In hysteroscopy, the inside of the uterus is viewed by passing an endoscope up through the cervix. The female reproductive organs can also be examined through an endoscope inserted through the abdominal wall (laparoscopy). CT scanning and MRI may be used to image the reproductive system in both sexes.

Uterus

Fallopian tube

Ovary

ENDOSCOPIC VIEW OF THE FEMALE REPRODUCTIVE SYSTEM

This endoscopic view shows the main female reproductive structures: the uterus (at the top), the two fallopian tubes, and the two ovaries.

IMAGING THE ARMS AND LEGS

BONES, JOINTS, AND MUSCLES OCCUPY MOST OF
THE ARMS AND LEGS. CONSEQUENTLY PLAIN X-RAYS,
CT SCANNING, AND MRI, WHICH SHOW THESE PARTS
OF THE BODY WELL, ARE THE IMAGING TECHNIQUES
MOST COMMONLY USED TO VISUALIZE THE LIMBS.

The arms and legs join the trunk at
the shoulder and hip joints, which are
supported by the bony structures of
the shoulder girdle (comprising the
clavicle and scapula) and pelvic girdle
(comprising the sacrum, coccyx, and
hip bones). Each arm has 30 bones,
as does each leg. As well as the bones,
the limbs contain a large number of
muscles, ligaments, and tendons,
which make possible the complex
movements the limbs can perform,
together with nerves and blood
vessels. X-rays, and CT and MRI
scans, as well as radionuclide
scanning (see p.13), can be used to
investigate injuries and abnormalities,
such as tumors, and endoscopy may
sometimes be used to visualize the
inside of joints. Angiography and
ultrasound may be used to image
blood vessels—for example, to
detect blocked or damaged areas.

LEVELS OF SCANS

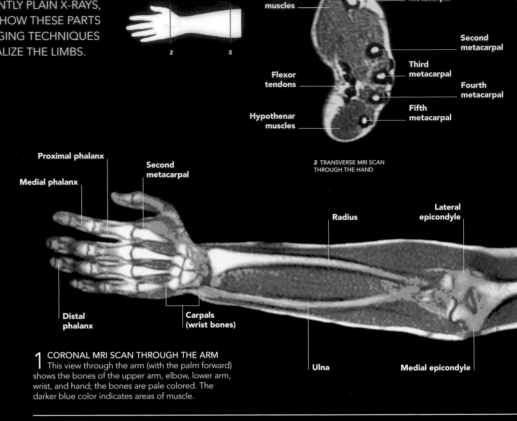

2 TRANSVERSE MRI SCAN THROUGH THE HAND

Thenar muscles / First metacarpal / Second metacarpal / Third metacarpal / Fourth metacarpal / Fifth metacarpal / Flexor tendons / Hypothenar muscles

Proximal phalanx / Medial phalanx / Second metacarpal / Radius / Lateral epicondyle / Distal phalanx / Carpals (wrist bones) / Ulna / Medial epicondyle

1 CORONAL MRI SCAN THROUGH THE ARM
This view through the arm (with the palm forward)
shows the bones of the upper arm, elbow, lower arm,
wrist, and hand; the bones are pale colored. The
darker blue color indicates areas of muscle.

LEVELS OF SCANS

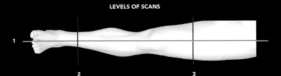

1 SAGITTAL MRI SCAN THROUGH THE LEG
This image shows some of the major bones and muscles in the
leg, ankle, and foot, including a toe bone (phalanx); one of the three
cuneiforms; the navicular and talus (ankle bones); the calcaneus
(heel bone); and the gastrocnemius and quadriceps muscles.

2,3 TRANSVERSE MRI SCANS THROUGH THE LEG
The scan through the lower leg (2)
shows the tibia and fibula, as well
as muscles of the shin and calf. The
image at mid-thigh level (3) shows
the femur and surrounding muscles.

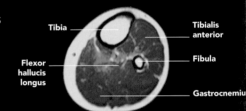

Tibia / Tibialis anterior / Flexor hallucis longus / Fibula / Gastrocnemius

2 TRANSVERSE MRI SCAN THROUGH THE LOWER LEG

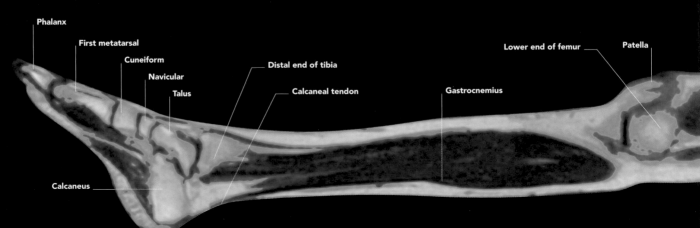

Phalanx / First metatarsal / Cuneiform / Navicular / Talus / Distal end of tibia / Calcaneal tendon / Lower end of femur / Patella / Gastrocnemius / Calcaneus

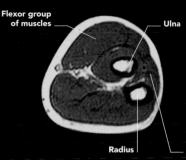

Flexor group of muscles
Ulna
Radius
Extensor group of muscles

3 TRANSVERSE MRI SCAN THROUGH THE LOWER ARM

2, 3 TRANSVERSE MRI SCANS THROUGH THE ARM AND HAND

The section through the palm of the hand (2) shows the small muscles that act on the thumb (thenar muscles) and the flexor tendons that lie in the carpal tunnel at the front of the wrist, together with the metacarpal bones. The scan through the forearm (3) shows the radius and ulna bones and flexor and extensor muscles.

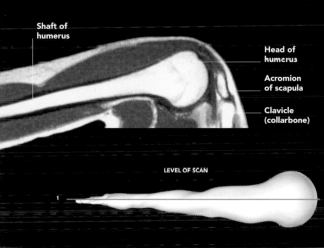

Shaft of humerus
Head of humerus
Acromion of scapula
Clavicle (collarbone)

LEVEL OF SCAN

1

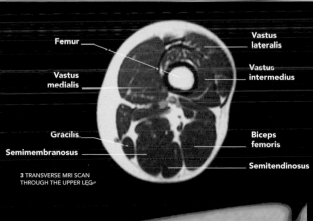

Femur
Vastus lateralis
Vastus intermedius
Vastus medialis
Gracilis
Semimembranosus
Biceps femoris
Semitendinosus

3 TRANSVERSE MRI SCAN THROUGH THE UPPER LEG

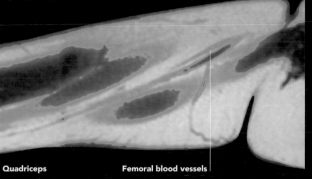

Quadriceps
Femoral blood vessels

SKELETAL SYSTEM

Plain X-rays are probably the most commonly used technique for examining the bones and joints, to assess injuries such as fractures and damage from diseases such as osteoporosis. CT scanning and MRI may be used to view the bones and the interior of joints, and other tissues such as the muscles, tendons, and ligaments, in greater detail. The inside of joints may also be viewed directly by arthroscopy. The cell activity in bones and bone marrow can be assessed by means of radionuclide scanning techniques such as PET (see p.13). In such scans, areas with overactive cells, such as tumors, show up as "hot spots" that are lighter than the surrounding bone.

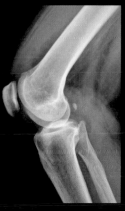

X-RAY OF THE KNEE
This side view of a bended knee clearly shows the femur (above), patella, tibia, and fibula. Plain X-rays are still widely used to visualize bones.

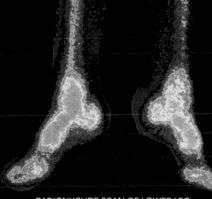

RADIONUCLIDE SCAN OF LOWER LEG
In this scan of the lower legs and feet, the bone (blue) has absorbed the radionuclide more than the other tissues. This method of scanning can reveal increased cell activity that could indicate cancer.

CARDIOVASCULAR SYSTEM

The blood vessels of the arms and legs can be viewed by angiography (sometimes defined as extremity angiography), in which a radiopaque contrast medium is injected into the bloodstream so that the vessels will show up on X-rays or on CT scans. This technique is useful for detecting problems such as blood clots in the leg veins (deep vein thrombosis) or a buildup of fatty deposits in the arteries (atherosclerosis), as well as signs of inflammation or injury. CT or MRI scanning without using a contrast medium may also sometimes be used to show the blood vessels. Blood flow through the vessels can be assessed by means of Doppler ultrasound, which can reveal obstructed or abnormal blood flow in real time, and enable investigation of conditions such as varicose veins in the legs or Raynaud's phenomenon in the fingers.

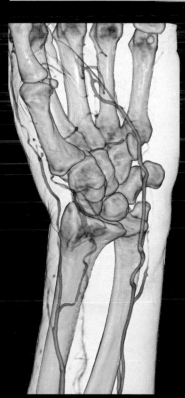

ANGIOGRAM OF THE WRIST
In this colored image, a contrast medium has been used to highlight the major arteries of the lower forearm, wrist, and palm, and a CT scan has then been taken.

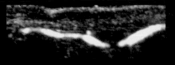

DOPPLER ULTRASOUND SCAN OF BLOOD FLOW THROUGH A FINGER
Doppler ultrasound can detect the movement of fluids. This image shows blood flow (in orange) through a healthy finger.

SKELETAL

EXPLORED ON PAGES 48–69

The skeleton is a solid, movable framework that supports the body. Its bones work as levers and anchor plates to allow for movement. Bones also work for other body systems—blood cells develop in the fatty inner tissue of bones (red marrow), for example. The body draws from mineral stores in bones during times of shortage, such as when calcium is needed for healthy nerve function.

COMPONENTS

- Skull, spine, ribs, and breastbone (axial skeleton)
- Limb bones, shoulders, and hips (appendicular skeleton)
- Cartilage and ligaments

MUSCULAR

EXPLORED ON PAGES 70–81

Muscles work with the skeleton, providing the pulling force for movement, from powerful to intricate. Involuntary muscles work largely automatically to control internal processes, such as blood distribution and digestion. Muscles rely on nerves to control them and blood to supply them with oxygen and energy.

COMPONENTS

- Skeletal muscles (attached to bones)
- Smooth muscle within organs
- Tendons
- Cardiac muscle of heart

NERVOUS

EXPLORED ON PAGES 82–119

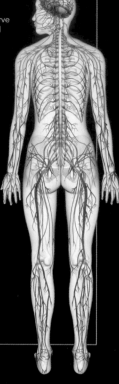

The brain is both the seat of consciousness and creativity and, through the spinal cord and nerve branches, it controls all body movements with its motor output. The brain also receives sensory information from outside the body and within. Yet much of the brain's second-by-second activity is carried out unconsciously as it works with endocrine glands to monitor and maintain other body systems.

COMPONENTS

- Brain
- Spinal cord
- Peripheral nerves
- Sense organs

ENDOCRINE

EXPLORED ON PAGES 120–29

The glands and cells of the endocrine system produce chemical messengers called hormones, which circulate in blood and other fluids. In response to physiological feedback, they maintain an optimal internal environment. Hormones also govern long-term processes such as growth, the changes that take place during puberty, and reproductive activity. The endocrine system is linked closely to the nervous system via the brain, allowing dual monitoring and control of all other systems.

COMPONENTS

- Pituitary gland
- Hypothalamus
- Thyroid gland
- Thymus gland
- Heart
- Stomach
- Pancreas
- Intestines
- Adrenal glands
- Ovaries (in female)
- Testes (in male)

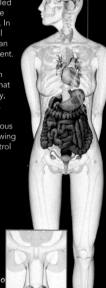

MALE

CARDIOVASCULAR

EXPLORED ON PAGES 130–45

The most basic function of the cardiovascular, or circulatory, system is to pump blood around the body. It supplies all organs and tissues with freshly oxygenated, nutrient-rich blood. Any waste products are removed with the blood as it leaves. The circulatory system also transports other vital substances, such as nutrients, hormones, and immune cells.

COMPONENTS

- Heart
- Blood
- Major vessels (arteries and veins)
- Minor vessels (arterioles and venules)
- Microscopic vessels (capillaries)

RESPIRATORY

EXPLORED ON PAGES 146–61

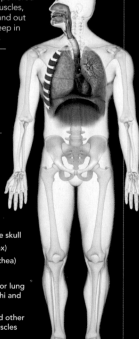

The respiratory tract and its movements, powered by breathing muscles, carries air into and out of the lungs. Deep in the lungs gases are exchanged—vital oxygen is absorbed from the air and carbon dioxide is passed into it before the air is carried back out of the body. A secondary function of this system is vocalization.

COMPONENTS

- Nasal and other air passages in the skull
- Throat (pharynx)
- Windpipe (trachea)
- Lungs
- Major and minor lung airways (bronchi and bronchioles)
- Diaphragm and other respiratory muscles

BODY SYSTEMS

THE HUMAN BODY'S SYSTEMS WORK TOGETHER AS A TRUE COOPERATIVE—EACH ONE FULFILLS ITS OWN VITAL FUNCTION, BUT ALL WORK TOGETHER TO MAINTAIN HEALTH AND EFFICIENCY.

Just like every other living thing, the prime biological aim of the human body is to replace itself with viable offspring. However, it is far from being simply a gene-carrier, a reproductive system with extra supporting parts "added on." In fact, and somewhat ironically, the reproductive system is the only one that is not required for basic survival. The exact number and extent of the body's systems is debated—the muscles, bones, and joints are sometimes combined as the musculoskeletal system, for instance. Although these systems can be described as separate entities, each depends on all others for physical and physiological support. Most systems have some "general" body tissues such as the connective tissues that delineate, support, and cushion many organs.

SKIN, HAIR, AND NAILS

EXPLORED ON PAGES 162–71

The skin, hair, and nails form the body's outer protective covering and are together termed the integumentary system. They repel physical damage and hazards such as microorganisms and radiation. The skin also regulates body temperature by sweating when too hot. The layer of subcutaneous fat under the skin acts as an insulator, an energy store, and a physical shock absorber.

COMPONENTS

- Skin
- Hair
- Nails
- Subcutaneous fat layer

LYMPHATIC AND IMMUNE

EXPLORED ON PAGES 172–87

The immune system's intricate interrelationships of physical, cellular, and chemical defenses provide vital resistance to many threats, including infectious diseases and malfunctions of internal processes. The slowly circulating lymph fluid helps distribute nutrients and collect waste. It also delivers immunity-providing white blood cells when needed.

COMPONENTS

- White blood cells (such as lymphocytes)
- Antibodies
- Spleen
- Tonsils and adenoids
- Thymus gland
- Lymph fluid
- Lymph vessels, nodes ("glands"), and ducts

DIGESTIVE

EXPLORED ON PAGES 188–209

The digestive tract's 30 ft (9 m) or so of tubing, which varies in size between the mouth and the anus, has a complex range of functions. It chops and chews food, stores and then digests it. eliminates waste products, and passes the nutrients to the major gland, the liver, which makes optimal use of the various digestive products. Healthy digestion depends on the proper functioning of the immune and nervous systems, and psychological state also greatly affects digestion.

COMPONENTS

- Mouth and throat (pharynx)
- Esophagus
- Stomach
- Pancreas
- Liver
- Gallbladder
- Small intestine (duodenum, jejunum, and ileum)
- Large intestine (colon, appendix, and rectum)
- Anus

URINARY

EXPLORED ON PAGES 210–19

The formation of urine by the kidneys eliminates wastes and excess substances from the blood, helping maintain the body's correct balance of water, fluids, salts, and minerals. Urine production is controlled by several hormones and is influenced by blood flow and pressure, the quantities of incoming water and nutrients, fluid loss (through sweating and bleeding, for instance), external conditions (especially temperature), and regular bodily cycles (such as sleeping and waking).

COMPONENTS

- Kidneys
- Ureters
- Bladder
- Urethra

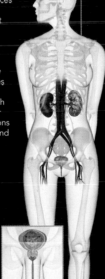

MALE

REPRODUCTIVE

EXPLORED ON PAGES 220–45

Unlike any other system, the reproductive system differs dramatically between female and male, it functions for only part of the human life span, and it can be surgically removed without threatening life. The production of sperm in the male is continual while the female production of ripe eggs is cyclical. In the male, both sperm and urine use the urethra as an exit tube but at different times.

COMPONENTS

Female:
- Ovaries, fallopian tubes, and uterus
- Vagina and external genitalia
- Breasts

Male:
- Testes, spermatic ducts, seminal vesicles, urethra, and penis
- Prostate and bulbourethral glands

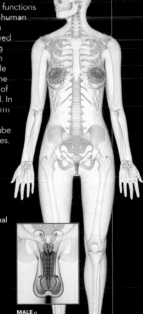

MALE

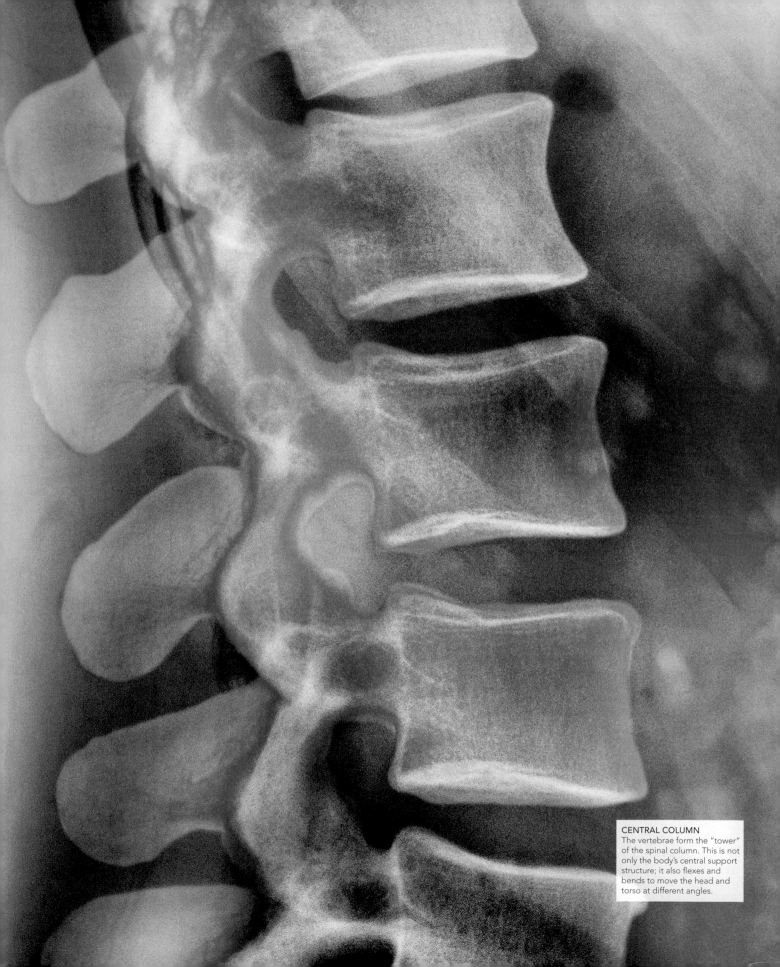

CENTRAL COLUMN
The vertebrae form the "tower" of the spinal column. This is not only the body's central support structure; it also flexes and bends to move the head and torso at different angles.

SUPPORT AND MOVEMENT

THE BODY'S MUSCLES, BONES, AND JOINTS PROVIDE A SUPPORTIVE FRAMEWORK CAPABLE OF AN ENORMOUS RANGE OF DYNAMIC MOTION. MUSCLES AND BONES ALSO HAVE NUMEROUS INTERACTIONS WITH OTHER BODILY SYSTEMS, ESPECIALLY THE NERVES FOR CONTROL AND COORDINATION, AND THE BLOOD, WHICH SUPPLIES THE ENERGY-HUNGRY MUSCLES WITH THEIR ESSENTIAL REQUIREMENTS.

The body's muscular system is never still. Even as the body sleeps, breathing continues, the heart beats, and the intestines squirm. Most muscles relax during sleep, but some contract occasionally to shift the body into a new position. This avoids squashing nerves and vessels, which could cause blood deprivation and damage, and thereby safeguards all body systems.

MUSCLE TEAMWORK

Apart from simple movements, such as the blink of an eye, actions are the result of multiple muscle contractions. A subtle movement such as a smile, for example, involves 20 facial muscles. Writing utilizes more than 60 muscles in the arm, hand,

STAYING SUPPLE
Our potential for movement, and the health of the skeletal and muscular systems, is maximized by regular "3 S" exercises, for strength, stamina, and suppleness. Warm-ups and cool-downs avoid sudden strain that could cause injury.

and wrist. As the arm moves, muscles in the shoulder come into play, while the shifting weight load on the main torso uses yet more muscles to balance the body. The other muscles do not simply relax. They maintain tension, so that their opposing partners have some resistance to pull against, in a continuing sequence of split-second give-and-take.

STRESS AND FLEXIBILITY

Bones are slightly flexible so that they can absorb normal stress without cracking or snapping. Sensory systems built into

muscles, bones, and joints also protect against injury. Microsensors within them, and inside associated parts, such as tendons and ligaments, gauge tensions and pressures. Nerve messages feed the brain and provide warning, registering stress to the bones as discomfort or pain. Awareness of the pain stimulates action by the body.

POSTURE AND FEEDBACK

Feedback signals also provide the brain with information about the body's posture and the detailed positions of its parts, which is known as the proprioceptive sense. In this way we "know," without having to look, or feel, that the fingers are clenched or a knee is bent. When learning a new motor skill, the eyes watch the movement's progress, and the skin feels it, as the brain adjusts its muscle control through trial and error. With practice, the motor nerve patterns and their proprioceptive feedback become well tuned and established. Eventually the movement becomes automatic. It is organized by part of the lower rear brain called the cerebellum, and we no longer need to concentrate on it.

MUTUAL HEALTH

The interdependence of muscles, bones, and joints not only allows them to function but also keeps them healthy. During vigorous exercise, two-thirds of the heart's output of

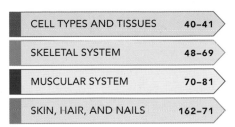

Sensory cortex
Part of brain that monitors sensory information from the body

Sensory nerve
Muscle-stretch information travels to brain

Biceps muscle
Moves arm to a flexed position

Sensory neuron
Nerve cell that carries sensory nerve impulses

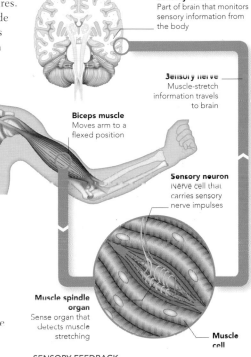

Muscle spindle organ
Sense organ that detects muscle stretching

Muscle cell

SENSORY FEEDBACK
Nerve endings form tiny sense organs (muscle spindle organs) within the muscles. Specialized to respond to tension or stretching, they fire signals along nerve fibers to the brain. The signals arrive at the sensory cortex, where they inform the brain about what is happening.

blood goes to the muscles, compared to only one-fifth at rest, giving the heart muscle a workout. Muscles exercised to extremes can exert so much force on a bone that it breaks. Conversely, weak muscles do not put bones under regular pressure, so the bone starts to weaken and waste.

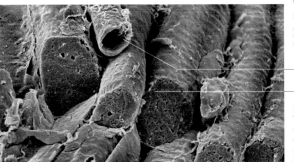

Blood vessel

Muscle fiber

MUSCLE FIBERS
This color-enhanced electron micrograph of muscle tissue shows the cut ends of several myofibers, which are large cells. Each fiber has within it bundles of even slimmer myofilaments.

INFORMATION PROCESSING

THE HUMAN BODY IS ALIVE WITH INFORMATION. BEING A COMPLEX, DYNAMIC MECHANISM, ITS INTERACTING AND INTERDEPENDENT PARTS REQUIRE CONTROL AND COORDINATION. THIS IS DONE BY PASSING INFORMATION BETWEEN THEM. TWO BODY SYSTEMS ARE RESPONSIBLE FOR COMMAND-CONTROL AND DATA MANAGEMENT—THE NERVOUS AND ENDOCRINE SYSTEMS.

| NERVOUS SYSTEM | 82–119 |
| ENDOCRINE SYSTEM | 120–29 |

Information processing involves inputs, evaluation, and decision-making, followed by outputs. The body has inputs from the various senses such as sight and hearing. Its brain is the "CPU" (central processing unit), whose outputs control the physical actions of muscles and the chemical responses of glands. Both nerves and hormones are involved in data management.

ELECTRICAL AND CHEMICAL PATHWAYS

The "language" of the nervous system is tiny electrical impulses. They are small and fast—each just one-tenth of a volt in strength and lasting a few one-thousandths of a second—and numerous. Every second, millions pass through the network of long, pale, stringlike pathways we call nerves. Information from the senses flows to the brain as electrical impulses. Here, it is sifted, analyzed, and evaluated, causing millions more signals to pass around and within the brain between its numerous, complex areas. Decisions are reached and command messages are produced in the form of electrical impulses. The brain's electrical output travels along motor nerves to the muscles to stimulate and coordinate their contractions for movements. Different information carriers called hormones are secreted by endocrine glands into the bloodstream to stimulate distant tissues to action. More than 50 hormones circulate in the bloodstream. The specific molecular structure of each hormone stimulates only

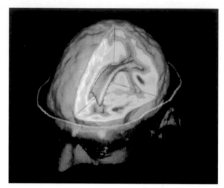

BRAIN ACTIVITY
This image is a three-dimensional, functional MRI scan showing brain activity during speech. Red indicates areas of high activity, yellow indicates medium activity, while blue indicates low activity.

cells with suitable receptors on their surface, instructing the cells to carry out certain procedures. In general, nerves work quickly (within fractions of a second). Most hormones function over longer times, within minutes, days, or even months. Long-lasting effects, as in growth hormone for example, occur because the hormone is continuously secreted over many years; an individual dose would last only a few days.

BODY-CLOCK INPUT

The body has built-in rhythms of activity. People in experimental "timeless" surroundings (of constant light, temperature, food availability, and other conditions) still tend to sleep, wake, eat, become alert, and move about in a roughly 24-hour cycle. A small part of the brain known as the

suprachiasmatic nucleus, located just above the place where the visual, or optic, nerves meet (see p.95), is the "body clock." It is continually adjusted by external cues, such as light levels and temperature fluctuations, and our mental acknowledgment of clock times. In turn, it feeds information to many brain parts that deal with cyclical activities, such as hormone release, tissue repair, body temperature control, urine production, and digestive matters. In this way, the natural rhythms of the body are coordinated.

THE IMPORTANCE OF INPUT

As can be seen by the workings of the body clock, feeding information into the brain's processing centers relies on more than the five senses. The continuing environmental adjustment of the body clock is one example

SELECTIVE FOCUS
The nose detects smells continuously and sends endless streams of nerve signals to the brain. However, we can choose to ignore this information, or to focus on it, as part of the mind's selective awareness of incoming data.

of more subtle and complex sensory input. Within the body, there are thousands of microreceptors that continually monitor variables, such as blood pressure, body temperature, and levels of important chemicals, for example oxygen, waste carbon dioxide, and blood glucose. These data feed to "automatic" or subconscious parts of the brain, which make decisions that do not register in the conscious mind. In this way a huge amount of information processing occurs of which we are hardly ever aware.

KEY
- ■ Aldosterone
- ■ Melatonin
- ■ Cortisol

DAILY CYCLES
Hormone levels follow a 24-hour cycle. Melatonin, the "sleep hormone," both maintains and is affected by the rhythm control system. Aldosterone affects urine production. Cortisol is involved in many tasks, from influencing glucose levels to promoting healing and alleviating stress.

Chart: HORMONE CONCENTRATION vs TIME (HOURS)
DAY | NIGHT | DAY | NIGHT
12:00 18:00 00:00 06:00 12:00 18:00 00:00 06:00 12:00

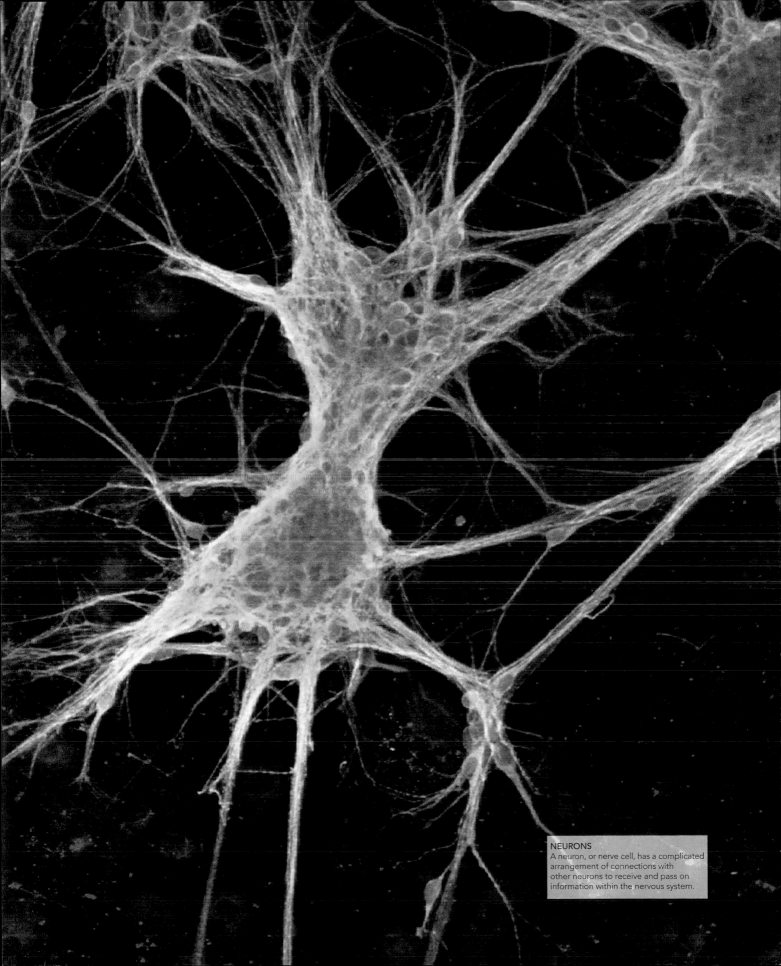

NEURONS
A neuron, or nerve cell, has a complicated arrangement of connections with other neurons to receive and pass on information within the nervous system.

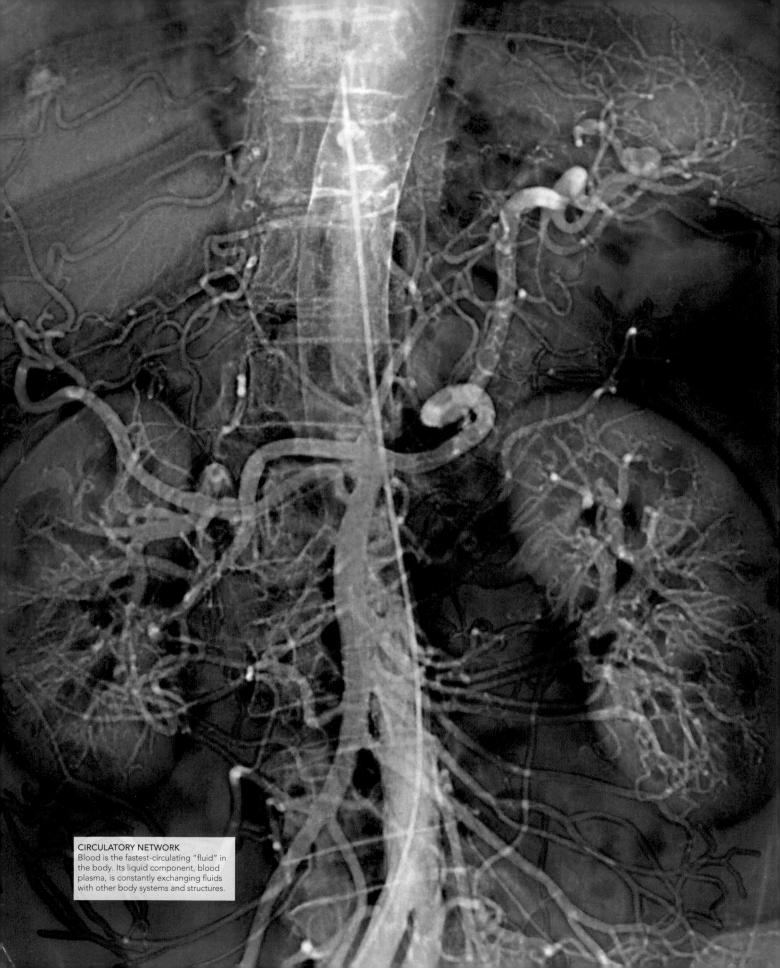

CIRCULATORY NETWORK
Blood is the fastest-circulating "fluid" in
the body. Its liquid component, blood
plasma, is constantly exchanging fluids
with other body systems and structures.

THE FLUID BODY

ROUGHLY TWO-THIRDS OF THE BODY IS COMPOSED OF WATER AND THE VARIOUS ESSENTIAL SUBSTANCES DISSOLVED WITHIN IT. THESE FLUIDS HAVE INNUMERABLE VITAL ROLES WITHIN MANY BODY SYSTEMS. THEY ARE FOUND IN CELLS, AROUND THE BODY'S TISSUES AND, MOST OBVIOUSLY, IN BLOOD AND LYMPH.

| CARDIOVASCULAR SYSTEM | 130–45 |
| LYMPH AND IMMUNITY | 172–87 |

Most body parts are largely composed of water. Tissues are 70–80 percent fluid, which means that organs such as the brain and intestines are typically three-quarters water. Blood plasma is over 90 percent water, while bones contain almost 25 percent. Fat has 10–15 percent water in its composition.

FLUID COMPARTMENTS

The body's different fluids can be grouped into physiological categories that are known as compartments. There are two major fluid compartments: intracellular and extracellular. Intracellular fluid (also known as cytoplasm) is found within the body's cells. Extracellular fluid accounts for all other fluids in the body. Its subcompartments are: interstitial fluid, which occupies the spaces between cells and tissues; blood plasma and lymph; the fluids found in bones, joints, and dense connective tissue; and transcellular fluid, which includes saliva and other digestive juices, mucus, sweat, and urine.

FUNCTIONS OF FLUIDS

Water is an excellent solvent. Thousands of substances that are dissolved in it are used in the body's biochemical reactions. These reactions are the very basis of life. Water is

Blood plasma
The pressure produced by the heart's pumping squeezes blood plasma through capillary walls.

BLOOD PLASMA AND LYMPH CYCLE
Blood plasma leaks out from capillaries to become interstitial fluid. Some of this drains into lymph vessels and becomes lymph fluid. Eventually this fluid is returned to blood circulation, as lymphatic vessels empty into large veins.

Lymph
Lymph vessels collect and circulate the fluid, then route it back into the blood circulation.

Interstitial fluid
The fluid, now under little pressure, flows randomly and slowly around cells and tissues.

also an effective transport system. It moves around the body distributing nutrients and collecting and delivering waste materials. Fluids spread heat from active parts of the body, such as exercising muscles, to cooler areas, and in doing so they aid in thermoregulation. The body uses fluids as shock absorbers to cushion sensitive areas such as the brain, the eyes, and the spinal cord. Fluids also work as lubricants within the body, so that tissues and organs slip past each other with minimal friction. Small amounts of specific fluids that specialize in this role include the pleural fluid around the lungs, the pericardial fluid around the heart, and the synovial fluid inside joints.

BLOOD AND LYMPH

The blood and lymph circulatory systems are closely linked because they are constantly swapping fluids. Blood plasma, the fluid in which blood cells are suspended, transports red blood cells (which carry oxygen and remove carbon dioxide) around the body. Blood plasma leaks from capillaries into the tissues around them, becoming interstitial fluid. Most of this leaked fluid is reabsorbed into the blood, but some of it is drawn into the capillaries of the lymphatic system where it becomes

lymph fluid. This transports white blood cells (which produce antibodies to fight infection and disease) around the body. After flowing through the lymphatic system, lymph drains back into the blood stream once again to be used as blood plasma.

BALANCE AND RECYCLING

The average adult body contains about 70 pints (40 liters) of water. Every day, water is lost from the body in the form of urine, sweat, water vapor from the lungs, and in feces. Water is used up and also produced by the body in biochemical reactions. For example, the glands producing saliva and digestive juices. To maintain a healthy balance of fluids, we need to drink at least 3½ pints (2 liters) of water per day. But if it were not for the body's amazing water conservation and recycling measures (such as recycling blood plasma as lymph, and vice versa), we would need to consume at least 100 times more water than the recommended amount every day.

COMPONENTS OF BLOOD PLASMA

Blood cells are transported by blood plasma, a liquid that accounts for about 55 percent of blood volume and is itself about 90 percent water. It contains many important substances.

PLASMA PROTEINS	Such as albumins (stop water leaking into the tissues), fibrinogen (involved in blood clotting), and globulins (such as antibodies).
ELECTROLYTES	Mainly mineral salts, that form ions when dissolved, principally sodium, chloride, potassium, calcium, and phosphate.
HORMONES	Like insulin and glucagon (regulate blood-glucose levels), thyroid hormones (control rate of cell metabolism), and sex hormones.
NUTRIENTS	Such as glucose (for energy), amino acids, and lipids, such as cholesterol and triglycerides (for cellular components and energy).
WASTES	Such as carbon dioxide, lactic acid, creatinine, and uric acid. These are transported out of the blood circulation by the kidneys.

2½ PINTS (1.4 L) LYMPH

29¼ PINTS (16.6 L) INTERSTITIAL FLUIDS

35¾ PINTS (20.3 L) INTRACELLULAR FLUIDS

7¼ PINTS (4.15 L) BLOOD IN VEINS

2 PINTS (1.1 L) BLOOD IN ARTERIES

½ PINT (0.28 L) BLOOD IN CAPILLARIES

VOLUMES OF MAJOR BODY FLUIDS
The fluids within cells (intracellular fluids) and around cells and tissues (interstitial fluids) form the greatest proportion of body liquids. This chart ignores many other fluids, such as saliva and other secretions, and fluid in bones, joints, and connective tissues.

EQUILIBRIUM

THE BODY'S CELLS AND TISSUES ARE DELICATE AND EASILY DISRUPTED. THEY ONLY FUNCTION WELL IF ALL ASPECTS OF THEIR CHEMICAL AND PHYSICAL ENVIRONMENT ARE CONTINUOUSLY ADJUSTED TO KEEP THEM STABLE AND IN EQUILIBRIUM. SEVERAL BODY SYSTEMS WORK TOGETHER TO MAINTAIN A BALANCED ENVIRONMENT, A PROCESS OR STATE CALLED HOMEOSTASIS.

Chemical changes that occur inside every cell are attuned to specific conditions: body fluid concentration, oxygen levels, glucose and vital supplies, acid-alkali balance, and external circumstances, such as temperature and pressure. The body must maintain these internal conditions within certain limits, or its biochemical pathways go awry, waste builds up, energy runs out, and the resulting adverse effects rapidly spread.

HOMEOSTATIC SYSTEMS

Several body systems contribute to homeostasis. The respiratory system ensures that the body has a constant supply of oxygen, which is consumed when liberating energy

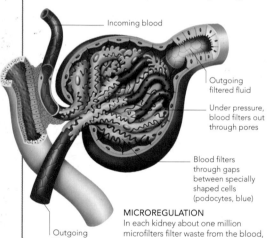

Incoming blood

Outgoing filtered fluid

Under pressure, blood filters out through pores

Blood filters through gaps between specially shaped cells (podocytes, blue)

Outgoing blood

MICROREGULATION
In each kidney about one million microfilters filter waste from the blood, and regulate the amount of water, salts, and minerals it contains.

from nutrients, and cannot be stored in the body in any quantity. The digestive system takes in and processes nutrients, some of which are used for the repair and maintenance of old cells and tissues. The circulatory system ensures that oxygen and nutrients are distributed throughout the body, as well as gathering waste products, which are removed by the urinary and respiratory systems. The integumentary system (skin, hair, and nails) buffers the body's interior against the ever-varying external environment and fluctuations in temperature, moisture levels, and radiation.

CONTROL AND FEEDBACK

The body's two major control systems, nerves and hormones, are mainly responsible for coordinating homeostatic mechanisms using feedback loops. For example, if water levels fall slightly in the tissues, blood and other fluids become more concentrated. Various sensors monitor this, switch on, and feed back information to alert the brain. The brain's homeostatic centers trigger a sequence of regulating actions. Hormonal control of urinary excretion is adjusted to conserve water and nervous activity produces a thirst urge in our conscious awareness, so we take a drink to replenish water levels. The sensors detect the changes as fluid concentrations return toward normal, then they switch off until the next time they are needed. In this way, constant monitoring of internal conditions maintains a stable internal environment, so that cells and tissues can function with maximum efficiency.

THERMOREGULATION

One facet that demonstrates the intricacy of internal homeostasis and is obvious from the outside is thermoregulation (the maintenance of an approximately constant body temperature). The principle is much the same as a thermostat-equipped heater. When the thermostat sensor detects a fall

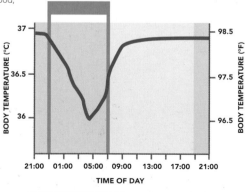

SLEEP

BODY TEMPERATURE (°C)

37

36.5

36

BODY TEMPERATURE (°F)

98.5

97.5

96.5

21:00 01:00 05:00 09:00 13:00 17:00 21:00

TIME OF DAY

DYNAMIC EQUILIBRIUM
The body's normal core temperature is about 98.6°F (37°C) when awake and active, falling to about 96.8°F (36°C) when asleep. Because the equilibrium point alters according to circumstances, it is called dynamic equilibrium.

in temperature, it switches on the heating; as the temperature reaches its required set point, it is switched off. In the body, active muscles generate heat, which is dissipated to all parts by blood flow. But a temperature variation of much more than 2°F (1°C) starts to affect the chemical reactions inside cells. Protein molecules, in particular, which include enzymes that control rates of reactions, are very heat-sensitive. They begin to distort and lose their complex three-dimensional structure when too warm. The body's temperature-sensing nerve endings trigger the thermostatic processes.

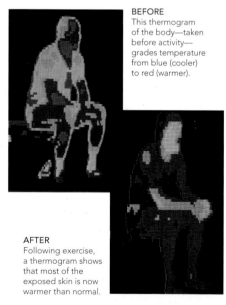

BEFORE
This thermogram of the body—taken before activity—grades temperature from blue (cooler) to red (warmer).

AFTER
Following exercise, a thermogram shows that most of the exposed skin is now warmer than normal.

Blood vessels in the skin widen to allow greater blood flow, which leads to increased heat loss to the surrounding air, while perspiration also occurs, so that warmth is drawn from the body by evaporation of watery sweat. In these ways, both the physical and chemical conditions inside the body are kept relatively stable, and an ongoing equilibrium is maintained.

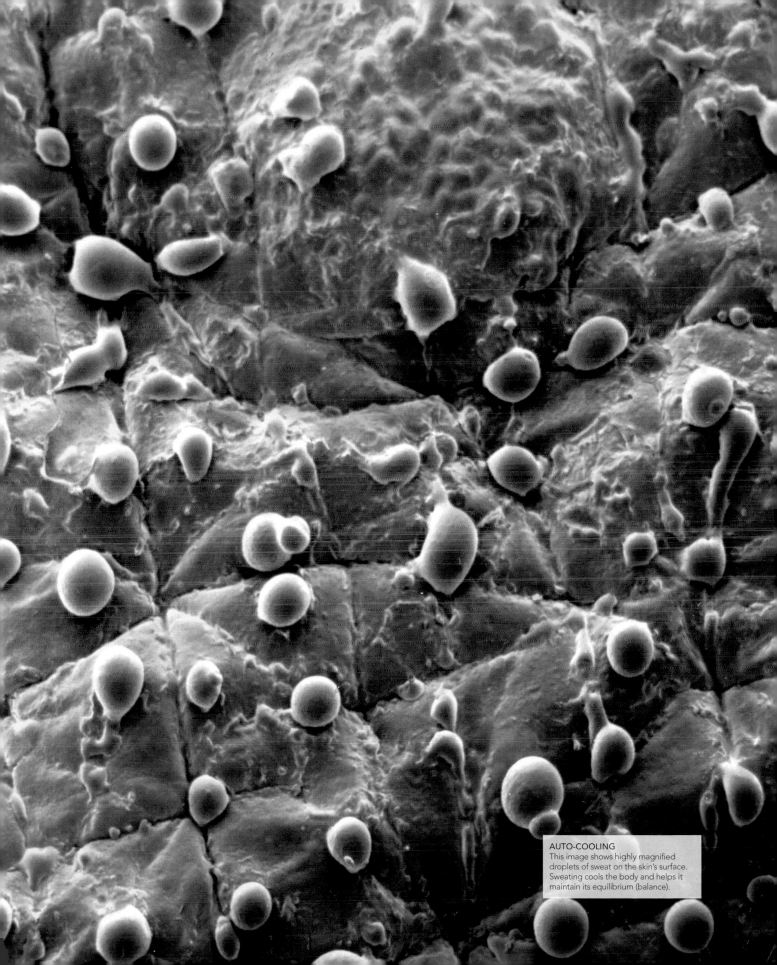

AUTO-COOLING
This image shows highly magnified droplets of sweat on the skin's surface. Sweating cools the body and helps it maintain its equilibrium (balance).

ADAPTING TO EXTREMES

THE HUMAN BODY IS LEAST STRESSED AND MOST HEALTHY IF IT RECEIVES ADEQUATE FOOD AND FLUID INTAKE, AND EXISTS IN A COMFORTABLE ENVIRONMENT WITH EQUITABLE TEMPERATURE, NORMAL AIR PRESSURE, AND STANDARD GRAVITY. YET THE BODY HAS AN AMAZING ABILITY TO COPE WITH EXTREME CONDITIONS, BOTH IN THE SHORT TERM USING EMERGENCY RESPONSES, AND BY GRADUAL ADAPTATION TO A NEW ENVIRONMENT.

TEMPERATURE RANGE

The body's myriad inner processes and chemical changes, collectively known as metabolism, have evolved to work most efficiently at a core temperature of 97.7–99.5°F (36.5–37.5°C). As part of its ability to balance and maintain internal conditions, called homeostasis, the body has automatic reactions known as thermoregulation (see p.30). These include dilating skin blood vessels when warm or constricting them if cold. We also have conscious behavioral responses, such as putting on extra clothes when chilled or seeking shade if hot. But once body core temperature strays by more than about 2.7°F (1.5°C) for an appreciable time, further and more drastic reactions start to occur, and damage may follow. Hypothermia is usually regarded as a core temperature of 95°F (35°C) or below, which is inadequate for normal metabolism. Likewise, hyperthermia begins as core temperature rises more than about 1.8°F (1°C) above what would be normal body temperature in the 24-hour cycle.

COOLING EFFECT
This "heat image" (thermogram), showing temperature from yellow as hottest to indigo as coldest, reveals a cooling drink that will help counteract heat stress by chilling the mouth, throat, and body core.

THE BODY'S TEMPERATURE SPECTRUM

Biological reactions such as shivering and sweating, and behavioral actions can protect the body from temperature extremes for a short time. However, in freezing conditions, the body's extremities are at risk of serious problems such as frostbite within a few minutes, with consequent permanent loss of tissue. Nevertheless, even if this happens, core temperature in the vital organs may be maintained long enough for survival.

CORE BODY TEMPERATURE

95°F ◄———————————— 98.6°F ————————————► 104°F

EXTREME COLD	MODERATE COLD	MILD COLD	MILD HEAT	MODERATE HEAT	EXTREME HEAT
Hypothermic state sets in, breathing rate slows even further but heart rate may increase due to arrhythmia	Heart (pulse) rate and breathing rate slow further to save energy for brain	Heart (pulse) rate and breathing rate slow slightly	Sensation of thirst increases, to replenish water lost in sweat and hotter, more humid exhaled air	Headache and nausea are more likely, followed by vomiting and diarrhea	Hyperthermia established, with confusion and perhaps aggression, then fainting; tissues suffer lasting damage
Surface tissues suffer freezing and irreversible frostbite; core organs begin to shut down	Blood circulation to skin and extremities is restricted even more, leading to pale, "waxy" appearance and increased risk of frostbite	Skin blood vessels constrict, limiting blood flow and heat loss from skin, thereby conserving core warmth and turning skin pale	Skin blood vessels dilate, increasing blood flow and heat loss, and turning skin redder or "flushed"	If dehydration occurs, blood thickens and is more likely to clot, so chances of stroke or heart attack rise	Great thirst may fade ("paradoxical replenishment") even as body's water content falls to critical levels
Memory loss and confusion may lead to irrational removal of clothing ("paradoxical undressing")	Feeling and dexterity are lost, and sensitivity to pain is dulled; movement becomes slow and labored	Muscles linked to skin hairs constrict, raising the hairs to trap insulating air next to the skin (causing "goose bumps")	Tiny muscles attached to skin hairs relax, so hairs lie flat and heat from skin can pass through easily to air	Blood pressure drops, affecting ability to stand up and move about; falls are more likely	Low blood pressure and lack of oxygen turn skin blue; muscle cramps set in
Brain is usually the last organ to cease functioning	If violent shivering does not help, the body becomes still to conserve remaining energy and heat in vital organs	Muscles tremble or shiver involuntarily, causing them to produce heat that is distributed by the blood	Sweat glands in skin secrete sweat, which evaporates and cools body surface	Sweating may cease so skin becomes hot and dry, a major sign of hyperthermia	Convulsions or seizures; unconsciousness; heartbeat becomes weaker and more irregular; vital organs fail

COLD WARNING SIGNS
Shivering and loss of skin sensitivity are early signs that warn the affected person to put on more clothing, especially on the head and neck, and to move into a sheltered place where wind does not constantly blow away body heat (the effect known as wind chill).

HEAT WARNING SIGNS
Very hot conditions bring on sweating, faintness, and feelings of thirst. These are signals to take in fluids, and to rest in a cool place such as in the shade, especially where a breeze can continuously carry away evaporated sweat to increase its cooling effect.

PRESSURE AND ALTITUDE

Water resists being compressed under pressure exceptionally well. Much the same applies to its expansion as pressure falls. So most body tissues, being water-based, are little affected by external pressure changes. The chief problem with increased external pressure, as when diving deeper in water, is air being compressed in breathing passages and lungs. Diving equipment delivers air at the same pressure as the surrounding water. Decreased external air pressure is usually due to increasing altitude. The main problem here is not so much the pressure but less oxygen in the air. Above 8,000 ft (2,400 m), oxygen depletion can cause altitude sickness, with raised breathing (hyperventilation) and heart rates, headache, fatigue, nausea, dizziness, faintness, and collapse.

ALTITUDE ACCLIMATIZATION
Over days and weeks, breathing air with lowered oxygen content causes the kidneys and liver to release more erythropoietin. This hormone raises production of red blood cells in the bone marrow, which then enter the bloodstream, to enhance oxygen absorption and transport.

Normal blood oxygen level

More red blood cells enable blood to carry more oxygen

Blood oxygen level falls due to reduced oxygen availability at altitude (or fewer red blood cells, or increased oxygen demand from tissues)

Erythropoietin stimulates bone marrow to raise rate of red blood cell manufacture

Kidney and liver release more erythropoietin

FREE DIVING
Free divers do not use breathing equipment. The body's air spaces are compressed, but its natural "mammalian diving reflex" minimizes harm by forcing extra blood into the lungs' vessels, pushing out most of the air so the lungs are less prone to collapse.

HIGH AND LOW GRAVITY

The body's physical structure, and especially its bones, joints, and muscles, are adapted to withstand the downward pull of Earth's gravity—usually called g force. Increased g force is encountered during rapid acceleration in vehicles like cars and space rockets, and also during fast ups, downs, and turns, as on a roller-coaster. "High g" can bring on dizziness and disorientation, motion sickness, and reflexes such as throwing out limbs. If prolonged, the person feels faint and blood pools in some areas as the heart struggles against the force preventing its return. Breathing difficulty, joint stress, and unconsciousness may also occur. Zero g or weightlessness, such as that experienced in space, also produces disorientation and physical reflexes, accompanied by nausea and blood pooling. Over time, muscles deteriorate and waste, joints weaken, and bones lose mineral content.

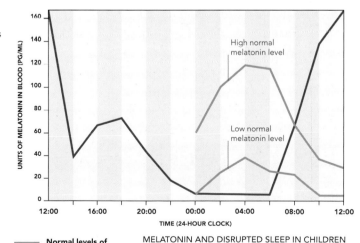

TRAINING FOR LOW GRAVITY
The nausea and disorientation that may accompany weightlessness are called space sickness. Astronauts can adapt to this condition by training in modified aircraft.

LACK OF SLEEP

Good health depends not only the total length of sleep in each 24 hours but also its quality and phase, such as REM or NREM (see p.95). Sleep deprivation or frequent changes in sleep times disrupt the late-night rise in melatonin production by the pineal gland that usually brings on sleep (see p.126). At first, this leads to fatigue, yawning, irritability, and clumsiness. If it continues, further effects are memory problems, confusion, depressed mood, emotional swings, and even hallucinations, coupled with headaches, trembling, and unsteadiness. There are also increased risks of blood-pressure problems, diabetes, gastric ulcers, and accidents.

IN THE SLEEP LAB
EEGs (electroencephalogram recordings of brain activity) may show underlying reasons why and how a person's sleep cycle is disrupted. One cause is abnormal melatonin production (see right).

— Normal levels of melatonin at night

— Levels of melatonin in SMS

MELATONIN AND DISRUPTED SLEEP IN CHILDREN
In the genetic disorder Smith–Magenis Syndrome (SMS), the melatonin production cycle is shifted by 6–12 hours, causing wakefulness at night and sleepiness by day.

DEHYDRATION

Lack of fluids can harm the body more, and faster, than lack of nutrients. Water output in urine is essential to remove dissolved wastes that would otherwise build up in blood and become toxic. During short-term water deprivation, for several hours to a couple of days (depending greatly on environmental conditions) the body adjusts. For example, the tubules in kidney nephrons respond to increased antidiuretic hormone (ADH or vasopressin) by increasing the water they take in from urine so this becomes more concentrated (see p.215). Also, feelings of thirst originating from the hypothalamus increase greatly. If water is still not available, tissues start to suffer damage from generalized effects such as thickened blood.

EFFECTS OF DEHYDRATION ON THE BODY

A convenient way to measure dehydration is by percentage weight loss compared to average healthy hydrated weight. Effects of dehydration are due not only to less water, but also to the loss of dissolved essential salts and minerals that are carried away by urine. For example, falling sodium levels affect the way nerve signals pass around the brain and along nerves.

PERCENTAGE WEIGHT LOSS	EFFECTS ON THE BODY
1–2	Feelings of thirst; dry "cotton" mouth due to less saliva; dry, flushed skin due to less sweating; reduced urine output; constipation; fatigue
3–4	Greater thirst; lowered appetite; lack of sweating allows core body temperature to rise, especially after activity; dizziness or light-headedness; muscle weakness
5–6	Body temperature continues to rise; heart and breathing rates increase; severe headaches; muscle cramps; increasing risk of heat exhaustion and hyperthermia
MORE THAN 6	Muscles uncontrollable; skin shrivels; vision problems; painful swallowing; confusion, memory lapse, hallucinations; collapse and unconsciousness

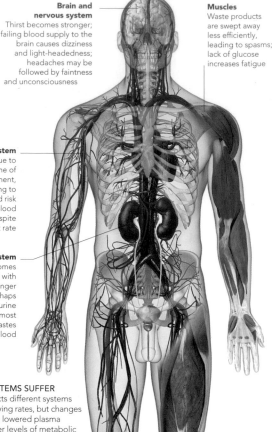

Brain and nervous system
Thirst becomes stronger; failing blood supply to the brain causes dizziness and light-headedness; headaches may be followed by faintness and unconsciousness

Muscles
Waste products are swept away less efficiently, leading to spasms; lack of glucose increases fatigue

Circulatory system
Blood thickens due to decreased volume of its liquid component, plasma, leading to increased risk of clotting; blood pressure falls, despite rising heart rate

Urinary system
Urine initially becomes more concentrated with darker color, stronger odor, and perhaps cloudiness; urine production then almost shuts down, so wastes build up in blood

HOW BODY SYSTEMS SUFFER
Dehydration affects different systems and organs at varying rates, but changes in the blood, with lowered plasma volume and higher levels of metabolic wastes, gradually damage all parts.

Head and brain
Mental problems such as confusion and irritability; reduced or absent appetite; fatigue; early loss and delayed eruption of teeth; scalp hair becomes thinned, dry and brittle

Chest and abdomen
Flattened chest with weak breathing muscles; weak heartbeat; distended abdomen due to enlarged, fatty liver; weakened immune system with increased risk of infection

Skin and limbs
Skin sores, rashes and peeling; loss of skin pigmentation; wasted muscles; unstable joints; thinned bones.

Feet
Swelling of the feet, known as pedal edema, is one of the key signs of kwashiorkor; puffy skin due to edema may occur elsewhere.

KWASHIORKOR
This condition, which mainly affects children, is caused by severe lack of protein in the diet, despite possibly adequate intake of energy and other dietary requirements.

MALNUTRITION AND STARVATION

A healthy body has sufficient stores of most essential nutrients to last for several days. An urgent need is readily available energy, usually provided by starchy foods digested to yield sugars, chiefly glucose. If these are unavailable, glycogen stocks in the liver are broken apart and mobilized as sugars; this supply lasts up to one day. Then body fats (lipids) and proteins, especially in muscle, begin to be metabolized into energy. Stocks of essential vitamins (most of which cannot be manufactured or recycled in the body) and minerals also begin to run low. This affects processes such as digestion, nerve-signal transmission, and hormone production; the maintenance of healthy bones, muscles, and skin; resistance to infection; and vision, taste, and other senses. These are the widespread, longer-term effects of general malnutrition. Unless they are relieved by a careful, graded reintroduction to a healthier diet, progress to true starvation will result.

MALNUTRITION
When the body's supply of food is cut off, it draws on energy from its own stores, starting with resources that are most easily converted into energy.

Energy from dietary glucose
Obtained by breaking down carbohydrates, glucose is the body's main source of energy, but freely circulating supplies in the blood last for less than one hour.

Energy from glycogen
Stores of glycogen (animal starch) in the liver and muscles are broken down by a process called glycogenolysis into glucose and distributed by the blood.

Energy from fats
First free blood fatty acids then, over hours, adipose tissue fats are split into energy-yielding products such as ketones, and glycerol (used to make glucose by gluconeogenesis).

Energy from proteins
Amino acids derived from muscle during starvation are used by the liver (for gluconeogenesis) and by the kidneys (for ammoniagenesis, which also yields glucose).

HUMAN PERFORMANCE

FACTORS SUCH AS "PREADAPTED" NATURAL PHYSIQUE, MATCHLESS TRAINING REGIMES, THE BEST FACILITIES MONEY CAN BUY, AND UNFLINCHING DETERMINATION ALL CONTRIBUTE TO THE BODIES OF HUMAN BEINGS WHO EXCEL IN SPORTS AND OTHER PHYSICAL PERFORMANCES.

NATURAL PHYSIQUE AND ABILITY

Some people are born with a particular body size and shape, and seeming built-in ability, that suits a certain sport or competition. For example, great height gives a head start in basketball, although ball-handling skills, court mobility, and tactical awareness are also important. Sprinters tend to be tall, muscular, and long-legged, while middle and long distance runners are generally smaller, slimmer, and more sparely built. However, genetic research indicates that, for example, the dominance of East African runners in long-distance events cannot be assigned solely to inherited traits. Also vital in this case are social and economic factors. These include plentiful walking or running in daily life due to habit or lack of vehicle transport, a traditional diet based on carbohydrates (see below), plus intense drive to succeed, since winning a competition may mean rewards equivalent to many lifetimes' earnings.

TYPICALLY TALL
Average height in the US's National Basketball Association is 6 ft 7 in (2 m), but smaller players can thrive if they are exceptionally fast.

DIET AND LIFESTYLE

Modern sports science pays huge attention to food and fluid intakes at all times, from leisure periods to training and competition. As well as sufficient overall energy to fuel physical activity, the form of energy is also key. Endurance performers tend to focus on carbohydrates, which are eaten as pasta, rice, bread, potatoes and similar starchy foods, and are digested and stored as glycogen. This can be readily broken down as a convenient source of glucose energy over one or a few hours. Eating especially plentiful carbohydrates, a practice known as "carb-loading," during training and before an event helps prepare the body for this lengthy exertion.

EXERCISE AND TRAINING

Each aspect of performance relies on specific sets of muscle groups, and training regimens aim to work particular groups, both individually and in combination. Some sports, like weightlifting, have relatively narrow requirements, while activities such as swimming, rowing, and cross-country skiing use more varied patterns of activity and widespread muscle groups. Even so, all training aims to develop general posture, coordination, balance, bone strength, joint flexibility, and overall fitness—including the heart and breathing muscles, since good respiratory and cardiovascular condition is so vital for general health. Training also includes warm up and cool-down (warm-down) routines, so muscles and joints are not suddenly strained.

NORDIC WALKING
This relatively new activity uses poles to involve the arms and shoulders, as well as the lower body. It can be tailored to people of most ages and levels of fitness.

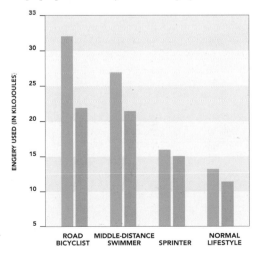

ATHLETES' ENERGY USE
During competition, top-class athletes need perhaps four or five times the amount of energy required by a semisedentary lifestyle. Road bicycling, which typically involves 4–5 hours in the saddle every day, is especially arduous.

ENERGY USED (IN KILOJOULES)

ROAD BICYCLIST | MIDDLE-DISTANCE SWIMMER | SPRINTER | NORMAL LIFESTYLE

Men
Women

BREAKING RECORDS

Since accurate record-keeping began, generally over the past century, almost all human performance records have improved. The causes are multiple— better diet and health care, especially during developmental years, improved training and equipment, and greater understanding of sports and the body. Rarely, a truly outstanding feat sets a record for years, perhaps partly due to chance factors. For example, in the 1968 Summer Olympics, with the high-altitude "thin air" of Mexico City presenting less resistance, and with a maximum allowable following wind, US long-jumper Bob Beamon increased the world record by 22 in (55 cm) when he leaped 29 ft 2.5 in (8.90 m). The record stood for nearly 23 years.

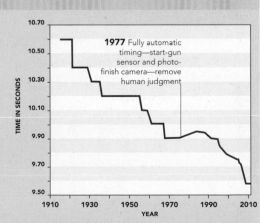

1977 Fully automatic timing—start-gun sensor and photo-finish camera—remove human judgment

TIME IN SECONDS

YEAR

100 METERS RECORDS
The trend is for athletic records to be broken by gradually smaller increments. But occasionally an exceptional competitor brings a "quantum leap," as with Jamaican sprinter Usain Bolt from 2008.

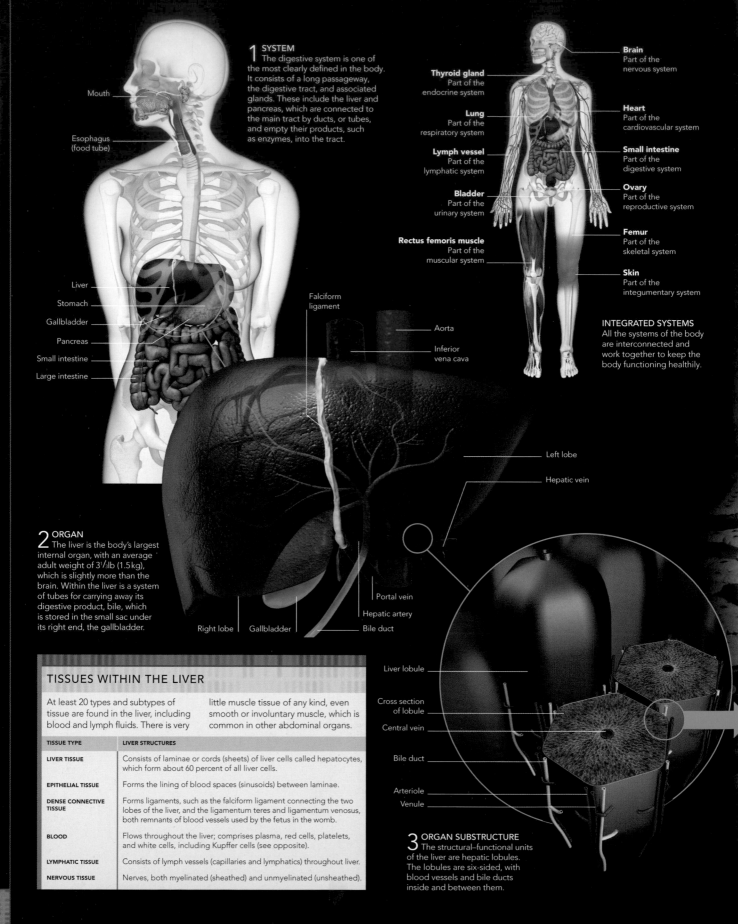

1 SYSTEM

The digestive system is one of the most clearly defined in the body. It consists of a long passageway, the digestive tract, and associated glands. These include the liver and pancreas, which are connected to the main tract by ducts, or tubes, and empty their products, such as enzymes, into the tract.

Mouth

Esophagus (food tube)

Liver

Stomach

Gallbladder

Pancreas

Small intestine

Large intestine

Falciform ligament

Aorta

Inferior vena cava

Thyroid gland
Part of the endocrine system

Lung
Part of the respiratory system

Lymph vessel
Part of the lymphatic system

Bladder
Part of the urinary system

Rectus femoris muscle
Part of the muscular system

Brain
Part of the nervous system

Heart
Part of the cardiovascular system

Small intestine
Part of the digestive system

Ovary
Part of the reproductive system

Femur
Part of the skeletal system

Skin
Part of the integumentary system

INTEGRATED SYSTEMS

All the systems of the body are interconnected and work together to keep the body functioning healthily.

Left lobe

Hepatic vein

2 ORGAN

The liver is the body's largest internal organ, with an average adult weight of 3¹/₃lb (1.5 kg), which is slightly more than the brain. Within the liver is a system of tubes for carrying away its digestive product, bile, which is stored in the small sac under its right end, the gallbladder.

Right lobe

Gallbladder

Portal vein

Hepatic artery

Bile duct

Liver lobule

Cross section of lobule

Central vein

Bile duct

Arteriole

Venule

3 ORGAN SUBSTRUCTURE

The structural–functional units of the liver are hepatic lobules. The lobules are six-sided, with blood vessels and bile ducts inside and between them.

TISSUES WITHIN THE LIVER

At least 20 types and subtypes of tissue are found in the liver, including blood and lymph fluids. There is very little muscle tissue of any kind, even smooth or involuntary muscle, which is common in other abdominal organs.

TISSUE TYPE	LIVER STRUCTURES
LIVER TISSUE	Consists of laminae or cords (sheets) of liver cells called hepatocytes, which form about 60 percent of all liver cells.
EPITHELIAL TISSUE	Forms the lining of blood spaces (sinusoids) between laminae.
DENSE CONNECTIVE TISSUE	Forms ligaments, such as the falciform ligament connecting the two lobes of the liver, and the ligamentum teres and ligamentum venosus, both remnants of blood vessels used by the fetus in the womb.
BLOOD	Flows throughout the liver; comprises plasma, red cells, platelets, and white cells, including Kupffer cells (see opposite).
LYMPHATIC TISSUE	Consists of lymph vessels (capillaries and lymphatics) throughout liver.
NERVOUS TISSUE	Nerves, both myelinated (sheathed) and unmyelinated (unsheathed).

BODY SYSTEMS TO CELLS

BROADLY SPEAKING, EACH SYSTEM CAN BE SEEN AS A HIERARCHY OF LARGE COMPONENTS COMPOSED OF SMALLER ONES. THE SYSTEM ITSELF IS AT THE TOP OF THE HIERARCHY; NEXT ARE ITS ORGANS; BELOW THESE ARE THE TISSUES THAT MAKE UP THE ORGANS; AND AT THE BOTTOM OF THE HIERARCHY ARE THE CELLS OF WHICH THE TISSUES ARE MADE.

A system of the body is usually regarded as a collection of organs and parts designed for one important task. The systems are integrated and interdependent, but each has its own identifiable components and boundaries. The main parts of a system are its organs and tissues. (In the circulatory system, the heart is the essential organ and pumps the liquid tissue, blood, around the body.) Most organs are composed of different tissues. The brain, for example, contains not only nervous tissue but also connective and epithelial (covering or lining) tissues. In turn, a tissue is a group of microscopic cells, all similar in structure and carrying out the same specialized function (see pp.40–41).

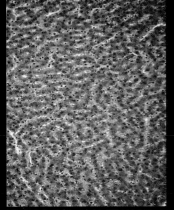

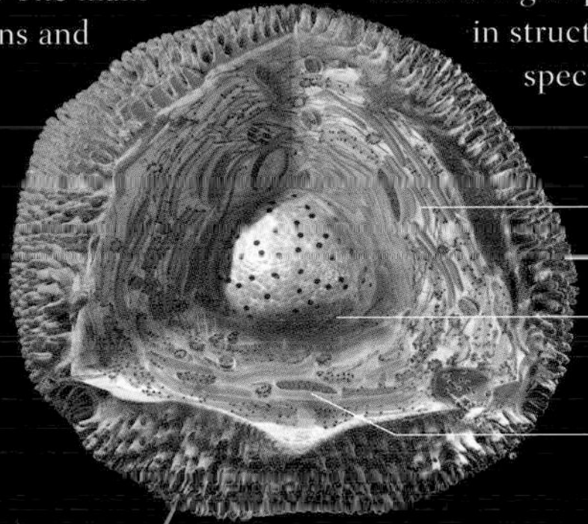

MICROSECTION
In this magnified section of liver tissue, the cells (pink) and their nuclei (dark purple) are visible. Blood cells lie in the lighter areas between the cells (hepatic sinusoids).

Kupffer cell
Also known as a hepatic macrophage; a type of white blood cell specific to the liver that engulfs and digests old worn-out blood cells and other debris

4 TISSUE
The unique tissue of the liver consists of branching sheets, or laminae, of liver cells (hepatocytes) arranged at angles. These are permeated by fluids and microscopic branches of two main kinds of tubes: blood vessels and bile ducts.

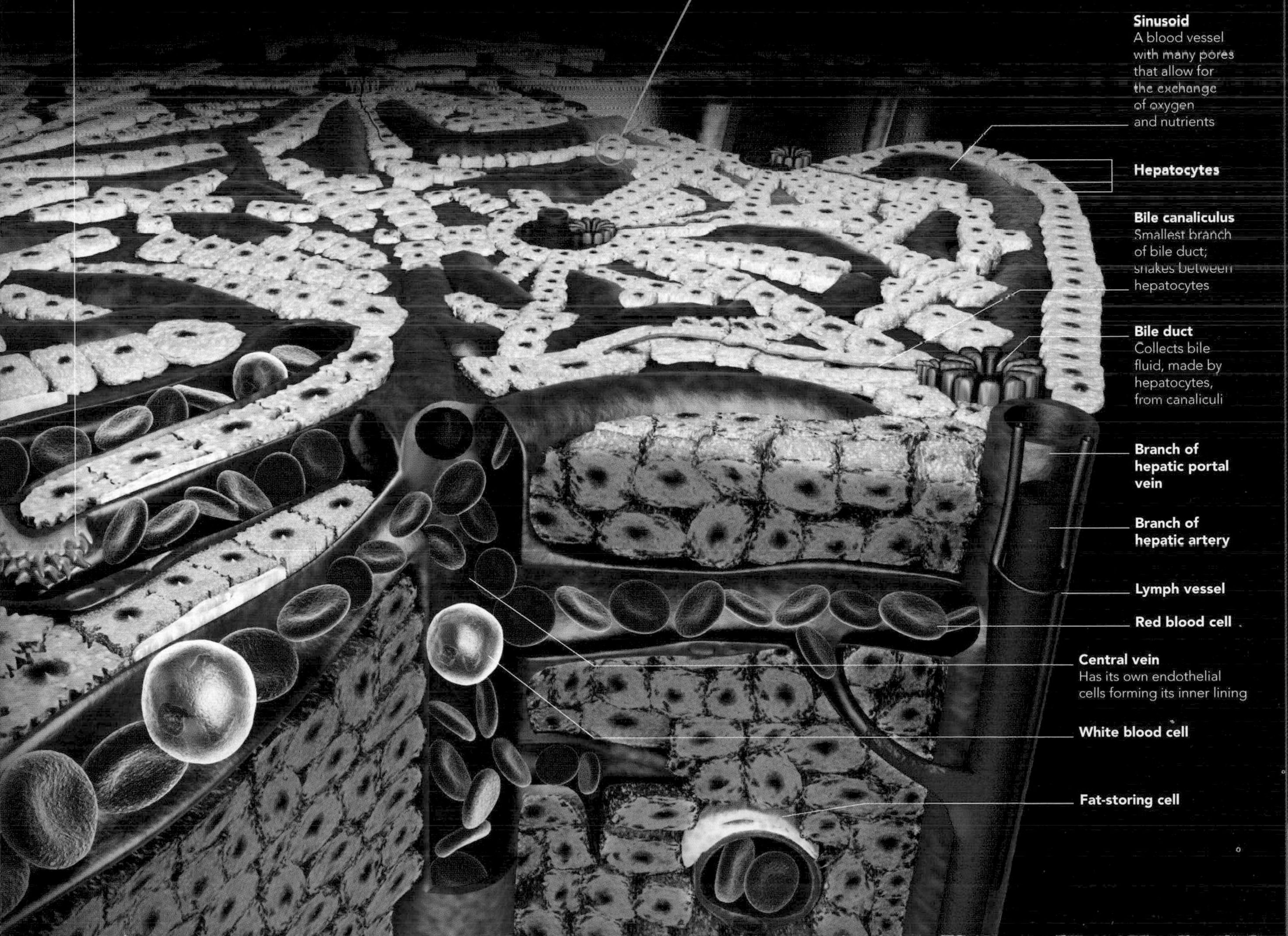

Cytoplasm
Cell membrane
Nucleus
Mitochondrion

5 CELL
The fundamental living unit of all tissues, a typical cell is capable of obtaining energy and processing nutrients. The hepatocytes of the liver are an example of body cells, containing most types of the miniature structures, called organelles, inside them.

Sinusoid
A blood vessel with many pores that allow for the exchange of oxygen and nutrients

Hepatocytes

Bile canaliculus
Smallest branch of bile duct; snakes between hepatocytes

Bile duct
Collects bile fluid, made by hepatocytes, from canaliculi

Branch of hepatic portal vein

Branch of hepatic artery

Lymph vessel

Red blood cell

Central vein
Has its own endothelial cells forming its inner lining

White blood cell

Fat-storing cell

Nucleolus
The region at the center of the nucleus that plays an important role in ribosome production

Nucleus
The cell's control center, containing most of the cell's DNA

Nuclear membrane
A two-layered membrane with pores through which substances enter and leave the nucleus

Nucleoplasm
The fluid within the nucleus in which nucleolus and chromosomes float

Cytoskeleton
Internal framework of the cell, comprised of microfilaments and hollow microtubules

Microfilament
Provides support for the cell; sometimes linked to the cell's outer membrane

Vacuole
A sac that stores and transports ingested materials, waste products, and water

Cytoplasm
Jellylike fluid in which organelles float; primarily water but also contains enzymes and amino acids

Mitochondrion
The site of fat and sugar digestion in the cell; produces energy

Microtubules
Part of the cell's cytoskeleton; aid movement of substances through the watery cytoplasm

Centriole
Composed of two cylinders of tubules; essential to cell reproduction

Microvilli
Projections found on some cells; they increase the cell's surface area, helping absorption of nutrients

Released secretions
Secretions are released from the cell by exocytosis— a vesicle merges with the cell membrane and releases its contents

Secretory vesicle
Sac that contains various substances, such as enzymes, that are produced by the cell and secreted at the cell membrane

Golgi complex
Organelle that processes and repackages proteins produced in rough endoplasmic reticulum for release at cell membrane

Lysosome
Produces powerful enzymes that aid in digestion and excretion of substances and worn-out organelles

Smooth endoplasmic reticulum
Network of tubes and flat, curved sacs that helps transport materials through cell; site of calcium storage; main location of fat metabolism

Cell membrane
Encloses contents of the cell, regulating the flow of substances into and out of the cell

Peroxisome
Makes enzymes that oxidize some toxic chemicals

Ribosome
Small structure that functions in protein assembly

Rough endoplasmic reticulum
Folded membranes extending throughout cell, studded with ribosomes; help transport materials through cell; site of much protein manufacture

INSIDE A CELL
This generalized body cell shows all the tiny structures (organelles), each with a particular task. Liver cells may be the closest equivalent to such a "general" cell. A large number of an organelle indicates the cell's chief role.

THE CELL

THE CELL IS THE BASIC STRUCTURAL AND FUNCTIONAL UNIT OF THE BODY. IT IS THE SMALLEST PART CAPABLE OF THE PROCESSES THAT DEFINE LIFE, INCLUDING REPRODUCTION, MOVEMENT, RESPIRATION, DIGESTION, AND EXCRETION— ALTHOUGH NOT EVERY CELL HAS ALL OF THESE ABILITIES.

CELL ANATOMY

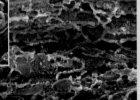

EMBRYONIC STEM CELL
All cells develop from one of two types of undifferentiated cell (stem cells), somatic or embryonic.

Most cells are microscopic; a typical cell is 20–30 μm in diameter, which means that 40 in a row would stretch across the period at the end of this sentence. Very specialized, long, thin cells include neurons (nerve cells) and muscle fiber cells (myofibers), which may extend more than 12 in (30 cm) but are incredibly thin. Most cells are bounded by an outer flexible "skin," the cell, or plasma, membrane. Inside is an array of structural components known as organelles, each with a characteristic shape, size, and function. These organelles do not float around at random. The cell is highly organized, with many interior chambers and compartments linked by sheets and membranes and held in place by a flexible, latticelike, ever-changing "skeleton" of even tinier tubules and filaments.

CELL MEMBRANE

Several features allow the membrane to fulfill its dual responsibilities of protecting the cell within and permitting movement of materials into and out of the cell. The primary component of this membrane is a double layer of phospholipid molecules. Each phospholipid has a water-loving (hydrophilic) head group and two hydrophobic tails. The two layers are arranged with the heads on both the outside and inside of the cell membrane, and the tails in between. The phospholipids are interspersed with protein molecules and carbohydrate chains that allow the cell to be recognized by other body cells.

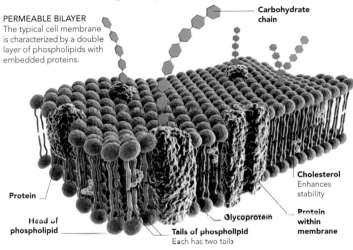

PERMEABLE BILAYER
The typical cell membrane is characterized by a double layer of phospholipids with embedded proteins.

Carbohydrate chain

Cholesterol
Enhances stability

Protein within membrane

Glycoprotein

Tails of phospholipid
Each has two tails

Head of phospholipid

Protein

MEMBRANES OF ORGANELLES

Membranes abound in the cell. They separate and divide the cytoplasm into sections and control the passage of materials between these regions; act as attachment points for ribosomes and other structures and as storage areas; and shape channels along which substances move. Several important organelles are enclosed in their own membrane.

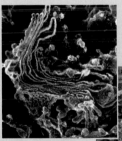

GOLGI COMPLEX
Within this stack of flattened membranous sacs, protein from the endoplasmic reticulum is modified and repackaged.

ENDOPLASMIC RETICULUM (ER)
A series of highly folded and curved ER membranes usually encloses one continuous labyrinthine space.

MITOCHONDRION
The inner membrane is folded in shelflike, incomplete partitions to increase the surface area for releasing energy from sugars and fats. The outer membrane of a mitochondrion is smooth and featureless.

TRANSPORT

The transfer of materials through the cell membrane occurs by one of three processes. Small molecules such as glycerol, water, oxygen, and carbon dioxide cross the membrane by diffusion. Molecules that cannot cross the phospholipid layer must cross by facilitated diffusion. When substances (including minerals and nutrients) are in lower concentration on the outside of the cell than on the inside, they must be conveyed by active transport, requiring energy.

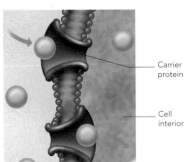

FACILITATED DIFFUSION
A carrier protein binds with a specific molecule, such as glucose, outside the cell, then changes shape and ejects the molecule into the cell.

Carrier protein

Cell interior

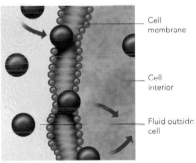

DIFFUSION
Many molecules naturally move from an area where they are in high concentration to one in which their numbers are fewer. This process is known as diffusion.

Cell membrane

Cell interior

Fluid outside cell

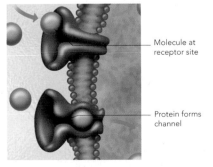

ACTIVE TRANSPORT
Molecules bind to a receptor site on the cell membrane, triggering a protein to change into a channel, through which molecules travel.

Molecule at receptor site

Protein forms channel

CELL TYPES AND TISSUES

MORE THAN 200 TYPES OF SPECIALIZED CELLS POPULATE THE HUMAN BODY. THEY DEVELOP WITH THEIR OWN KIND TO FORM CLOSELY KNIT CONFIGURATIONS, WHICH ARE CLEARLY RECOGNIZABLE AS SPECIFIC TYPES OF TISSUE. IN SOME CASES, TISSUES ARE MADE OF SEVERAL TYPES OF CELLS.

TISSUE TYPES

The cells that form tissue all have much the same structure and perform the same function. There are four primary tissue types, derived from specific cell layers in the early embryo: epithelial, connective, muscle, and nerve. Blood, bone, cartilage, tendons, and ligaments are forms of connective tissue. The epidermis and the tissues that line almost every organ are all types of epithelial tissues. Muscle and nerve tissues, of course, form muscles and nerves.

CELL TYPES

Cells come in many shapes and sizes, depending on their specialized functions within tissues. Speed of cell division also varies. It is most rapid in epithelial (covering and lining) cells, which are subjected to physical abrasion and wear, and which must continually replace themselves. It is slow or even nonexistent in structurally complex cells such as nerve cells (neurons).

Epithelial cells
These cells form skin, cover most organs, and line hollow cavities. The cells shown here are from the top surface of the intestinal tract.

Smooth muscle cell
The large, elongated, spindlelike cells of smooth muscle are known as muscle fibers. The shape allows for contraction by means of sliding strands of protein inside.

Photoreceptor cell
A cone cell is a type of light-sensitive neuron that is found in the retina of the eye. Cone cells are activated by bright light and are responsible for color perception.

Nerve cell
Each cell has a configuration of short extensions (dendrites) to receive nerve signals, and a long "wire" (axon) to send signals to other cells.

Red blood cell
The red cell (erythrocyte) is a bag of oxygen-carrying hemoglobin molecules. Its biconcave (double-dished) shape allows rapid, maximal oxygen absorption.

Sperm cell
Each sperm has a head that carries the paternal set of genetic material, and a long, whiplike tail that propels it toward the egg.

Adipose (fat) cell
The main adipose cells, adipocytes, are bulky and crammed with droplets of fat (lipids), which store energy in case the diet cannot meet requirements.

Ovum (egg) cell
These giant cells contain the maternal complement of genetic material and energy resources for the first cell divisions that shape the early embryo.

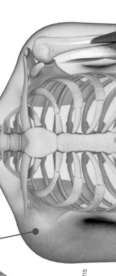

ELASTIC CARTILAGE
Cartilaginous tissues (types of connective tissue) have varying properties depending on cell proportions and type of tissue structure. This microscope image of an epiglottis reveals rounded cells (chondrocytes) in fibers of elastin, making it lightweight, flexible, and strong.

LARYNX

Elastic cartilage
Light and bendy; holds the larynx open

Hyaline cartilage
Tough yet flexible; the most common type of cartilage

White matter
Contains long, wirelike insulated nerve fibers

Gray matter
Contains nerve cell bodies and support cells

CROSS SECTION OF BRAIN FROM THE FRONT

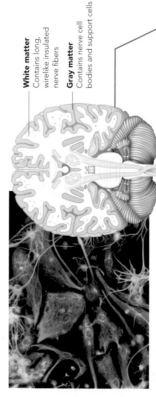

NERVE TISSUE
An immunofluorescent microscope image showing glial cells that support message-conducting nerve cells (neurons) in nerve tissue, and are known as neuroglia. Among these are astrocytes (light green spiderlike forms). These supply neurons with nutrients.

Connective dermal tissue
Connects skin (pictured) to underlying organs

LOOSE CONNECTIVE TISSUE
Some connective tissue (in parts of the lower layer of skin, for example) is made up of cells that are loosely embedded in fibers. Here, the nuclei of fibroblast cells (dark spots) can be seen among fibers of elastin (dark lines) and collagen (broad purple stripes).

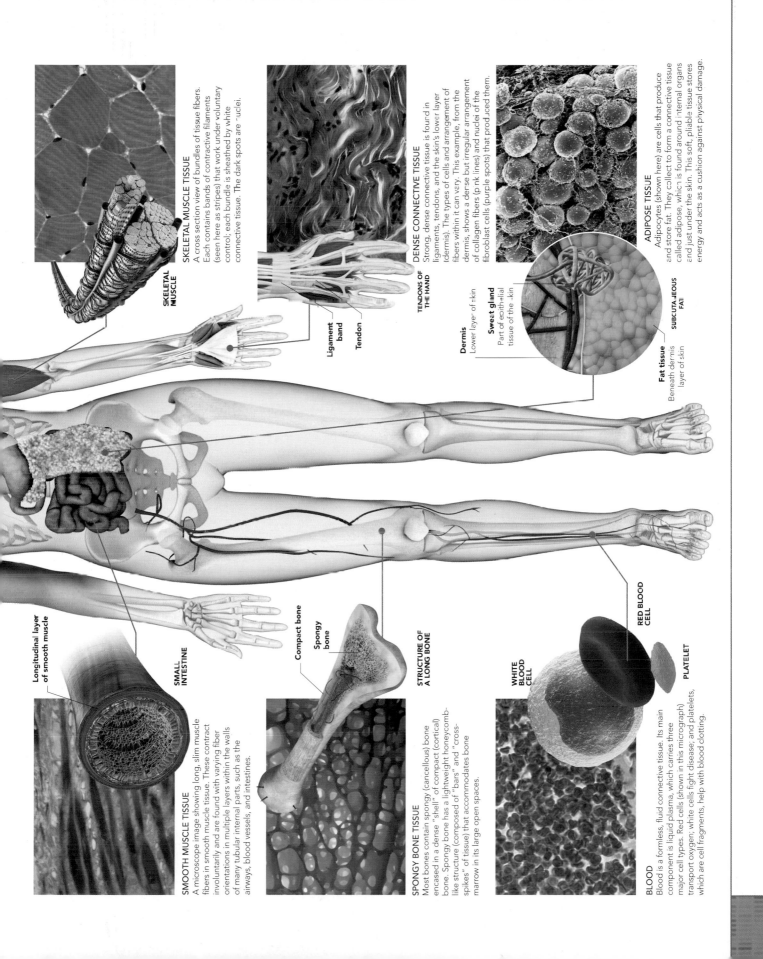

SKELETAL MUSCLE TISSUE

A cross section view of bundles of tissue fibers. Each contains bands of contractile filaments (seen here as stripes) that work under voluntary control; each bundle is sheathed by white connective tissue. The dark spots are nuclei.

SKELETAL MUSCLE

Ligament band

Tendon

TENDONS OF THE HAND

DENSE CONNECTIVE TISSUE

Strong, dense connective tissue is found in ligaments, tendons, and the skin's lower layer (dermis). The types of cells and arrangement of fibers within it can vary. This example, from the dermis, shows a dense but irregular arrangement of collagen fibers (pink lines) and nuclei of the fibroblast cells (purple spots) that produced them.

Dermis
Lower layer of skin

Sweat gland
Part of epithelial tissue of the skin

Fat tissue
Beneath dermis layer of skin

SUBCUTANEOUS FAT

ADIPOSE TISSUE

Adipocytes (shown here) are cells that produce and store fat. They collect to form a connective tissue called adipose, which is found around internal organs and just under the skin. This soft, pliable tissue stores energy and acts as a cushion against physical damage.

Longitudinal layer of smooth muscle

SMOOTH MUSCLE TISSUE

A microscope image showing long, slim muscle fibers in smooth muscle tissue. These contract involuntarily and are found with varying fiber orientations in multiple layers within the walls of many tubular internal parts, such as the airways, blood vessels, and intestines.

SMALL INTESTINE

Compact bone

Spongy bone

STRUCTURE OF A LONG BONE

SPONGY BONE TISSUE

Most bones contain spongy (cancellous) bone encased in a dense "shell" of compact (cortical) bone. Spongy bone has a lightweight honeycomb-like structure (composed of "bars" and "cross-spikes" of tissue) that accommodates bone marrow in its large open spaces.

WHITE BLOOD CELL

RED BLOOD CELL

PLATELET

BLOOD

Blood is a formless, fluid connective tissue. Its main component is liquid plasma, which carries three major cell types. Red cells (shown in this micrograph) transport oxygen; white cells fight disease; and platelets, which are cell fragments, help with blood clotting.

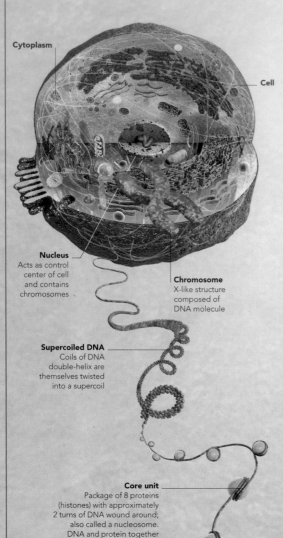

Cytoplasm

Cell

Nucleus
Acts as control center of cell and contains chromosomes

Chromosome
X-like structure composed of DNA molecule

Supercoiled DNA
Coils of DNA double-helix are themselves twisted into a supercoil

Core unit
Package of 8 proteins (histones) with approximately 2 turns of DNA wound around; also called a nucleosome. DNA and protein together like this is called chromatin

DNA

OFTEN REFERRED TO AS THE MOLECULE OF LIFE, DNA (DEOXYRIBONUCLEIC ACID) IS FOUND IN ALMOST ALL LIVING THINGS. IT ACTS AS A TYPE OF CHEMICAL CODE THAT CONTAINS INSTRUCTIONS, KNOWN AS GENES, FOR HOW THE BODY AND ALL ITS DIFFERENT PARTS GROW, DEVELOP, FUNCTION, AND MAINTAIN THEMSELVES.

In nearly all human cells, DNA is packaged into 46 X-shaped elements called chromosomes, which are situated in the cell's nucleus. DNA's enormous list of instructions takes the form of long, thin molecules, one per chromosome, each taking the shape of a double-helix. Each double-helix has two long, corkscrewlike strands, which act as "backbones" for the molecule, twining around each other. These are held together by rungs, like a twisted ladder. The rungs are made of pairs of chemicals called bases: adenine (A), guanine (G), thymine (T), and cytosine (C). In each rung, A always pairs with T, and G with C. This structure gives DNA its two key features: the order of the bases contains the chromosome's genetic code, while the way the bases cross-link enables DNA to make exact copies of itself.

DNA UNDER THE MICROSCOPE
This scanning tunneling micrograph (STM) of DNA, magnified about one million times, shows the twists of the helix as a series of yellow peaks on the left.

DNA backbone
Constructed of alternating units of deoxyribose (a form of sugar) and phosphate chemicals

Base pairs
Base pairs form cross-linked "rungs" on the DNA "twisted ladder"

BASE PAIRS

The four bases can pair in only two configurations due to their chemical structures. Adenine and thymine each have two positions for forming hydrogen bonds and so fit together, while guanine and cytosine each have three hydrogen-bond locations.

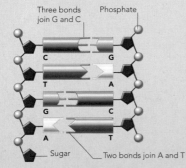

Three bonds join G and C

Phosphate

C — G

T — A

G — C

A — T

Sugar

Two bonds join A and T

At certain stages between cell divisions, the chromosome unravels, enabling the DNA molecule to expose information for making proteins and to allow replication

DOUBLE HELIX
A DNA molecule is coiled and supercoiled in order to fit into the chromosomes (see panel, opposite). It also loops and twists. The molecule is accompanied by various proteins, particularly histones.

Base pair sequence
Base pairs are arranged in a specific order; some sections of the sequence act as sets of coded instructions for making proteins

COILS AND SUPERCOILS

DNA's multicoiled structure allows an incredible length to be packed into a tiny space. The single length of DNA in a typical chromosome, if unwound, would stretch about 2 in (5 cm). There are 46 chromosomes in the nucleus of each cell (except mature red blood cells, which have no nucleus or DNA). When cells are not dividing, the DNA (wrapped around protein to form what is called chromatin) is relatively loosely coiled. This allows portions to be available for protein assembly and other functions. As a cell prepares to divide, its DNA coils into supercoils, which are shorter and denser, and visible as the typical chromosome "X" shapes.

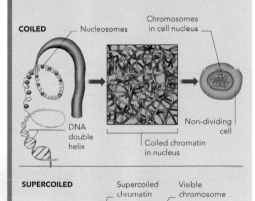

COILED
- Nucleosomes
- Chromosomes in cell nucleus
- DNA double helix
- Coiled chromatin in nucleus
- Non-dividing cell

SUPERCOILED
- Coiled chromatin
- Supercoiled chromatin
- Visible chromosome
- Cell prepared for divison

WHAT ARE GENES?

A gene is generally regarded as a unit of DNA needed to construct one protein. It consists of all the sections of DNA that code for all the amino acids for that protein. Usually one gene is located on one chromosome. However, it may have several sections on different regions of the DNA molecule, each containing the code for one portion of the protein. Typically, lengths of DNA called introns and exons are both transcribed (see pp.44–45) to form immature mRNA. The parts of mRNA made from the introns are then stripped out by the cell's molecular machinery, leaving mature mRNA for translation. There are also regulatory DNA sequences that code for their own proteins, affecting the gene transcription rate.

EYE COLOR
Iris color is affected by at least 15 genes, including OCA2 and HERC2, both sited on chromosome 15.

PARTS OF A GENE
Regions called introns and exons both transcribe to form mRNAs for different portions of a protein. The lengths made from introns are then spliced out chemically, to leave exon-only portions, which go on to make the protein.

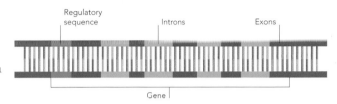

- Regulatory sequence
- Introns
- Exons
- Gene

RANGE OF GENE SIZE
Genes vary enormously in their size, which is usually measured in numbers of base pairs. Small genes may be just a few hundred base pairs long, while others are measured in millions of base pairs. The gene for beta globin is one of the smallest. It codes for part of the hemoglobin molecule. It is compared, right, with a larger gene.

LARGE GENE F8 (FOR BLOOD CLOTTING FACTOR VIII) ON X CHROMOSOME
- Exons
- Code for protein
- 186,935 base pairs

SMALL GENE (FOR BETA GLOBIN) ON CHROMOSOME 11
- 1,605 base pairs
- Exons

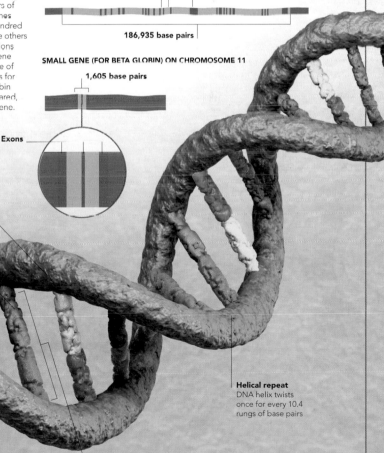

Adenine–thymine link
Adenine always forms a base pair with thymine

Strands coil around each other to form double helix

- Guanine
- Thymine
- Cytosine
- Adenine

Helical repeat
DNA helix twists once for every 10.4 rungs of base pairs

Guanine–cytosine link
Guanine always forms a base pair with cytosine

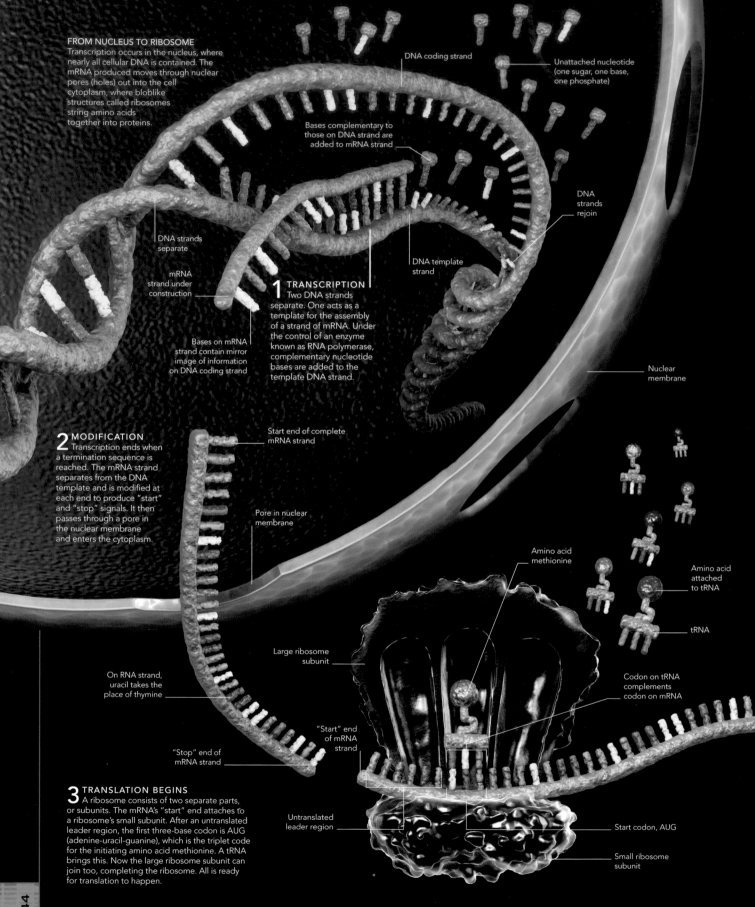

FROM NUCLEUS TO RIBOSOME
Transcription occurs in the nucleus, where nearly all cellular DNA is contained. The mRNA produced moves through nuclear pores (holes) out into the cell cytoplasm, where bloblike structures called ribosomes string amino acids together into proteins.

DNA coding strand

Unattached nucleotide (one sugar, one base, one phosphate)

Bases complementary to those on DNA strand are added to mRNA strand

DNA strands rejoin

DNA strands separate

mRNA strand under construction

DNA template strand

Bases on mRNA strand contain mirror image of information on DNA coding strand

1 TRANSCRIPTION
Two DNA strands separate. One acts as a template for the assembly of a strand of mRNA. Under the control of an enzyme known as RNA polymerase, complementary nucleotide bases are added to the template DNA strand.

Nuclear membrane

2 MODIFICATION
Transcription ends when a termination sequence is reached. The mRNA strand separates from the DNA template and is modified at each end to produce "start" and "stop" signals. It then passes through a pore in the nuclear membrane and enters the cytoplasm.

Start end of complete mRNA strand

Pore in nuclear membrane

Amino acid methionine

Amino acid attached to tRNA

tRNA

On RNA strand, uracil takes the place of thymine

Large ribosome subunit

Codon on tRNA complements codon on mRNA

"Start" end of mRNA strand

"Stop" end of mRNA strand

3 TRANSLATION BEGINS
A ribosome consists of two separate parts, or subunits. The mRNA's "start" end attaches to a ribosome's small subunit. After an untranslated leader region, the first three-base codon is AUG (adenine-uracil-guanine), which is the triplet code for the initiating amino acid methionine. A tRNA brings this. Now the large ribosome subunit can join too, completing the ribosome. All is ready for translation to happen.

Untranslated leader region

Start codon, AUG

Small ribosome subunit

FROM DNA TO PROTEIN

DNA PROVIDES INFORMATION TO ASSEMBLE PROTEINS—THE "BUILDING BLOCKS" OF ALMOST EVERY BODY PART. THERE ARE TWO STAGES TO THE PROCESS. IN ADDITION TO DNA, IT INVOLVES TWO MORE KINDS OF NUCLEIC ACIDS, mRNA AND tRNA.

Manufacture of proteins occurs in two main phases, transcription and translation. In transcription, information is taken from the DNA and copied to an intermediate type of molecule called mRNA (messenger ribonucleic acid). This is built from nucleotide units in a similar way to DNA. The mRNA moves out of the cell's nucleus to protein-building units known as ribosomes. In the translation phase,

the mRNA acts as a template for the assembly of units of protein called amino acids. There are about 20 different amino acids. Their order is specified by lengths of mRNA three bases long, called triplet codons. The order of bases in each codon is the code for a particular amino acid (hence the term genetic code). The mRNA carries instructions to make a specific protein from a sequence of amino acids.

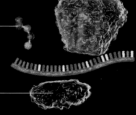

Completed amino acid chain (protein)

Small ribosome subunit separates from large subunit and mRNA strand

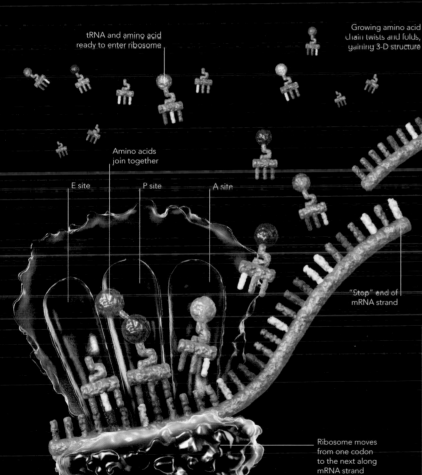

tRNA and amino acid ready to enter ribosome

Growing amino acid chain twists and folds, gaining 3-D structure

Amino acids join together

E site

P site

A site

Stop codon

Untranslated trailer region

"Stop" end of mRNA strand

5 TERMINATION
The amino acid chain bends and folds according to the angles and links between its amino acids. When the ribosome reaches the mRNA's "stop" signal, the chain of amino acids—now a large part of a protein or even a whole protein—detaches and the two subunits of the ribosome separate.

Ribosome moves from one codon to the next along mRNA strand

4 TRANSLATION IN MIDFLOW
Each mRNA codon attracts a particular tRNA with the right amino acid for that position in the sequence. Within the ribosome are three binding sites identified by the letters A, P, and E. As the ribosome moves along the mRNA, a tRNA joins at its A site; the amino acid detaches from it at the P site and joins the growing chain; and the tRNA leaves at the E site.

RNA

RNA (ribonucleic acid) is like DNA, but with essential differences. RNA has ribose sugars in its "backbone," not deoxyribose; it has a single helix strand rather than double; its sequences are generally shorter than DNA; and the base uracil, rather than thymine, links to adenine. However, like DNA, RNA carries coded information. Examples are mRNA and tRNA (see opposite), and the ribosome itself is partly made of rRNA (ribosomal RNA). Lengths of RNA called ribozymes work as enzymes, while other RNAs include DNA-splicing controllers and regulatory "gene switches" to turn genes on and off.

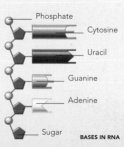

Phosphate

Cytosine

Uracil

Guanine

Adenine

Sugar

BASES IN RNA

THE GENOME

A GENOME IS THE FULL SET OF GENETIC INSTRUCTIONS FOR A LIVING THING, CONTROLLING ITS DEVELOPMENT FROM A SINGLE CELL INTO A COMPLEX ADULT BODY. THE HUMAN GENOME CONSISTS OF AN ESTIMATED 20,000 GENES FOR MAKING PROTEINS, CARRIED ON THE DOUBLE SET OF 46 CHROMOSOMES IN MOST BODY CELLS.

CHROMOSOMES AND DNA

The Human Genome Project, a multinational effort to map the sequence of the human genome, was completed in 2003. This research led to the identification of more than 20,000 individual human genes within 46 chromosomes that collectively include 3.2 billion base pairs. Although much DNA does not provide codes for individual genes, known as noncoding and "junk" DNA, it may regulate their function. No one knows for sure yet how much DNA really is "junk"—completely non-functional. Mapping of the genome makes it possible for medical researchers to know which genes are involved in certain metabolic processes.

98%
Non-coding DNA

2%
Coding DNA (genes)

CODING AND NON-CODING
Only about 2% of the genome's DNA is estimated to carry codes that are used to make proteins. Some of the rest is involved in complex switching and control systems for gene activation, many using RNA.

KARYOTYPE
A "group photograph" of all chromosomes from a cell, arranged in their pairs in a standard order, is known as a karyotype. It reveals extra, missing, broken, or oddly banded chromosomes. This example is from a female (note the two, equal-sized X chromosomes at the bottom right).

CHROMOSOMES
This scanning electron microscope image shows the coils and supercoils of a DNA double helix within each chromosome, shaped like a large fluffy brush.

CHROMOSOME COMPLEMENT
The full set of chromosomes in a human cell numbers 46. These consist of 22 equivalent pairs, one of each pair derived from the mother and one from the father. They are numbered from 1 (largest) to 22 (smallest). The 23rd pair is the sex chromosomes, XX signifying female and XY (as here), male. When colored by chemical stains, dark and pale stripes called banding patterns show up on each chromosome. These allow researchers to "map" the locations of particular genes within the chromosome.

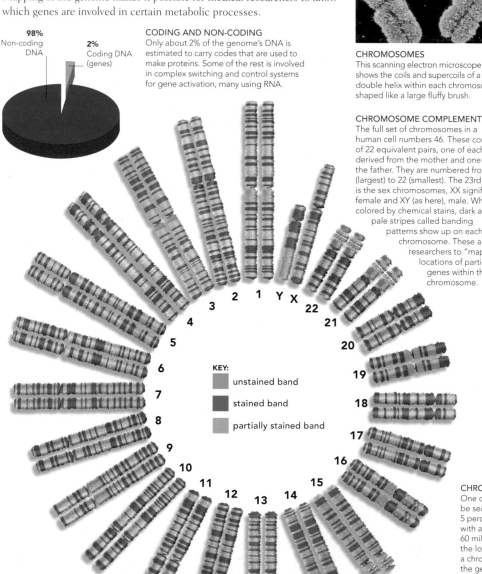

KEY:
- unstained band
- stained band
- partially stained band

p22.2
p21.3
p21.1
p15.2
p14.3
p14.1
p12.3
p12.1
q11.22
q11.23
q21.11
q21.2
q22.1
q23.3
q31.2
Cystic fibrosis
q31.32
q32.1
q33
q36.1
q36.3

CHROMOSOME SEVEN
One of the first chromosomes to be sequenced, it contains more than 5 percent of the genome's total DNA, with about 159 million pairs of bases. Almost 60 million are in the short arm, 7p, with the rest in the longer arm, 7q. The conventions of labeling a chromosome make it possible to find the site of the gene if you know its "address." The cystic fibrosis gene (CFTR), for example, is located at 7q31.2.

MITOCHONDRIAL GENES

Mitochondria, the powerhouses of the cell, have their own DNA. Unlike DNA in the nucleus, mitochondrial DNA (mtDNA) is circular, not linear. It contains just 37 genes which code for the proteins and RNA the mitochondrion needs for its functions. Mitochondrial DNA is unique in being inherited only from the maternal line, via the mitochondria present in the egg at fertilization. This type of DNA has been used to study genetic relationships and reunite families, as the high rate of mutation of mtDNA means unrelated people have very different mtDNA. Certain rare diseases are associated with changes in the mtDNA.

MITOCHONDRIAL DNA
This electron microscope image shows that mitochondrial DNA forms a closed loop, unlike the DNA in the nucleus of the cell, which is linear.

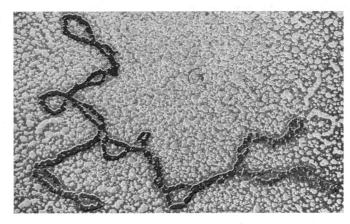

GENETIC CONTROL OF CELLS

Not all genes are active and working in all cells. The process by which a gene is able to make its protein or other substance is known as gene expression. The expression of each gene is controlled according to exposure to chemicals such as growth factors and regulators—products of other genes. Some types of genes are "switched on" and express themselves in most cells. These are genes concerned with basic life processes and "housekeeping," such as breaking down glucose sugar for energy. Other genes are "switched off" unless they are needed; these are for making specialized products, such as hormones, or particular proteins, like the actin and myosin filaments that pack muscle cells. As cells' genes are switched on and off in different circumstances they differentiate, or become different.

CHEMICAL ADDITION
A methyl group (one carbon and three hydrogen atoms) attaches to a gene to switch it off

EPIGENETICS
Expression of genes can be regulated by the attachment of small chemical compounds to DNA or to the histones (proteins) DNA wraps around. These modifications are called epigenetic changes. They do not change the DNA sequence (genetic code), but do affect the way the genes are expressed. Scientists are starting to realize that some epigenetic changes that build up in an individual's DNA during his or her life can actually be inherited by the next generation.

STEM CELLS

A stem cell is an undifferentiated cell, which retains the ability both to keep dividing for self-renewal of its population, and to become specialized in certain conditions. Embryonic stem cells occur in the early embryo, and have the ability to differentiate into any of the 200-plus types of specialized cells in the human body. Adult stem cells occur in certain tissues, where they multiply rapidly as part of ongoing maintenance. In the bone marrow, they produce millions of different blood cells every second.

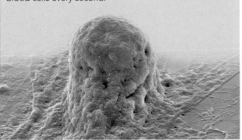

CELLULAR DIFFERENTIATION
The first cells produced by divisions of a fertilized egg are "generalized" stem cells. As they build in numbers, pre-programmed instructions begin to act. Patterns of intercellular contacts and the chemical environment influence genes in cells lying in certain parts of an embryo, making the cell differentiate, or become different. In differentiating, stem cells become nerve, muscle, and skin cells, and all other necessary cell types.

PRECURSOR CELL

PRECURSOR CELL
This can become any of a variety of cells. Some lines of offspring cells retain the ability to specialize, while others go on to become specialists

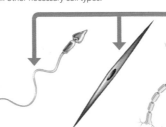

SPERM CELL
Packed with mitochondria to supply fuel

MUSCLE CELL
Long, thin cells packed with contractile proteins

NERVE CELL
Extreme specialization in shape and connections

EPITHELIAL CELL
Programmed to multiply rapidly and then die

FAT CELL
Stores energy in case diet does not meet energy requirements

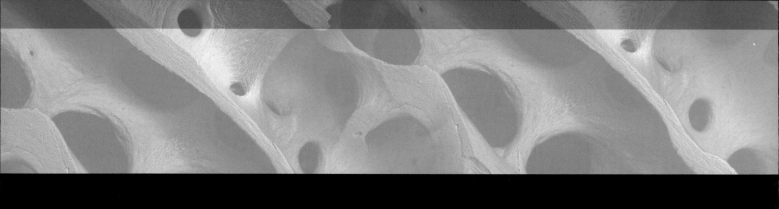

WITH ITS HIGHLY ENGINEERED JOINTS, THE LIVING SKELETON
PROVIDES A FRAMEWORK OF STIFF LEVERS AND STABLE PLATES
THAT PERMITS A MULTITUDE OF MOVEMENTS. INTIMATELY
CONNECTED WITH THE MUSCULAR SYSTEM, THE SKELETON
ALSO INTEGRATES FUNCTIONALLY WITH THE CARDIOVASCULAR
SYSTEM—EVERY SECOND, MILLIONS OF FRESH BLOOD CELLS
POUR OUT OF THE BONE MARROW. A HEALTHY DIET THAT

SKELETAL SYSTEM

SKELETON

THE SKELETON MAKES UP ALMOST ONE-FIFTH OF A HEALTHY BODY'S WEIGHT. THIS FLEXIBLE INNER FRAMEWORK SUPPORTS ALL OTHER PARTS AND TISSUES, WHICH WOULD COLLAPSE WITHOUT SKELETAL REINFORCEMENT. THE SKELETON ALSO PROTECTS CERTAIN ORGANS, SUCH AS THE DELICATE BRAIN INSIDE THE SKULL. IN ADDITION, BONES ARE RESERVOIRS OF IMPORTANT MINERALS, ESPECIALLY CALCIUM, AND THE SITE OF BLOOD CELL PRODUCTION.

The average skeleton has 206 bones. There are natural variations: about one individual in 200 has an extra rib. The number of small bones fused into the skull also varies. Bone is an active tissue, and even though it is about 22 percent water, it has an extremely strong yet lightweight and flexible structure. A similar frame made of high-technology composite materials could not match the skeleton's weight, strength, and durability. The skeleton also has the advantage of being able to repair itself if damaged. It can remodel its bones to thicken and strengthen them in areas of extra stress, as seen in some activities such as horseback riding and weight lifting. There are two major divisions: the axial and appendicular skeletons. The axial skeleton consists of the skull, vertebral (spinal) column, ribs, and sternum. The appendicular skeleton includes the bones of the shoulder, arm, wrist, and hand, and the hips, legs, ankles, and feet. Of the 206 bones, 80 are in the axial skeleton, with 64 in the upper appendicular and 62 in the lower appendicular skeleton.

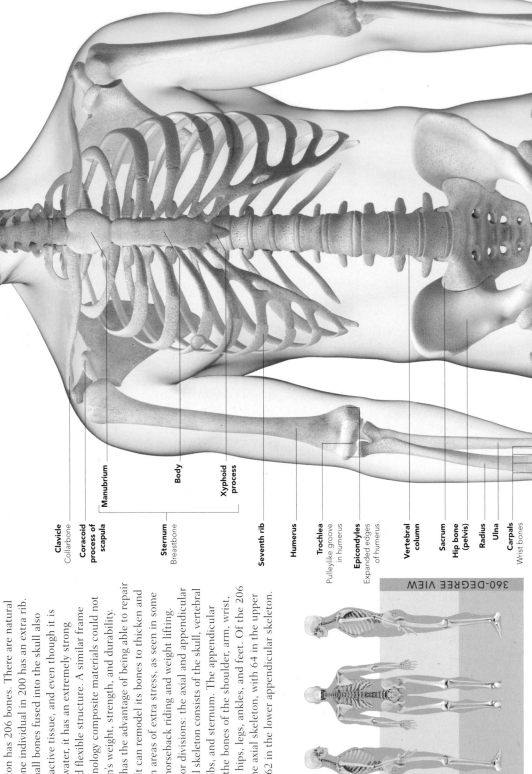

Cranium
Skull

Mandible
Jawbone

Clavicle
Collarbone

**Coracoid
process of
scapula**

Manubrium

Sternum
Breastbone

Body

**Xyphoid
process**

Seventh rib

Humerus

Trochlea
Pulleylike groove
in humerus

Epicondyles
Expanded edges
of humerus

**Vertebral
column**

Sacrum

**Hip bone
(pelvis)**

Radius

Ulna

Carpals
Wrist bones

360-DEGREE VIEW

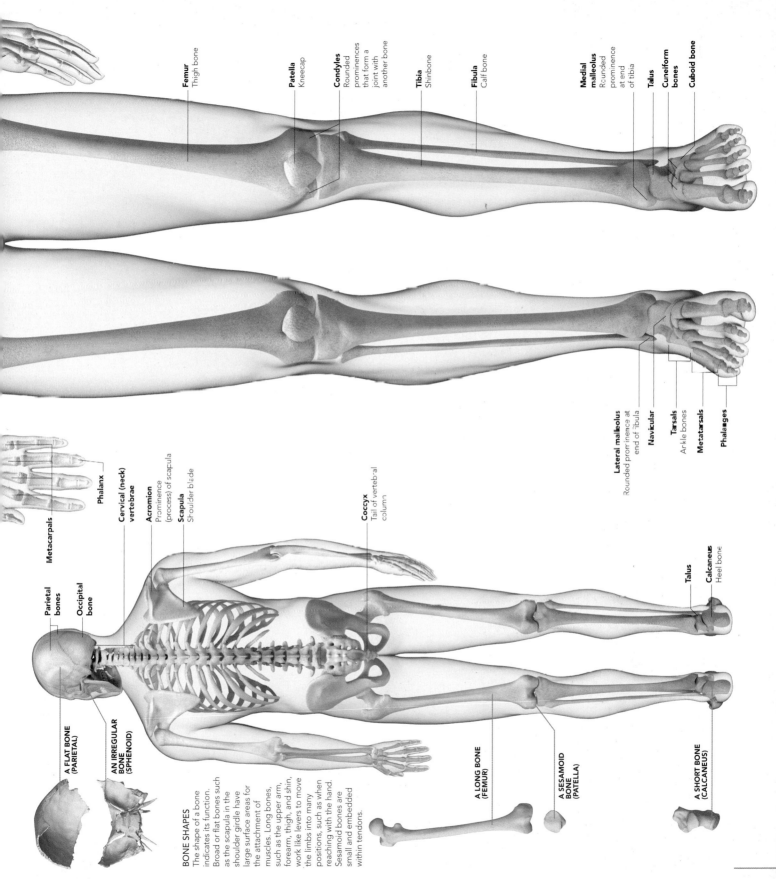

Femur
Thigh bone

Patella
Kneecap

Condyles
Rounded prominences that form a joint with another bone

Tibia
Shinbone

Fibula
Calf bone

Medial malleolus
Rounded prominence at end of tibia

Talus

Cuneiform bones

Cuboid bone

Lateral malleolus
Rounded prominence at end of fibula

Navicular

Tarsals
Ankle bones

Metatarsals

Phalanges

Metacarpals

Phalanx

Cervical (neck) vertebrae

Acromion
Prominence (process) of scapula

Scapula
Shoulder blade

Coccyx
Tail of vertebral column

Talus

Calcaneus
Heel bone

Parietal bones

Occipital bone

BONE SHAPES

The shape of a bone indicates its function. Broad or flat bones such as the scapula in the shoulder girdle have large surface areas for the attachment of muscles. Long bones, such as the upper arm, forearm, thigh, and shin, work like levers to move the limbs into many positions, such as when reaching with the hand. Sesamoid bones are small and embedded within tendons.

A FLAT BONE (PARIETAL)

AN IRREGULAR BONE (SPHENOID)

A LONG BONE (FEMUR)

A SESAMOID BONE (PATELLA)

A SHORT BONE (CALCANEUS)

BONE STRUCTURE

BONE IS A TYPE OF CONNECTIVE TISSUE THAT IS AS STRONG AS STEEL BUT AS LIGHT AS ALUMINUM. IT IS MADE OF SPECIALIZED CELLS AND PROTEIN FIBERS. NEITHER IMMOBILE NOR DEAD, BONE CONSTANTLY BREAKS DOWN AND REBUILDS ITSELF. EACH BONE ADJUSTS ITS SIZE AND SHAPE DURING THE GROWING PROCESS, AFTER AN INJURY, AND IN RESPONSE TO STRESS.

STRUCTURE OF A BONE

Along the central shaft of a long bone (such as the femur, tibia, or humerus) is the medullary canal or marrow cavity. This contains red bone marrow, which produces blood cells; yellow marrow, which is mostly fatty tissue; and plentiful blood vessels. Surrounding the marrow cavity is a layer of spongy (cancellous) bone, the honeycomb-like cavities of which also contain marrow. Around this is a shell-like layer of compact (cortical) bone, which is hard, dense, and strong. Small canals connect the marrow cavity with the periosteum, a membrane that covers the bone surface. Bone tissue is made of specialized cells and protein fibers, chiefly collagen, woven into a matrix of water, mineral crystals and salts, carbohydrates, and other substances. Bone cells include osteoblasts, which calcify bone as it forms; osteocytes, which maintain healthy bone structure; and osteoclasts, which absorb bone tissue where it is degenerating or not needed.

COMPACT BONE
Compact, or cortical, bone is composed of tiny rodlike cells called osteons. Under a microscope, these can be seen bundled tightly together, an arrangement that provides great strength.

Blood vessel
Rich network of blood vessels nourishes bone

INSIDE A BONE
Long bones, for example those in the leg, have most types of bone tissue. The ratio of compact to spongy bone varies with age and activity, reflecting the physical stresses placed on the bone.

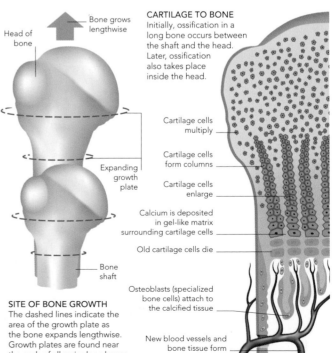

Periosteum
Thin, fibrous membrane covering entire bone surface (except in joints)

Compact bone
Bone gets its strength from this hard, shell-like tissue

BONE CELLS

Healthy bones depend on three types of cells, formed (along with blood cells) in the marrow. Osteoblasts at first produce bone as the skeleton grows. They then turn into osteocytes, which sustain surrounding bone tissue. Osteoclasts are large cells with several nuclei that break down unwanted or unhealthy bone (see pp.54–55).

OSTEOCYTE IN BONE
This greatly magnified picture shows an osteocyte bone cell within a tiny cavity (lacuna) in compact bone.

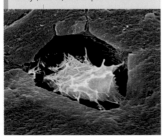

BONE GROWTH

During development in the womb and infancy, most bones develop from structures made of cartilage. Ossification is the process by which the cartilage tissue of these structures is converted into bone tissue by the deposition of mineral salts and crystals, mainly phosphates and carbonates of calcium. Much of the increase in height during childhood comes from lengthening long bones. Near each end of a long bone is an area known as the growth plate, where lengthening and ossification both occur. Cartilage cells (see opposite) multiply here and form columns toward the bone shaft. As the cartilage cells enlarge and die, the space they occupied is filled by new bone cells. In this way the growth plate moves along the lengthening bone shaft, remaining between the shaft and the head of the bone.

Bone grows lengthwise

Head of bone

Expanding growth plate

Bone shaft

SITE OF BONE GROWTH
The dashed lines indicate the area of the growth plate as the bone expands lengthwise. Growth plates are found near the ends of all major long bones.

CARTILAGE TO BONE
Initially, ossification in a long bone occurs between the shaft and the head. Later, ossification also takes place inside the head.

Cartilage cells multiply

Cartilage cells form columns

Cartilage cells enlarge

Calcium is deposited in gel-like matrix surrounding cartilage cells

Old cartilage cells die

Osteoblasts (specialized bone cells) attach to the calcified tissue

New blood vessels and bone tissue form

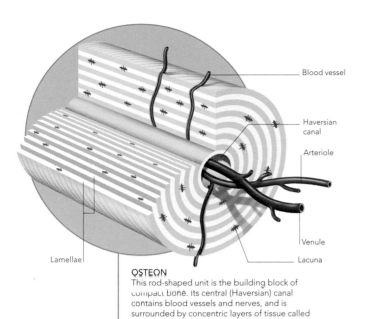

Lamellae

Blood vessel

Haversian canal

Arteriole

Venule

Lacuna

OSTEON
This rod-shaped unit is the building block of compact bone. Its central (Haversian) canal contains blood vessels and nerves, and is surrounded by concentric layers of tissue called lamellae. Gaps (lacunae) in the tissue contain osteocyte cells, which keep bones healthy.

CARTILAGE
Cartilage is a tough, highly adaptable form of connective tissue. It consists of a gel-like matrix containing many chemicals, such as proteins and carbohydrates. Chondrocytes (cells that make and maintain the whole tissue) and various types of fibers are embedded in this matrix. The chondrocytes occupy small cavities or spaces, known as lacunae. Cartilage does not usually have any blood vessels but instead receives nutrients and oxygen by diffusion, and wastes move by the same process in the opposite direction. There are several kinds of cartilage, including hyaline cartilage, fibrocartilage, and elastic cartilage, which are classified according their proportion of matrix jelly, chondrocytes, and fiber. The most flexible is elastic cartilage due to its high proportion of elastin fibers and relatively little matrix. It provides lightweight, flexible support at sites such as the outer ear flap, epiglottis, and larynx.

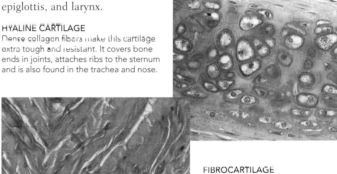

HYALINE CARTILAGE
Dense collagen fibers make this cartilage extra tough and resistant. It covers bone ends in joints, attaches ribs to the sternum and is also found in the trachea and nose.

FIBROCARTILAGE
This is mostly dense bundles of collagen fibers, with little gel-like matrix. It is found in the jaw, knee meniscus, and intervertebral disks.

Osteon

Bone marrow
Tissue filling a bone's central cavity; at first (as shown here) long bones have red marrow—later this turns into yellow marrow

Spongy bone
Latticework structure consisting of bony spikes (trabeculae), arranged along lines of greatest stress

Vein

Epiphysis
Expanded head of bone containing mainly spongy bone tissue

Bone shaft
Long shaft is mostly marrow and compact bone

Artery

BLOOD FACTORY
Red bone marrow contains hematopoietic tissue, the chief function of which is to produce all three main kinds of blood cell: red; white; and platelets. At birth, red marrow is present in all bones, but with increasing age, in the long bones it gradually becomes yellow marrow and loses its blood-making capacity.

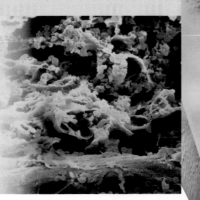

BLOOD CELL FORMATION
This microscopic image shows red marrow dotted with red blood cells destined for the bloodstream.

NEW BONE FOR OLD
Bone remodeling involves the breakdown of old bone, the removal of waste products, and the building of new bone tissue. The balance between these processes is controlled by a series of interacting substances that includes hormones, steroids, vitamin D, and signaling proteins called cytokines.

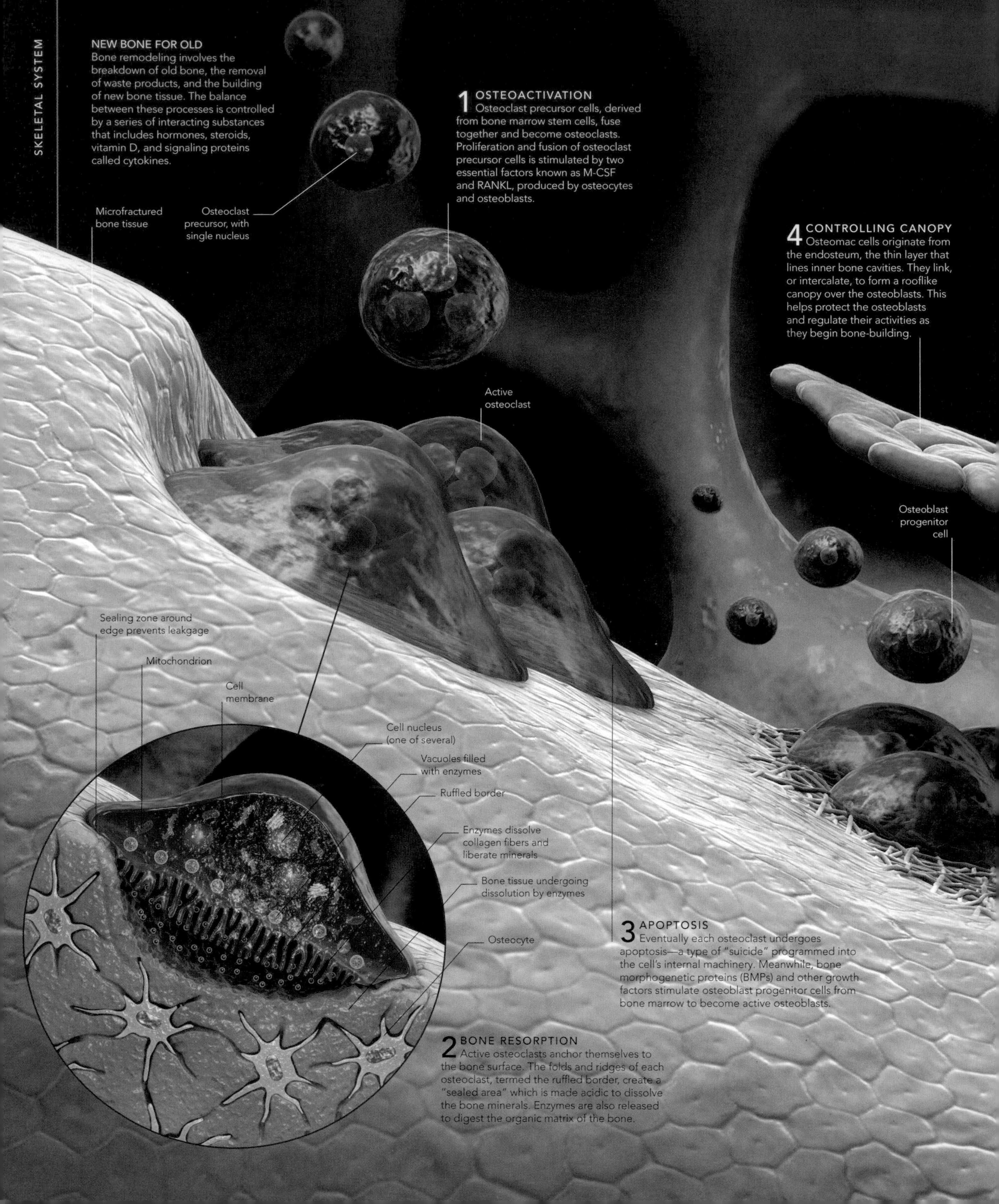

Microfractured bone tissue

Osteoclast precursor, with single nucleus

1 OSTEOACTIVATION
Osteoclast precursor cells, derived from bone marrow stem cells, fuse together and become osteoclasts. Proliferation and fusion of osteoclast precursor cells is stimulated by two essential factors known as M-CSF and RANKL, produced by osteocytes and osteoblasts.

4 CONTROLLING CANOPY
Osteomac cells originate from the endosteum, the thin layer that lines inner bone cavities. They link, or intercalate, to form a rooflike canopy over the osteoblasts. This helps protect the osteoblasts and regulate their activities as they begin bone-building.

Active osteoclast

Osteoblast progenitor cell

Sealing zone around edge prevents leakgage

Mitochondrion

Cell membrane

Cell nucleus (one of several)

Vacuoles filled with enzymes

Ruffled border

Enzymes dissolve collagen fibers and liberate minerals

Bone tissue undergoing dissolution by enzymes

Osteocyte

3 APOPTOSIS
Eventually each osteoclast undergoes apoptosis—a type of "suicide" programmed into the cell's internal machinery. Meanwhile, bone morphogenetic proteins (BMPs) and other growth factors stimulate osteoblast progenitor cells from bone marrow to become active osteoblasts.

2 BONE RESORPTION
Active osteoclasts anchor themselves to the bone surface. The folds and ridges of each osteoclast, termed the ruffled border, create a "sealed area" which is made acidic to dissolve the bone minerals. Enzymes are also released to digest the organic matrix of the bone.

REMODELING BONE

DESPITE THEIR INERT APPEARANCE, BONES ARE IN AN ONGOING STATE OF FLUX. AT THE MICROSCOPIC SCALE, MILLIONS OF CELLS BUSILY MAINTAIN, REPAIR, AND RESHAPE THEIR TISSUES.

Like all tissues, bone accumulates wear and tear. Some bone cells die when the tiny blood vessels supplying them with oxygen and nutrients suffer microtears. Knocks cause tiny fractures, splinters, and other damage. Inadequate diet may withdraw bone's calcium and other minerals are withdrawn for more urgent tasks elsewhere, such as transmitting nerve signals, then restored as the supply of dietary minerals resumes. Bone also responds gradually to physical strains, especially regular activity, from long-distance running to weightlifting,

tennis, or horseback riding. The stressed bones develop extra thickness and mineral density along lines of deformation, making them better able to withstand forces and pressures. This overall process is called bone remodeling, and about ten percent of the adult skeleton is remodeled each year. Four main kinds of cells are involved (see p.52): destructive osteoclasts, constructive osteoblasts, maintaining osteocytes, and enabling "osteomacs" (see panel, right). An imbalance in remodeling may lead to conditions such as osteoporosis (see p.48).

see p.52); see p.48).

MACROPHAGES

As types of white blood cells, macrophages play vital roles in body defenses. Some undertake a germ-fighting response (see p.177). Others roam around, engulfing debris and invading microbes (see p.158). "Macrophage" means "big eater," and over its life, one macrophage can consume up to 200 bacteria or thousands of viruses. Each tissue has its own community of resident macrophages. Those in bone are called osteal tissue resident macrophages, or osteomacs.

(see p.177). (see p.158).

MACROPHAGES ON THE MOVE
A macrophage's flexible frills and folds find and infiltrate gaps between other cells, membranes, and similar structures, then the whole macrophage squeezes through.

Osteoclast precursors and osteoblast progenitors ready for recruitment

Canopy structure

Mature osteoblast

Osteomac

Osteoid, made of collagen and other fibers

Calcium and other minerals are added to osteoid

Remodeled bone

5 BONE FORMATION
Under the influence of osteomacs, active osteoblasts arrive at the area digested by osteoclasts. The osteoblasts lay down osteoid, made of collagen and other bone-tissue fibers, then add calcium, phosphate, and other minerals to make fully formed or ossified new bone.

6 MAINTENANCE
Some osteoblasts are imprisoned in their new bone tissue, within tiny chambers called lacunae. Now known as osteocytes (see opposite and p.52), they survive perhaps for decades as they maintain surrounding bone tissue—until they are attacked by osteoclasts, as the remodeling cycle begins again.

(see opposite and p.52),

JOINTS

THE SITE AT WHICH TWO BONES MEET IS CALLED A JOINT OR AN ARTICULATION. JOINTS CAN BE CLASSIFIED ACCORDING TO THEIR STRUCTURE AND BY THE TYPES OF MOVEMENT THAT THEY ALLOW. THE BODY HAS MORE THAN 300 DIFFERENT JOINTS.

SYNOVIAL JOINTS

The body's most numerous, versatile, and freely moving joints are known as synovial joints. They can work well for many decades if used well and often, but not overused. Synovial joints are enclosed by a protective outer covering known as the joint capsule. The capsule's inner lining, the synovial membrane, produces a slippery, viscous synovial fluid that keeps the joint well lubricated so that the joint surfaces in contact slide with minimal friction and wear. There are around 230 synovial joints in the body.

TYPES OF SYNOVIAL JOINT
A synovial joint's range of movement is determined by the shape of its articular cartilage surfaces (see p.57) and how they fit together.

SEMIMOVABLE AND FIXED JOINTS

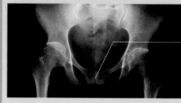

Not all joints have a wide range of movement. Some allow for growth or for greater stability. The bones in these joints are usually linked by cartilage or tough fibers made of substances such as the protein collagen. In the fixed joints of the skull, once growth is complete, the separate bone plates are securely connected by interlocking fibrous tissue, forming suture joints.

Suture

FIXED JOINT
The adult skull's suture joints show up as wiggly lines. In infancy, these joints are loosely attached to allow for expansion of the rapidly growing brain.

Pubic symphysis

SEMIMOVABLE JOINT
In partly flexible joints bones are linked by fibrous tissue or cartilage, as in the pubic symphysis.

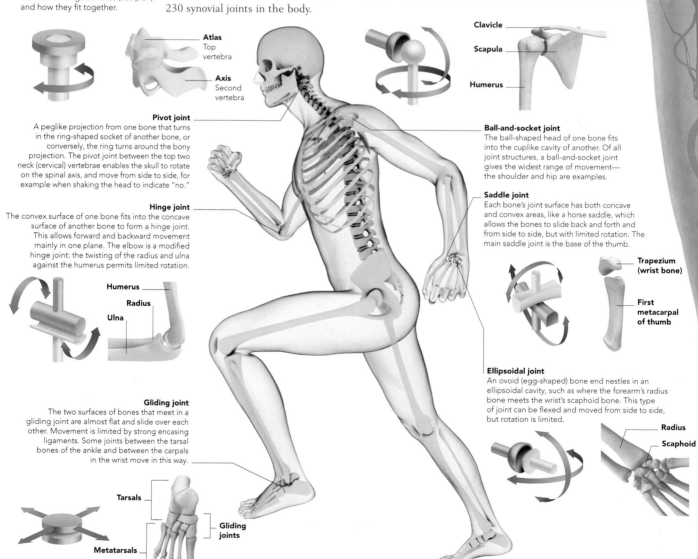

Atlas
Top vertebra

Axis
Second vertebra

Pivot joint
A peglike projection from one bone that turns in the ring-shaped socket of another bone, or conversely, the ring turns around the bony projection. The pivot joint between the top two neck (cervical) vertebrae enables the skull to rotate on the spinal axis, and move from side to side, for example when shaking the head to indicate "no."

Hinge joint
The convex surface of one bone fits into the concave surface of another bone to form a hinge joint. This allows forward and backward movement mainly in one plane. The elbow is a modified hinge joint: the twisting of the radius and ulna against the humerus permits limited rotation.

Humerus
Radius
Ulna

Gliding joint
The two surfaces of bones that meet in a gliding joint are almost flat and slide over each other. Movement is limited by strong encasing ligaments. Some joints between the tarsal bones of the ankle and between the carpals in the wrist move in this way.

Tarsals
Gliding joints
Metatarsals

Clavicle
Scapula
Humerus

Ball-and-socket joint
The ball-shaped head of one bone fits into the cuplike cavity of another. Of all joint structures, a ball-and-socket joint gives the widest range of movement—the shoulder and hip are examples.

Saddle joint
Each bone's joint surface has both concave and convex areas, like a horse saddle, which allows the bones to slide back and forth and from side to side, but with limited rotation. The main saddle joint is the base of the thumb.

Trapezium (wrist bone)

First metacarpal of thumb

Ellipsoidal joint
An ovoid (egg-shaped) bone end nestles in an ellipsoidal cavity, such as where the forearm's radius bone meets the wrist's scaphoid bone. This type of joint can be flexed and moved from side to side, but rotation is limited.

Radius
Scaphoid

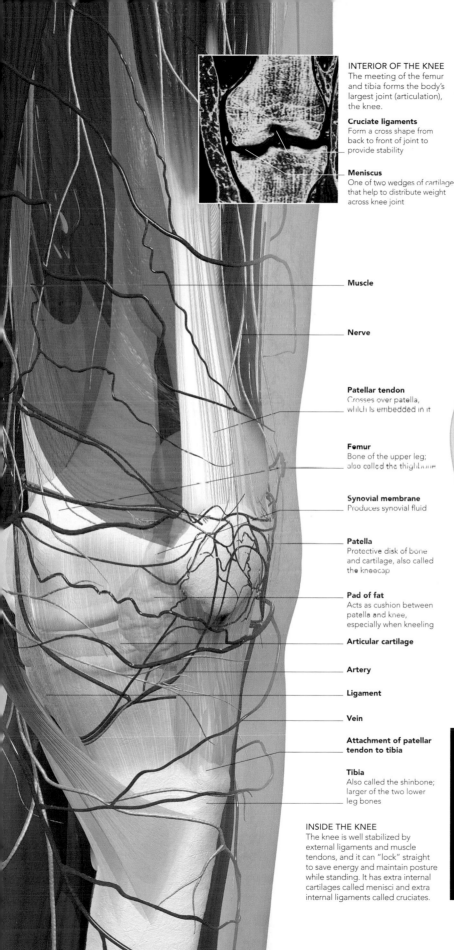

INTERIOR OF THE KNEE
The meeting of the femur and tibia forms the body's largest joint (articulation), the knee.

Cruciate ligaments
Form a cross shape from back to front of joint to provide stability

Meniscus
One of two wedges of cartilage that help to distribute weight across knee joint

Muscle

Nerve

Patellar tendon
Crosses over patella, which is embedded in it

Femur
Bone of the upper leg; also called the thighbone

Synovial membrane
Produces synovial fluid

Patella
Protective disk of bone and cartilage, also called the kneecap

Pad of fat
Acts as cushion between patella and knee, especially when kneeling

Articular cartilage

Artery

Ligament

Vein

Attachment of patellar tendon to tibia

Tibia
Also called the shinbone; larger of the two lower leg bones

INSIDE THE KNEE
The knee is well stabilized by external ligaments and muscle tendons, and it can "lock" straight to save energy and maintain posture while standing. It has extra internal cartilages called menisci and extra internal ligaments called cruciates.

INSIDE A JOINT

The bone ends in a synovial joint are covered and protected by a type of cartilage called articular cartilage, which is smooth and slightly compressible. Surrounding the joint is the joint capsule, which is made of strong connective tissue and is attached to the bone ends. Its delicate inner lining, the synovial membrane, continuously secretes viscous synovial fluid into the synovial cavity to keep the joint well oiled. The fluid also nourishes the cartilage with fats and proteins, and is constantly reabsorbed. Ligaments, the fibrous thickenings of the capsule, are anchored to bones at each end and prevent the bones from moving too far or in unnatural directions. Muscles around the joint, connected to its bones by tendons, tense for stability and contract to produce movement.

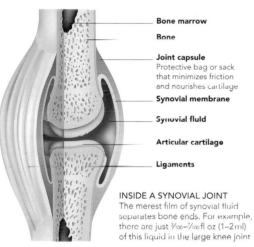

Bone marrow

Bone

Joint capsule
Protective bag or sack that minimizes friction and nourishes cartilage

Synovial membrane

Synovial fluid

Articular cartilage

Ligaments

INSIDE A SYNOVIAL JOINT
The merest film of synovial fluid separates bone ends. For example, there are just 3/100–7/100 fl oz (1–2 ml) of this liquid in the large knee joint

CARTILAGE AS A SHOCK ABSORBER

The articular cartilage that coats the bone ends in a synovial joint is also known as hyaline cartilage (see p.53). When sudden knocks or vibrations jolt the joint, this cartilage works as a shock absorber to dissipate some of the force of the impact, preventing jarring damage to the much stiffer bones. In certain joints the cartilage has tougher fibers. Examples include the fibrocartilaginous pads, called intervertebral disks, between the vertebrae of the backbone. Fibrocartilage also occurs in the jaw and wrist joints and the menisci in the knee.

SPINAL CARTILAGE
The fibrocartilage disks between the vertebrae play an important role in stabilizing and cushioning the spinal column.

SKULL

THERE ARE, IN TOTAL, 29 BONES IN THE HUMAN HEAD—22 BONES FORM THE SKULL ITSELF, WITH 21 OF THESE, EXCLUDING THE LOWER JAW (MANDIBLE), FUSED INTO A SINGLE, FIRM STRUCTURE. THE REMAINING BONES ARE THE HYOID BONE IN THE UPPER FRONT OF THE NECK AND THREE PAIRS OF TINY EAR BONES, CALLED OSSICLES, ONE SET LOCATED IN EACH MIDDLE EAR.

SKULL

Two groups of bones make up the skull. The upper set of eight bones forms the domelike cranium (cranial skull or cranial vault), which encloses and protects the brain. The other 14 bones make the skeleton of the face. Twenty-one of the 22 bones become strongly fused during growth at faint joint lines, called sutures. The lower jaw, or mandible, remains unfixed and is linked to the rest of the skull at the two jaw, or temporomandibular, joints.

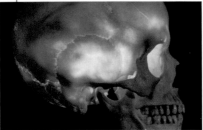

SKULL SUTURES
Lines on the skull's surface, highlighted here by backlighting, are the fused margins of the skull bones.

SINUSES

The four pairs of sinuses, known as paranasal sinuses, are air-filled cavities within the skull bones. They are named after the bones in which they are located: maxillary, frontal, sphenoidal, and ethmoidal sinuses. The first three pairs have fairly well-defined shapes. The ethmoidal sinuses are more honeycomb-like and variable.

Sphenoidal sinus   Frontal sinus   Maxillary sinus   Ethmoidal sinuses

RESONANT WEIGHT-SAVERS
The sinuses help lighten the skull's overall weight, and also act as resonating chambers to give each person's voice an individual character.

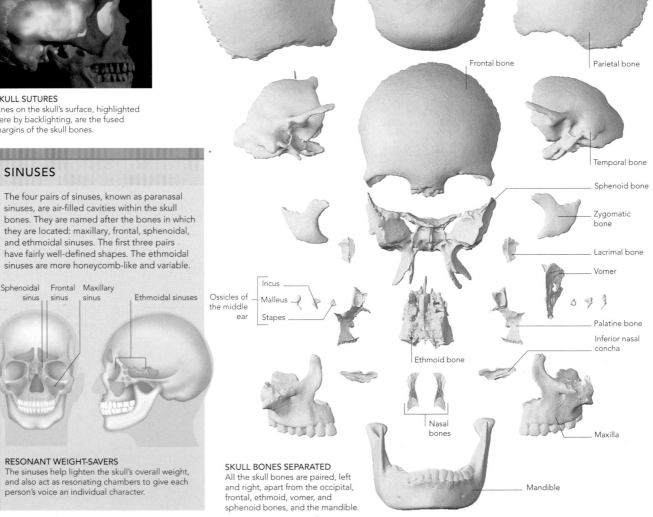

SKULL AND HEAD REGIONS
Two sets of bones form the structure of the skull. The eight bones that enclose the skull are called the cranial vault.

Parietal bone — Frontal bone
— Ethmoid bone
Temporal bone — Lacrimal bone
— Nasal bone
— Zygomatic bone
— Sphenoid bone
— Maxilla
Occipital bone — Mandible

Occipital bone
Frontal bone — Parietal bone
Temporal bone
Sphenoid bone
Zygomatic bone
Lacrimal bone
Vomer
Ossicles of the middle ear — Incus / Malleus / Stapes — Palatine bone
— Inferior nasal concha
Ethmoid bone
Nasal bones — Maxilla

SKULL BONES SEPARATED
All the skull bones are paired, left and right, apart from the occipital, frontal, ethmoid, vomer, and sphenoid bones, and the mandible.

Mandible

SPINE

THE SPINE IS ALSO KNOWN AS THE SPINAL OR VERTEBRAL COLUMN, OR SIMPLY "THE BACKBONE." THIS STRONG BUT FLEXIBLE CENTRAL SUPPORT HOLDS THE HEAD AND TORSO UPRIGHT, YET IT ALLOWS THE NECK AND BACK TO BEND AND TWIST.

SPINE FUNCTION

The spine consists of 33 ringlike bones called vertebrae. The bottom nine vertebrae are fused into two larger bones termed the sacrum and the coccyx, leaving 26 movable components within the spine. These components are linked by a series of mobile joints. Sandwiched between the bones in each joint is the intervertebral disk, a springy pad of tough, fibrous cartilage that squashes slightly under pressure to absorb shocks. Strong ligaments and many sets of muscles around the spine stabilize the vertebrae and help control movement. The spinal column also protects the spinal cord and allows nerve roots to exit through spaces in the vertebrae (see p.97).

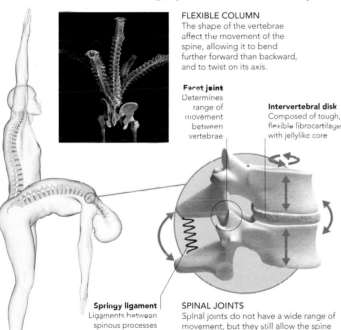

FLEXIBLE COLUMN
The shape of the vertebrae affect the movement of the spine, allowing it to bend further forward than backward, and to twist on its axis.

Facet joint
Determines range of movement between vertebrae

Intervertebral disk
Composed of tough, flexible fibrocartilage with jellylike core

Springy ligament
Ligaments between spinous processes limit movement and store energy for recoil

SPINAL JOINTS
Spinal joints do not have a wide range of movement, but they still allow the spine great flexibility, letting it arch backward, twist, and curve forward. Two facet joints help prevent slippage and torsion.

HYOID BONE

The single U-shaped hyoid bone is located at the root of the tongue, just above the larynx. It is one of the few bones in the body that does not join directly to another bone. It is held in position by muscles and by the strong stylohyoid ligament on each side of the bone, which links to the styloid process of the skull's temporal bone. The hyoid stabilizes several sets of muscles used in swallowing and speech.

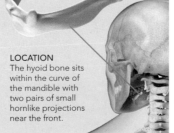

LOCATION
The hyoid bone sits within the curve of the mandible with two pairs of small hornlike projections near the front.

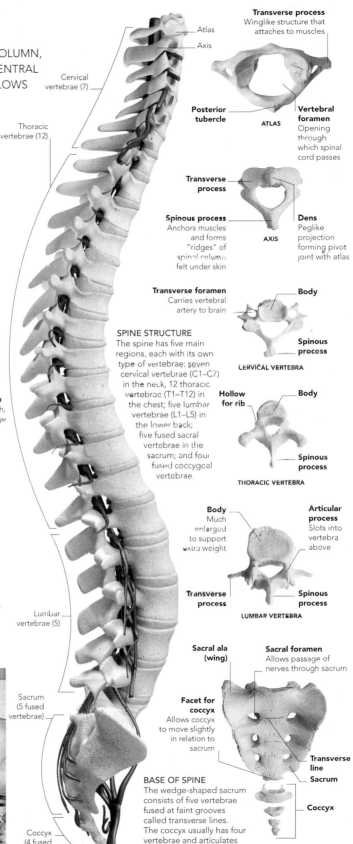

Atlas
Axis

Cervical vertebrae (7)

Thoracic vertebrae (12)

Transverse process
Winglike structure that attaches to muscles

Posterior tubercle

ATLAS

Vertebral foramen
Opening through which spinal cord passes

Transverse process

Spinous process
Anchors muscles and forms "ridges" of spinal column felt under skin

AXIS

Dens
Peglike projection forming pivot joint with atlas

Transverse foramen
Carries vertebral artery to brain

Body

Spinous process

CERVICAL VERTEBRA

SPINE STRUCTURE
The spine has five main regions, each with its own type of vertebrae: seven cervical vertebrae (C1–C7) in the neck; 12 thoracic vertebrae (T1–T12) in the chest; five lumbar vertebrae (L1–L5) in the lower back; five fused sacral vertebrae in the sacrum; and four fused coccygeal vertebrae.

Hollow for rib

Body

Spinous process

THORACIC VERTEBRA

Body
Much enlarged to support extra weight

Articular process
Slots into vertebra above

Transverse process

Spinous process

LUMBAR VERTEBRA

Lumbar vertebrae (5)

Sacral ala (wing)

Sacral foramen
Allows passage of nerves through sacrum

Facet for coccyx
Allows coccyx to move slightly in relation to sacrum

Transverse line

Sacrum

Coccyx

Sacrum (5 fused vertebrae)

BASE OF SPINE
The wedge-shaped sacrum consists of five vertebrae fused at faint grooves called transverse lines. The coccyx usually has four vertebrae and articulates with the sacrum.

SACRUM AND COCCYX

Coccyx (4 fused vertebrae)

RIBS, PELVIS, HANDS, AND FEET

THE RIBS AND HIPBONE (PELVIS) GUARD VITAL CHEST AND ABDOMINAL ORGANS, AND THEY DEMONSTRATE THE SKELETON'S TWIN FUNCTIONS OF SUPPORT AND PROTECTION. THE PELVIS PROVIDES SURFACES FOR ANCHORING THE POWERFUL HIP AND THIGH MUSCLES. THE WRISTS, HANDS, ANKLES, AND FEET, WHICH TOGETHER CONTAIN MORE THAN HALF OF ALL THE BONES IN THE BODY, ARE VITAL FOR COORDINATED MOVEMENT.

RIB CAGE

Most people have 12 pairs of ribs, but about 1 in 200 is born with one or more extra pairs. All ribs attach to the spinal column at the rear. The upper seven pairs of "true ribs" link directly to the sternum (breastbone) by their cartilage extensions (costal cartilages). The next two or three pairs of "false ribs" connect to the cartilages of the ribs above. The remaining "floating ribs" do not link to the sternum. The whole rib cage is flexible because the ribs tilt.

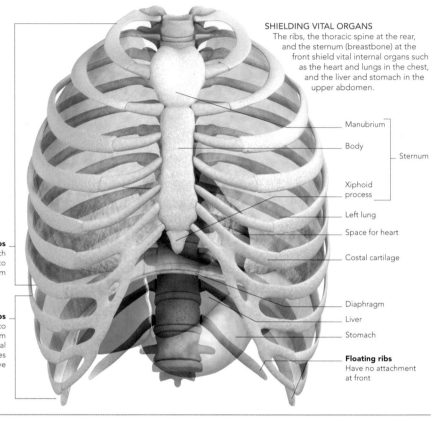

SHIELDING VITAL ORGANS
The ribs, the thoracic spine at the rear, and the sternum (breastbone) at the front shield vital internal organs such as the heart and lungs in the chest, and the liver and stomach in the upper abdomen.

Manubrium

Body — Sternum

Xiphoid process

Left lung

Space for heart

Costal cartilage

Diaphragm

Liver

Stomach

Floating ribs
Have no attachment at front

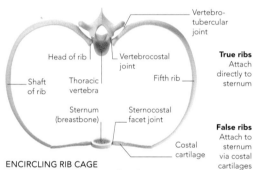

Vertebro-tubercular joint

Head of rib

Vertebrocostal joint

Shaft of rib

Thoracic vertebra

Fifth rib

Sternum (breastbone)

Sternocostal facet joint

Costal cartilage

True ribs
Attach directly to sternum

False ribs
Attach to sternum via costal cartilages above

ENCIRCLING RIB CAGE
Each rib links to its corresponding chest (thoracic) vertebra at two points. Flexible costal cartilage attaches ribs to the sternum, allowing the rib cage to change volume during breathing.

PELVIS

Often referred to as the hip bone, the pelvis is a bowl-like structure consisting of the left and right innominate bones or ossa coxae, and the wedge-shaped sacrum and coccyx, which makes up a "tailbone" at the rear. Each innominate bone has three fused bony elements: the large, flaring ilium at the rear, which forms the hip bone you can feel under the skin; the ischium at the lower front; and the pubis above it. There are paired sacroiliac joints at the rear and the pubic symphysis, a semimovable joint made of fibrocartilage, at the front. The shape of the pelvis is shallower and wider in females than in males, with a larger gap, or pelvic inlet, and a greater pelvic outlet, to allow a baby to pass through at birth.

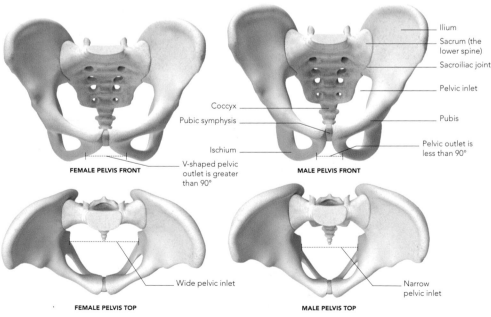

Ilium

Sacrum (the lower spine)

Sacroiliac joint

Pelvic inlet

Coccyx

Pubic symphysis

Ischium

Pubis

Pelvic outlet is less than 90°

FEMALE PELVIS FRONT

V-shaped pelvic outlet is greater than 90°

MALE PELVIS FRONT

Wide pelvic inlet

FEMALE PELVIS TOP

Narrow pelvic inlet

MALE PELVIS TOP

WRIST AND HAND

The wrist is made up of the eight carpal bones, arranged approximately in two rows of four. They are linked to each other chiefly by plane or gliding joints (see p.56), and to the forearm bones by the radiocarpal joint. The palm of the hand contains five metacarpal bones. Each of these joins at its outer end to the finger bone (phalanx), of which there are two in the thumb (first digit, or pollex) and three each in the other four digits. The entire structure is moved by more than 50 muscles, including some in the forearm, to provide great flexibility and delicate manipulation.

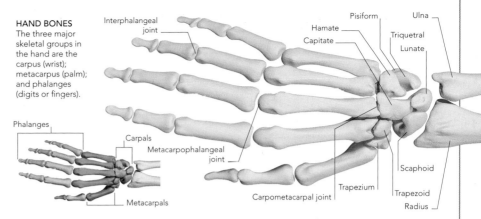

HAND BONES
The three major skeletal groups in the hand are the carpus (wrist); metacarpus (palm); and phalanges (digits or fingers).

Phalanges

Carpals

Metacarpophalangeal joint

Metacarpals

Interphalangeal joint

Pisiform
Hamate
Capitate
Ulna
Triquetral
Lunate

Scaphoid

Trapezium

Carpometacarpal joint

Trapezoid

Radius

ANKLE AND FOOT

The ankle and foot have a similar bone arrangement to the wrist and hand (see above), except there are only seven tarsal (ankle) bones. The build of the ankle and foot bones is heavier, for strength and weight-bearing stability at the expense of precision and mobility. The sole of the foot is supported by the five metatarsal bones. As in the hand, the hallux (first digit or big toe) has two phalanges (toe bones), and the others have three each. The bony prominence commonly called the "heel bone" is formed by the calcaneus.

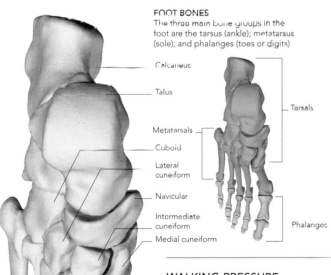

FOOT BONES
The three main bone groups in the foot are the tarsus (ankle); metatarsus (sole); and phalanges (toes or digits)

Calcaneus

Talus

Metatarsals

Cuboid

Lateral cuneiform

Navicular

Intermediate cuneiform

Medial cuneiform

Tarsals

Phalanges

LIGAMENTS

Ligaments are strong bands or straps of fibrous tissue that provide support to bones and link bone ends together in and around joints. They are made of collagen—a tough, elastic protein. A large number of ligaments bind together the complex wrist and ankle joints. Each one is named after the bones it links; for example, the calcaneofibular ligament links the calcaneus and the fibula. The foot ligaments store energy as they stretch when the foot is planted, and then release energy as they recoil and shorten to put a "spring in the step." This saves an enormous amount of energy when walking. Ligaments are susceptible to a wide range of injuries as a result of the stresses and strains placed on them, especially during sports.

ANKLE LIGAMENTS
More than a dozen ligaments bind the foot's tarsal bones to each other, and ligaments run from the tarsals to the fibula, tibia, and metatarsals (viewed here from outer side).

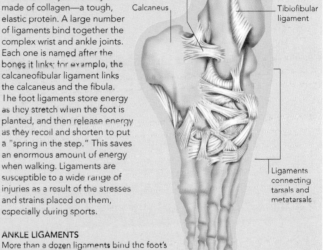

Fibula

Calcaneofibular ligament

Calcaneus

Tibia

Tibiofibular ligament

Ligaments connecting tarsals and metatarsals

WALKING PRESSURE

With each step, the main weight of the body moves from the rear to the front of the foot. The heel region bears initial pressure as the foot is put down. The force passes along the arch, which flattens slightly, then recoils to transfer the energy and pressure to the ball of the foot, and finally to the big toe for the push-off.

LOAD AREAS ON THE FOOT
These footprint impressions show (from left to right) how the body's weight transfers from the heel to the ball to the big toe when walking.

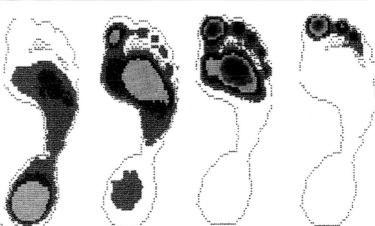

BONE DISORDERS

BONE STRENGTH GRADUALLY DECREASES WITH AGE AND, TOGETHER WITH INCREASED RISK OF FALLS, MAKES FRACTURES MORE COMMON IN THE ELDERLY. HOWEVER, FRACTURES ARE ALSO COMMON EARLY IN LIFE BECAUSE CHILDREN ARE LESS AWARE OF RISKS. OTHER FACTORS THAT INFLUENCE BONE HEALTH ARE NUTRITIONAL AND HORMONE DEFICIENCIES, LACK OF EXERCISE, AND BEING OVERWEIGHT.

FRACTURE

BROKEN BONES—FRACTURES—RANGE FROM A MINOR CRACK IN THE BONE SURFACE, TO A SPLIT PARTWAY THROUGH THE BONE, TO A COMPLETE BREAK.

Fractures may be caused by a sudden impact, by compression, or by repeated stress. A displaced fracture occurs when the broken surfaces of bone are forced from their normal positions. There are various types of displaced fracture, depending on the angle and strength of the blow. A compression fracture occurs when spongy bone, such as in the vertebrae, is crushed. Stress fractures are caused by prolonged or repeated force straining the bone; they occur in long-distance runners and in the elderly, in whom minor stress, such as coughing, may cause a fracture. Nutritional deficiencies or certain chronic diseases such as osteoporosis, which can weaken bone, may increase the likelihood of fractures. If a broken bone remains beneath the skin, the fracture is described as closed or simple, and there is a low risk of infection. If the ends of the fractured bones project through the skin, the injury is described as open or compound, and there is a danger of dirt entering the bone tissue and causing microbial contamination.

Bone repair

Despite its image as dry, brittle, and even lifeless, bone is an active tissue with an extensive blood supply and its own restorative processes (see pp.54–55). After a fracture, blood clots as it does elsewhere in the body. Fibrous tissue, and then new bone growth, bridge the break and eventually restore strength. However, medical treatment is often required to ensure that the repair process is effective and the result is not misshapen. If the bones are displaced, manipulation to restore their normal position—known as reduction—may be performed under anesthesia. The bone will also be immobilized to allow the ends to heal correctly.

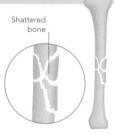 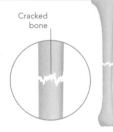

COMMINUTED FRACTURE
A direct impact can shatter a bone into several fragments or pieces. This type of fracture is likely to occur during a traffic accident.

Shattered bone

Cracked bone

TRANSVERSE FRACTURE
A powerful force may cause a break across the bone width. The injury is usually stable; the broken surfaces are unlikely to move.

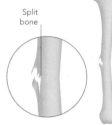 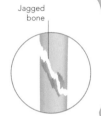

Split bone

Jagged bone

GREENSTICK FRACTURE
A crack on one side of a long bone caused by bending under force. This is common in children, whose bones are flexible.

SPIRAL FRACTURE
A sharp, twisting force may break a bone diagonally across the shaft. The jagged ends may be difficult to reposition.

COMMONLY INJURED BONES

Typical breaks vary depending on age and activity levels. Elbow fracture is common in childhood; the humerus (upper arm bone) breaks just above the elbow joint, often as a result of a fall during play. A young person is likely to injure a lower leg bone during activity, especially team sports. With age, bones naturally become "thinner"—weaker and more brittle—and they are more likely to fracture with only minimal force. The hip joint is especially vulnerable, and fracture is often the consequence of a fall. Another common injury in the elderly is Colles' fracture, which affects the wrist; it usually occurs when an outstretched arm tries to break a fall.

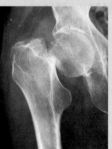 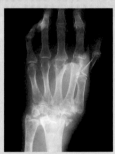

HIP FRACTURE
Common in the elderly, this is a break in the femur just below its ball-shaped head.

COLLES' FRACTURE
Flexing a hand to cushion a fall may break the end of the radius and the tip of the ulna.

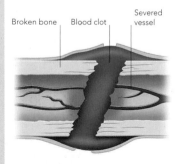 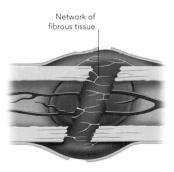 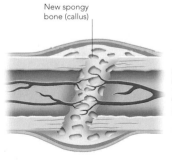 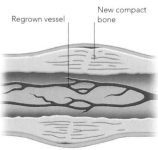

Broken bone Blood clot Severed vessel

Network of fibrous tissue

New spongy bone (callus)

Regrown vessel New compact bone

IMMEDIATE RESPONSE
Blood leaks from the blood vessel and clots. White blood cells gather at the area to scavenge damaged cells and debris.

AFTER SEVERAL DAYS
Fibroblast cells construct new fibrous tissue across the break. The limb is immobilized, usually in a plaster cast or splint.

AFTER 1–2 WEEKS
Bone-building cells (osteoblasts) multiply and form new bone tissue. Initially spongy, the tissue infiltrates the site as a callus.

AFTER 2–3 MONTHS
Blood vessels reconnect across the break. The callus reshapes while new bone tissue is "remodeled" into dense, compact bone.

SPINAL FRACTURES

MOST MAJOR SPINAL INJURIES OCCUR AS A RESULT OF SEVERE FORCES OF COMPRESSION, ROTATION, OR FLEXING BEYOND THE SPINE'S NORMAL RANGE OF MOVEMENT.

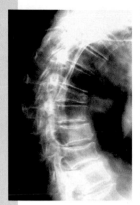

Many injuries to the spine are minor and cause only slight bruising. However, a severe fall or an accident may dislocate or fracture one or more of the vertebrae. If the spinal cord or nerves are damaged, a loss of sensation or function may result; paralysis may occur if the damage is severe, particularly in the neck region. Bone disease, for example osteoporosis, can affect the spine and increase the likelihood of fractures. The outcome of a spinal fracture depends on whether it is stable (unlikely to shift) or unstable, in which case damage to the spinal cord or nerves is more likely.

COMPRESSION FRACTURE
The area in red on this X-ray shows a fractured vertebra that has collapsed. This type of fracture often occurs in the elderly.

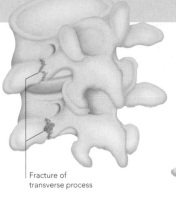

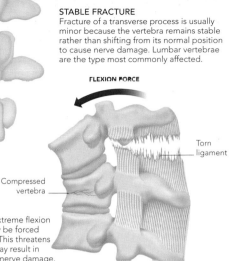

STABLE FRACTURE
Fracture of a transverse process is usually minor because the vertebra remains stable rather than shifting from its normal position to cause nerve damage. Lumbar vertebrae are the type most commonly affected.

FLEXION FORCE

Fracture of transverse process

Torn ligament

Compressed vertebra

UNSTABLE FRACTURE
If ligaments tear during extreme flexion or rotation, vertebrae may be forced out of normal alignment. This threatens the spine's stability and may result in permanent spinal cord or nerve damage.

SCIATICA

PRESSURE ON THE ROOTS OF THE SCIATIC NERVE CAUSES PAIN IN THE BUTTOCK AND THE BACK OF THE THIGH.

The sciatic nerve is the largest nerve in the body, and pressure on its roots may cause pain to radiate down the entire leg. In severe cases, pain may be accompanied by weakness of the leg muscles. The source of the pressure on the sciatic nerve roots (junctions with the spinal cord) is usually a prolapsed intervertebral disk. Other causes include muscle spasm, sitting awkwardly for a long time, and, in older people, osteoarthritis. Rarely, a tumor may be the cause.

Spinal cord

Sciatic nerve roots

Sciatic nerve

Tibial nerve

Peroneal nerve

SCIATIC NERVE
The large sciatic nerve in the thigh sends branches down the leg and into the foot from its roots within the spinal cord.

WHIPLASH

SUDDEN BENDING OF THE SPINE CAUSES INJURY TO THE CERVICAL VERTEBRAE.

Whiplash injury is usually the result of a car accident. If hit from behind, a vehicle jerks forward, causing a rapid motion of the head first backward and then forward. The whiplike backward motion hyperextends the cervical vertebrae, and this movement is quickly followed by flexion of the vertebrae as the head's momentum carries it forward and causes the chin to arc down to the chest. The effect of this violent motion is spraining of the ligaments attached to the cervical vertebrae, or partial dislocation of a cervical joint, or both.

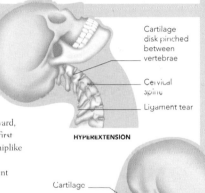

Cartilage disk pinched between vertebrae

Cervical spine

Ligament tear

HYPEREXTENSION

Cartilage

Ligament tear

FLEXION

DISK PROLAPSE

A PROLAPSED (ALSO KNOWN AS HERNIATED OR "SLIPPED") DISK IS A PROTRUSION FROM ONE OF THE SHOCK-ABSORBING PADS BETWEEN THE VERTEBRAE.

The cushionlike cartilage disks or pads that separate adjacent vertebrae have a hard outer covering and a jellylike center. An accident, wear and tear, or excessive pressure when lifting awkwardly may rupture the outer layer. This forces some of the core material to bulge out or prolapse. The prolapsed (or herniated) portion may cause pressure on the nearby spinal nerve root. Symptoms of disk prolapse include dull pain, muscle spasm and stiffness in the area of the back affected, and pain, tingling, numbness, or weakness in the body part supplied by the nerve—usually the leg or, with a prolapse higher in the spinal column, the arm. The term "slipped disk" is misleading because it is not the whole disk that slides out of position.

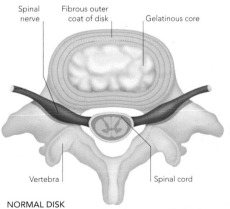

Spinal nerve

Fibrous outer coat of disk

Gelatinous core

Vertebra

Spinal cord

Protruding core pressing on nerve

Fibrous outer coat of disk

Spinal nerve

Compressed spinal cord

NORMAL DISK
The outer casing or capsule of the intervertebral disk is intact and completely encloses its gelatinous core. The disk sits between the bodies, or centra, of adjacent vertebrae.

PROLAPSED DISK
A weak site in the outer casing allows the gelatinous core material to bulge through as the disk is compressed. The resulting pressure on the spinal nerve causes pain.

SPINAL CURVATURE

KYPHOSIS AND LORDOSIS INVOLVE EXAGGERATED CURVATURE OF THE UPPER AND LOWER PARTS OF THE SPINE.

The spinal column, or backbone, has two main natural curves. These are the thoracic curve to the rear in the chest region, and the lumbar curvature to the front in the lower back. Increased thoracic curvature, causing a rounded, or humped, upper back, is called kyphosis. Lordosis is an exaggerated lumbar curvature that produces a hollow in the small of the back. The conditions may occur together since one tends to compensate for the other. Causes include bone or joint problems, such as osteoarthritis or osteoporosis, poor posture, and being overweight.

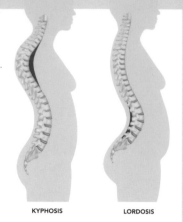

KYPHOSIS　　**LORDOSIS**

TYPES OF CURVATURE OF THE SPINE
Kyphosis accentuates the upper spinal column while lordosis affects the lower region (normal curvatures are shown in red).

OSTEOMYELITIS

INFECTION OF A BONE, USUALLY BY BACTERIA, CAN LEAD TO PAINFUL, WEAK, AND DAMAGED BONE TISSUE.

Osteomyelitis generally affects young and elderly people, although it can occur in those with reduced immunity, for example people on immunosuppressive drugs, or with a condition such as sickle-cell anemia. In children, the vertebrae or long limb bones are most frequently affected and, in adults, the vertebrae or pelvis. In acute osteomyelitis, the causative bacteria may be *Staphylococcus aureus*. Symptoms include swelling, pain, and a fever. The chronic form may be caused by tuberculosis, which does not produce swelling or fever.

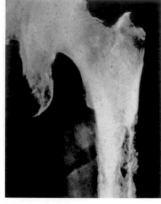

INFECTED FEMUR
A region of the leg infected by osteomyelitis (darker area, lower right) can be seen clearly in the shaft of the femur (thighbone).

OSTEOPOROSIS

MORE COMMON WITH INCREASED AGE, OSTEOPOROSIS IS LOSS OR THINNING OF BONE TISSUE THAT MAKES BONES WEAKER, MORE BRITTLE, AND MORE LIKELY TO BREAK.

In order for healthy bone growth and repair to occur, bone tissue is continually being broken down and replaced. Sex hormones are essential to initiate and maintain this process and with the decline in production of sex hormones in both sexes after middle age, bones become notably thinner and more porous. Estrogen levels fall rapidly in women after menopause, which can lead to severe thinning, or osteoporosis. The decline in testosterone in men is gradual and, in general, males are less prone to osteoporosis. Exercise is an essential component in maintaining bone health, and a lack of activity is a predisposing factor to developing osteoporosis. The decreased density of osteoporotic bones makes them more likely to fracture. Crush fractures in the spine can cause spinal curvature; hip or wrist fractures may occur after minor falls. Other factors that influence the development of osteoporosis include smoking, corticosteroid treatment, rheumatoid arthritis, an overactive thyroid, and long-term kidney failure.

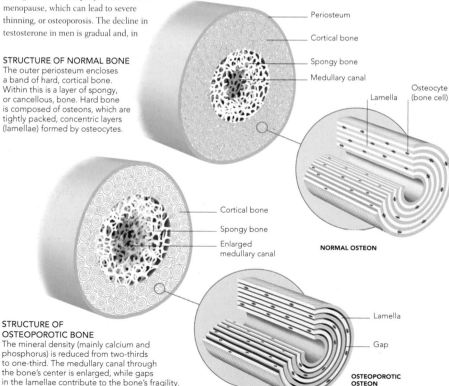

STRUCTURE OF NORMAL BONE
The outer periosteum encloses a band of hard, cortical bone. Within this is a layer of spongy, or cancellous, bone. Hard bone is composed of osteons, which are tightly packed, concentric layers (lamellae) formed by osteocytes.

- Periosteum
- Cortical bone
- Spongy bone
- Medullary canal
- Lamella
- Osteocyte (bone cell)

NORMAL OSTEON

- Cortical bone
- Spongy bone
- Enlarged medullary canal

STRUCTURE OF OSTEOPOROTIC BONE
The mineral density (mainly calcium and phosphorus) is reduced from two-thirds to one-third. The medullary canal through the bone's center is enlarged, while gaps in the lamellae contribute to the bone's fragility.

- Lamella
- Gap

OSTEOPOROTIC OSTEON

WHY OSTEOPOROSIS OCCURS

Bone tissue is built up by the deposition of minerals (mainly calcium salts) on a framework of collagen fibers. It is continually broken down and rebuilt in order to allow growth and repair. Osteoporosis develops when the rate at which fibers, minerals, and cells are broken down becomes much greater than the formation of new tissue.

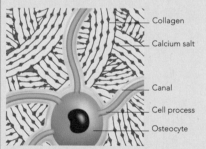

- Collagen
- Calcium salt
- Canal
- Cell process
- Osteocyte

NORMAL BONE
Osteocytes (bone-maintaining cells) form collagen fibers and aid calcium deposition. Calcium moves in canals between bone and blood in response to hormones.

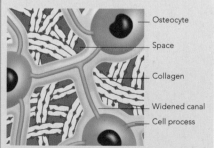

- Osteocyte
- Space
- Collagen
- Widened canal
- Cell process

OSTEOPOROTIC BONE
In osteoporosis, the collagen framework and deposited minerals are broken down faster than they form. The canals widen, new spaces appear, and bone weakens.

OSTEOMALACIA

A LOSS OF CALCIUM AND PHOSPHORUS, OFTEN AS THE RESULT OF VITAMIN D DEFICIENCY, CAN CAUSE WEAK BONES.

In osteomalacia, bones are weakened by a loss of minerals, most notably calcium. Other symptoms include bone tenderness and deformity. The main cause is a shortage of vitamin D, which is essential to enable the body to absorb calcium and phosphorus. Vitamin D is obtained from food and also by the action of sunlight on skin. Inadequate supplies can be due to lack of sunlight, an unbalanced diet, or disorders that affect absorption of the vitamin, such as celiac disease, and it can also occur as a result of some kidney diseases. In children, the condition is known as rickets.

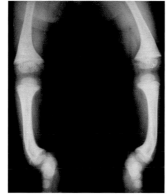

RICKETS
This X-ray shows the legs of a child diagnosed with rickets. The characteristic bowing of the legs at the knees can become a permanent disability if the disorder occurs early in the child's development.

PAGET'S DISEASE

THIS ABNORMALITY IN THE BALANCE OF BONE FORMATION AND BREAKDOWN CAUSES BONE DISTORTION.

Paget's disease, also known as osteitis deformans, can affect any bone in the skeleton, although it occurs most commonly in the pelvis, collarbone, vertebrae, skull, and leg bones. The bone tissue is broken down at an increased rate and is replaced rapidly by abnormal bone. The affected bone becomes weakened and distorted, is often painful, and may become more liable to fracture. If the enlarged bone presses on a nerve, there may be numbness, tingling, weakness, and loss of function. Rare in young people, the condition becomes increasingly common over the age of 50 years.

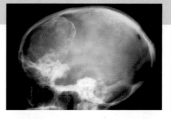
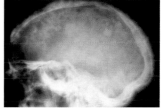

BONE THICKENING
A normal skull (top) is compared with an affected skull (bottom). The bone of the skull has a typical "cottony" appearance. The bone distortion may cause hearing loss if the auditory nerve is compressed.

BONE CANCER

CANCER IN A BONE MAY BE PRIMARY, THAT IS FORMING IN THE BONE ITSELF, BUT MORE OFTEN IT IS SECONDARY, HAVING SPREAD FROM ELSEWHERE IN THE BODY.

Primary cancer

A malignant, or cancerous, tumor that originates within a bone is described as primary. Cancers that start in bone are most likely to occur in children and adolescents. Osteosarcoma, which affects long bones, such as the femur (thighbone), is the most common type of primary bone cancer. The affected leg may be painful and swollen and is susceptible to fracture. Another primary bone cancer, chondrosarcoma, occurs mainly in the pelvis, ribs, and breastbone.

Secondary cancer

More frequent than primary bone cancers, secondary tumors in bone are the result of cancer cells spreading from a primary tumor elsewhere in the body. This type of bone cancer is known as metastatic. Secondary bone cancer is more likely to occur in older people, mainly because this age group is likely to have cancer elsewhere. Cancers most likely to spread to bone are breast, lung, thyroid, kidney, and prostate cancer, but sometimes the primary site is unknown. Symptoms include gnawing pain that is worse at night, and swelling and tenderness at the site. The most commonly affected areas are the skull, sternum, pelvis, vertebrae, ribs and, less often, the top of the femur and humerus.

SECONDARY CANCER
Cancerous cells travel to bones through the blood circulation. The breast, lung, thyroid, kidney, bladder, and (in males) prostate gland are the most common primary sites linked to bone metastases.

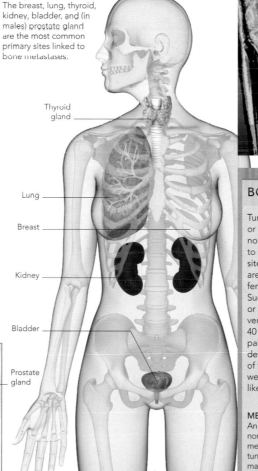

Thyroid gland

Lung

Breast

Kidney

Bladder

Prostate gland

PROSTATE GLAND
In men, the prostate gland at the base of the bladder makes secretions for sperm. Prostate cancer often spreads to bones around the body.

Tumor

OSTEOSARCOMA
This primary bone tumor is seen just above the knee at the lower end of the femur (the dark blue area at the upper left of this scan). Externally, the leg would appear swollen and distorted.

BONE TUMORS

Tumors in bone are either benign (noncancerous) or malignant (cancerous). Benign tumors and noninvasive malignant tumors do not spread to other parts of the body. The most common sites for these noncancerous growths or tumors are the long bones of the limbs, such as the femur (thighbone), and the bones in the hands. Such tumors tend to occur during childhood or adolescence, and are very rare after the age of 40 years. There may be pain, enlargement, and deformity at the location of the tumor, and the weakened bone is more likely to fracture.

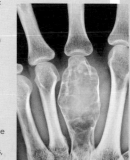

METACARPAL TUMOR
An X-ray showing a large, noncancerous tumor on a metacarpal (hand bone). The tumor causes swelling and may press on nearby nerves, blood vessels, and tendons.

JOINT DISORDERS

JOINTS ARE DESIGNED TO WORK IN SPECIFIC WAYS, AND ANY MOVEMENT BEYOND THE NORMAL
RANGE, OR IN AN UNNATURAL DIRECTION, CAN RESULT IN INJURY. COMMON CAUSES INCLUDE A DIRECT
BLOW OR FALL, AND INJURY DURING PHYSICAL ACTIVITIES SUCH AS SPORTS. THE PROBLEMS CAN ALSO
STEM FROM OVERUSE. CONGENITAL DEFECTS MAY CAUSE JOINT PROBLEMS (SEE ALSO PP.68–69).

LIGAMENT INJURIES

IF A JOINT IS FORCED BEYOND ITS NATURAL RANGE,
THE LIGAMENTS THAT ARE NORMALLY ABLE TO PREVENT
EXCESSIVE MOVEMENT MAY BE STRAINED OR TORN.

Ligaments are strong, flexible bands of fibrous tissue,
linking bone ends together around a joint. If the bones
within a joint are pulled too far apart, often as a result of
a sudden, unexpected, or forceful movement, the fibers
of the ligaments may overstretch or tear. This commonly
results in swelling, pain, and muscle spasm. A joint
"sprain" is usually due to partial tearing of a ligament.
Rest, ice, compression, and elevation of the joint are
the usual treatments if a sprain is not serious. If the
injury is severe, it may result in joint instability or
dislocation, which requires medical intervention.

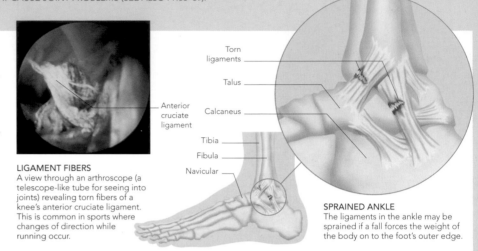

LIGAMENT FIBERS
A view through an arthroscope (a
telescope-like tube for seeing into
joints) revealing torn fibers of a
knee's anterior cruciate ligament.
This is common in sports where
changes of direction while
running occur.

Torn
ligaments

Talus

Calcaneus

Tibia

Fibula

Navicular

Anterior
cruciate
ligament

SPRAINED ANKLE
The ligaments in the ankle may be
sprained if a fall forces the weight of
the body on to the foot's outer edge.

TORN CARTILAGE

CARTILAGE COVERS THE BONE ENDS IN MANY
JOINTS, BUT THE TERM "TORN CARTILAGE"
USUALLY REFERS TO THE KNEE IN PARTICULAR.

The knee joint contains pad-like curved "disks"
of cartilage called menisci. These are almost
C-shaped and made of tough fibrous cartilage.
The disks are sited between the lower end of
the femur and upper end of the tibia, with the
medial cartilage on the knee's inner side and
the lateral cartilage on the outside. These
menisci stabilize the joint, helping it "lock"
straight while standing, and cushion the
bones. A meniscus may be crushed or torn
by rapid twisting of the knee, often while
playing sport. If painful, surgery can
remove the damaged piece of cartilage.

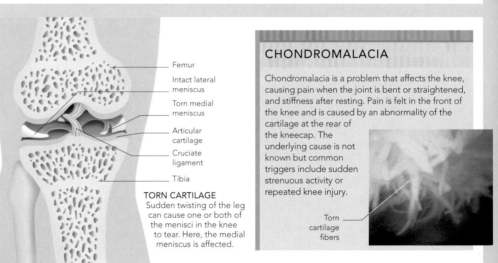

Femur

Intact lateral
meniscus

Torn medial
meniscus

Articular
cartilage

Cruciate
ligament

Tibia

TORN CARTILAGE
Sudden twisting of the leg
can cause one or both of
the menisci in the knee
to tear. Here, the medial
meniscus is affected.

CHONDROMALACIA

Chondromalacia is a problem that affects the knee,
causing pain when the joint is bent or straightened,
and stiffness after resting. Pain is felt in the front of
the knee and is caused by an abnormality of the
cartilage at the rear of
the kneecap. The
underlying cause is not
known but common
triggers include sudden
strenuous activity or
repeated knee injury.

Torn
cartilage
fibers

FROZEN SHOULDER

FROZEN SHOULDER REFERS TO PAIN
AND RESTRICTED MOVEMENT BROUGHT
ON BY INFLAMMATION IN THE JOINT.

The cause of frozen shoulder, or
adhesive capsulitis, may be linked
to injury or overuse of the joint, or
immobilization after an arm bone
fracture or a stroke, but sometimes
there is no obvious cause. The pain
can be severe and may result in loss
of all arm and shoulder movements.
Analgesics and anti-inflammatories
may ease the condition, along with
physical therapy, but the condition
usually gets better with time.

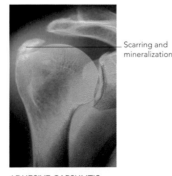

Scarring and
mineralization

ADHESIVE CAPSULITIS
Scar tissue and mineral deposits, a typical
sign of frozen shoulder, can be seen in the
right shoulder joint shown in this X-ray.

BUNION

A BUNION CONSISTS OF INFLAMED, THICKENED
SOFT TISSUE AND BONY OVERGROWTHS AT THE
BASE OF THE BIG TOE.

A bunion is usually caused by hallux
valgus, in which the big toe bends in
toward the other toes. The condition
is more common in women and tends
to run in families. The metatarsal (foot
bone) of the big toe angles toward the
body's midline but the phalanges
(toe bones) angle the other way. A
bunion can make walking painful.
If severe, it is corrected by surgery,
in which some bone is removed to
realign the toe.

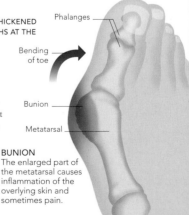

Phalanges

Bending
of toe

Bunion

Metatarsal

BUNION
The enlarged part of
the metatarsal causes
inflammation of the
overlying skin and
sometimes pain.

DISLOCATED JOINTS

WHERE BONES ARE FORCED OUT OF THEIR USUAL POSITIONS WITHIN A JOINT, THIS IS CALLED A DISLOCATION.

Often painful, a dislocation can be partial, in which only part of the bone is misplaced, or complete, such as in a dislocated shoulder where the humerus is totally out of its socket. A dislocation is often the result of a fall or sports injury. Rarely, the dislocation may also damage nerves, adjacent blood vessels, and other soft tissues, which rapidly swell and become painful. The affected area may have a different appearance from the normal joint on the other side of the body. Some people have

joints that are prone to dislocation because of slight natural variations in the shapes of the bone ends, or laxity of the ligaments, which can be inherited.

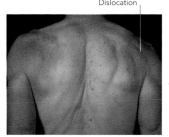
Dislocation

DISLOCATED SHOULDER
The area around the right shoulder joint appears swollen and misshapen compared with the normal shoulder on the left.

BURSITIS

INFLAMMATION OF THE BURSA, THE CUSHIONING PAD AT OR NEAR A JOINT, CAUSES PAIN, REDNESS, AND SWELLING.

A bursa is a fluid-filled sac that acts as a cushioned lubricating pad around a joint. It reduces the effect of friction and wear between muscle, tendon, and bone. Prolonged or repeated pressure, or sudden excessive stress at a joint, can cause a bursa to become inflamed and swollen. This may occur at various spots in the body but is most common at the knee and elbow. Predisposing factors include rheumatoid arthritis, gout, previous joint injury, or repetitive motion of a joint. In rare cases, it is due to bacterial infection. Treatment includes

rest and anti-inflammatory medication, and possibly draining excess synovial fluid from the bursa by aspiration. Sometimes a corticosteroid drug may be injected into the site.

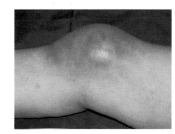

SWOLLEN KNEE
Swelling and tenderness in the bursa of the knee joint is often due to repeated kneeling and traditionally called "housemaid's knee."

HIP DISORDERS IN CHILDREN

Bone and joint abnormalities in children are often caused by injuries. But a painful or misshapen hip may be due to a transient "irritable hip" condition, a congenital defect, a problem with the blood supply to the hip, an infection, or juvenile rheumatoid arthritis (Still's disease).

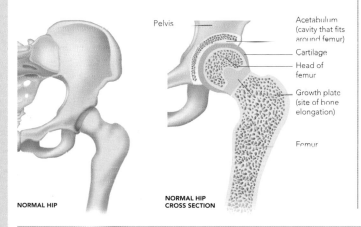
NORMAL HIP

Pelvis
Acetabulum (cavity that fits around femur)
Cartilage
Head of femur
Growth plate (site of bone elongation)
Femur
NORMAL HIP CROSS SECTION

DEVELOPMENTAL DYSPLASIA OF THE HIP

THIS CONDITION RESULTS FROM A FLATTENED OR MISPLACED SOCKET IN THE PELVIS FAILING TO HOLD THE FEMUR.

Previously known as congenital dislocation of the hip, or CDH, this problem is usually detected during the postnatal check given to babies soon after birth. The condition may cause mild looseness of the joint; occasional dislocation if manipulated; or complete displacement of the femoral head outside the socket in the hip bone and formation of a false joint (see right). If detected at birth, it may simply be monitored as the infant grows or treated with splints, a harness, or a cast, or even surgery. However, the dysplasia may be missed if it is very slight. It can then become apparent when the child begins to walk with a limp.

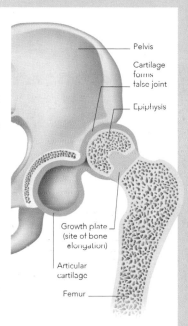
Pelvis
Cartilage forms false joint
Epiphysis
Growth plate (site of bone elongation)
Articular cartilage
Femur

PERTHES' DISEASE

THIS DISORDER IS THOUGHT TO BE DUE TO INSUFFICIENT BLOOD CIRCULATION IN THE HEAD OF THE FEMUR.

In Perthes' disease, the rounded head of the femur softens and becomes deformed, leading to pain in the thigh and groin, which may cause limping. The disease often affects only one hip. Perthes' disease is more common in boys than girls, and tends to occur around the ages of 4 to 10 years. It is thought to be caused by decreased blood circulation. Treatment is required, including rest, splinting, and perhaps traction and surgery, to help prevent osteoarthritis later in life.

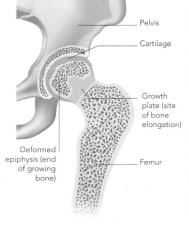
Pelvis
Cartilage
Growth plate (site of bone elongation)
Deformed epiphysis (end of growing bone)
Femur

SLIPPED EPIPHYSIS

THE HEAD OF THE FEMUR, OR PROXIMAL EPIPHYSIS, MAY SLIP THROUGH INJURY OR GRADUALLY BECOME DISPLACED.

The bony ball-shaped head (epiphysis) of the femur is separated from its shaft by a soft, cartilaginous region, known as the growth plate, where bone growth occurs. This is the usual site of slippage. Whether the displacement is slow or sudden, it is associated with obesity and tends to happen during rapid growth phases, often at puberty, when growth hormone may cause the tissues to soften. Surgical repair repositions the displaced bone, which may then be secured with metal pins.

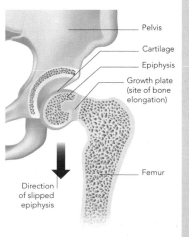
Pelvis
Cartilage
Epiphysis
Growth plate (site of bone elongation)
Femur
Direction of slipped epiphysis

ARTHRITIS

"Arthritis" is a collective term used to describe several different disorders that damage joints, causing pain, swelling, and restricted movement. The most common disorder in this group is osteoarthritis, which is widespread among older people. Rheumatoid arthritis can occur at any age, including during childhood, but usually begins after age 40.

OSTEOARTHRITIS

IN OSTEOARTHRITIS, THE CARTILAGE COVERING THE BONE ENDS (ARTICULAR CARTILAGE) INSIDE A JOINT BEGINS TO DEGENERATE, CAUSING PAIN AND SWELLING.

Osteoarthritis is often confused with rheumatoid arthritis (see opposite), but the two disorders have different causes and progressions. Osteoarthritis may affect only a single joint and can be triggered by localized "wear and tear," resulting in painful inflammation from time to time. Joint degeneration may be hastened by a congenital defect, injury, infection, or obesity. Because cartilage normally wears away as the body ages, a mild form of osteoarthritis affects many people after about the age of 60 years. Typical symptoms are pain and swelling in the joint that worsen with activity and fade with rest; stiffness for a short time after rest; restricted movement; crepitus (crackling noises) when moving the joint; and referred pain (in areas remote from the site of damage but on the same nerve pathway as the affected joint). Symptomatic treatment and lifestyle changes are effective in many milder cases.

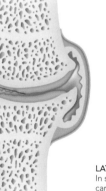

OSTEOARTHRITIS OF THE HIP
The right hip, on the left of this X-ray, is badly eroded by osteoarthritis. The head of the femur, which is normally round, is flattened.

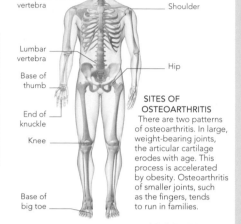

Cervical vertebra
Shoulder
Lumbar vertebra
Base of thumb
Hip
End of knuckle
Knee
Base of big toe

SITES OF OSTEOARTHRITIS
There are two patterns of osteoarthritis. In large, weight-bearing joints, the articular cartilage erodes with age. This process is accelerated by obesity. Osteoarthritis of smaller joints, such as the fingers, tends to run in families.

Bone
Joint capsule
Synovial membrane
Synovial fluid
Articular cartilage

HEALTHY JOINT
The articular cartilages coating the ends of the bones are smooth and compressible. They are lubricated by synovial fluid and slip past each other with minimal friction.

Inflamed synovial membrane
Osteophyte
Reduced joint space
Excess synovial fluid
Thinned articular cartilage

EARLY OSTEOARTHRITIS
In early osteoarthritis, articular cartilage becomes thin and rough with fissures in its surface. Bony outgrowths (osteophytes) form, and the synovial lining is inflamed, producing excess fluid.

Tight thickened capsule
Inflamed synovial membrane
Thickened bone
Bone surfaces in contact
Osteophyte
Cyst forming in bone

LATE OSTEOARTHRITIS
In severe cases of osteoarthritis, cartilage and underlying bone crack and erode. The bones rub together, thicken, and overgrow, causing extreme discomfort. The joint capsule thickens.

JOINT REPLACEMENT

When the symptoms of an osteoarthritic hip cannot be controlled with drug treatment the hip may be replaced by an artificial joint, or prosthesis. Joint replacement may also be used to treat hip fractures. A hip prosthesis is made of metal, ceramic, or plastic, and comprises a shaft with a ball-shaped head, and a cuplike pelvic socket, cemented in place. Other joints that can be replaced by prostheses include the knee, shoulder, and small joints in the hand. After the operation the joint is no longer painful, but physical therapy is needed to strengthen the muscles and restore full function.

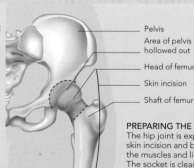

Pelvis
Area of pelvis hollowed out
Head of femur
Skin incision
Shaft of femur

PREPARING THE HIP
The hip joint is exposed by a skin incision and by moving the muscles and ligaments. The socket is cleared and the head of the femur removed.

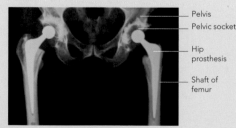

Pelvis
Pelvic socket
Hip prosthesis
Shaft of femur

DOUBLE HIP REPLACEMENT
This X-ray shows a hip prosthesis (light blue) in each leg. The ball-shaped head and an anchoring "spike" are clearly visible.

RHEUMATOID ARTHRITIS

IN THIS AUTOIMMUNE FORM OF ARTHRITIS, THE IMMUNE SYSTEM DAMAGES THE BODY'S OWN TISSUES, IN THIS CASE THE JOINTS. IT CAN AFFECT SEVERAL BODY SYSTEMS.

Rheumatoid arthritis develops when the immune system produces antibodies that attack its own body tissues, especially the synovial membranes inside joints. The joints become swollen and deformed, with painful and restricted movement. Early general symptoms include fever, fatigue, and weakness. Characteristically, many of the small joints are affected in a symmetrical pattern; for example, the hands and feet may become inflamed to the same degree on both sides. Stiffness is often worse in the mornings but eases during the day. Painless small lumps or nodules (clusters of inflamed tissue cells) may form in areas of pressure, commonly on the forearms, and the skin over the joint is thin and fragile. The condition may flare up then fade for a time. The diagnosis is supported if a blood test detects certain antibodies (RF and anti-CCP) associated with the problem. The disease can also affect the tissues of the eyes, skin, heart, nerves, and lungs. Anemia may also develop.

Treatment includes simple anti-inflammatory drugs and disease-modifying anti-rheumatic drugs (DMARDs).

JOINT INFLAMMATION

In this X-ray, the middle knuckles of the hands are severely damaged by rheumatoid arthritis (red). Inflammation of the joints causes abnormal bending of the fingers.

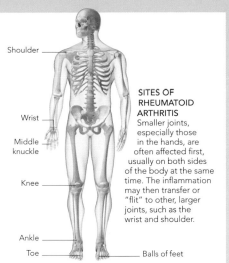

SITES OF RHEUMATOID ARTHRITIS

Smaller joints, especially those in the hands, are often affected first, usually on both sides of the body at the same time. The inflammation may then transfer or "flit" to other, larger joints, such as the wrist and shoulder.

Shoulder

Wrist

Middle knuckle

Knee

Ankle

Toe

Balls of feet

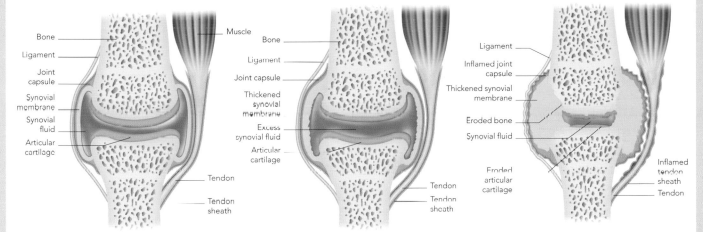

HEALTHY JOINT
Cartilage is smooth and intact in a healthy joint. Ligaments aid stability, and tendons slide in sheaths as muscles pull on them.

Bone
Ligament
Joint capsule
Synovial membrane
Synovial fluid
Articular cartilage
Muscle
Tendon
Tendon sheath

EARLY RHEUMATOID ARTHRITIS
The synovial membrane becomes inflamed and thickens, spreading across the joint. Excess synovial fluid accumulates.

Bone
Ligament
Joint capsule
Thickened synovial membrane
Excess synovial fluid
Articular cartilage
Tendon
Tendon sheath

LATE RHEUMATOID ARTHRITIS
As the synovial membrane thickens, the cartilage and bone ends are eroded. The joint capsule and tendon sheath become inflamed.

Ligament
Inflamed joint capsule
Thickened synovial membrane
Eroded bone
Synovial fluid
Eroded articular cartilage
Inflamed tendon sheath
Tendon

GOUT

IN GOUT, CRYSTALS OF URIC ACID FORM WITHIN A JOINT, CAUSING INTENSELY PAINFUL ARTHRITIS. IT CAN AFFECT ANY JOINT BUT COMMONLY OCCURS IN THE BIG TOE.

Gout is a type of crystal-induced arthritis that may cause sudden and severe pain, swelling, and redness in one or more joints. The problem is more common in men than in women, and when it occurs in women, it is usually after menopause. Owing to a problem of metabolism, with causes linked to certain genes and also lifestyle, excess uric acid accumulates in the body. Normally, uric acid stays in dissolved form, is collected by the blood, and then excreted in urine. In gout, however, this uric acid comes out of solution in the synovial fluid of a joint, forming needlelike crystals. The affected joint becomes red, hot, swollen, and very painful. Gout can occur spontaneously or be linked with drinking alcohol, certain forms of surgery, or some medications such as diuretics or chemotherapy. Drug treatment can relieve the pain of an attack and help prevent recurrence.

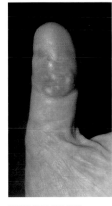

SWOLLEN THUMB
Here, uric acid crystals (pale yellow) have been deposited in the soft tissues of the thumb. They can eventually discharge through the skin as a chalky substance.

JOINT ASPIRATION

In this procedure, excess fluid is removed from a swollen joint by sucking it out with a needle and syringe, possibly under local anesthetic. The procedure can be used for diagnosis, treatment, or both. For example, the fluid may be examined for characteristic contents, such as the uric acid crystals of gout, while removal of the fluid eases the joint's swelling and pain. A similar procedure is used to inject medication directly into the joint.

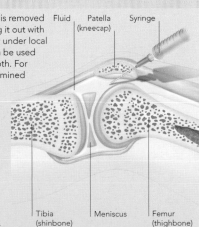

Fluid
Patella (kneecap)
Syringe
Tibia (shinbone)
Meniscus
Femur (thighbone)

ASPIRATING THE KNEE
The patella is held still with the knee relaxed. A needle is inserted into the space under the patella to withdraw the fluid.

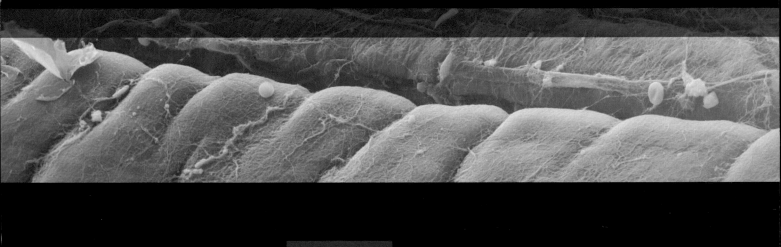

THE MUSCULAR SYSTEM PRODUCES AN ENDLESS VARIETY
OF ACTIONS BY USING MUSCLES AS COORDINATED TEAMS.
MUSCLE TISSUE CREATES BODY MOVEMENTS AND IT
ALSO POWERS INTERNAL PROCESSES, FROM THE HEARTBEAT
AND THE MOVEMENT OF FOOD THROUGH THE INTESTINES
TO THE ADJUSTMENT OF ARTERY DIAMETER AND FOCUSING
THE EYE. THE MUSCULAR SYSTEM LEADS A VERY PHYSICAL
EXISTENCE, IN WHICH REGULAR USE PREVENTS WASTING,
AND INJURY IS MORE COMMON THAN DISEASE. HOWEVER,
MUSCLES ARE HELPLESS WITHOUT THE NERVOUS SYSTEM
TO STIMULATE AND INTEGRATE THEIR ACTIVITIES.

MUSCULAR SYSTEM

MUSCLES OF THE BODY

MUSCLES ARE THE BODY'S "FLESH." THEY BULGE AND RIPPLE JUST UNDER THE SKIN, AND ARE ARRANGED IN CRISSCROSSING LAYERS DOWN TO THE BONES. THEIR JOB IS TO CONTRACT AND PULL THE BONES TO WHICH THEY ARE ANCHORED. RARELY WORKING ALONE, THEY USUALLY CONTRACT IN GROUPS, MOVING BONES AT ACCURATE ANGLES AND BY PRECISE DISTANCES.

The typical male body contains approximately 640 muscles, which compose around two-fifths of its weight. The same number in a female body make up a slightly smaller proportion of the total weight. A typical muscle spans a joint and tapers at each end into a fibrous tendon anchored to a bone. The more stable attachment of a muscle, usually nearer the center of the body, is known as its origin. The other end, the insertion, is toward the body's periphery and moves more. Some muscles divide to attach to different bones. The names of some muscles reflect their shape: the deltoid in the shoulder, for example, is triangular. Superficial muscles, those just under the skin, are pictured here on the left side of a male body. On the right of this body are the deeper layers—the intermediate muscles and deep muscles.

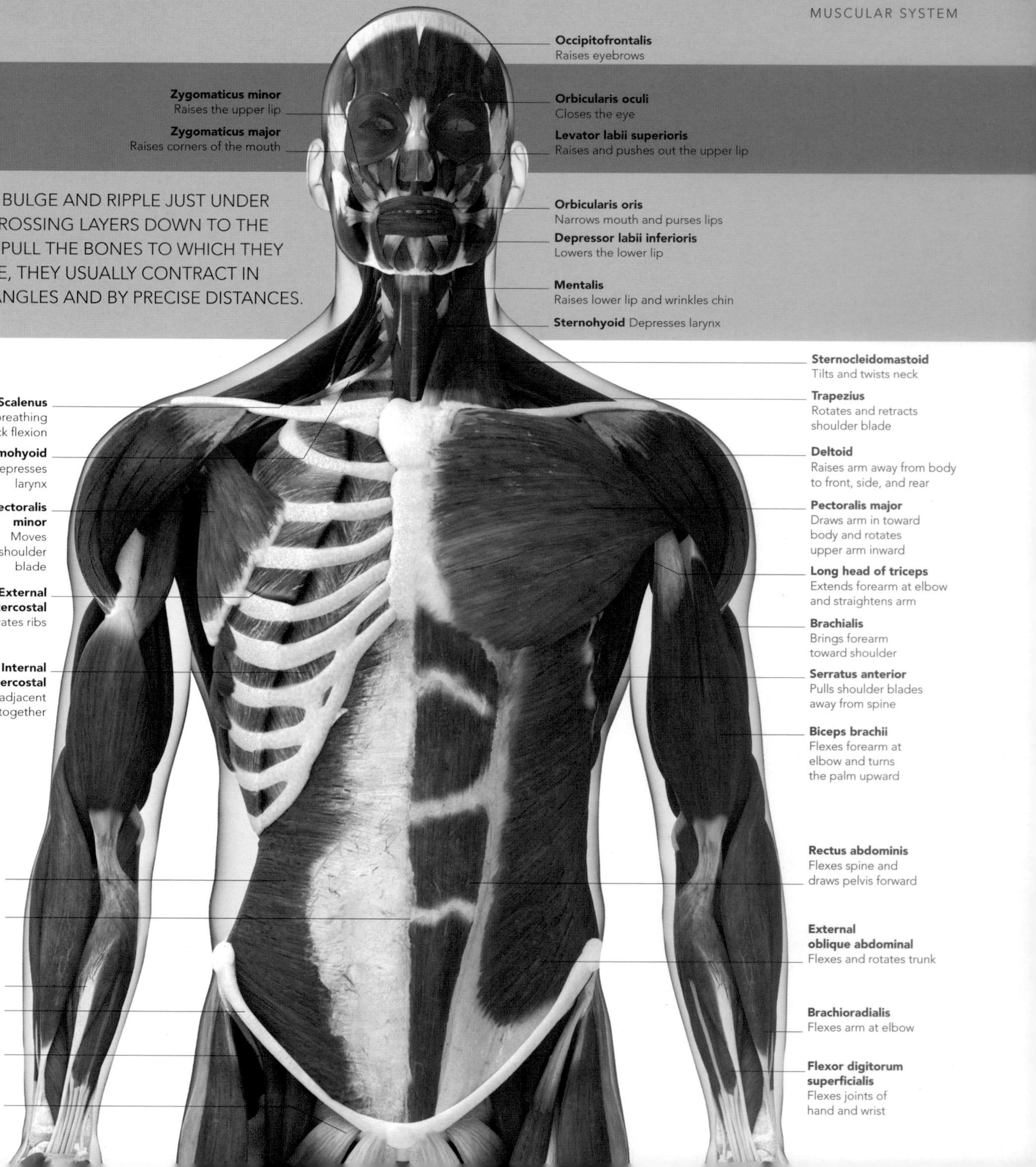

Occipitofrontalis
Raises eyebrows

Orbicularis oculi
Closes the eye

Levator labii superioris
Raises and pushes out the upper lip

Orbicularis oris
Narrows mouth and purses lips

Depressor labii inferioris
Lowers the lower lip

Mentalis
Raises lower lip and wrinkles chin

Sternohyoid Depresses larynx

Zygomaticus minor
Raises the upper lip

Zygomaticus major
Raises corners of the mouth

Sternocleidomastoid
Tilts and twists neck

Trapezius
Rotates and retracts shoulder blade

Deltoid
Raises arm away from body to front, side, and rear

Pectoralis major
Draws arm in toward body and rotates upper arm inward

Long head of triceps
Extends forearm at elbow and straightens arm

Brachialis
Brings forearm toward shoulder

Serratus anterior
Pulls shoulder blades away from spine

Biceps brachii
Flexes forearm at elbow and turns the palm upward

Rectus abdominis
Flexes spine and draws pelvis forward

External oblique abdominal
Flexes and rotates trunk

Brachioradialis
Flexes arm at elbow

Flexor digitorum superficialis
Flexes joints of hand and wrist

Scalenus
Aids breathing and neck flexion

Omohyoid
Depresses larynx

Pectoralis minor
Moves shoulder blade

External intercostal
Elevates ribs

Internal intercostal
Pulls adjacent ribs together

Internal oblique abdominal
Flexes and rotates trunk

Linea alba
Tendinous structure dividing left and right abdominal muscles

Flexor carpi radialis
Flexes hand at wrist

Inguinal ligament

Iliopsoas
Flexes thigh at hip

Pectineus
Flexes and draws thigh in toward body

Abductor pollicis brevis
Pulls thumb in toward palm

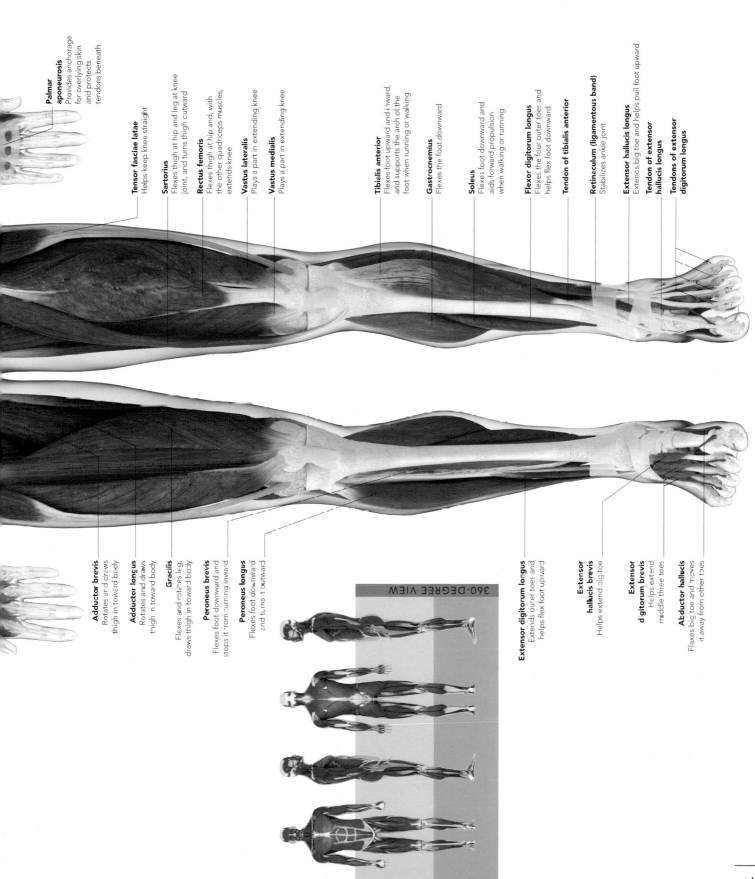

Palmar aponeurosis
Provides anchorage for overlying skin and protects tendons beneath

Tensor fasciae latae
Helps keep knee straight

Sartorius
Flexes thigh at hip and leg at knee joint, and turns thigh outward

Rectus femoris
Flexes thigh at hip and, with the other quadriceps muscles, extends knee

Vastus lateralis
Plays a part in extending knee

Vastus medialis
Plays a part in extending knee

Tibialis anterior
Flexes foot upward and inward, and supports the arch of the foot when running or walking

Gastrocnemius
Flexes the foot downward

Soleus
Flexes foot downward and aids forward propulsion when walking or running

Flexor digitorum longus
Flexes the four outer toes and helps flex foot downward

Tendon of tibialis anterior

Retinaculum (ligamentous band)
Stabilizes ankle joint

Extensor hallucis longus
Extends big toe and helps pull foot upward

Tendon of extensor hallucis longus

Tendons of extensor digitorum longus

Adductor brevis
Rotates and draws thigh in toward body

Adductor longus
Rotates and draws thigh in toward body

Gracilis
Flexes and rotates leg, draws thigh in toward body

Peroneus brevis
Flexes foot downward and stops it from turning inward

Peroneus longus
Flexes foot downward and turns it outward

Extensor digitorum longus
Extends outer toes and helps flex foot upward

Extensor hallucis brevis
Helps extend big toe

Extensor digitorum brevis
Helps extend middle three toes

Adductor hallucis
Flexes big toe and moves it away from other toes

360-DEGREE VIEW

73

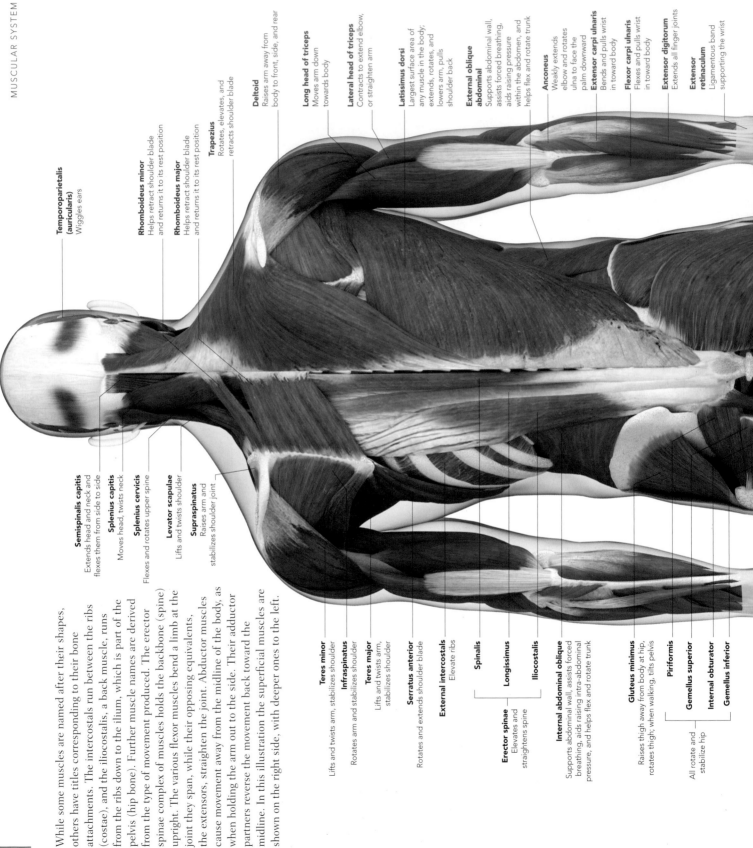

While some muscles are named after their shapes, others have titles corresponding to their bone attachments. The intercostals run between the ribs (costae), and the iliocostalis, a back muscle, runs from the ribs down to the ilium, which is part of the pelvis (hip bone). Further muscle names are derived from the type of movement produced. The erector spinae complex of muscles holds the backbone (spine) upright. The various flexor muscles bend a limb at the joint they span, while their opposing equivalents, the extensors, straighten the joint. Abductor muscles cause movement away from the midline of the body, as when holding the arm out to the side. Their adductor partners reverse the movement back toward the midline. In this illustration the superficial muscles are shown on the right side, with deeper ones to the left.

Temporoparietalis (auricularis)
Wiggles ears

Rhomboideus minor
Helps retract shoulder blade and returns it to its rest position

Rhomboideus major
Helps retract shoulder blade and returns it to its rest position

Trapezius
Rotates, elevates, and retracts shoulder blade

Deltoid
Raises arm away from body to front, side, and rear

Long head of triceps
Moves arm down towards body

Lateral head of triceps
Contracts to extend elbow, or straighten arm

Latissimus dorsi
Largest surface area of any muscle in the body; extends, rotates, and lowers arm, pulls shoulder back

External oblique abdominal
Supports abdominal wall, assists forced breathing, aids raising pressure within the abdomen, and helps flex and rotate trunk

Anconeus
Weakly extends elbow and rotates ulna to face the palm downward

Extensor carpi ulnaris
Bends and pulls wrist in toward body

Flexor carpi ulnaris
Flexes and pulls wrist in toward body

Extensor digitorum
Extends all finger joints

Extensor retinaculum
Ligamentous band supporting the wrist

Semispinalis capitis
Extends head and neck and flexes them from side to side

Splenius capitis
Moves head, twists neck

Splenius cervicis
Flexes and rotates upper spine

Levator scapulae
Lifts and twists shoulder

Supraspinatus
Raises arm and stabilizes shoulder joint

Teres minor
Lifts and twists arm, stabilizes shoulder

Infraspinatus
Rotates arm and stabilizes shoulder

Teres major
Lifts and twists arm, stabilizes shoulder

Serratus anterior
Rotates and extends shoulder blade

External intercostals
Elevate ribs

Spinalis

Longissimus

Iliocostalis

Erector spinae
Elevates and straightens spine

Internal abdominal oblique
Supports abdominal wall, assists forced breathing, aids raising intra-abdominal pressure, and helps flex and rotate trunk

Gluteus minimus
Raises thigh away from body at hip, rotates thigh; when walking, tilts pelvis

Piriformis

Gemellus superior

Internal obturator

Gemellus inferior

All rotate and stabilize hip

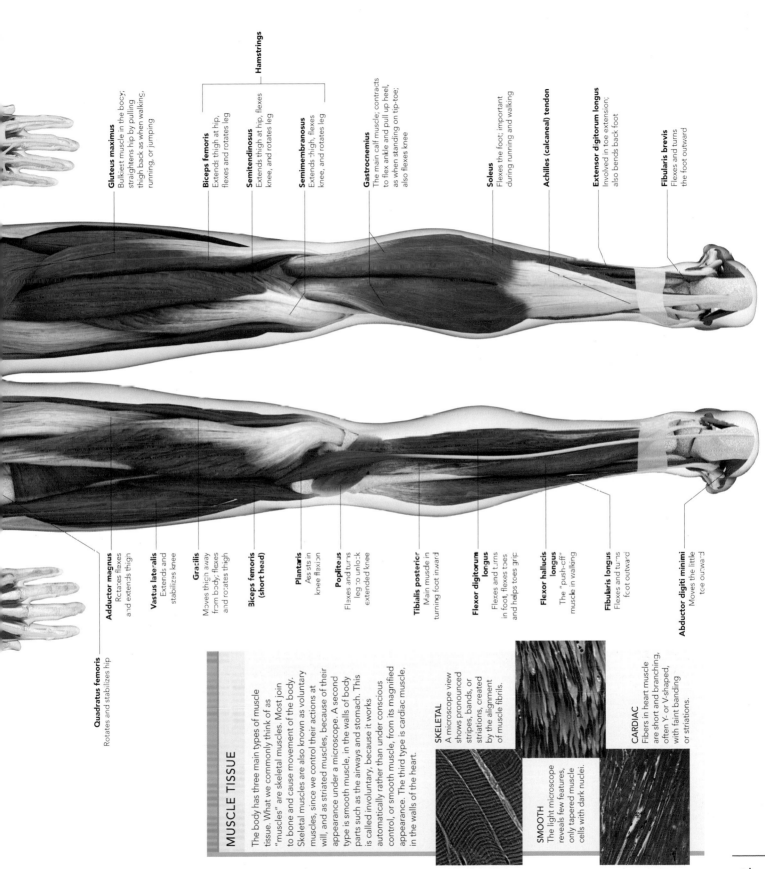

Quadratus femoris
Rotates and stabilizes hip

Adductor magnus
Rotates flexes and extends thigh

Vastus lateralis
Extends and stabilizes knee

Gracilis
Moves thigh away from body; flexes and rotates thigh

Biceps femoris (short head)
Extends thigh

Plantaris
Assists in knee flexion

Popliteus
Flexes and turns leg to unlock extended knee

Tibialis posterior
Main muscle in turning foot inward

Flexor digitorum longus
Flexes and turns in foot, flexes toes and helps toes grip

Flexor hallucis longus
The "push-off" muscle in walking

Fibularis longus
Flexes and turns foot outward

Abductor digiti minimi
Moves the little toe outward

Gluteus maximus
Bulkiest muscle in the body; straightens hip by pulling thigh back as when walking, running, or jumping

Biceps femoris
Extends thigh at hip, flexes and rotates leg

Semitendinosus
Extends thigh at hip, flexes knee, and rotates leg

Semimembranosus
Extends thigh, flexes knee, and rotates leg

Hamstrings

Gastrocnemius
The main calf muscle; contracts to flex ankle and pull up heel, as when standing on tip-toe; also flexes knee

Soleus
Flexes the foot; important during running and walking

Achilles (calcaneal) tendon

Extensor digitorum longus
Involved in toe extension; also bends back foot

Fibularis brevis
Flexes and turns the foot outward

MUSCLE TISSUE

The body has three main types of muscle tissue. What we commonly think of as "muscles" are skeletal muscles. Most join to bone and cause movement of the body. Skeletal muscles are also known as voluntary muscles, since we control their actions at will, and as striated muscles, because of their appearance under a microscope. A second type is smooth muscle, in the walls of body parts such as the airways and stomach. This is called involuntary, because it works automatically rather than under conscious control, or smooth muscle, from its magnified appearance. The third type is cardiac muscle, in the walls of the heart.

SKELETAL
A microscope view shows pronounced stripes, bands, or striations, created by the alignment of muscle fibrils.

SMOOTH
The light microscope reveals few features, only tapered muscle cells with dark nuclei.

CARDIAC
Fibers in heart muscle are short and branching, often Y- or V-shaped, with faint banding or striations.

MUSCLES OF THE FACE, HEAD, AND NECK

TO STEADY AND MOVE THE HEAD AND TO MOVE FACIAL FEATURES SUCH AS THE EYEBROWS, EYELIDS, AND LIPS, THE MUSCLES OF THE FACE, HEAD, AND NECK INTERACT. THE MUSCULATURE INVOLVED IS HIGHLY COMPLEX, ALLOWING FOR A HUGE RANGE OF FACIAL EXPRESSIONS.

FACIAL MUSCLES

Some facial muscles are anchored to bones. Others are joined to tendons or to dense, sheetlike clusters of fibrous connective tissue called aponeuroses. This means that some facial muscles are joined to each other. Many of these muscles have their other end inserted into deeper layers of the skin. The advantage of this complex system is that even a slight degree of muscle contraction produces movement of the face's skin, which reveals itself as a show of expression or emotion. Almost all facial muscles are controlled by the facial nerve called cranial VII (see p.98). Damage or disease of this nerve results in loss of facial mobility and expression, reducing the ability to communicate.

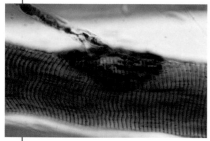

NERVE-MUSCLE JUNCTION
In this microscope image a nerve cell (top left) joins a facial muscle fiber. At the point of contact is the motor end plate (center), an area of highly excitable muscle fiber.

LAUGHTER LINES

Healthy young skin contains resilient fibers made of the protein elastin, which help it return to its original position, for example, after smiling. With increasing age, the elastin degenerates and the skin's dermis (see pp.164–65) becomes more loosely attached to the muscle beneath. This causes wrinkles since the skin can no longer stretch or shrink easily. Initially "crow's feet" radiate from the corners of the eyes. These are followed by lines around the brow and mouth, in front of the ears, between the eyebrows, on the chin and bridge of the nose. Facial wrinkles are always at right angles to the muscle fibers so they reveal the pattern of facial muscles. Exposure to excessive sunlight and temperature hastens wrinkling.

Forehead wrinkles

Crow's foot

Cheek crease

SIGNS OF AGING
Wrinkles and furrows tend to appear around often-used muscles (see pp.258–61). Crow's feet are associated with the orbicularis oculi, forehead wrinkles with the frontalis, and creases in the cheeks with the levator labii superioris.

FACE AND NECK MUSCLES
Intermeshing muscles around the lips are involved in speech, nonverbal expression, eating, and drinking. Some facial muscles act as sphincters to open and close orifices, such as the eyelids, nostrils, and lips.

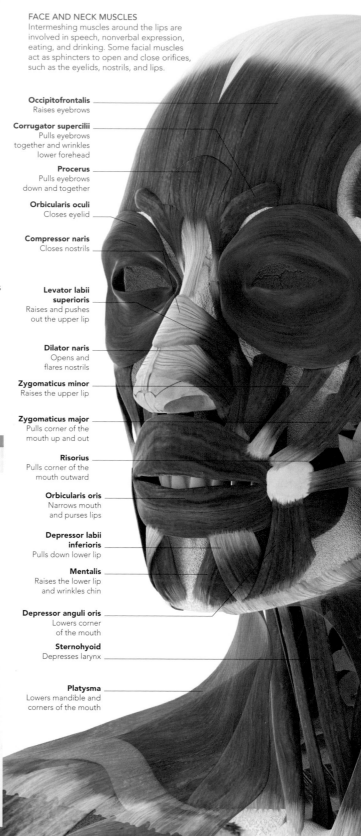

Occipitofrontalis
Raises eyebrows

Corrugator supercilii
Pulls eyebrows together and wrinkles lower forehead

Procerus
Pulls eyebrows down and together

Orbicularis oculi
Closes eyelid

Compressor naris
Closes nostrils

Levator labii superioris
Raises and pushes out the upper lip

Dilator naris
Opens and flares nostrils

Zygomaticus minor
Raises the upper lip

Zygomaticus major
Pulls corner of the mouth up and out

Risorius
Pulls corner of the mouth outward

Orbicularis oris
Narrows mouth and purses lips

Depressor labii inferioris
Pulls down lower lip

Mentalis
Raises the lower lip and wrinkles chin

Depressor anguli oris
Lowers corner of the mouth

Sternohyoid
Depresses larynx

Platysma
Lowers mandible and corners of the mouth

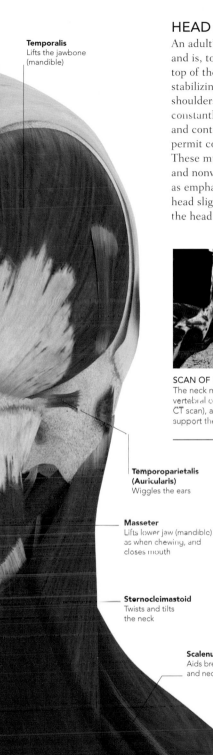

Temporalis
Lifts the jawbone
(mandible)

HEAD AND NECK MUSCLES

An adult's head weighs over 11 lb (5 kg) and is, to some extent, "balanced" on top of the vertebral column. Strong, stabilizing muscles in the neck, inner shoulders, and upper back are active constantly, tensing to steady the head and contracting in coordinated teams to permit complex movements of the neck. These muscles assist facial expressions and nonverbal communication, such as emphasizing doubt by cocking the head slightly to one side or moving the head to indicate "yes" or "no."

SCAN OF NECK MUSCLES
The neck muscles move the skull on the vertebral column (shown in green in this CT scan), and also wrap around and support the trachea and esophagus.

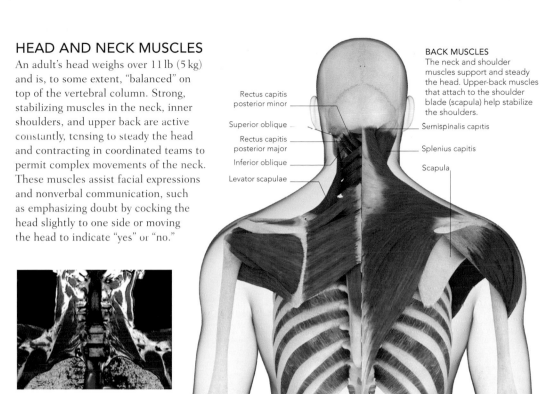

BACK MUSCLES
The neck and shoulder muscles support and steady the head. Upper-back muscles that attach to the shoulder blade (scapula) help stabilize the shoulders.

Rectus capitis
posterior minor

Superior oblique

Rectus capitis
posterior major

Inferior oblique

Levator scapulae

Semispinalis capitis

Splenius capitis

Scapula

**Temporoparietalis
(Auricularis)**
Wiggles the ears

Masseter
Lifts lower jaw (mandible)
as when chewing, and
closes mouth

Sternocleimastoid
Twists and tilts
the neck

Scalenus
Aids breathing
and neck flexion

RELAXED "NEUTRAL" EXPRESSION

FACIAL EXPRESSIONS

Facial expressions are among our most important methods of nonverbal communication. The facial musculature enables us to make many subtle nuances of appearance that convey an enormous variety of emotions. A smile often indicates pleasure, and a frown the opposite—but not always. The smile is a highly ambiguous and versatile expression, which can also convey relief or pity, or widen into a grin for sarcastic disapproval. Likewise, a frown can articulate various feelings, including disappointment and confusion. In addition to the mouth, other regions of the face are involved to add shades of meaning.

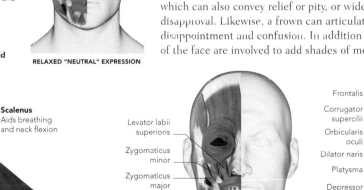

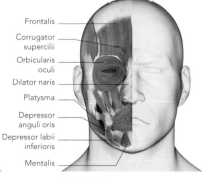

Levator labii
superioris

Zygomaticus
minor

Zygomaticus
major

Risorius

Frontalis

Corrugator
supercilii

Orbicularis
oculi

Dilator naris

Platysma

Depressor
anguli oris

Depressor labii
inferioris

Mentalis

SMILING
The levator labii superioris lifts the upper lip, while the zygomaticus major and minor and the risorius pull the angle of the mouth and the lip corners up and sideways.

FROWNING
The platysma and depressor muscles pull the mouth and corners of the lips down, and the mentalis wrinkles the chin. The corrugator supercilii furrows the brow, the dilator naris flares the nostrils, and the orbicularis oculi narrows the eyes.

MUSCLES AND TENDONS

MUSCLES CAN ONLY CONTRACT AND SHORTEN. TO RETURN
TO THEIR ORIGINAL SHAPE, THEY RELAX AND LENGTHEN
PASSIVELY AS OTHER MUSCLES CONTRACT.
CONTRACTION OF SKELETAL MUSCLES AND
TENDONS GENERATES BODY MOTION.

MUSCLE STRUCTURE

Skeletal (striated or voluntary) muscle consists
of densely packed groups of hugely elongated
cells known as myofibers. These are grouped
into bundles (fascicles). A typical myofiber
is ³/₄–1¹/₅ in (2–3 cm) long and ¹/₅₀₀ in (0.05 mm)
in diameter and is composed of narrower
structures called myofibrils. These contain
thick and thin myofilaments made up mainly
of the proteins actin and myosin. Numerous
capillaries keep the muscle supplied with the
oxygen and glucose needed to fuel contraction.

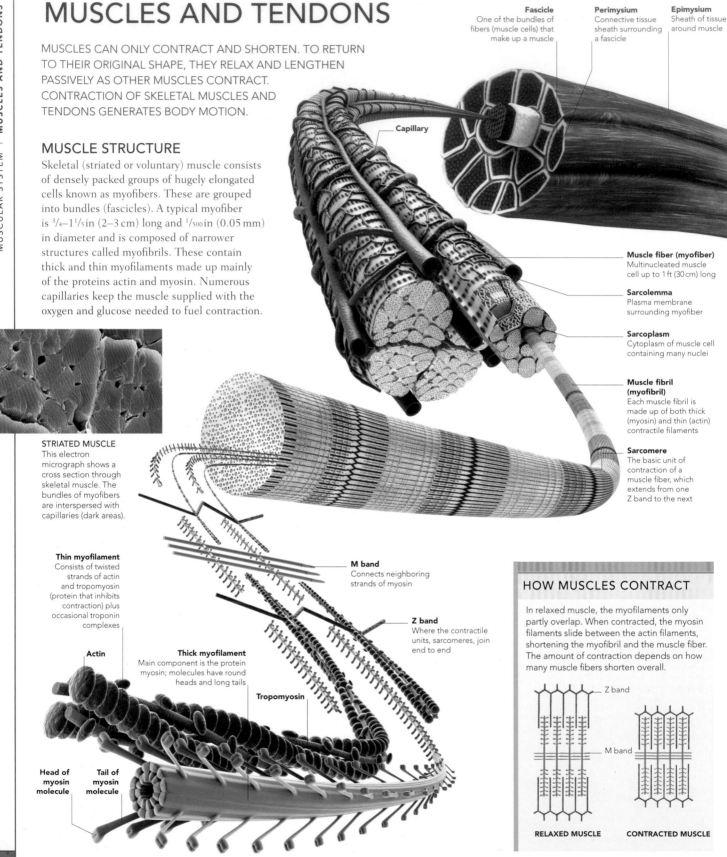

STRIATED MUSCLE
This electron
micrograph shows a
cross section through
skeletal muscle. The
bundles of myofibers
are interspersed with
capillaries (dark areas).

Fascicle
One of the bundles of
fibers (muscle cells) that
make up a muscle

Perimysium
Connective tissue
sheath surrounding
a fascicle

Epimysium
Sheath of tissue
around muscle

Capillary

Muscle fiber (myofiber)
Multinucleated muscle
cell up to 1 ft (30 cm) long

Sarcolemma
Plasma membrane
surrounding myofiber

Sarcoplasm
Cytoplasm of muscle cell
containing many nuclei

**Muscle fibril
(myofibril)**
Each muscle fibril is
made up of both thick
(myosin) and thin (actin)
contractile filaments

Sarcomere
The basic unit of
contraction of a
muscle fiber, which
extends from one
Z band to the next

Thin myofilament
Consists of twisted
strands of actin
and tropomyosin
(protein that inhibits
contraction) plus
occasional troponin
complexes

Actin

Thick myofilament
Main component is the protein
myosin; molecules have round
heads and long tails

Tropomyosin

M band
Connects neighboring
strands of myosin

Z band
Where the contractile
units, sarcomeres, join
end to end

**Head of
myosin
molecule**

**Tail of
myosin
molecule**

HOW MUSCLES CONTRACT

In relaxed muscle, the myofilaments only
partly overlap. When contracted, the myosin
filaments slide between the actin filaments,
shortening the myofibril and the muscle fiber.
The amount of contraction depends on how
many muscle fibers shorten overall.

Z band

M band

RELAXED MUSCLE　　**CONTRACTED MUSCLE**

BODY PARTS AS LEVERS

Movements in the body, such as nodding and walking, employ the mechanical principles of applying a force to one part of a rigid lever, which tilts at a pivot point (fulcrum) to move a weight (load) elsewhere on the lever. The muscles apply force, the bones serve as levers, and the joints function as fulcrums. A whole range of lever systems exist in the body and among them they allow a wide range of movement as well as providing a means to lift and carry things.

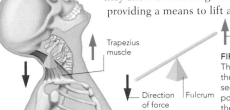

Trapezius muscle

Movement of load

Direction of force   Fulcrum

FIRST-CLASS LEVER
The fulcrum is positioned between the force and the load, like a seesaw. An example is seen in the posterior neck muscles that tilt back the head on the cervical vertebrae.

SECOND-CLASS LEVER
The load lies between the force and the fulcrum. Standing on tiptoe, the calf muscles provide the force, the heel and foot form the lever, and the toes provide the fulcrum.

Movement of load

Direction of force

Gastrocnemius muscle

Tendon

Fulcrum

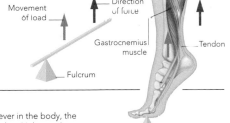

THIRD-CLASS LEVER
The most common type of lever in the body, the force is applied between load and fulcrum. An example is flexing the elbow joint (the fulcrum) by contracting the biceps brachii muscle.

Biceps brachii muscle

Tendon

Movement of load

Direction of force

Fulcrum

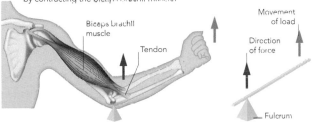

POSITIONAL SENSE

Muscles contain many tiny sensors, known as neuromuscular spindles. These are modified muscle fibers with a spindle-shaped sheath or capsule and several types of nerve supply. The sensory or afferent nerve fibers, which are wrapped around the modified muscle fibers, relay information to the spinal cord and brain about muscle length and tension as the muscle stretches. Signals are then sent back through motor neurons to the muscle to tell it to contract, thus restoring muscle tension to normal. Similar receptors are found in ligaments and tendons. Together they provide the body's innate sense of its own position and posture, called proprioception.

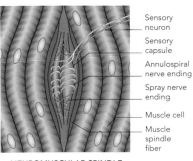

Sensory neuron

Sensory capsule

Annulospiral nerve ending

Spray nerve ending

Muscle cell

Muscle spindle fiber

NEUROMUSCULAR SPINDLE
These stretch sensors lie between and in parallel with skeletal muscle fibers; information from them allows the brain to gauge the muscle's tension and elongation.

TENDONS

Tendons are tough, fibrous cords of connective tissue that link skeletal muscles to bones. Within them, Sharpey's fibers pass through the bone covering (periosteum) to embed in the bone. Tendons in the hands and feet are enclosed in self-lubricating sheaths to protect them from rubbing against the bones. From the hand bones, tendons extend upward to muscles near the elbow.

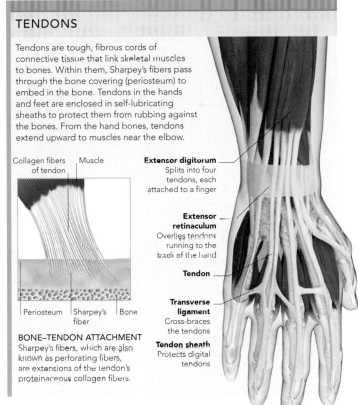

Collagen fibers of tendon   Muscle

Periosteum   Sharpey's fiber   Bone

BONE–TENDON ATTACHMENT
Sharpey's fibers, which are also known as perforating fibers, are extensions of the tendon's proteinaceous collagen fibers.

Extensor digitorum
Splits into four tendons, each attached to a finger

Extensor retinaculum
Overlies tendons running to the back of the hand

Tendon

Transverse ligament
Cross-braces the tendons

Tendon sheath
Protects digital tendons

HOW MUSCLES WORK TOGETHER

Muscles can only pull, not push, and are arranged in pairs that act in opposition to one other. The movement produced by one muscle can be reversed by its opposing partner. When a muscle contracts to produce movement, it is called the agonist, while its opposite partner, the antagonist, relaxes and is passively stretched. In reality, few movements are achieved by a single muscle contraction. Usually, whole teams of muscles act as agonists to give the precisely required degree and direction of motion, while the antagonists tense to prevent the movement from overextending.

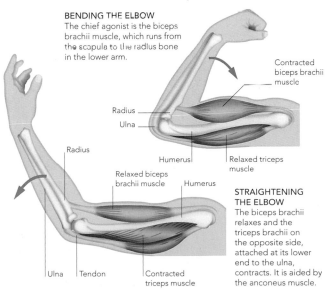

BENDING THE ELBOW
The chief agonist is the biceps brachii muscle, which runs from the scapula to the radius bone in the lower arm.

Contracted biceps brachii muscle

Radius
Ulna

Humerus   Relaxed triceps muscle

Radius

Relaxed biceps brachii muscle   Humerus

Ulna   Tendon   Contracted triceps muscle

STRAIGHTENING THE ELBOW
The biceps brachii relaxes and the triceps brachii on the opposite side, attached at its lower end to the ulna, contracts. It is aided by the anconeus muscle.

MUSCLE AND TENDON DISORDERS

INJURIES TO MUSCLES AND THEIR TENDON ATTACHMENTS ARE USUALLY THE RESULT OF PHYSICAL EXERTION DURING DAILY ACTIVITIES, OR DUE TO SUDDEN PULLING OR TWISTING MOVEMENTS, SUCH AS THOSE OCCURRING IN SPORTS OR AN ACCIDENT. REPETITIVE ACTIONS, FOR EXAMPLE AS PART OF EMPLOYMENT, CAN ALSO DAMAGE MUSCLES AND TENDONS OVER TIME. A NUMBER OF RARE MUSCLE DISORDERS MAY BE RESPONSIBLE FOR MUSCLE WEAKNESS AND PROGRESSIVE DEGENERATION.

MUSCLE STRAINS AND TEARS

A MILD INJURY RESULTING FROM AN OVERSTRETCHED MUSCLE IS CALLED A STRAIN; MORE SEVERE DAMAGE IS A TEAR.

Muscle strain is the term used for a moderate amount of soft-tissue damage to muscle fibers, usually caused by sudden, strenuous movements. Limited bleeding inside the muscle causes tenderness and swelling, which may be accompanied by painful spasms or contractions. Visible bruising may follow. More serious damage, involving a larger number of torn or ruptured fibers, is called a muscle tear. A torn muscle causes severe pain and swelling. Following a medical check to gauge severity, the usual treatment is rest, anti-inflammatory medication, and perhaps physical therapy. Rarely, surgery may be needed to repair a muscle that has been badly torn. The risk of muscle strains and tears can be reduced by warming up adequately before exercise.

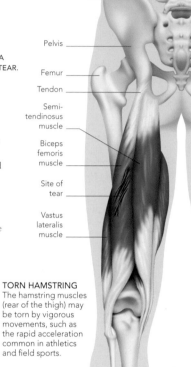

Pelvis
Femur
Tendon
Semi-tendinosus muscle
Biceps femoris muscle
Site of tear
Vastus lateralis muscle

TORN HAMSTRING
The hamstring muscles (rear of the thigh) may be torn by vigorous movements, such as the rapid acceleration common in athletics and field sports.

SOFT-TISSUE INFLAMMATION

THE BODY'S OWN DEFENSES CAUSE MUSCLE TISSUE TO BECOME INFLAMED AS THE HEALING PROCESS BEGINS.

Like any soft tissue, muscle reacts to damage (such as that from a physical blow) with inflammation (see pp.178–79). The affected area becomes hot, red, and swollen as blood and fluids accumulate from ruptured cells and capillaries. Blood vessels widen (dilate) as white blood cells congregate, attracted by the leaking debris from muscle fibers (cells) and other tissues. Moving the muscle causes discomfort or pain. Longer-term causes of muscle tissue inflammation are the group of disorders called repetitive strain injuries (RSIs). The basic cause is a particular movement or action repeated often over a long period. Movements that are rapid and forceful increase the risk. RSIs are linked to many and varied daily activities, from working on assembly lines or with computers, to sports or playing a musical instrument.

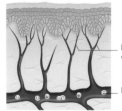

Blood vessels
Neutrophil

NORMAL TISSUE
Blood flows through undamaged vessels, where occasional white cells, such as neutrophils, scavenge debris and attack microorganisms that have managed to enter.

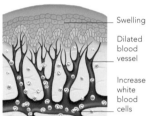

Swelling
Dilated blood vessel
Increased white blood cells

INFLAMED TISSUE
Blood vessels enlarge in diameter, bringing increased numbers of white blood cells, as fluid leaks from disrupted cells and tissues. Heat, pain, and redness result.

TENDINITIS AND TENOSYNOVITIS

INFLAMMATION CAN AFFECT THE TENDON ITSELF, AS TENDINITIS, OR THE LININGS OF THE TENDON SHEATHS THAT ENCLOSE THEM, AS TENOSYNOVITIS.

Tendinitis may occur when strong or repeated movement creates excessive friction between the tendon's outer surface and an adjacent bone. Tenosynovitis may be the result of overstretching or repeated movement causing inflammation of the lubricating sheaths that enclose some tendons. Both of these problems can occur together and may be part of the group of disorders known as repetitive strain injuries (RSIs), as described in soft-tissue inflammation, above. Areas affected include the shoulder, elbow, wrist, fingers, knee, and the back of the heel. Symptoms of both tendinitis and tenosynovitis are stiffness, swelling, and pain, with hot, reddened skin at the site.

Tendon sheaths

Tendons

Clavicle (collarbone)
Inflamed supraspinous tendon
Humerus

TENDINITIS
Repeated arm lifting, such as in racquet sports, may force the supraspinous tendon to rub against the shoulder blade's acromion process, causing tendinitis.

Acromion process of shoulder blade
Supraspinatus muscle

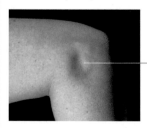

Inflammation
Tendon sheath

TENOSYNOVITIS
The complex, weight-bearing nature of the foot makes it susceptible to tendon damage. Activities that involve running or kicking, and awkward movements, such as dancing, may cause inflammation.

TENNIS ELBOW

TENNIS AND GOLFER'S ELBOW ARE COMMON NAMES FOR TENDON DAMAGE IN THE AREA WHERE THE ARM MUSCLES ATTACH TO BONES NEAR THE ELBOW JOINT.

Most cases of tennis elbow involve the common extensor tendon, which anchors several forearm muscles involved in wrist and hand movements to the lateral epicondyle, a knoblike projection on the upper arm bone (humerus). Golfer's elbow is a similar type of injury but the pain is at the site of the medial epicondyle on the elbow's inner side.

Lateral epicondyle—the site of pain from tennis elbow

ELBOW INFLAMMATION
In tennis elbow, vigorous, repeated use of the forearm against resistance causes small tears in the tendon, leading to tenderness and pain on the outer side of the joint.

RUPTURED TENDON

A SUDDEN, POWERFUL MUSCLE CONTRACTION OR WRENCHING INJURY CAN COMPLETELY TEAR A TENDON.

Playing sports and unaccustomed lifting of heavy weights may result in torn, or ruptured, tendons. Examples are tearing of the tendons attached to the biceps brachii muscle in the upper arm, or of the quadriceps tendon at the front of the thigh that stretches over the knee. A sudden impact that bends a fingertip toward the palm may snap the extensor tendon on the back of the finger. In severe

cases, the tendon may even be torn away from the bone. Main symptoms include a snapping or twanging sensation, pain, swelling, and impaired movement. Some injuries, such as a ruptured Achilles tendon (at the back of the heel), may require immobilizaton of the affected area with a cast to prevent the tendon from stretching in the early stages of healing.

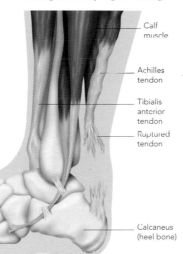

TORN ACHILLES TENDON
The Achilles (long calcaneal) tendon attaches the calf muscle to the heel bone (calcaneus). It can snap after sudden exertion and may need to be treated by surgery and immobilization in a cast.

- Calf muscle
- Achilles tendon
- Tibialis anterior tendon
- Ruptured tendon
- Calcaneus (heel bone)

MYASTHENIA GRAVIS

THIS AUTOIMMUNE DISORDER CAUSES CHRONIC MUSCLE WEAKNESS; EYE AND FACIAL MUSCLES ARE AFFECTED MOST.

Myasthenia gravis is caused by antibodies that attack and gradually destroy the receptors in muscle fibers that receive nerve signals. As a result, muscles are not stimulated to contract, or respond only very weakly. Affected muscles

include those of the face, throat, and eyes, which can lead to problems with speech and vision. Arm, leg, and respiratory muscles are more rarely affected. A thymus disorder may trigger the disease, so treatment may involve immunosuppressant and other medications and removal of the thymus gland.

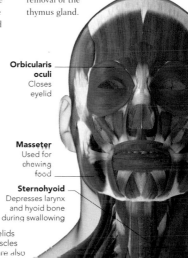

Orbicularis oculi
Closes eyelid

Masseter
Used for chewing food

Sternohyoid
Depresses larynx and hyoid bone during swallowing

EFFECTS OF MYASTHENIA GRAVIS
Early symptoms include drooping eyelids (above) as facial muscles weaken. Muscles involved in chewing and swallowing are also affected, so eating can become difficult.

MUSCULAR DYSTROPHY

MUSCULAR DYSTROPHIES ARE A GROUP OF INHERITED DISORDERS THAT CAUSE DEGENERATION OF MUSCLE, LEADING TO WEAK AND IMPAIRED MOVEMENTS.

Common symptoms of various types of muscular dystrophy (MD) are progressive wasting of muscles and loss of movement. There is no effective treatment to halt the underlying process. However, stretching exercises and surgery to release shortened muscles and tendons can benefit people with muscular dystrophy by improving mobility. Among the well-known forms are Duchenne and Becker MD, in which the genetic abnormality is carried on the X chromosome; they almost always affect boys.

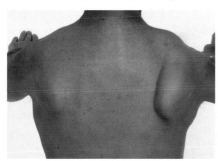

EFFECT OF MUSCULAR DYSTROPHY
In facioscapulohumeral muscular dystrophy (FSH), the muscles of the face, shoulder, and upper arm become weak. Holding the arm out forward causes "winging" of the shoulder blade (scapula), where the bone's inner edge protrudes rearward.

CARPAL TUNNEL SYNDROME

COMPRESSION OF A NERVE IN THE WRIST LEADS TO SYMPTOMS SUCH AS TINGLING AND PAIN IN THE HAND, WRIST, AND FOREARM, AND WEAKENED GRIP.

The carpal tunnel is a narrow passageway formed by the carpal ligament (flexor retinaculum), on the inside of the wrist and the underlying wrist bones, the carpals. Long tendons run through the passage from the muscles in the forearm to the bones of the hand and fingers. The median nerve also passes through the carpal tunnel, to control hand muscles

and convey sensations from the fingers. In carpal tunnel syndrome (CTS) the median nerve is compressed by swelling of the tissues around it in the tunnel. Causes include diabetes mellitus, pregnancy, a wrist injury, rheumatoid arthritis, and repetitive movements; in some cases the cause is not clear. CTS tends to affect women aged 40–60 and can occur in both wrists. The nerve compression causes numbness and pain, especially in the thumb to middle fingers and one side of the ring finger. Anti-inflammatory drugs and perhaps surgery to loosen the ligament can bring relief.

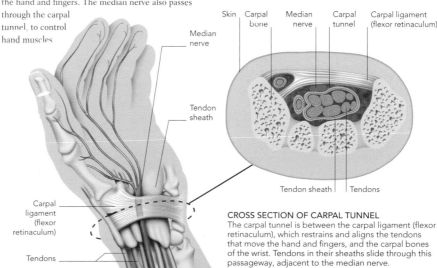

Median nerve

Tendon sheath

Carpal ligament (flexor retinaculum)

Tendons

Skin — Carpal bone — Median nerve — Carpal tunnel — Carpal ligament (flexor retinaculum)

Tendon sheath | Tendons

CROSS SECTION OF CARPAL TUNNEL
The carpal tunnel is between the carpal ligament (flexor retinaculum), which restrains and aligns the tendons that move the hand and fingers, and the carpal bones of the wrist. Tendons in their sheaths slide through this passageway, adjacent to the median nerve.

IN SOME WAYS, THE HUMAN BRAIN RESEMBLES A
COMPUTER. BUT IN ADDITION TO LOGICAL PROCESSING,
IT IS CAPABLE OF COMPLEX DEVELOPMENT, LEARNING,
SELF-AWARENESS, EMOTION, AND CREATIVITY. EVERY

NERVOUS SYSTEM
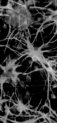

NERVOUS SYSTEM

CONSTANTLY ALIVE WITH ELECTRICITY, THE NERVOUS SYSTEM IS THE BODY'S PRIME COMMUNICATION AND COORDINATION NETWORK. IT IS SO VAST AND COMPLEX THAT, AT A CAUTIOUS ESTIMATE, ALL THE INDIVIDUAL NERVES FROM ONE BODY JOINED END TO END COULD REACH AROUND THE WORLD TWO AND A HALF TIMES.

The nervous system actually comprises three systems or components, defined by both anatomy and function. The central nervous system, CNS, is central to the body's structure and workings. It is composed of the brain and its chief nerve, the spinal cord, which runs along the inside of the backbone (spinal or vertebral column). From the CNS branch 43 pairs of nerves: 12 from the brain and 31 from the cord. As these divide, snake among organs and tissue, and infiltrate every tiny nook and cranny, they form the network of the peripheral nervous system, PNS. The CNS can be viewed as the coordinator and decision-maker, with the PNS sending information as sensory input, and receiving instructions as motor output to muscles and glands. The third component is the autonomic nervous system, ANS. This has some elements located in the CNS and shares some nerves with the PNS; it also has its own nerve chains alongside the spinal cord. Its work is primarily "automatic" in that it deals with activities such as blood pressure control and heart rate adjustment, of which we are rarely aware.

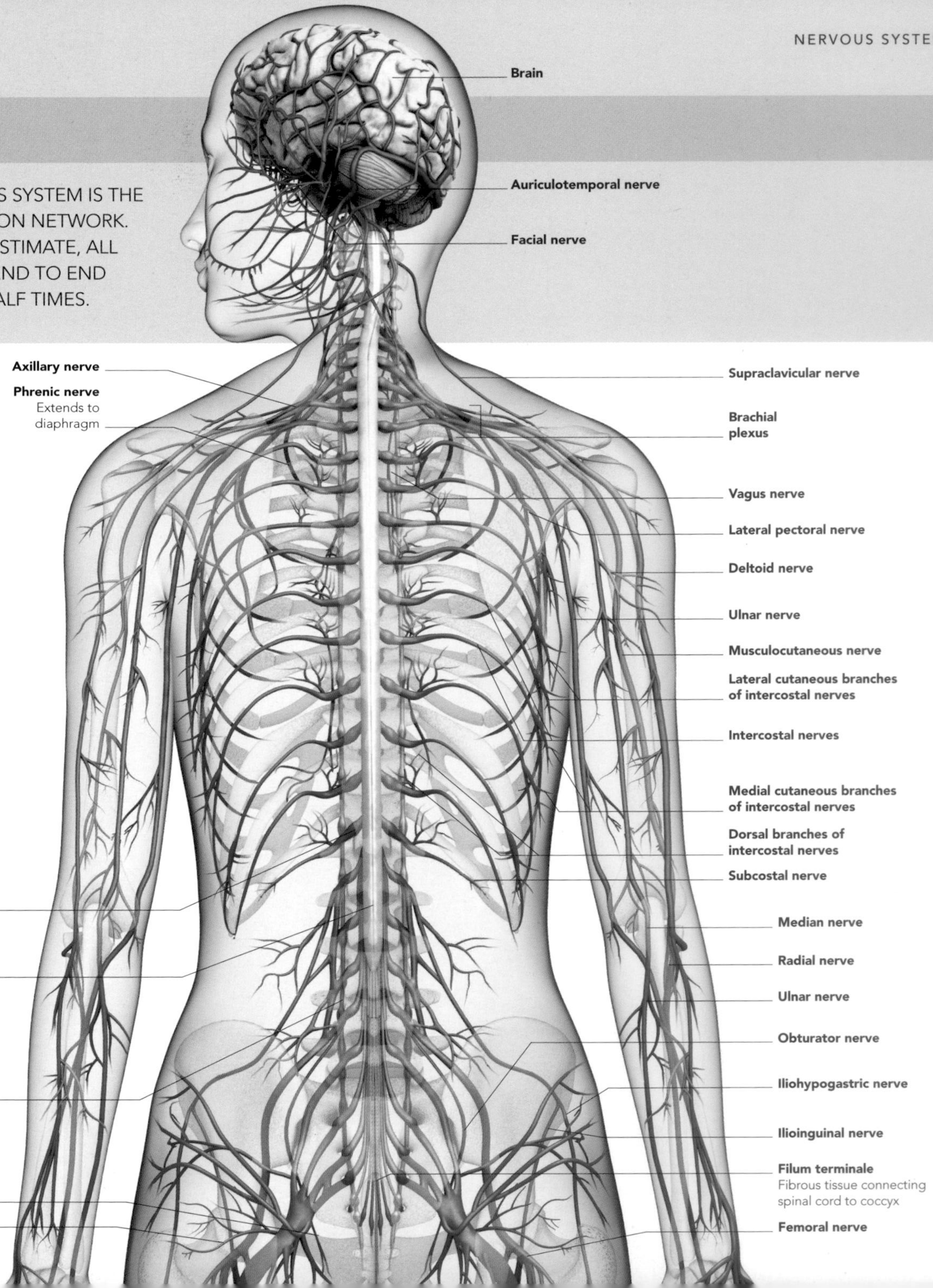

Brain

Auriculotemporal nerve

Facial nerve

Supraclavicular nerve

Brachial plexus

Vagus nerve

Lateral pectoral nerve

Deltoid nerve

Ulnar nerve

Musculocutaneous nerve

Lateral cutaneous branches of intercostal nerves

Intercostal nerves

Medial cutaneous branches of intercostal nerves

Dorsal branches of intercostal nerves

Subcostal nerve

Median nerve

Radial nerve

Ulnar nerve

Obturator nerve

Iliohypogastric nerve

Ilioinguinal nerve

Filum terminale
Fibrous tissue connecting spinal cord to coccyx

Femoral nerve

Axillary nerve

Phrenic nerve
Extends to diaphragm

Spinal ganglion
One of many nodules that send sensory information to brain via spinal cord

Spinal cord
Part of central nervous system, extends from brain down the back, protected by vertebral column

Sympathetic ganglia chain
Part of sympathetic nervous system, also called paravertebral ganglia; conveys stress signals to body

Pudendal nerve

Gluteal nerve

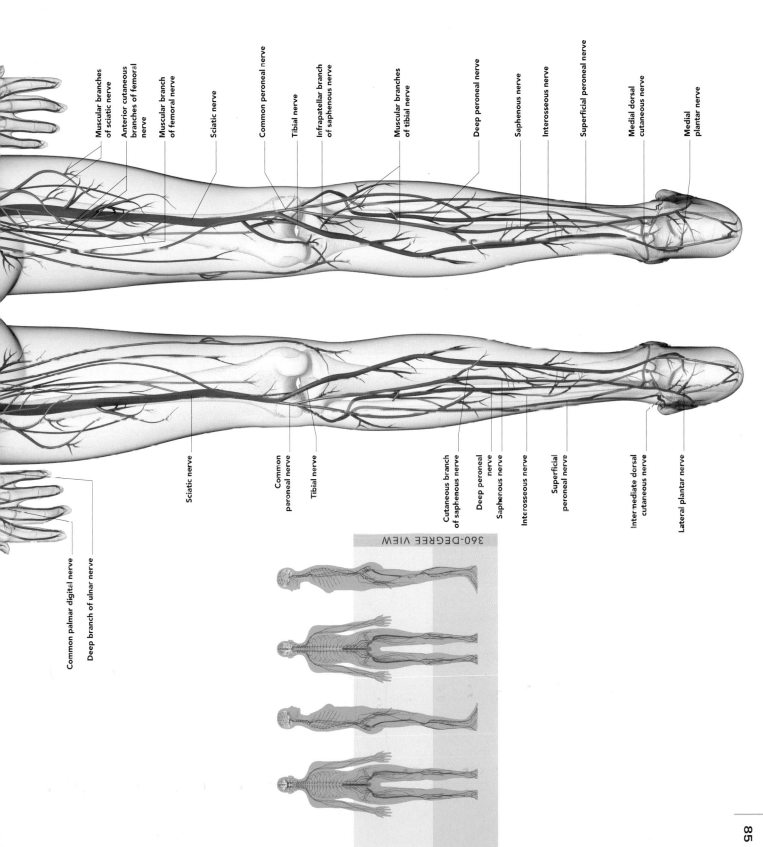

Muscular branches of sciatic nerve

Anterior cutaneous branches of femoral nerve

Muscular branch of femoral nerve

Sciatic nerve

Common peroneal nerve

Tibial nerve

Infrapatellar branch of saphenous nerve

Muscular branches of tibial nerve

Deep peroneal nerve

Saphenous nerve

Interosseous nerve

Superficial peroneal nerve

Medial dorsal cutaneous nerve

Medial plantar nerve

Sciatic nerve

Common peroneal nerve

Tibial nerve

Cutaneous branch of saphenous nerve

Deep peroneal nerve

Saphenous nerve

Interosseous nerve

Superficial peroneal nerve

Intermediate dorsal cutaneous nerve

Lateral plantar nerve

Common palmar digital nerve

Deep branch of ulnar nerve

360-DEGREE VIEW

NERVES AND NEURONS

THE BRAIN HAS OVER 100 BILLION NERVE CELLS, OR NEURONS, AND THE BODY CONTAINS MILLIONS MORE. BUNDLES OF NERVE FIBERS PROJECTING FROM NEURONS FORM A BODY-WIDE NETWORK OF NERVES. NEURONS ARE HIGHLY SPECIALIZED IN THEIR STRUCTURE, FUNCTION, AND THE WAY THEY LINK TOGETHER TO COMMUNICATE.

NEURON STRUCTURE

Like all other cells, a typical neuron has a main cell body with a nucleus. But a neuron also has long, wirelike processes that connect the neuron to others, allowing messages to be passed at junctions called synapses.

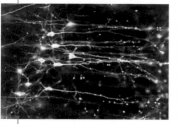

MICROSCOPE VIEW
Nerve cells under the microscope display their cell bodies with nuclei (left) and processes (right).

These processes are of two main kinds. Dendrites receive messages from other neurons, or from nervelike cells in sense organs, and conduct them toward the cell body of the neuron. The axon conveys messages away from the cell body, to other neurons or to muscle or gland cells. Dendrites tend to be short and have many branches, while axons are usually longer and branch less along their length. Neurons in the brain and spinal cord are protected and nurtured by supporting nerve cells known as glial cells.

Axon terminal fiber

Schwann cell
Produces myelin

Schwann cell nucleus

NEURONAL NETWORK
The snaking dendrites and axons of a neural net, which are reaching out to communicate, are clearly visible in this image. These neurons are of the multipolar type, found especially in the cortex of the brain. A single neuron can correspond via its processes with tens of thousands of others.

Dendrite process
Receives messages from other neurons

Axon process
Transmits messages from the nerve cell body to other tissues

TYPES OF NEURON

The shapes and sizes of the bodies of neuron cells vary greatly, as do the type, number, and length of their projections. Neurons are classified according to the number of processes that extend from the cell body. Bipolar neurons are the "original" neuronal design in the embryo, but by adulthood, they are found in only a few locations, such as the retina in the eye and the olfactory nerve in the nose. Most neurons in the brain and spinal cord are multipolar. Unipolar neurons are present mainly in the sensory nerves of the peripheral nervous system.

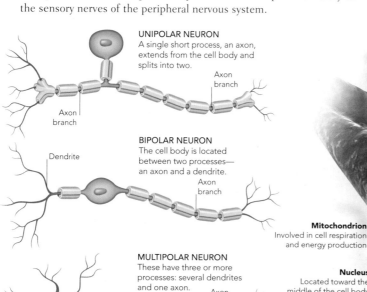

UNIPOLAR NEURON
A single short process, an axon, extends from the cell body and splits into two.

Axon branch

Axon branch

BIPOLAR NEURON
The cell body is located between two processes— an axon and a dendrite.

Dendrite

Axon branch

MULTIPOLAR NEURON
These have three or more processes: several dendrites and one axon.

Axon branch

Dendrite

Mitochondrion
Involved in cell respiration and energy production

Nucleus
Located toward the middle of the cell body

Cell body

SUPPORT CELLS

Supporting nerve cells, known as glial cells or neuroglia, protect and nourish the neurons. Several types of glial cell exist. The smallest are microglia, which destroy microorganisms, foreign particles, and cell debris from disintegrating neurons. Ependymal cells line the cavities filled with cerebrospinal fluid, which surrounds the brain and spinal cord. Other glial cells insulate the axons and dendrites or regulate the flow of cerebrospinal fluid.

ASTROCYTE
Named for their starlike appearance, these nerve cells provide support and nutrition.

OLIGODENDROCYTE
These cells provide a support framework and produce and nourish myelin sheath segments for certain axons.

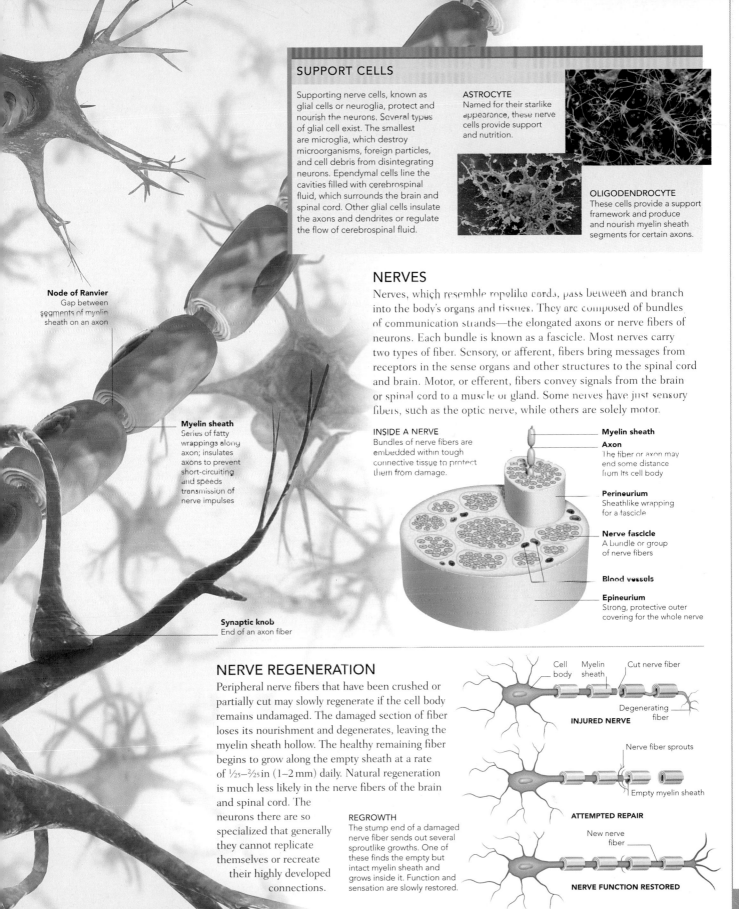

Node of Ranvier
Gap between segments of myelin sheath on an axon

Myelin sheath
Series of fatty wrappings along axon; insulates axons to prevent short-circuiting and speeds transmission of nerve impulses

Synaptic knob
End of an axon fiber

NERVES

Nerves, which resemble ropelike cords, pass between and branch into the body's organs and tissues. They are composed of bundles of communication strands—the elongated axons or nerve fibers of neurons. Each bundle is known as a fascicle. Most nerves carry two types of fiber. Sensory, or afferent, fibers bring messages from receptors in the sense organs and other structures to the spinal cord and brain. Motor, or efferent, fibers convey signals from the brain or spinal cord to a muscle or gland. Some nerves have just sensory fibers, such as the optic nerve, while others are solely motor.

INSIDE A NERVE
Bundles of nerve fibers are embedded within tough connective tissue to protect them from damage.

Myelin sheath

Axon
The fiber or axon may end some distance from its cell body

Perineurium
Sheathlike wrapping for a fascicle

Nerve fascicle
A bundle or group of nerve fibers

Blood vessels

Epineurium
Strong, protective outer covering for the whole nerve

NERVE REGENERATION

Peripheral nerve fibers that have been crushed or partially cut may slowly regenerate if the cell body remains undamaged. The damaged section of fiber loses its nourishment and degenerates, leaving the myelin sheath hollow. The healthy remaining fiber begins to grow along the empty sheath at a rate of 1/25–2/25 in (1–2 mm) daily. Natural regeneration is much less likely in the nerve fibers of the brain and spinal cord. The neurons there are so specialized that generally they cannot replicate themselves or recreate their highly developed connections.

REGROWTH
The stump end of a damaged nerve fiber sends out several sproutlike growths. One of these finds the empty but intact myelin sheath and grows inside it. Function and sensation are slowly restored.

Cell body
Myelin sheath
Cut nerve fiber

Degenerating fiber

INJURED NERVE

Nerve fiber sprouts

Empty myelin sheath

ATTEMPTED REPAIR

New nerve fiber

NERVE FUNCTION RESTORED

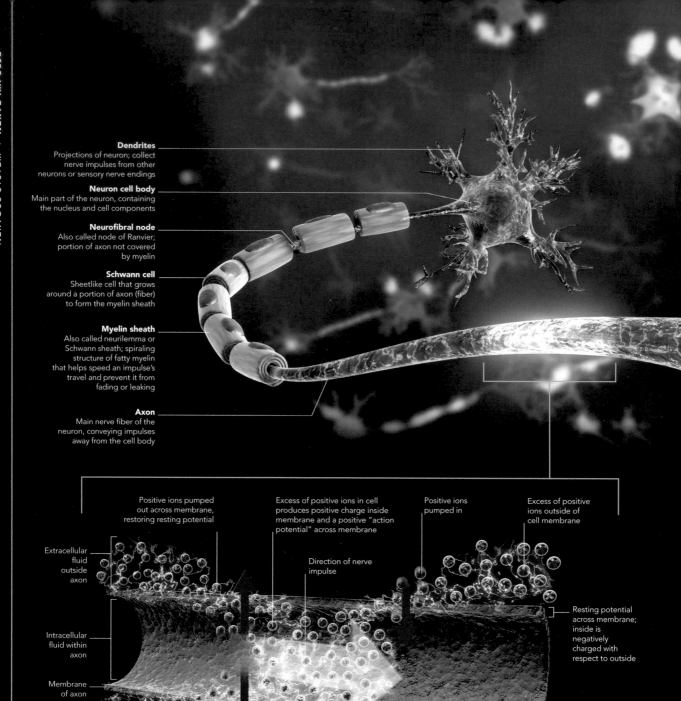

Dendrites
Projections of neuron; collect nerve impulses from other neurons or sensory nerve endings

Neuron cell body
Main part of the neuron, containing the nucleus and cell components

Neurofibral node
Also called node of Ranvier; portion of axon not covered by myelin

Schwann cell
Sheetlike cell that grows around a portion of axon (fiber) to form the myelin sheath

Myelin sheath
Also called neurilemma or Schwann sheath; spiraling structure of fatty myelin that helps speed an impulse's travel and prevent it from fading or leaking

Axon
Main nerve fiber of the neuron, conveying impulses away from the cell body

Positive ions pumped out across membrane, restoring resting potential

Excess of positive ions in cell produces positive charge inside membrane and a positive "action potential" across membrane

Positive ions pumped in

Excess of positive ions outside of cell membrane

Direction of nerve impulse

Extracellular fluid outside axon

Intracellular fluid within axon

Membrane of axon

Resting potential across membrane; inside is negatively charged with respect to outside

Action potential across membrane

3 REPOLARIZATION
Positively charged potassium ions flow in the opposite direction, restoring the charge balance. The change in electrical charge stimulates an adjacent area of membrane, and the next, and so on. The impulse moves along the membrane as a wave of depolarization and repolarization.

2 DEPOLARIZATION
During this phase (depolarization), positive sodium ions rush in through ion channels in a patch of neuron membrane. The membrane is first depolarized, then its polarity is reversed to become slightly positive, resulting in an "action potential" of +30 millivolts on the inside.

1 RESTING POTENTIAL
With no impulse, there are more positively charged ions, particularly sodium ions, outside the cell membrane and more negative ions inside. This produces an electrical "resting potential" of –70 millivolts. The membrane is polarized, with the inside negative.

IMPULSE MOVEMENT WITHIN A NERVE CELL
The nerve impulse is based chiefly on movement of positively charged sodium and potassium ions through the neuron's cell membrane. Impulses travel at speeds of between 3–400 ft/s (1 and 120 m/s), depending on the type of nerve. Movement is much faster in sheathed (myelinated) axons, in which the action potential jumps along successive myelin-coated sections from one node to the next (see above).

NERVE IMPULSE

NERVE CELLS, OR NEURONS, ARE EXCITABLE. WHEN STIMULATED, THEY UNDERGO CHEMICAL CHANGES THAT PRODUCE TINY TRAVELING WAVES OF ELECTRICITY—NERVE SIGNALS, OR IMPULSES. THESE PASS TO OTHER NEURONS, ELICITING SIMILAR RESPONSES FROM THEM.

Throughout the nervous system, information is conveyed as tiny electrical signals called nerve impulses, or action potentials. These impulses are the same all over the body, about 100 millivolts (0.1 volts) in strength and lasting just 1 millisecond (1/1000s). The information carried depends on their position in the nervous system, and their frequency, from one impulse every few seconds to several hundreds per second. When a neuron receives enough impulses from other neurons it fires one of its own, as wavelike movements of ions (electrically charged particles). Impulses jump from one neuron to another at junctions known as synapses.

EXCITEMENT AND INHIBITION

When neurotransmitters land on their receptor sites, they can either excite or inhibit the receiving cell. Both responses are equally valuable in relaying messages through the nervous system. To excite a receiving cell, positive sodium ions flow into it, depolarizing the membrane in a similar way to a nerve impulse (see opposite). The depolarizing effect spreads through the membrane for a few milliseconds, fading as it does so. If further signals enter the cell, they may become strong enough to fire a new nerve impulse. To inhibit a cell, negatively charged chloride ions rush into a cell. The negative effect spreads through the cell membrane and prevents its excitement.

CROSSING THE GAP BETWEEN NEURONS

When an electrical impulse arrives at the junction (synapse), it triggers the release of chemicals called neurotransmitters. They cross the incredibly thin gap (synaptic cleft) between the membranes of the presynaptic (sending) and postsynaptic (receiving) neurons. They either trigger a new impulse in the receiving neuron or actively inhibit it from firing.

Microfilament
Thinnest element of the flexible, supporting scaffolding found in most cells

Mitochondrion
Standard cellular component that provides energy

Synaptic vesicle
Package of neurotransmitter molecules that fuses with the cell membrane when an impulse arrives, releasing the molecules

Neurotransmitter
Molecule that flows across the synaptic cleft in about 1 millisecond, passing on the nerve impulse in chemical form

Presynaptic membrane
Membrane of sending cell's axon

Postsynaptic membrane
Membrane of receiving cell's dendrite

Neurotubule
Specialized microtubule that works as a conveyor belt to bring synaptic vesicles from the cell body to the axon terminal

Positive ion

Synaptic knob
Enlarged end of axon terminal

Membrane channel protein
Complex protein embedded in cell membrane; when enough ions flood through the channel, they cause a response in the receiving cell

Receptor
Site in membrane channel into which neurotransmitter molecules slot, altering the shape of the channel to admit charged ions

Synaptic cleft
Fluid-filled gap between the sending and receiving neuron, just 1/1,000,000 in (25 nanometers, or 25 billionths of a meter) wide

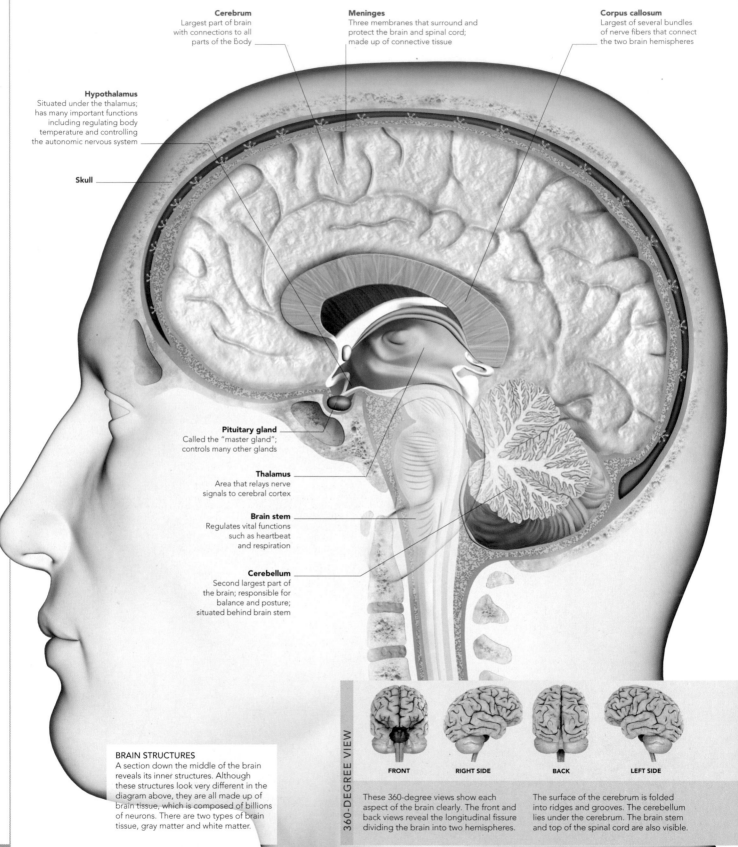

Cerebrum
Largest part of brain
with connections to all
parts of the body

Meninges
Three membranes that surround and
protect the brain and spinal cord;
made up of connective tissue

Corpus callosum
Largest of several bundles
of nerve fibers that connect
the two brain hemispheres

Hypothalamus
Situated under the thalamus;
has many important functions
including regulating body
temperature and controlling
the autonomic nervous system

Skull

Pituitary gland
Called the "master gland";
controls many other glands

Thalamus
Area that relays nerve
signals to cerebral cortex

Brain stem
Regulates vital functions
such as heartbeat
and respiration

Cerebellum
Second largest part of
the brain; responsible for
balance and posture;
situated behind brain stem

BRAIN STRUCTURES
A section down the middle of the brain
reveals its inner structures. Although
these structures look very different in the
diagram above, they are all made up of
brain tissue, which is composed of billions
of neurons. There are two types of brain
tissue, gray matter and white matter.

360-DEGREE VIEW

FRONT **RIGHT SIDE** **BACK** **LEFT SIDE**

These 360-degree views show each
aspect of the brain clearly. The front and
back views reveal the longitudinal fissure
dividing the brain into two hemispheres.

The surface of the cerebrum is folded
into ridges and grooves. The cerebellum
lies under the cerebrum. The brain stem
and top of the spinal cord are also visible.

BRAIN

THE BRAIN, IN CONJUNCTION WITH THE SPINAL
CORD, REGULATES NONCONSCIOUS PROCESSES
AND COORDINATES MOST VOLUNTARY MOVEMENT.
FURTHERMORE, THE BRAIN IS THE SITE OF CONSCIOUSNESS,
ALLOWING HUMANS TO THINK AND LEARN.

BRAIN STRUCTURE

The largest part of the brain is the cerebrum, which has a heavily
folded surface, the pattern of which is unique in each person.
Shallow grooves are called sulci and deep grooves are called fissures.
The fissures and some of the large sulci outline four functional areas
called lobes: frontal, parietal,
occipital, and temporal (see
p.92). A ridge on the surface of
the brain is called a gyrus. The
center of the brain contains
the thalamus, which acts as the
brain's information relay station.
Surrounding this is a group of
structures known as the limbic
system (see p.94), which is
involved in survival instincts,
behavior, and emotions. Closely
linked with the limbic system
is the hypothalamus, which
receives sensory information.

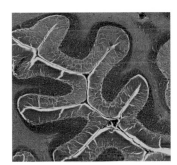

CEREBELLUM
The cerebellum (section shown above)
contains billions of neurons that link up
with other regions of the brain and spinal
cord to facilitate precise movement.

BLOOD SUPPLY TO THE BRAIN

The brain accounts for 2 percent of total body weight, yet it
requires 20 percent of the body's blood. Both oxygen and glucose
are transported by blood; without these essential elements, brain
function quickly deteriorates and dizziness, confusion, and loss
of consciousness may occur. Within only four to eight minutes of
oxygen depravation, brain damage or death results. The brain has
an abundant supply of blood from a vast network of blood vessels
that stem from the carotid arteries,
which run up each side of the neck,
and from two vertebral arteries
that run alongside the spinal cord.

CIRCLE OF WILLIS
A ring of communicating arteries, known as the
Circle of Willis, encircle the base of the brain.
This arterial ring provides multiple pathways to
supply oxygenated blood to all parts of the brain.
If one pathway becomes blocked, blood can be
supplied from an alternative artery in the circle.

BLOOD SUPPLY
The brain has an extensive blood
supply from two front and two
rear arteries, as illustrated in
this color, three-dimensional
magnetic resonance angiography
(MRA) scan. The blood vessels
are colored red; here they are
seen supplying oxygenated blood
to various parts of the brain, which
is shown as the blue area.

PROTECTION

The brain has several forms of protection. It is shielded by the three
protective membranes (meninges) that envelop it, and the ventricles
(chambers) in the brain produce a watery medium within the
skull known as cerebrospinal fluid (CSF, see below) that absorbs
and disperses excessive mechanical forces that might otherwise
cause serious injury. An
analysis of the chemical
constituents and flow
pressure of CSF has
offered vital clues in
the diagnosis of many
diseases and disorders
of the brain and spinal
cord, such as meningitis.

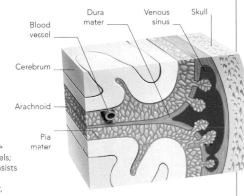

MENINGES
The tough outermost membrane
dura mater, contains blood vessels;
the middle layer, arachnoid, consists
of connective tissue; and the
innermost membrane, pia mater,
lies closest to the brain.

CEREBROSPINAL FLUID FLOW

The soft tissue of the brain floats in cerebrospinal fluid (CSF) within
the bony casing of the skull. CSF is a clear liquid that is renewed four
to five times a day. It contains proteins and glucose that provide energy
for brain cell function as well as lymphocytes that guard against
infection. The CSF protects and nourishes both the brain and spinal
cord as it flows around them. The fluid is produced by the choroid
plexuses in the lateral ventricles, which drains into the third ventricle.
It then flows into the fourth ventricle, located in front of the cerebellum.
Circulation of the fluid is aided by pulsations of the cerebral arteries.

**1 Site of fluid production
(choroid plexuses)**
The CSF found in the ventricles
in the brain is produced in
clusters of thin-walled capillaries,
known as choroid plexuses.
These capillaries line the walls
of the ventricles.

2 Direction of flow
Fluid moves from the
brain's lateral ventricles into
the third and fourth ventricles.
The fluid then flows up the
back of the brain, down around
the spinal cord, and up to the
front of the brain, as indicated
by the arrows.

**3 Circulation around
spinal cord**
CSF ebbs and flows
around the spinal cord and
through its central canal.

**4 Site of reabsorption
(arachnoid granulations)**
After circulating around the brain,
CSF is reabsorbed into the blood
via structures known as arachnoid
granulations, which are projections
of the arachnoid layer into the large
sagittal sinus, or cerebral vein.

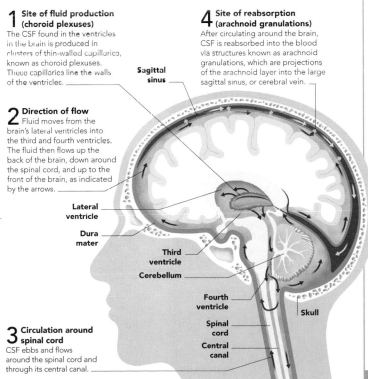

BRAIN STRUCTURES

THE BRAIN COMPRISES ABOUT ONE-FIFTIETH OF THE WEIGHT OF THE WHOLE
BODY, AVERAGING 3LB 2OZ (1.4KG) IN ADULTS. ANATOMICALLY, IT HAS FOUR
MAIN STRUCTURES: THE LARGE, DOMED CEREBRUM; THE DEEPER, INNER
DIENCEPHALON (CONSISTING OF THE THALAMUS AND NEARBY STRUCTURES);
THE CEREBELLUM TO THE LOWER REAR; AND THE BRAIN STEM AT THE BASE.

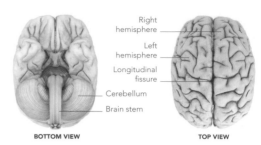

Right
hemisphere

Left
hemisphere

Longitudinal
fissure

Cerebellum

Brain stem

BOTTOM VIEW

TOP VIEW

EXTERNAL BRAIN FEATURES

The brain's most obvious feature is the cerebrum, which
makes up more than four-fifths of all its tissues. It has a
grooved appearance due to its heavily folded surface, which
is called the cerebral cortex. The cerebrum partly envelops
the thalamus and nearby structures (diencephalon) and the
brain stem below this (see opposite). The smaller cerebellum
forms about one-tenth of the brain's whole volume; it is mainly
concerned with the organization of motor information being
sent to muscles to make movements smooth and coordinated.

OUTER BRAIN STRUCTURES

The cerebrum is partly separated into two halves
(cerebral hemispheres) by the deep longitudinal fissure.
The cerebellum is the smaller bulbous structure
responsible for muscle control. Beneath the cerebellum
is the brain stem, which controls basic life processes.

BRAIN-PRINT
A scan reveals a unique
"brain-print"—the pattern of
cerebral grooves and bulges
that is different in each person.

Parietal lobe
Area in which bodily
sensations such as
touch, temperature,
pressure, and pain
are perceived and
interpreted, in
the region called the
somatosensory cortex

Postcentral gyrus
A ridge, or bulge, on the
brain's surface is called a
gyrus; the postcentral gyrus
(just behind midpoint from
front to rear) is an important
anatomical landmark

Frontal lobe
Speech production,
movement
initiation, and
aspects of
"personality" are
based in this lobe

Parietal occipital fissure
Fissure (deep groove)
that demarcates border
between parietal and
occipital lobes

Lateral sulcus
Groove running
along upper part of
temporal lobe

**Superior
temporal sulcus**
Upper of two
main sulci (shallow
grooves) that divide
chief gyri (bulges) of
temporal lobe

Temporal lobe
Recognition of sounds,
their tones, and
loudness takes place in
temporal lobes; they
also play a role in
storage of memory

Pons
Upper portion
of brain stem

Occipital lobe
This area is mainly
concerned with
analyzing and
interpreting visual
information from
sensory nerve signals
sent by the eyes

Inferior temporal sulcus
Lower of two main sulci (shallow
grooves) that divide gyri (bulges)
of temporal lobe

LOBES OF THE BRAIN
Traditionally the cerebral surface is
divided into four major lobes, partly by
patterns of deeper grooves, or fissures,
and partly by functional significance—the
roles each fulfills. The names of some of
the lobes parallel the names of the skull
bones that overlie them (see p.58).

Brain stem
Lowest, mainly
"automatic" region
of brain
(see opposite)

Cerebellum
This "little brain" is involved
with timing and accuracy of
skilled movements, and controls
balance and posture

THE HOLLOW BRAIN

The brain is, in a sense, hollow: it contains four chambers known as ventricles, filled with cerebrospinal fluid, or CSF (see p.91). There are two lateral ventricles, one in each hemisphere, and the fluid is produced here. It then drains via the interventricular foramen into the third ventricle, which is situated close to the thalamus. From here it flows through the cerebral aqueduct and into the fourth ventricle, which extends down between the pons and cerebellum into the medulla. The total volume of CSF in the ventricles is about ⁹⁄₁₀ fl oz (25 ml). Circulation is aided by head movements and pulsations of the cerebral arteries.

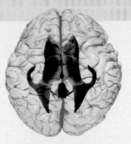

VIEW FROM ABOVE
The lateral ventricles have frontward-, backward-, and side-facing horns, or cornua. Seen between them in this view is the central third ventricle.

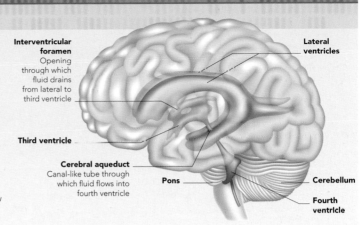

Interventricular foramen
Opening through which fluid drains from lateral to third ventricle

Third ventricle

Cerebral aqueduct
Canal-like tube through which fluid flows into fourth ventricle

Pons

Lateral ventricles

Cerebellum

Fourth ventricle

GRAY AND WHITE MATTER

The bulk of the cerebrum has two main layers. The outer, pale-gray layer, often known as "gray matter," is the cerebral cortex. It follows the folds and bulges of the cerebrum to cover its entire surface. Its average thickness is ¹⁄₁₀–²⁄₁₀ in (3–5 mm), and spread out flat, it would cover about the same area as a pillowcase. Deeper within the cerebrum are small islands of gray matter. These and the cerebral cortex are composed chiefly of the cell bodies and impulse-collecting projections (dendrites) of nerve cells (neurons). Beneath the cortex's gray matter is the paler "white matter," forming the bulk of the cerebrum's interior. It is composed mainly of nerve fibers.

Gray matter
Outermost layer of cerebral cortex containing an estimated 50 billion neurons and perhaps 10 times as many supporting cells

White matter interior
Here axons, or fibers, of neurons run up from lower areas and project down from neuron cell bodies of cortex

Corpus callosum
Largest of several bundles of nerve fibers, called commissures, which connect specific areas of the two halves, or cerebral hemispheres, of the upper brain

Basal ganglia
"Islands" of gray matter deep in cerebrum

Motor nerve tracts
Large fiber bundles that carry instructions for movements down to the spinal cord, and that cross over in lower brain stem

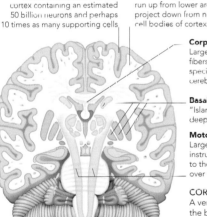

Brain stem

CORONAL SECTION
A vertical "slice" through the middle of the brain reveals the paired structures, outer gray layer, and inner white matter. The corpus callosum contains more than 100 million nerve fibers and is the main "bridge" between the two hemispheres.

BASAL GANGLIA
These structures include the lentiform nucleus (putamen and globus pallidus), caudate nucleus, subthalamic nucleus, and substantia nigra (the latter two not seen in this view). They are a complex interface between sensory inputs and motor skills, especially for semiautomatic movements, such as walking.

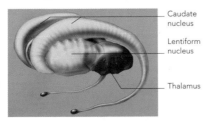

Caudate nucleus

Lentiform nucleus

Thalamus

VERTICAL LINKS

Sheathed (myelinated) nerve fibers, organized into bundles known as projection tracts, transmit impulses between the spinal cord and lower brain areas and the cerebral cortex above. These nerve tracts pass through a communication link called the internal capsule and also intersect the corpus callosum. In addition, similar bundles pass through the upper, outer zones of the white matter, from one area of the cerebral cortex to another. These association tracts convey nerve signals directly between different regions or centers of the cortex.

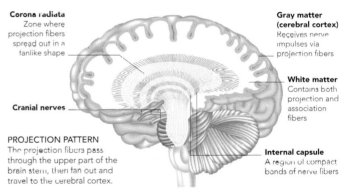

Corona radiata
Zone where projection fibers spread out in a tanlike shape

Cranial nerves

PROJECTION PATTERN
The projection fibers pass through the upper part of the brain stem, then fan out and travel to the cerebral cortex.

Gray matter (cerebral cortex)
Receives nerve impulses via projection fibers

White matter
Contains both projection and association fibers

Internal capsule
A region of compact bands of nerve fibers

THE THALAMUS AND BRAIN STEM

The thalamus sits on top of the brain stem. Shaped like two eggs side by side, it lies almost at the "heart" of the brain. It is a major relay station that monitors and processes incoming information before this is sent to the upper regions of the brain. The brain stem contains centers that regulate several functions vital for survival: these include heartbeat, respiration, blood pressure, and some reflex actions, such as swallowing and vomiting.

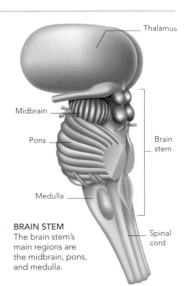

Thalamus

Midbrain

Pons

Medulla

Brain stem

Spinal cord

BRAIN STEM
The brain stem's main regions are the midbrain, pons, and medulla.

THE PRIMITIVE BRAIN

HUMAN BEHAVIOR IS NOT ALWAYS RATIONAL. IN TIMES OF STRESS OR CRISIS, DEEP-SEATED INSTINCTS WELL UP FROM DEEP WITHIN AND TAKE OVER OUR AWARENESS. SUCH EVENTS INVOLVE THE "PRIMITIVE BRAIN," WHICH IS BASED MAINLY IN A SERIES OF PARTS KNOWN AS THE LIMBIC SYSTEM.

THE LIMBIC SYSTEM

The limbic system influences subconscious, instinctive behavior, similar to animal responses that relate to survival and reproduction. In humans, many of these innate, early-evolved "primitive" behaviors are modified by conscious, thoughtful considerations based in upper regions of the brain, as we take into account moral, social, and cultural codes, and the results of our actions. However, primal urges sometimes prevail, and this is when the limbic system and associated structures take over. At other times they play lesser, but still complex and important, roles in the expression of instincts, drives, and emotions.

FRONT   **RIGHT SIDE**   **BACK**   **LEFT SIDE**

The limbic system involves parts situated between the "automatic" centers of the mid and lower brain stem, and the "thinking" regions associated with higher mental functions in the cortex. The limbic cortex is on the inner sides of the cortical lobes where they fold up against the midbrain (see the Back view).

LIMBIC STRUCTURES

The components of the ring-shaped limbic system, located in the lower center of the brain, mediate the effects of innermost moods on external behavior. They also influence changes in bodily functions, such as those involving digestion and urination. The association of emotions with sensory inputs is also influenced by this system.

Cingulate gyrus
With the parahippocampal gyrus and olfactory bulbs, comprises the limbic cortex, which modifies behavior and emotions

Fornix
Pathway of nerve fibers that transmits information from hippocampus and other limbic areas to the mamillary bodies

Column of fornix

Mamillary body
Tiny lump of neurons that acts as a relay station, transmitting information mainly between fornix and thalamus; involved in memory processes

Olfactory bulbs
The brain's "smell processors"; they are "hard-wired" into the limbic system, which helps explain why the sense of smell can evoke such strong memories and emotional responses

Midbrain
Uppermost part of brain stem; limbic areas in the midbrain connect to the cortex and to the thalamus, and also link to the clusters of nerve cell bodies known as basal ganglia

Pituitary gland

Pons
Part of the brain stem; not part of the limbic system

Hippocampus
Curved band of gray matter involved with recognizing new experiences, and with learning and memory, especially short-term memory and information relating to recent events

Amygdala
Double-almond-shaped structure that influences behavior and activities so that they are directed toward the body's needs; also concerned with emotions such as anger and jealousy, and drives such as hunger, thirst, and sexual desire

Parahippocampal gyrus
Helps modify expression of forceful emotions; also forms and recalls topographic memories of scenes and views (rather than objects, faces, or facts)

Mamillary bodies   Corpus callosum

Fornix

RINGED AND ARCHED
The limbic system—"limbic" meaning linked—encompasses parts in the cerebrum, diencephalon, and midbrain, and links the cortical and midbrain areas with lower centers that control automatic functions.

THE HYPOTHALAMUS

The hypothalamus ("below the thalamus") is about the size of a sugar cube and contains numerous tiny clusters of neurons called nuclei. It is usually regarded as the vital integrating center of the limbic system. A stalk below links it to the pituitary, the chief gland of the hormonal system. In addition to this major endocrine connection, the hypothalamus also has complex associations with the rest of the limbic system around it, and with the autonomic parts of the general nervous system. Hypothalamic functions include monitoring and regulating vital internal conditions such as body temperature, nutrient levels, water-salt balance, blood flow, the sleep-wake cycle, and the levels of hormones. The hypothalamus initiates feelings, actions, and emotions such as hunger, thirst, rage, and terror.

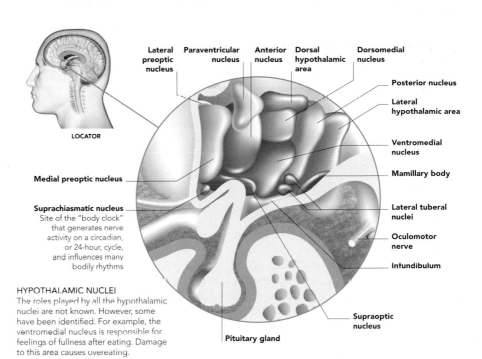

LOCATOR

Lateral preoptic nucleus • Paraventricular nucleus • Anterior nucleus • Dorsal hypothalamic area • Dorsomedial nucleus • Posterior nucleus • Lateral hypothalamic area • Ventromedial nucleus • Mamillary body • Lateral tuberal nuclei • Oculomotor nerve • Infundibulum

Medial preoptic nucleus

Suprachiasmatic nucleus
Site of the "body clock" that generates nerve activity on a circadian, or 24-hour, cycle, and influences many bodily rhythms

Supraoptic nucleus

Pituitary gland

HYPOTHALAMIC NUCLEI

The roles played by all the hypothalamic nuclei are not known. However, some have been identified. For example, the ventromedial nucleus is responsible for feelings of fullness after eating. Damage to this area causes overeating.

THE RETICULAR FORMATION

The reticular formation is a structure containing various clusters of neurons (nuclei) together with a series of long, slim nerve tracts located in much of the length of the brain stem. It comprises several distinct neural systems, each with its own neurotransmitter (the chemical that passes on nerve signals at the junctions, or synapses, between neurons). One of its many functions is to operate an arousal system, known as the reticular activating system (RAS), that keeps the brain awake and alert. The reticular formation also includes the cardioregulatory and respiratory centers that control heart rate and breathing, and other essential centers.

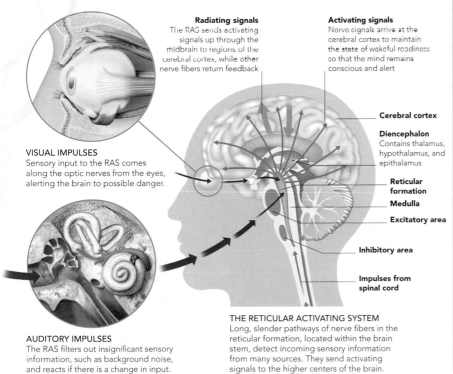

Radiating signals
The RAS sends activating signals up through the midbrain to regions of the cerebral cortex, while other nerve fibers return feedback

Activating signals
Nerve signals arrive at the cerebral cortex to maintain the state of wakeful readiness so that the mind remains conscious and alert

Cerebral cortex

Diencephalon
Contains thalamus, hypothalamus, and epithalamus

Reticular formation

Medulla

Excitatory area

Inhibitory area

Impulses from spinal cord

VISUAL IMPULSES
Sensory input to the RAS comes along the optic nerves from the eyes, alerting the brain to possible danger.

AUDITORY IMPULSES
The RAS filters out insignificant sensory information, such as background noise, and reacts if there is a change in input.

THE RETICULAR ACTIVATING SYSTEM
Long, slender pathways of nerve fibers in the reticular formation, located within the brain stem, detect incoming sensory information from many sources. They send activating signals to the higher centers of the brain.

SLEEP CYCLES

During sleep, much of the body rests, but not the brain. Its billions of neurons continue to send signals, as shown by EEG traces. Sleep occurs in cycles, made up of lengthening phases of REM (rapid eye movement) sleep, when dreaming occurs, and four stages of NREM (nonrapid eye movement sleep), which is dreamless. In stage 1, sleep is light: people wake relatively easily and brain waves are active. In stage 2, brain waves begin to slow down. In stage 3, fast and slow waves are interspersed, and finally in stage 4, the deepest stage, there are slow waves only.

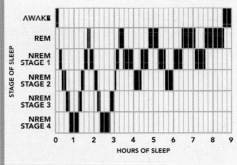

AWAKE
REM
NREM STAGE 1
NREM STAGE 2
NREM STAGE 3
NREM STAGE 4

STAGE OF SLEEP

0 1 2 3 4 5 6 7 8 9
HOURS OF SLEEP

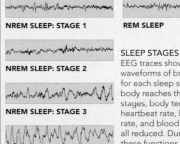

NREM SLEEP: STAGE 1

REM SLEEP

NREM SLEEP: STAGE 2

NREM SLEEP: STAGE 3

NREM SLEEP: STAGE 4

SLEEP STAGES
EEG traces show different waveforms of brain activity for each sleep stage. As the body reaches the deeper stages, body temperature, heartbeat rate, breathing rate, and blood pressure are all reduced. During REM sleep these functions rise slightly and dreams usually occur.

SPINAL CORD

THE NERVE FIBERS OF THE SPINAL CORD LINK THE BRAIN WITH THE TORSO, ARMS, AND LEGS. THE BRAIN IS DIRECTLY CONNECTED TO THE SENSE ORGANS IN THE HEAD BY THE CRANIAL NERVES (SEE P.98) BUT NEEDS THE SPINAL CORD TO CARRY INFORMATION TO AND FROM THE REST OF THE BODY. THE CORD IS MORE THAN A PASSIVE CONDUIT FOR NERVE SIGNALS—WHEN NECESSARY, IT CAN BYPASS THE BRAIN, FOR EXAMPLE IN REFLEX ACTIONS.

SPINAL CORD ANATOMY

The spinal cord is a complex bundle of nerve fibers (axons) that is about 16–18 in (40–45 cm) long. It extends from the base of the brain, down to the lower (lumbosacral) part of the spinal column. It is shaped like a flattened cylinder and is only slightly wider than a pencil for most of its length, tapering to a threadlike tail at the base. Branching out from the spinal cord are 31 pairs of spinal nerves, which connect it to the skin, muscles, and other parts of the limbs, chest, and abdomen. The nerves carry sensory information to the cord about conditions within the body and transmit the sense of touch from the skin. They also convey motor information to muscles throughout the body and to glands within the chest and abdomen.

(SEE P.98)

NERVE CROSSOVER

Bundles of nerve fibers (axons) in the left and right sides of the spinal cord do not all pass straight up into the left and right sides of the brain. In the uppermost portion of the cord and the lower brain stem (medulla), many of the fibers cross over, or decussate, to the other side—left to right, and right to left. As a result, nerve signals about, for example, touch sensations on the left side of the body reach the touch center (somatosensory cortex) on the right side of the brain. Likewise, motor signals from the right motor cortex and right side of the cerebellum travel to the muscles on the left side of the body. Different major bundles, or tracts, of fibers decussate at slightly different levels. About one-tenth of those that cross over do so in the upper spinal cord, and the remainder cross over in the medulla.

SPINAL GRAY MATTER
This microscope view of a cross section through the spinal cord shows one brown-stained "wing" of the butterfly-shaped gray matter, which lies at the cord's center.

Nerve fiber tract
Bundle of nerve fibers (axons) that carries signals to and from spinal cord and specific areas of brain

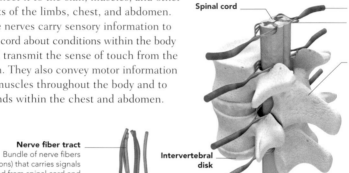

Spinal cord
Spinal nerve
Spinal nerve root
Vertebra
Intervertebral disk
REAR OF BODY

HOW SPINAL NERVES ATTACH
The spinal nerves reach the cord through gaps between vertebrae, which are held apart by pads of cartilage, known as intervertebral disks. The nerves divide and enter the back and front of the spinal cord as spinal nerve roots, each composed of many rootlets.

Central canal
Cerebrospinal fluid fills the narrow central canal and provides nourishment and waste collection for neurons and tissues around it

Spinal nerve
Sensory and motor nerve rootlets merge to form spinal nerve

Motor nerve rootlets (ventral)
Bundles of fibers that emerge from front (ventral side) of spinal cord; carry signals to voluntary skeletal muscles and involuntary smooth muscles

Anterior fissure
Deep groove along front of spinal cord; almost reaches gray matter and central canal

Subarachnoid space

White matter
Gray matter

Sensory nerve rootlets (dorsal)
Bundles of fibers that enter spinal cord at rear (dorsal side); carry impulses of incoming information about touch sensations on skin, and conditions within the body

Sensory root ganglion (dorsal)
Cluster of nerve cell bodies on each spinal nerve; partially processes incoming information

Pia mater
Arachnoid
Dura mater

Meninges
Three layers of connective tissues that protect spinal cord; cerebrospinal fluid fills space under middle layer

SPINAL CORD
The inner organization of the spinal cord resembles an "inside-out" brain. The brain has gray matter outside and white within. The cord has an inner, butterfly-shaped core of gray matter. This is made up of neuron cell bodies and nonsheathed (unmyelinated) nerve fibers. Around this is the outer later of white matter, composed mainly of myelinated nerve fiber tracts that carry nerve impulses up and down the cord, between the brain and the rest of the body.

FRONT OF BODY

PROTECTION OF SPINAL CORD

The spinal cord is located inside the spinal canal, a long tunnel within the aligned column of backbones (vertebrae). The vertebral column, along with its strengthening ligaments and muscles, bends and flexes the cord, but also guards it from direct knocks and blows. Within the spinal canal the circulating cerebrospinal fluid acts as a shock-absorber and the epidural space provides a cushioning layer of fat and connective tissue. The epidural tissues lie between the periosteum (the membrane that lines the bone of the spinal canal) and the dura mater, the outer layer of the meninges.

INSIDE THE SPINAL CANAL

A cross section of the vertebral column in the neck (cervical) region shows how the spinal cord nestles in the well-padded bony cavity. Although the vertebrae shift position as the trunk of the body moves, the spinal cord remains well supported and protected.

EXTENT OF SPINAL CORD

During growth, the spinal cord does not continue to lengthen, unlike the spinal bones. By adulthood, it extends from the brain down to the first lumbar vertebra (L1) in the lower back. Here, it forms a conelike ending that tapers to a slender, tail-like filament, known as the filum terminale. This extends down through the lumbar and sacral vertebrae to the coccyx.

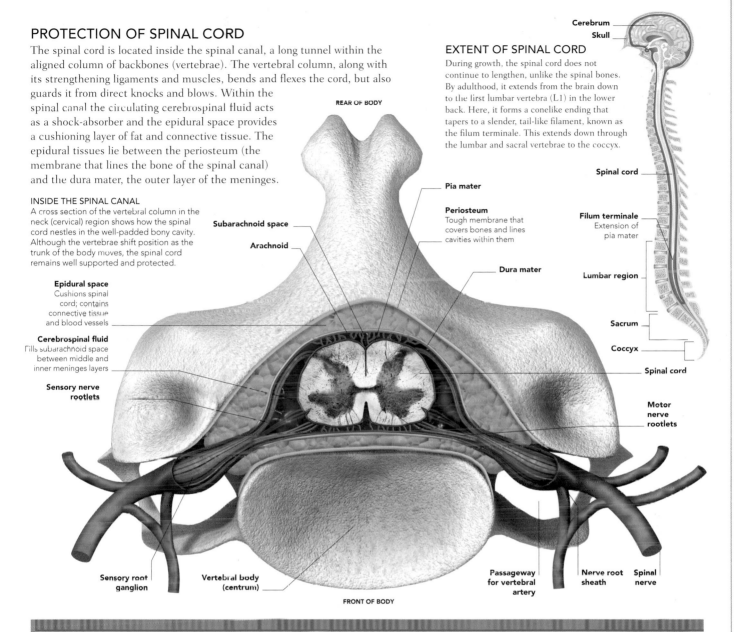

REAR OF BODY

Subarachnoid space

Arachnoid

Epidural space
Cushions spinal cord; contains connective tissue and blood vessels

Cerebrospinal fluid
Fills subarachnoid space between middle and inner meninges layers

Sensory nerve rootlets

Sensory root ganglion

Vertebral body (centrum)

FRONT OF BODY

Pia mater

Periosteum
Tough membrane that covers bones and lines cavities within them

Dura mater

Motor nerve rootlets

Passageway for vertebral artery

Nerve root sheath

Spinal nerve

Cerebrum

Skull

Spinal cord

Filum terminale
Extension of pia mater

Lumbar region

Sacrum

Coccyx

Spinal cord

NERVE TRACTS OF SPINAL CORD

In the white matter of the spinal cord, nerve fibers are grouped into main bundles, or tracts, according to the direction of the nerve signals they carry and the type of signals they transmit and respond to, such as pain or temperature. Some of these tracts connect and relay impulses between a few local pairs of spinal nerves, without sending fibers up to the brain. The central gray matter of the cord is organized into horns, or columns.

ASCENDING TRACTS
These bundles of nerve fibers relay impulses about bodily sensations and inner sensors like pain up the spinal cord to the brain.

DESCENDING TRACTS
These convey motor signals from the brain to skeletal muscles of the torso and limbs in order to bring about voluntary movements.

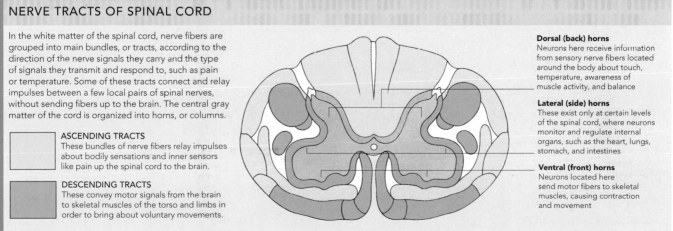

Dorsal (back) horns
Neurons here receive information from sensory nerve fibers located around the body about touch, temperature, awareness of muscle activity, and balance

Lateral (side) horns
These exist only at certain levels of the spinal cord, where neurons monitor and regulate internal organs, such as the heart, lungs, stomach, and intestines

Ventral (front) horns
Neurons located here send motor fibers to skeletal muscles, causing contraction and movement

PERIPHERAL NERVES

THE BODY'S NETWORK OF PERIPHERAL NERVES CONVEYS INFORMATION TO AND FROM THE BRAIN AND SPINAL CORD. SENSORY FIBERS IN THE NERVES CARRY MESSAGES FROM THE SENSE ORGANS, SUCH AS THE EYES, EARS, AND SKIN, AND FROM INTERNAL ORGANS. MOTOR FIBERS CONTROL MUSCLE MOVEMENT AND GLAND ACTIVITY.

CRANIAL NERVES

The 12 pairs of cranial nerves connect directly to the brain, rather than via the spinal cord. Some nerves perform sensory functions for organs and tissues in the head and neck, while others provide motor functions. The nerves with predominantly motor fibers also contain some sensory fibers that convey information to the brain about the amount of stretch and tension in the muscles they serve, as part of the proprioceptive sense (see p.79). Most of the cranial nerves are named according to the body parts they serve, such as the optic nerves (eyes). By convention, the nerves are also identified by Roman numerals, so the trigeminal nerve, for example, is cranial V (five).

Olfactory nerve
(I, sensory)
Relays information about smells from the olfactory epithelium inside the nose, just above the nasal chamber, via the olfactory bulbs and the olfactory tracts to the brain's limbic centers.

Trigeminal nerve
(V, two sensory and one mixed branch)
Ophthalmic and maxillary branches gather signals from the eye, face, and teeth; mandibular motor fibers control chewing muscles; and sensory fibers bring signals from the lower jaw.

Facial nerve
(VII, mixed)
Sensory branches come from the taste buds of the front two-thirds of the tongue; motor fibers run to the muscles of facial expression and to the salivary and lacrimal glands.

VIEW FROM BELOW
In this view of the brain's underside, the cranial nerves are seen joining mainly to the lower regions of the brain. Some of these nerves are sensory, taking impulses to the brain. Others are motor, bringing nerve signals from the brain to muscles and glands. Some are mixed, with both sensory and motor nerve fibers.

Optic nerve
(II, sensory)
The optic nerve brings visual information from the rod and cone cells in the retina to the visual cortex in the brain; parts of the two nerves cross at the optic chiasm (see p.109) where they form bands of nerve fibers, called optic tracts. Each nerve consists of a bundle of about one million sensory fibers—it carries the most information of any cranial nerve.

Oculomotor, trochlear, and abducens nerves
(III, IV, VI, mainly motor)
These three nerves regulate voluntary movements of the eye muscles, to move the eyeball and eyelids; the oculomotor also controls pupil constriction by the iris muscles and focusing changes in the lens by the ciliary muscles.

Vestibulocochlear nerve
(VIII, sensory)
The vestibular branch collects nerve signals from the inner ear about head orientation and balance; the cochlear branch brings signals from the ear concerning sound and hearing.

Glossopharyngeal and hypoglossal nerves
(IX, XII, both mixed)
Motor fibers of these nerves are involved in tongue movement and swallowing, while sensory fibers relay information about taste, touch, and temperature from the tongue and pharynx.

Spinal accessory nerve
(XI, mainly motor)
This nerve controls muscles and movements in the head, neck, and shoulders. It also stimulates the muscles of the pharynx and larynx, which are involved in swallowing.

Vagus nerve
(X, mixed)
The longest and most branched cranial nerve, the vagus (meaning "wanderer") has sensory, motor, and autonomic fibers that pass to the lower head, throat, neck, chest, and abdomen; these are involved in many vital body functions, including swallowing, breathing, heartbeat, and the formation of stomach acid.

SPINAL REFLEXES

A reflex is a rapid, involuntary, predictable response to a stimulus. Most reflexes are concerned with survival and defending the body against damage and harm, such as coughing to remove irritants from the lower airways and sneezing to clear the nasal airways. In general, a reflex occurs in a complete neural circuit that does not involve the higher regions of the brain, where consciousness and awareness occur; the mind usually becomes aware of the reflex response just after it has occurred, when it is too late to prevent. Spinal reflexes involve circuits of sensory nerve fibers that feed information to the spinal cord and then connect directly, or via an intermediate neuron, to motor nerve fibers, so that the resulting instructions for movement go directly out from the cord to the relevant muscles.

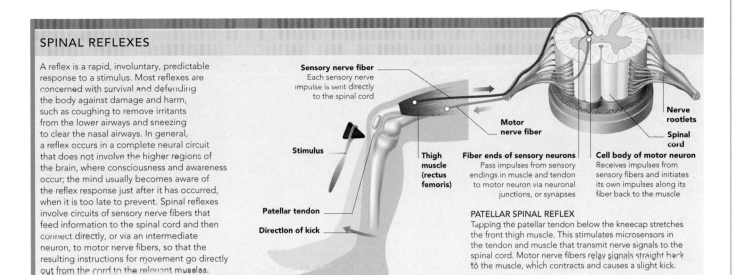

Sensory nerve fiber
Each sensory nerve impulse is sent directly to the spinal cord

Stimulus

Thigh muscle (rectus femoris)

Patellar tendon

Direction of kick

Motor nerve fiber

Nerve rootlets

Spinal cord

Fiber ends of sensory neurons
Pass impulses from sensory endings in muscle and tendon to motor neuron via neuronal junctions, or synapses

Cell body of motor neuron
Receives impulses from sensory fibers and initiates its own impulses along its fiber back to the muscle

PATELLAR SPINAL REFLEX
Tapping the patellar tendon below the kneecap stretches the front thigh muscle. This stimulates microsensors in the tendon and muscle that transmit nerve signals to the spinal cord. Motor nerve fibers relay signals straight back to the muscle, which contracts and causes a slight kick.

SPINAL NERVES

The 31 pairs of peripheral spinal nerves emerge from the spinal cord through spaces between the vertebrae. Each nerve divides and subdivides into a number of branches, the dorsal branches serve the rear portion of the body, while the ventral serve the front and sides. The branches of one spinal nerve may join with other nerves to form meshes called plexuses where nerves merge and intersect. Nerves leaving each plexus then go on to carry signals to and from that particular area of the body.

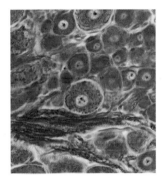

SPINAL NERVE GANGLION
This microscope image shows a section through a cluster of spinal nerve cells (ganglion), where nerve impulses are coordinated. Each neuron (purple) is surrounded by support cells (light blue).

Cervical region (C1–C8)
Eight pairs of cervical spinal nerves form two networks, the cervical (C1–C4) and brachial plexuses (C5–C8/T1). These run to the chest, head, neck, shoulders, arms, and hands, and to the diaphragm.

Thoracic region (T1–T12)
Apart from T1, which is considered part of the brachial plexus, thoracic spinal nerves are connected to the intercostal muscles between the ribs, the deep back muscles, and the abdominal muscles.

Lumbar region (L1–L5)
Four of the five pairs of lumbar spinal nerves (L1–L4) form the lumbar plexus, which supplies the lower abdominal wall and parts of the thighs and legs. L4 and L5 interconnect with the first four sacral nerves (S1–S4).

Sacral region (S1–S5)
Two nerve networks, the sacral plexus (L5–S3) and the coccygeal plexus (S4/S5/Co 1), send branches to the thighs, buttocks, muscles, and skin of the legs and feet, and anal and genital areas.

SPINAL REGIONS
The organization and naming of the four main spinal nerve regions reflect the regions of the spine itself—cervical or neck, thoracic or chest, lumbar or lower back, and sacral or base of spine.

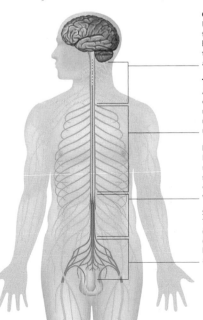

DERMATOMES

A dermatome is a region or zone of skin supplied by the dorsal (rear, sensory) nerve roots of one pair of spinal nerves. The nerve branches carry sensory information about touch, pressure, heat, cold, and pain from the skin microsensors within the zone, along the sensory nerve fibers of the branches of the spinal nerve, to the spinal nerve root and then into the spinal cord. A "skin map" delineates these zones, or dermatomes. In real life, the distribution of nerve roots, and sensations, overlaps slightly.

DERMATOME MAP
Spinal nerve C1, which lacks sensory fibers from the skin, is missing; the face and forehead send signals via the three branches of the trigeminal cranial (V) nerve, coded here as V1–V3.

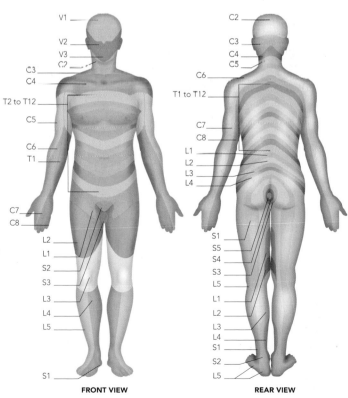

FRONT VIEW

REAR VIEW

AUTONOMIC NERVOUS SYSTEM

THE AUTONOMIC NERVOUS SYSTEM (ANS) HAS THE KEY TASK OF MAINTAINING CONSTANT CONDITIONS WITHIN THE BODY, WHICH IS KNOWN AS HOMEOSTASIS. MOST OF THE ACTIVITY IN THE ANS IS INDEPENDENT (AUTONOMIC) OF THE CONSCIOUS MIND, SO WE ARE RARELY AWARE OF ITS WORKINGS.

AUTOMATIC FUNCTIONS

The ANS is one of the three main components of the nervous system. The central and peripheral nervous systems share some nerve structures with the ANS, and it also has chains of ganglia (clusters of nerve cells where axons communicate) along each side of the spinal cord. The ANS works largely "automatically" to provide involuntary responses, both immediate and longer-term. Sensory nerve fibers send information about the organs and internal activities, such as heart rate. This information is integrated in the hypothalamus, brainstem, or spinal cord. The ANS then sends commands, as motor nerve signals, to three main destinations: the involuntary smooth muscles of many organs and blood vessels; cardiac muscle; and certain glands.

TWO DIVISIONS
The ANS has two divisions: sympathetic and parasympathetic. The ganglia of the sympathetic division are arranged into two ganglion chains, one on either side of the spinal column (only one shown here). The ganglia of the parasympathetic ANS are inside organs (see diagram). Only skin and blood vessels receive nerve messages from all positions on the cord.

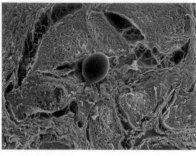

LACRIMAL GLAND
This electron microscope image shows the lacrimal gland, here producing tear fluid (red drop). It is one of many glands under autonomic control.

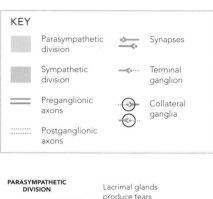

KEY

- Parasympathetic division
- Sympathetic division
- Preganglionic axons
- Postganglionic axons
- Synapses
- Terminal ganglion
- Collateral ganglia

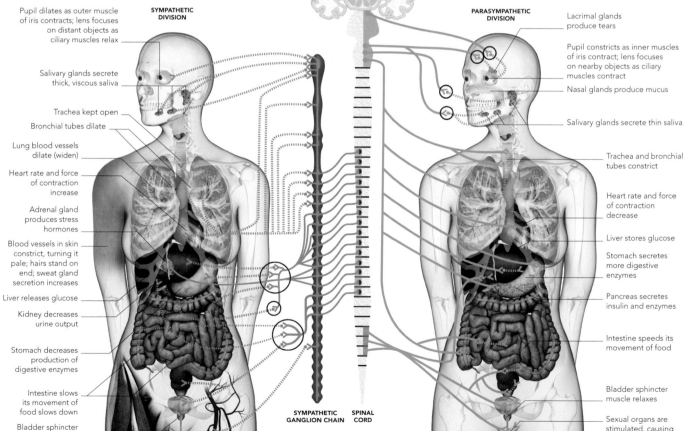

SYMPATHETIC DIVISION

Pupil dilates as outer muscle of iris contracts; lens focuses on distant objects as ciliary muscles relax

Salivary glands secrete thick, viscous saliva

Trachea kept open

Bronchial tubes dilate

Lung blood vessels dilate (widen)

Heart rate and force of contraction increase

Adrenal gland produces stress hormones

Blood vessels in skin constrict, turning it pale; hairs stand on end; sweat gland secretion increases

Liver releases glucose

Kidney decreases urine output

Stomach decreases production of digestive enzymes

Intestine slows its movement of food slows down

Bladder sphincter muscle constricts

Blood vessels dilate

SYMPATHETIC GANGLION CHAIN

SPINAL CORD

PARASYMPATHETIC DIVISION

Lacrimal glands produce tears

Pupil constricts as inner muscles of iris contract; lens focuses on nearby objects as ciliary muscles contract

Nasal glands produce mucus

Salivary glands secrete thin saliva

Trachea and bronchial tubes constrict

Heart rate and force of contraction decrease

Liver stores glucose

Stomach secretes more digestive enzymes

Pancreas secretes insulin and enzymes

Intestine speeds its movement of food

Bladder sphincter muscle relaxes

Sexual organs are stimulated, causing increased lubrication and erection of the clitoris in females and erection of the penis in males

SYMPATHETIC AND PARASYMPATHETIC FUNCTIONS

The sympathetic and parasympathetic division nerves produce contrasting responses. The sympathetic division prepares the body for action and stress. The parasympathetic division restores normal function to conserve energy.

AFFECTED ORGAN	SYMPATHETIC RESPONSE	PARASYMPATHETIC RESPONSE
EYES	Pupils dilate	Pupils constrict
LUNG	Bronchial tubes dilate (widen)	Bronchial tubes constrict
HEART	Rate and strength of heartbeat increase	Rate and strength of heartbeat decrease
STOMACH	Enzymes decrease	Enzymes increase

BALANCED COORDINATION

The ANS divisions primarily send out signals to muscles, creating a "push-pull" relationship between them. These opposing effects interact and balance. For example, involuntary changes in eye pupil size occur constantly. Smooth muscle fibers in the iris are arranged as a circular inner band and radial outer band. Sensory receptors in the eyes respond to light and send nerve signals to the brain, which sends messages to one or the other muscle band to adjust pupil size.

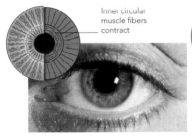
Inner circular muscle fibers contract

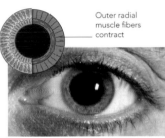
Outer radial muscle fibers contract

CONSTRICTED PUPIL
In bright light or to view nearby objects, the pupil constricts as parasympathetic influence stimulates muscle contraction.

DILATED PUPIL
Widening of the pupil signals the body's heightened awareness as sympathetic nerve messages signal muscle fibers to shorten.

RESPONSES UNDER VOLUNTARY CONTROL

Nervous responses under voluntary control are the opposite of reactions controlled by the ANS. Stimulated by incoming sensory nerve messages, or by conscious thought and intention, the brain's cerebral cortex (outer layer) formulates a central motor plan for a particular movement, and sends out instructions as motor nerve signals to voluntary muscles. As the movement progresses, it is monitored by sensory endings in the muscles, tendons, and joints. The sensory endings update the cerebellum, so that the cerebral cortex can send corrective nerve signals back to the muscles, to keep the movement coordinated and on course.

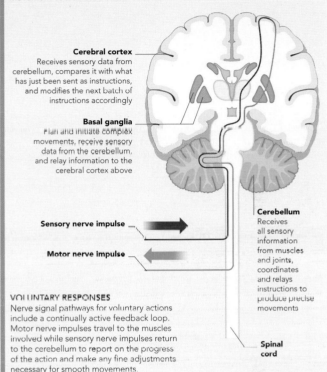

Cerebral cortex
Receives sensory data from cerebellum, compares it with what has just been sent as instructions, and modifies the next batch of instructions accordingly

Basal ganglia
Plan and initiate complex movements, receive sensory data from the cerebellum, and relay information to the cerebral cortex above

Sensory nerve impulse

Motor nerve impulse

Cerebellum
Receives all sensory information from muscles and joints, coordinates and relays instructions to produce precise movements

Spinal cord

VOLUNTARY RESPONSES
Nerve signal pathways for voluntary actions include a continually active feedback loop. Motor nerve impulses travel to the muscles involved while sensory nerve impulses return to the cerebellum to report on the progress of the action and make any fine adjustments necessary for smooth movements.

INVOLUNTARY RESPONSES

There are two main categories of involuntary, or automatic, responses, which do not usually involve conscious awareness. One category involves reflex actions (see p.99). Reflexes mainly affect muscles normally under voluntary control. The other type of response includes autonomic motor actions. The initial nerve pathways for these responses run along spinal nerves into the spinal cord, then up ascending nerve tracts to the lower autonomic regions of the brain, particularly the hypothalamus and parts of the limbic system. These regions analyze and process the information received and then use the autonomic pathways to send out motor impulses as instructions for the involuntary muscles and the glands. Parasympathetic and sympathetic response signals have separate pathways.

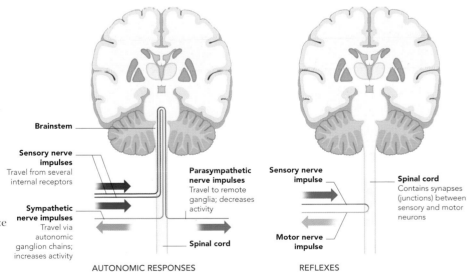

Brainstem

Sensory nerve impulses
Travel from several internal receptors

Sympathetic nerve impulses
Travel via autonomic ganglion chains; increases activity

Parasympathetic nerve impulses
Travel to remote ganglia; decreases activity

Spinal cord

AUTONOMIC RESPONSES
Nerve signals pass along spinal nerves and up the spinal cord to the lower autonomic regions of the brain, which output motor impulses in response.

Sensory nerve impulse

Motor nerve impulse

Spinal cord
Contains synapses (junctions) between sensory and motor neurons

REFLEXES
Sensory signals arrive, and motor signals depart, entirely within the spinal cord and without brain involvement—although the brain becomes aware soon after.

MEMORIES, THOUGHTS, AND EMOTIONS

THE BRAIN IS ENORMOUSLY COMPLEX AND INTEGRATED, AND MANY OF ITS MENTAL FACULTIES ARE NOT CONTROLLED BY ONE AREA. FOR EXAMPLE, THERE IS NO SINGLE "MEMORY CENTER." THOUGHTS, FEELINGS, AWARENESS, EMOTIONS, AND MEMORY INVOLVE MANY PARTS OF THE BRAIN.

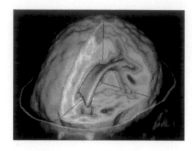

BRAIN ACTION IN SPEECH
This PET scan (see opposite) provides a "snapshot" of a brain taken while the person is talking. Red patches indicate areas of high activity. Here, they show Broca's area, which controls speech (bottom), and Wernicke's area, where language is analyzed (rear). In most people, speech and language comprehension centers are on the left side of the brain.

MAP OF THE CORTEX

Certain regions of the brain's cortex are called primary sensory areas. Each of these receives sensory information from a specific sense. The primary visual cortex, for instance, analyzes data from the eyes. Around each region are association areas, where data from the specific sense is integrated with data from other senses, compared with memories and knowledge, and associated with feelings and emotions. In this way, seeing a particular scene allows us to recognize, identify, and name the objects in it, remember where we saw them previously, recall related sensory data, such as a certain smell, and reexperience associated emotions.

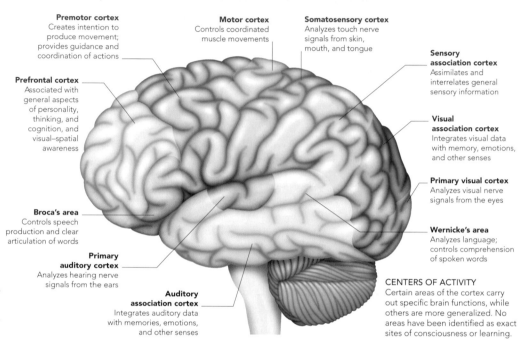

Premotor cortex
Creates intention to produce movement; provides guidance and coordination of actions

Prefrontal cortex
Associated with general aspects of personality, thinking, and cognition, and visual–spatial awareness

Broca's area
Controls speech production and clear articulation of words

Primary auditory cortex
Analyzes hearing nerve signals from the ears

Auditory association cortex
Integrates auditory data with memories, emotions, and other senses

Motor cortex
Controls coordinated muscle movements

Somatosensory cortex
Analyzes touch nerve signals from skin, mouth, and tongue

Sensory association cortex
Assimilates and interrelates general sensory information

Visual association cortex
Integrates visual data with memory, emotions, and other senses

Primary visual cortex
Analyzes visual nerve signals from the eyes

Wernicke's area
Analyzes language; controls comprehension of spoken words

CENTERS OF ACTIVITY
Certain areas of the cortex carry out specific brain functions, while others are more generalized. No areas have been identified as exact sites of consciousness or learning.

Putamen
Stores "subconscious" memories, such as motor skills gained by repetition

Prefrontal cortex
Controls grasp of passing situations such as visual–spatial awareness of current surroundings

Amygdala
Recalls powerful emotions associated with memorable events, such as fear

Hippocampus
Establishes long-term memories and knowledge linked with spatial awareness

Temporal lobe
Stores language, words, vocabulary, speech, and syntax

Cortex
Stores parts of a memory associated with specific senses and motor actions in their relevant areas

MEMORY AND RECALL

Memories are the brain's information storehouse. They include not only formally learned facts and data, but also informal information, such as likes and dislikes, and experiences encountered during emotionally significant events. No single region of the brain processes memories as they are being established, or acts as a storage site for all memories. These processes depend on several factors: the significance and time span of the memory, its depth of emotional impact, and its association with specific senses such as eyesight. For example, musical recall comes partially from the areas in the cortex that deal with auditory information.

AREAS INVOLVED IN MEMORY STORAGE
The outer gray layer of the cortex is heavily involved in memory. The hippocampus takes part in the transfer of immediate thoughts and sensory information into short- and long-term memory stores. If it is damaged, a person can recall events from long ago, before the damage, but not what happened a few hours, or even minutes, previously.

FORMING MEMORIES

To create memories, neurons (nerve cells) are thought to form new projections (axons) and interconnections. Information is constantly monitored for its significance by parts of the brain such as the thalamus and cortex. Certain facts, feelings, and sensory data, such as a smell, are selected for inclusion in the initial stages of memory formation by parts such as the amygdala and hippocampus. Various aspects of the memory are assigned to their relevant areas. Nerve cells form new links, or synapses, that create a new circuit, called a memory trace, or engram, for that particular aspect of the memory.

1 INPUT
A neuron receives a memory-associated input as nerve impulses collected by its dendrites. It outputs a corresponding series of impulses to a second neuron.

2 CIRCUIT FORMATION
The second neuron forms new links with a third, as does the first. The new synapses are established by growth of the axon terminals and dendrites.

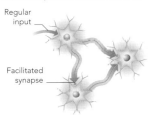
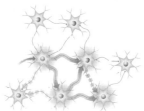

3 INCREASING ACTIVITY
Further activity creates more synapses, which gradually become established (facilitated). Recalling the memory "refreshes" the circuit to prolong retention.

4 INTEGRATION
Through continued activation, the memory circuit is assimilated into a network of surrounding neurons. The full web represents a single memory.

LONG- OR SHORT-TERM?

There are several systems for classifying memories and the processes by which they form. One is "durational" (based on time) and involves three basic stages. Sensory memory, such as the brief recognition of a sound, is fleeting and generally stored for up to half a second only. If consciously retained and interpreted, this sensory input may become short-term memory for a few minutes. The transfer of short-term to long-term memory is known as consolidation and requires attention, repetition, and associative ideas. How easily the information is recalled depends on how it was consolidated.

FROM INPUT TO MEMORY
Sensory input is monitored for significant information, which is then either retained briefly or, if focused upon, is consolidated and stored.

INPUT FROM SENSES → SENSORY MEMORY → ATTENTION NOT PAID → INFORMATION LOST

SENSORY MEMORY → ATTENTION PAID → SHORT-TERM MEMORY → MEMORIES NOT CONSOLIDATED → INFORMATION LOST

SHORT-TERM MEMORY → MEMORIES CONSOLIDATED → LONG-TERM MEMORY

THOUGHT IN ACTION

The advent of real-time scanning methods such as fMRI (functional magnetic resonance imaging) allows researchers to monitor activity levels in different parts of the brain. The fMRI scans reveal tiny localized increases in blood flow (hemodynamic activity). Scans showing hemodynamic activity form instantaneous "maps" that reveal which areas of the brain are busy during well-defined mental activities, such as studying the visual details of an image, listening to and understanding speech, or performing a particular set of movements. For many mental functions, several areas of the brain "light up" simultaneously, showing the complexity of the regional interactions that take place during thought.

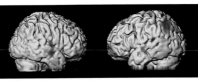

RIGHT SIDE OF BRAIN **LEFT SIDE OF BRAIN**

PLANNING A MOVEMENT
The subject of this fMRI scan was asked to think about performing a task during the scan. The image shows activity in the left and right prefrontal areas and also in the left and right auditory cortex.

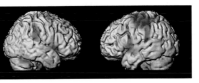

RIGHT SIDE OF BRAIN **LEFT SIDE OF BRAIN**

MAKING THAT MOVEMENT
When actually performing the task, large parts of the premotor and motor cortex show up on the brain's left side. The cerebellum (at the base of the brain) helps control precise muscle coordination.

EMOTIONAL MEMORY

Many memories include powerful emotions linked with an event, such as grief at the loss of a loved one, or great joy at hearing good news. Also, experiencing an associated situation can cause emotional recall. For example, a person witnessing a traffic accident may be reminded of an accident he or she personally experienced previously, and the strong feelings of fear and pain felt at the time may recur. One chief correlating structure in the storage and recall of such strong emotions is the amygdala, located within the temporal lobe on each side of the brain. Normally, the powerful emotional responses it can elicit are kept in check by other brain areas, especially the prefrontal cortex and thalamus.

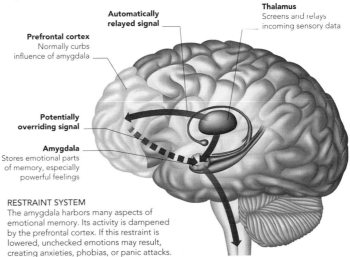

Prefrontal cortex
Normally curbs influence of amygdala

Automatically relayed signal

Thalamus
Screens and relays incoming sensory data

Potentially overriding signal

Amygdala
Stores emotional parts of memory, especially powerful feelings

RESTRAINT SYSTEM
The amygdala harbors many aspects of emotional memory. Its activity is dampened by the prefrontal cortex. If this restraint is lowered, unchecked emotions may result, creating anxieties, phobias, or panic attacks.

TOUCH, TASTE, AND SMELL

RECEPTORS THAT SENSE PRESSURE, PAIN, AND TEMPERATURE ARE WIDESPREAD IN THE BODY. TASTE AND SMELL, IN CONTRAST, ARE "SPECIAL SENSES" BECAUSE THEIR RECEPTORS ARE COMPLEX, LOCALIZED, AND DETECT SPECIFIC STIMULI.

SMELL

Smell (along with taste) is a chemosense, which is a sense that can detect chemical substances. The sense of smell detects molecules, or tiny particles, known as odorants floating in the air. In humans, smell is much more sensitive than taste and may be able to distinguish millions of odors. Specialized epithelial tissue provides a smelling zone, known as the olfactory epithelium, on the roof of the nasal chamber. In addition to warning of dangers, such as smoke and poisonous gas, smell makes an important contribution to the appreciation of food and drink. The sense of smell tends to deteriorate with age, so children and young adults are able to distinguish a wider range of odors and experience them more vividly than older people.

NASAL LINING
Epithelial cells lining the nasal chamber have tufts of hairlike cilia, which wave germ- and odorant-trapping mucus toward the back of the chamber to be swallowed.

HOW WE SMELL

Incoming odor molecules dissolve in the mucus lining the nasal chamber. In the roof of the chamber, they touch the cilia (microscopic hairlike endings of olfactory receptor cells). If the correct odor molecule slots into the same-shaped receptor on the cilial membrane, like a key in a lock, a nerve impulse is generated. The impulses are partly processed by intermediate neurons called glomeruli in the olfactory bulb.

LOCATION

Olfactory bulb
Dura matter
Ethmoid bone
Mucus-secreting gland
Basal cell
Olfactory receptor cell
Supporting cell
Airflow
Odor molecules
Glomerulus
Nerve fiber
Cilia

OLFACTORY EPITHELIUM
Nerve impulses are stimulated by odor molecules that make contact with the cilia on olfactory receptor cells. These impulses travel along nerve fibers and up through small holes in the ethmoid bone, which separates the nasal cavity from the brain.

TOUCH

The sense of touch is provided by microscopic sensory receptors (which are specialized endings of nerve cells) in the skin or in deeper tissues (see p.166). Some receptors are enclosed in capsules of connective tissue, while others remain uncovered. Different shapes and sizes of receptors detect a range of stimuli, such as light touch, heat, cold, pressure, and pain. These receptors relay their signals via the spinal cord and lower brain to a strip curving around the cerebral cortex, known as the somatosensory cortex, or "touch center."

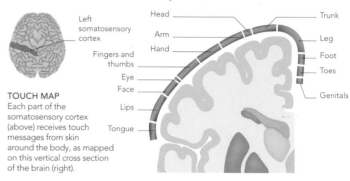

Left somatosensory cortex

Head
Arm
Hand
Fingers and thumbs
Eye
Face
Lips
Tongue

Trunk
Leg
Foot
Toes
Genitals

TOUCH MAP
Each part of the somatosensory cortex (above) receives touch messages from skin around the body, as mapped on this vertical cross section of the brain (right).

TASTE

Taste works in a similar way to smell. Its gustatory cell (taste) receptors detect specific chemicals dissolved in saliva by a "lock-and-key" method (see box, left). Groups of receptor cells are known as taste buds. A child has about 10,000 taste buds, but with age, the number may fall to fewer than 5,000. They are located mainly on and between the pimplelike protuberances (papillae) that dot certain regions of the tongue's upper surface. There are also some taste buds on the palate (roof of the mouth), throat, and epiglottis.

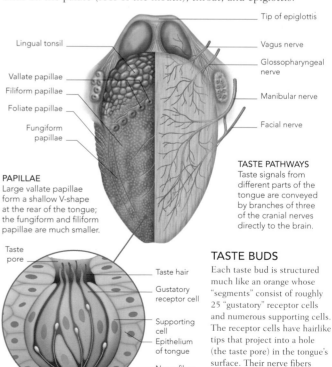

Lingual tonsil
Vallate papillae
Filiform papillae
Foliate papillae
Fungiform papillae

Tip of epiglottis
Vagus nerve
Glossopharyngeal nerve
Manibular nerve
Facial nerve

PAPILLAE
Large vallate papillae form a shallow V-shape at the rear of the tongue; the fungiform and filiform papillae are much smaller.

Taste pore
Taste hair
Gustatory receptor cell
Supporting cell
Epithelium of tongue
Nerve fiber

TASTE PATHWAYS
Taste signals from different parts of the tongue are conveyed by branches of three of the cranial nerves directly to the brain.

TASTE BUDS
Each taste bud is structured much like an orange whose "segments" consist of roughly 25 "gustatory" receptor cells and numerous supporting cells. The receptor cells have hairlike tips that project into a hole (the taste pore) in the tongue's surface. Their nerve fibers gather at the bud base.

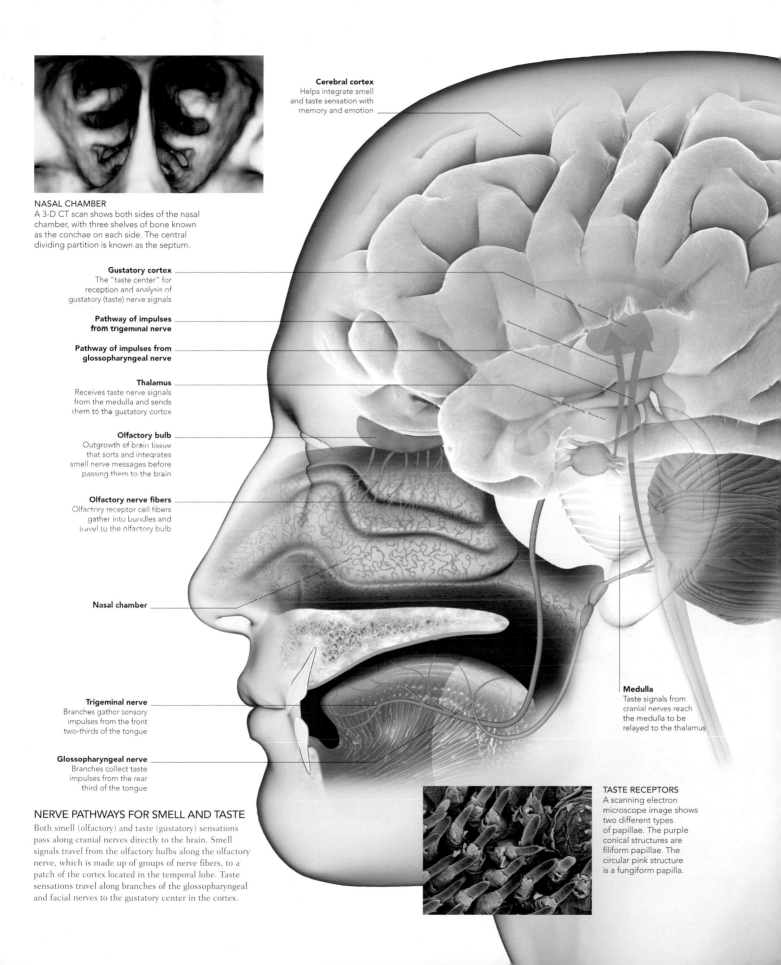

NASAL CHAMBER
A 3-D CT scan shows both sides of the nasal chamber, with three shelves of bone known as the conchae on each side. The central dividing partition is known as the septum.

Cerebral cortex
Helps integrate smell and taste sensation with memory and emotion

Gustatory cortex
The "taste center" for reception and analysis of gustatory (taste) nerve signals

Pathway of impulses from trigeminal nerve

Pathway of impulses from glossopharyngeal nerve

Thalamus
Receives taste nerve signals from the medulla and sends them to the gustatory cortex

Olfactory bulb
Outgrowth of brain tissue that sorts and integrates smell nerve messages before passing them to the brain

Olfactory nerve fibers
Olfactory receptor cell fibers gather into bundles and travel to the olfactory bulb

Nasal chamber

Trigeminal nerve
Branches gather sensory impulses from the front two-thirds of the tongue

Glossopharyngeal nerve
Branches collect taste impulses from the rear third of the tongue

Medulla
Taste signals from cranial nerves reach the medulla to be relayed to the thalamus

TASTE RECEPTORS
A scanning electron microscope image shows two different types of papillae. The purple conical structures are filiform papillae. The circular pink structure is a fungiform papilla.

NERVE PATHWAYS FOR SMELL AND TASTE

Both smell (olfactory) and taste (gustatory) sensations pass along cranial nerves directly to the brain. Smell signals travel from the olfactory bulbs along the olfactory nerve, which is made up of groups of nerve fibers, to a patch of the cortex located in the temporal lobe. Taste sensations travel along branches of the glossopharyngeal and facial nerves to the gustatory center in the cortex.

EARS, HEARING, AND BALANCE

THE EARS PROVIDE THE SENSE OF HEARING. THEY ALSO DETECT HEAD POSITION AND MOTION, SO THEY ARE ESSENTIAL TO BALANCE. THE PARTS CONCERNED WITH HEARING AND BALANCE ARE LOCATED IN DIFFERENT AREAS OF THE EAR, BUT THE FUNCTION OF BOTH IS BASED ON "HAIR CELL" RECEPTORS.

INSIDE THE EAR

The ear is divided into three parts. The outer ear comprises the ear flap (pinna) and the slightly S-shaped outer ear canal (external acoustic meatus). The ear canals guides sound waves to the second region, the middle ear. The elements of the middle ear amplify the sound waves and transfer them from the air into the fluid of the inner ear. They include the eardrum (tympanic membrane) and the three smallest bones in the body, the auditory ossicles, which span the air-filled middle-ear cavity (tympanic chamber). The fluid-filled inner ear changes sound waves to nerve signals inside the snail-shaped cochlea. The middle ear cavity is connected to the throat by the Eustachian tube, and thus to the air outside. This connection allows atmospheric pressure to transfer to the cavity, equalizing the air pressure on either side of the eardrum and preventing it from bulging as the outside pressure changes.

Scalp muscle

Auricular cartilage
Provides springy C-shaped framework to pinna

Temporal bone
Lower side bone of skull

Outer ear canal (external auditory meatus)

Pinna (ear flap)
Skin-covered flap with subcutaneous fat, cartilage, and connective tissue

OUTER EAR
The vaguely trumpet-shaped pinna helps funnel sound waves into the outer ear canal. The wax secreted continuously by its lining traps dirt and germs, and slowly flakes off to work its way out by jaw movements involved in chewing and talking.

Semicircular canals
Contain sense organs functioning in balance

Suspensory ligament
Ossicle ligaments keep the bones in position but free to vibrate

Tympanic chamber (middle-ear cavity)

Malleus (hammer)

Ear ossicles

Incus (anvil)

Stapes (stirrup)

Tympanum, or tympanic membrane (eardrum)
About the size of the owner's little fingernail; resembles thin skin

Canal lining
Secretes wax, which traps unwanted debris

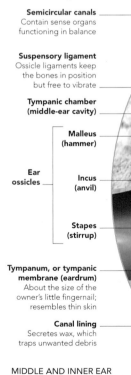

Vestibular nerve
Carries nerve signals from the balance organs to the brain

Vestibulocochlear (auditory) nerve
Conveys nerve signals from the vestibule and cochlea to the brain

Section cut from cochlea

Vestibular canal

Cochlear duct

Tympanic canal

Vestibule
Contains the utricle and saccule, organs of balance

Cochlea
Contains the organ of hearing; barely larger than the little finger's tip, it spirals for 2¾ turns

Oval window
Membrane in the cochlea wall receiving vibrations from the stapes

Round window
Pressure-relief membrane that allows cochlear fluid to bulge with vibrations

Eustachian tube
Runs to an opening in the side of the upper throat, level with the soft palate

MIDDLE AND INNER EAR
The cochlea, semicircular canals, and vestibule of the inner ear are linked. They are all filled with fluid and are encased and protected within the thickness of the skull's temporal bone, occupying a complex series of tunnels and chambers known as the osseous labyrinth. The ossicles are positioned and connected by miniature ligaments, tendons, and joints, just like larger bones.

HOW WE HEAR

Ears act as energy converters, changing pressure differences in air, known as sound waves, into electrochemical nerve impulses. Sound waves, which usually occur as a complex pattern of frequencies, set the eardrum vibrating in the same pattern. The vibrations are conducted along the ossicle chain, which rocks like a bent lever and forces the footplate of the stapes to act like a piston, pushing and pulling at the flexible oval window of the cochlea. The motions set off waves through the perilymph fluid inside the cochlea. These in turn transfer their vibrational energy to the tube-shaped organ of Corti (spiral organ), which coils within the cochlea.

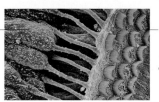

HAIR CELLS
Within the organ of Corti, shown with the tectorial membrane removed on the right, each hair cell is seen to have 40–100 hairs arranged in a curve. Nerve fibers run from the cell bases.

Hair cell — Hairs of hair cell

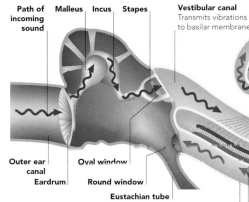

Organ of Corti
Central spiral element of the cochlea, composed of tectorial and basilar membranes linked by sensitive hair cells

Hair cells
Generate nerve signals in response to motion of basilar and tectorial membranes

Cochlear nerve
Carries nerve signals to brain

Vestibular canal
Transmits vibrations to basilar membrane

Path of incoming sound · **Malleus** · **Incus** · **Stapes**

Outer ear canal
Eardrum
Oval window
Round window
Eustachian tube

Tympanic canal
Conveys residual vibrations returning to round window

Tectorial membrane
Tips of hairs from hair cells embed in this

Basilar membrane
Supports the bases of hair cells and their nerve fibers

Nerve fiber
Nerve signal

Frequency response
Organ of Corti "shakes" at a particular point along its length, according to frequency of vibration

VIBRATION TRANSFER
Vibrations travel from the oval window, through the cochlea's fluid in the vestibular canal, and transfer to the organ of Corti. Here, hair cells on the basilar membrane have their microhair tips embedded in the jellylike tectorial membrane above. As this structure vibrates, various forces pull the hairs, stimulating their cells to produce nerve impulses. These travel via the cochlear nerve to the auditory cortex for interpretation. Residual vibrations from the vestibular canal pass down the tympanic canal to the round window.

HEARING RANGE

Human ears respond to a range of sound frequencies (pitches) from about 20 Hz (vibrations per second) to over 16,000 Hz. Pressure waves beyond this range (infrasound and ultrasound) cannot be heard. Hearing range varies among individuals and reduces with age, especially at the upper end.

AUDIOGRAM
A graph plotting the lowest audible pressure of sound waves (the hearing threshold) is called an audiogram, and it reveals that the human ear is most sensitive to sounds of medium frequency.

"Middle C" is at 262 Hz

Top of hearing range; above this is ultrasound

Bottom of hearing range; below this is infrasound

THRESHOLD OF HEARING (dB): 80 70 60 50 40 30 20 10 0 -10 -20

FREQUENCY (Hz): 7.8 15.6 31.2 62.5 125 250 500 1000 2000 4000 8000 16,000

THE PROCESS OF BALANCE

Balance is not a single sense, but a process involving a range of sensory inputs, analysis in the brain, and motor outputs. Inputs arrive from the eyes (see p.108), microreceptors in muscles and tendons (see p.79), and skin pressure sensors (see p.166), as in the soles of the feet. The inner ear's fluid-filled vestibule and semicircular canals also play a key role. They incorporate sensitive hair cells similar to the cochlea's (see above). The vestibule responds mainly to the head's position relative to gravity (static equilibrium), while the canals react chiefly to the speed and direction of head movements (dynamic equilibrium). In practice, both respond to most head positions and movements.

VESTIBULE
The vestibule's two parts, the utricle and saccule, each have a patch, the macula, containing hair cells. The tips of the cells extend into a membrane covered in heavy mineral crystals (otoliths). With head level, the saccule's macula is vertical and the utricle's horizontal. As the head bends forward, the hair cells monitor the head's position in relation to the ground.

MACULA ACTION
Mineral crystals (otoliths) cover membrane
Otolithic membrane
Hair of hair cell
Hair cell

Utricular macula rotated to vertical
Gravity pulls membrane
Hairs deflected
Hair cell stimulated

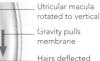

CANALS
Each semicircular canal has a bulge near one end called the ampulla. This houses a low mound of hair cells, their hair ends set into a taller jellylike mound, the cupula. As the head moves, fluid in the canal lags behind, swirls past the cupula and bends it. This pulls the hairs and triggers their cells to fire nerve signals.

AMPULLA ACTION
Cupula
Hairs of hair cells
Mound of hair cells (crista ampullaris)
Ampulla

Fluid swirls due to head motion
Cupula bends
Hair cells stimulated

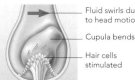

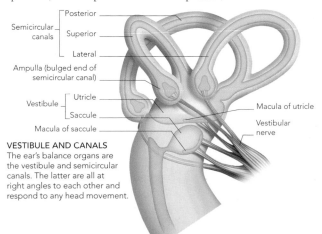

VESTIBULE AND CANALS
The ear's balance organs are the vestibule and semicircular canals. The latter are all at right angles to each other and respond to any head movement.

Semicircular canals — Posterior, Superior, Lateral
Ampulla (bulged end of semicircular canal)
Vestibule — Utricle, Saccule
Macula of saccule
Macula of utricle
Vestibular nerve

EYES AND VISION

EYESIGHT PROVIDES THE BRAIN WITH MORE INPUT THAN ALL OTHER SENSES COMBINED. EACH OPTIC NERVE CONTAINS ONE MILLION NERVE FIBERS, AND IT IS ESTIMATED THAT MORE THAN HALF THE INFORMATION IN THE CONSCIOUS MIND ENTERS THROUGH THE EYES.

THE SEQUENCE OF VISION

Rays of light enter the eye through the clear, domed front of the eyeball, the cornea, where they are partly bent (refracted). The rays then pass through the transparent lens, which changes shape to fine-focus the image, a mechanism known as accommodation. The light continues through the fluid, or vitreous humor, within the eyeball and shines an upside-down image onto the retina lining. The retina contains over 120 million cone cells and about 7 million rod cells. These convert the light energy falling onto them into nerve signals. Rods are scattered through the retina and respond to low levels of light, but do not differentiate colors. Cones, which are concentrated in the fovea, need brighter conditions to function, and distinguish colors and fine details. Nerve fibers from the rods and cones connect via intermediate retinal cells to neurons whose fibers form the optic nerve. Through this, the image is transmitted to the brain's visual cortex where it is turned upright.

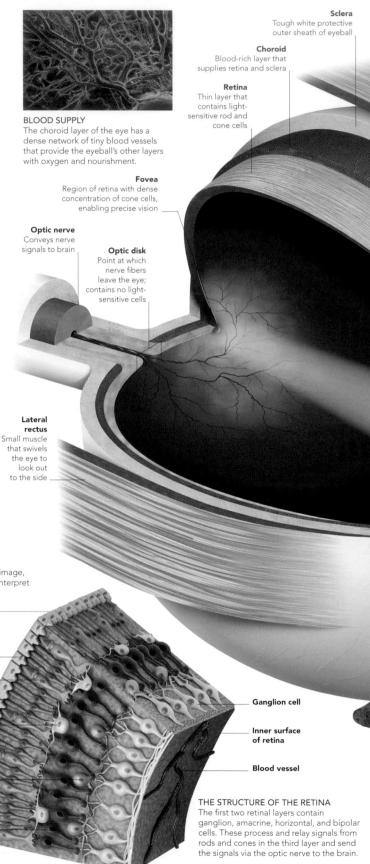

BLOOD SUPPLY
The choroid layer of the eye has a dense network of tiny blood vessels that provide the eyeball's other layers with oxygen and nourishment.

Sclera
Tough white protective outer sheath of eyeball

Choroid
Blood-rich layer that supplies retina and sclera

Retina
Thin layer that contains light-sensitive rod and cone cells

Fovea
Region of retina with dense concentration of cone cells, enabling precise vision

Optic nerve
Conveys nerve signals to brain

Optic disk
Point at which nerve fibers leave the eye; contains no light-sensitive cells

Lateral rectus
Small muscle that swivels the eye to look out to the side

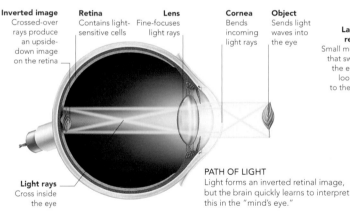

Inverted image
Crossed-over rays produce an upside-down image on the retina

Retina
Contains light-sensitive cells

Lens
Fine-focuses light rays

Cornea
Bends incoming light rays

Object
Sends light waves into the eye

Light rays
Cross inside the eye

PATH OF LIGHT
Light forms an inverted retinal image, but the brain quickly learns to interpret this in the "mind's eye."

THE RETINA

The retina contains three main layers of cells, each communicating with the next via junctions (synapses). The light-sensitive layer containing the rods and cones is at the back of the retina, and light must therefore pass through the two inner layers (and blood vessels on the retina's inner surface) before it reaches the rods and cones and stimulates them to produce nerve signals. The two inner layers of the retina contain ganglion, amacrine, horizontal, and bipolar cells. These cells provide initial processing of the nerve signals generated by the rods and cones and send the processed nerve signals to the visual cortex of the brain (see pp.110–11).

Pigmented epithelium

Cone cell

Horizontal cell

Rod cell

Amacrine cell

Bipolar cell

Ganglion cell

Inner surface of retina

Blood vessel

THE STRUCTURE OF THE RETINA
The first two retinal layers contain ganglion, amacrine, horizontal, and bipolar cells. These process and relay signals from rods and cones in the third layer and send the signals via the optic nerve to the brain.

ACCOMMODATION

The cornea provides most of the eye's focusing power, bending light waves so that they converge, which allows for sharper focus on the retina. Fine adjustment is carried out by the lens, which is altered in shape by the ringlike ciliary muscle around it. When the muscle contracts, the elastic lens bulges and thickens, providing greater focusing power to converge light waves from nearby objects. As the ciliary muscle relaxes, the lens becomes flatter and thinner, which allows the eyes to focus on more distant objects.

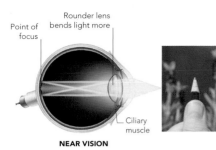

NEAR VISION
Light waves from close objects diverge more and need the extra focusing power of a fatter lens to bend the light waves so that they converge.

Point of focus / Rounder lens bends light more / Ciliary muscle

NEAR VISION

Point of focus / Flatter lens bends light less

DISTANT VISION
Light waves from distant objects are almost parallel and require less refracting power to focus, so the ciliary muscle relaxes to make the lens bulge less.

DISTANT VISION

Superior rectus
Small muscle that swivels eye to look up

Suspensory ligaments
Hold lens within the ring of ciliary muscle

Posterior chamber
Fluid-filled cavity behind the iris

Iris
Ring of muscle that changes size of pupil to regulate amount of light entering the eye

Anterior chamber
Between cornea and iris, filled with aqueous humor fluid

Pupil
Hole in iris that becomes wider in dim light

Cornea
Domed transparent "window" at front of eye

Conjunctiva
Delicate, sensitive covering of cornea and eyelid lining

AROUND THE EYE

The accessory structures around the eye are not involved directly in vision but help the eye function and remain healthy. The skin folds of the eyelids contain a ringlike (sphincter) muscle known as the orbicularis oculi. When this contracts during a normal blink, it narrows the gap in its center to shut the eyelids. This protects the eye and also smears lacrimal, or tear, fluid over the conjunctiva. The fluid flows from the lacrimal gland to wash dirt and dust off the surface and provides antimicrobial protection. Additionally, there are six small, straplike muscles attaching the eyeball to the rear of the eye socket (orbit) in the skull. Known as extraocular or extrinsic eye muscles, they swivel, or roll, the eyeball in its socket to look up or down, inward or out. These muscles are very fast-acting and are controlled by branches of the oculomotor, trochlear, and abducens nerves, which are cranial nerves (see p.98).

Lacrimal gland
Secretes tears to keep eye clean and moist

Lacrimal ducts
5–10 ducts convey fluid to eye surface

Lacrimal canals
Collect tears draining through small holes in the corner of the eyelid

Lacrimal sac
Channels tears toward nasal cavity

Nasolacrimal duct
Opens into nasal cavity

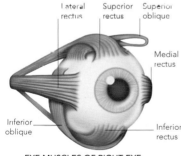

Lateral rectus / Superior rectus / Superior oblique / Medial rectus / Inferior oblique / Inferior rectus

EYE MUSCLES OF RIGHT EYE
The six extrinsic eye muscles are 1⅕–1⅖in (30–35 mm) long. They contract or relax in close coordination to move the eyeball within its socket.

Ciliary muscle
Ring of muscle that alters shape of lens

Lens
Transparent disk of tissue that changes shape for near or far vision

INSIDE THE EYE
An average eyeball is 1 in (25mm) in diameter and has three main outer layers: the sclera, choroid, and retina. Near the front, the sclera can be seen as the white of the eye, and at the front it becomes the clear cornea. The main bulk of the eye, between the lens and retina, is filled with a clear, jellylike fluid known as vitreous humor. This maintains the eyeball's spherical shape.

TEAR APPARATUS
The tear (lacrimal) gland is under the soft tissues of the upper eyelid's outer part and under the bony part of the orbit. It produces ⅓–⅔fl oz (1–2ml) of fluid daily.

HOW VISION WORKS

THE JOURNEY TAKEN BY NERVE SIGNALS FROM THE RETINA TO THE VISUAL CORTEX AT THE REAR OF THE BRAIN IS ONLY PART OF THE WAY WE PERCEIVE IMAGES IN THE MIND. FOR FULL VISUAL PROCESSING, OTHER BRAIN AREAS MUST ALSO COME INTO PLAY.

The multiphase process of visual perception becomes more complex and spread out as it proceeds. Rod and cone cells in the retina respond to light intensity, colors, and motion. Cells in the retina "preprocess" this information, then the resulting nerve signals travel along the optic nerve

to the optic chiasm, where some visual-field information is swapped. Next the optic radiation takes this information to the visual cortex. Here it is analyzed and passed along two main pathways, dorsal and ventral, where recognition and meaning are added, before conscious perception forms in the frontal lobes.

EYE TO BRAIN TO MIND

Information entering the eyes as light rays ends up being interpreted as images in the brain's frontal lobes. It can take less than one-fifth of a second for signals in the retina to be turned into a conscious perception in the mind.

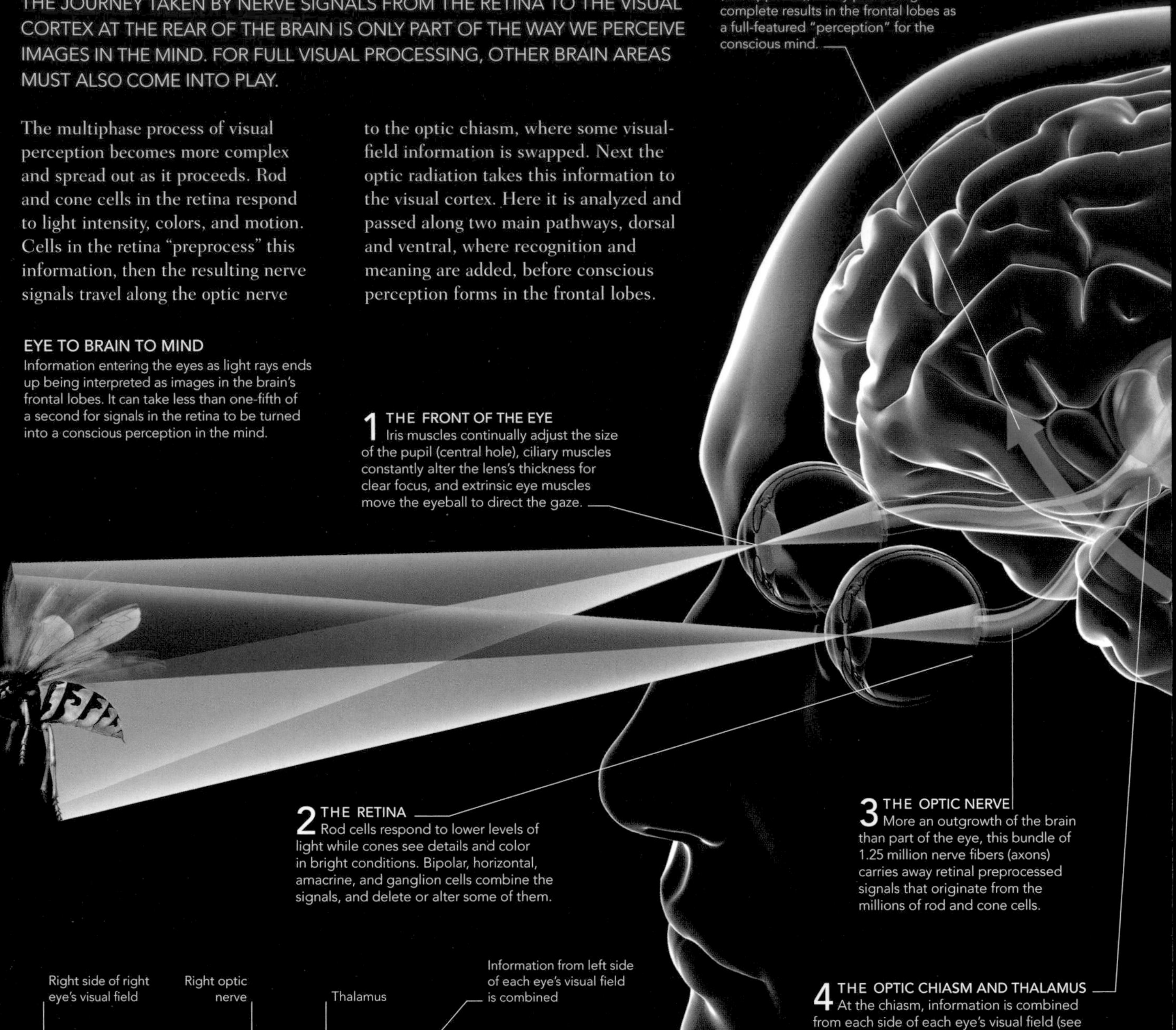

1 THE FRONT OF THE EYE
Iris muscles continually adjust the size of the pupil (central hole), ciliary muscles constantly alter the lens's thickness for clear focus, and extrinsic eye muscles move the eyeball to direct the gaze.

2 THE RETINA
Rod cells respond to lower levels of light while cones see details and color in bright conditions. Bipolar, horizontal, amacrine, and ganglion cells combine the signals, and delete or alter some of them.

3 THE OPTIC NERVE
More an outgrowth of the brain than part of the eye, this bundle of 1.25 million nerve fibers (axons) carries away retinal preprocessed signals that originate from the millions of rod and cone cells.

9 PERCEPTION
The dorsal and ventral routes send signals through various cortex lobes (see opposite), finally presenting complete results in the frontal lobes as a full-featured "perception" for the conscious mind.

4 THE OPTIC CHIASM AND THALAMUS
At the chiasm, information is combined from each side of each eye's visual field (see below left). Signals then proceed to a specialized part of the thalamus called the lateral geniculate nucleus (LGN). The LGN acts as a relay station that sorts information from the optic chiasm.

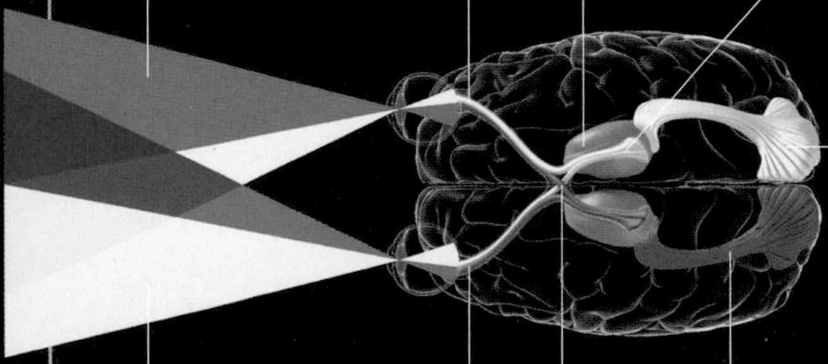

Right side of right eye's visual field

Right optic nerve

Thalamus

Information from left side of each eye's visual field is combined

Right visual cortex receives information from left sides of visual fields of both eyes

Left side of left eye's visual field

Left optic nerve

Optic chiasm

Optic radiation

SWAPPING FIELDS

At the optic chiasm, nerve signals from the right side of the right visual field swap to the left side of the brain, and vice versa. So data from the right visual field goes to the left visual cortex, where it is assessed for information about distance and direction.

5 THE OPTIC RADIATION
Signals from the LGN are sent onward, along a fanning-out, ribbon-like band, the optic radiation, to their main target, the primary visual cortex, where reconstruction of the image from the retina begins.

6 PRIMARY VISUAL CORTEX
The primary visual cortex is the central zone of the visual cortex—the surface layer of gray matter specializing in sight, at the lower rear of the brain. Analysis of visual stimuli begins here and in adjacent areas of the visual cortex. It continues along two pathways called the dorsal and ventral routes.

7 THE DORSAL ROUTE: "WHERE"
Starting from the primary visual cortex, the dorsal route contributes knowledge about an object's position in the visual field, its distance, motion, and direction in relation to the viewer, and also some aspects of its size and shape.

THE CORTEX
The visual cortex, located in the occipital lobe, takes the lead in processing sight signals including information about color, form, movement, and orientation. Visual information is then processed further in various other parts of the brain to add more information about recognition, relevance, and other details of context.

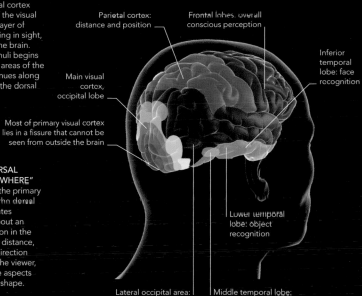

Parietal cortex: distance and position

Frontal lobes: overall conscious perception

Inferior temporal lobe: face recognition

Main visual cortex, occipital lobe

Most of primary visual cortex lies in a fissure that cannot be seen from outside the brain

Lower temporal lobe: object recognition

Lateral occipital area: perception of symmetry

Middle temporal lobe: motion perception

BINOCULAR VISION
The brain gets information from both eyes at the same time, but the image from each eye is slightly different. The brain combines the two-dimensional information from each eye to create a single three-dimensional image (stereoscopic vision). This allows depth perception and helps with shape recognition. It also helps a person to judge how quickly an object is moving toward or away from them. The full field of vision for humans is about 180 degrees, with the central binocular field of vision being around 120 degrees. Newborns do not have binocular vision; this develops by about four months of age.

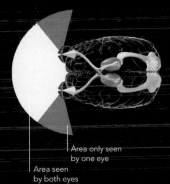

Area only seen by one eye

Area seen by both eyes

8 THE VENTRAL ROUTE: "WHAT"
Nerve signals from the primary visual cortex move through lower brain areas that add details of shape, color, and depth, and then recognition and further knowledge called up from memory, such as the name of the object.

CONE CELLS AND COLOR VISION
In the retina, the millions of light-sensitive cone cells are not all the same. There are three kinds or classes. Red-sensitive cones respond best to longer wavelengths of light rays, at the red-orange-yellow end of the spectrum. Green-sensitive cones react mainly to middle wavelengths, and blue-sensitive ones to the shortest waves at the blue-indigo-violet end of the spectrum. So, for example, yellow light stimulates red-sensitive cones most, green-sensitive ones less, and blue-sensitive cones hardly at all; this pattern of signals is interpreted by the brain as the color yellow.

PACKED TOGETHER
In the retina, color-sensing cone cells (yellow-green) are shorter and wider than light-intensity-detecting rods (white). In this microscope view, light rays would come from the right.

CEREBROVASCULAR DISORDERS

THE TERM "CEREBROVASCULAR DISORDERS" COVERS ANY PROBLEM THAT AFFECTS THE BLOOD VESSELS
SUPPLYING THE BRAIN. STROKE IS ONE OF THE MOST SERIOUS OF SUCH DISORDERS, WITH ONE IN SEVEN
PEOPLE DYING. ALSO SERIOUS IS BLEEDING INSIDE THE SKULL, WHICH MAY OCCUR SPONTANEOUSLY
DUE TO A DEFECT PRESENT FROM BIRTH OR AS A RESULT OF HEAD INJURY. MIGRAINE INVOLVES BLOOD
VESSELS IN THE SCALP AND BRAIN BUT DOES NOT CAUSE ANY PERMANENT LOSS OF FUNCTION.

STROKE

DAMAGE TO THE BRAIN OCCURS IF ITS BLOOD SUPPLY IS
INTERRUPTED AS A RESULT OF A BLOCKAGE OR BLEEDING
FROM ONE OF THE ARTERIES SUPPLYING THE BRAIN.

Any disruption of blood supply to the brain starves some
of the nerve cells of oxygen and nutrients. These affected
cells are unable to communicate with the parts of the
body they serve, which results in temporary or permanent
loss of function. In most people, symptoms of stroke
develop rapidly over seconds or minutes and may include
weakness or numbness on one side of the body, visual
disturbances, slurred speech, and difficulty maintaining
balance. Immediate admission to the hospital is essential
if there is to be a chance of preventing brain damage;
close monitoring is required. In some types of stroke,
drugs may be given to dissolve a blood clot.
Long-term treatment to reduce the risk of
further strokes depends on the cause of the
stroke but usually consists of drug treatment
and sometimes surgery; rehabilitative
treatments, such as physical therapy
and speech therapy, are often needed.
The aftereffects of a stroke are
variable and range from mild,
temporary symptoms, such as
slurred speech, to lifelong
disability or death.

Blockage of tiny vessels
Prolonged high blood pressure or
diabetes may damage tiny blood
vessels within the brain; this may
lead to localized blockages known
as lacunar strokes that sometimes
result in a form of dementia

**Branches of anterior
cerebral artery**

Thrombus
Buildup of fatty
deposits within
artery walls, called
atherosclerosis, narrows
blood vessels and may
encourage formation of a
blood clot, or thrombus;
if thrombus blocks off an
artery to the brain, a
stroke follows

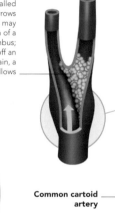

Hemorrhage   Blood vessel

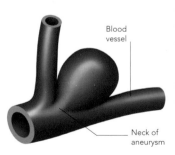

BLEEDING WITHIN THE BRAIN
An intracerebral hemorrhage,
bleeding within brain tissue, is a
main cause of stroke in older people
who have hypertension. High blood
pressure may put extra strain on
small arteries in the brain, which
causes them to rupture.

Embolus
Blockage of a cerebral
artery, resulting in a
stroke, can be caused by
a fragment of material,
called an embolus, that
has traveled through the
bloodstream and lodged
in the vessel

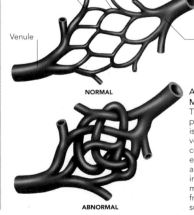

**Posterior
ceberal
artery**

**Basilar
artery**

**External
carotid
artery**

**Internal
carotid
artery**

**Vertebral
artery**

**Common cartoid
artery**

BLOCKED BLOOD VESSELS
Blocked arteries that cause a stroke can occur for several
reasons, ranging from localized blockages in tiny blood vessels
deep within the brain to a blockage caused by a fragment of
material that has traveled to the brain from elsewhere.

SUBARACHNOID
HEMORRHAGE

RARELY, AN ARTERY NEAR THE BRAIN RUPTURES
SPONTANEOUSLY AND LEAKS BLOOD INTO THE
SUBARACHNOID SPACE BETWEEN THE MIDDLE AND
INNERMOST OF THE MEMBRANES COVERING THE BRAIN.

The most common cause of subarachnoid hemorrhage
is rupture of a berry aneurysm, an abnormal, berrylike
swelling in a cerebral artery. Another major cause is
rupture of an arteriovenous malformation, an abnormal
tangle of blood vessels. Subarachnoid hemorrhage
is life-threatening and needs emergency medical
treatment. To stop bleeding from a berry aneurysm,
a clip is applied around its neck. Distended or knotted
vessels can sometimes be made safe without major
surgery either by insertion of wire coils using a
catheter, or by radiation therapy.

Blood
vessel

Neck of
aneurysm

BERRY ANEURYSM
A berry aneurysm usually forms at an
arterial bifurcation, often on a blood
vessel at the base of the brain (the
circle of Willis). Berry aneurysms are
thought to be present from birth,
and there may be one or several.

Capillaries

Venule

Arteriole

NORMAL

ABNORMAL

**ARTERIOVENOUS
MALFORMATION**
This defect, usually
present from birth,
is a tangle of blood
vessels. Fewer capillary
connections than normal
exist between arterioles
and venules. An increase
in pressure results, which
may cause blood to leak
from the vessels into the
subarachnoid space.

TRANSIENT ISCHEMIC ATTACK

PART OF THE BRAIN SUDDENLY AND BRIEFLY FAILS TO FUNCTION DUE TO BLOCKAGE OF ITS BLOOD SUPPLY.

A transient ischemic attack (TIA) produces temporary strokelike symptoms usually lasting anything from a few minutes to a few hours and has no or few after-effects. Symptoms lasting longer indicate a stroke (see opposite). The blockage may be due to an embolus or a thrombus (see below) and these may have many different underlying causes, including atherosclerosis (fatty deposits in artery walls), a previous heart attack, irregular heartbeat, and diabetes mellitus. Of people who have a TIA, about 1 in 3 will have a stroke within a year. The more frequently TIAs occur, the higher the risk of having a stroke in the future. The risk of TIAs is reduced by treating underlying causes; lifestyle changes such as a low-fat diet and giving up smoking are often helpful.

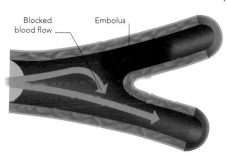

Blocked blood flow — **Embolus**

BLOCKAGE
An artery supplying the brain may become blocked if a fragment from a blood clot, called an embolus, detaches itself from elsewhere in the body and travels in the blood to block an artery. A blood clot, called a thrombus, may also develop in a cerebral artery itself, usually as a result of atherosclerosis (see p.140).

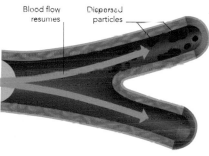

Blood flow resumes — **Dispersed particles**

DISPERSAL
As normal blood flow breaks up and disperses the blood clot, oxygenated blood again reaches the area of the brain that has been starved of oxygen due to disruption of its blood supply. Although clots usually disperse and symptoms then disappear, attacks tend to recur. People may have a number of attacks over one or several days. Sometimes several years elapse between attacks.

SUBDURAL HEMORRHAGE

A TORN VEIN CAUSES BLEEDING INSIDE THE SKULL BETWEEN THE TWO OUTER MEMBRANES SURROUNDING THE BRAIN.

Bleeding may occur suddenly (acute subdural hemorrhage) following a severe blow to the head or blood may build up slowly over days or weeks (chronic subdural hemorrhage), often following an apparently trivial head injury. Symptoms such as headache, confusion, and drowsiness may come on within minutes or over months depending on the type. The symptoms are caused by the formation of a blood clot, which enlarges and compresses the surrounding brain tissue. Surgery to drain the blood may be required. The outlook depends on the size and location of the clot. Many people recover quickly, but if the hemorrhage has affected a large area of the brain, the condition may be fatal.

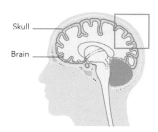

Skull — **Brain**

LOCATION

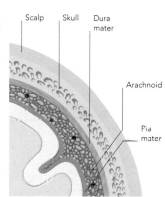

Scalp — **Skull** — **Dura mater** — **Arachnoid** — **Pia mater**

NORMAL
Three membranes, the meninges, cover the brain. The outermost membrane is the dura mater, which contains veins and arteries that nourish cranial bones. Next is the arachnoid, consisting of elastic tissue, and nearest the brain is the pia mater.

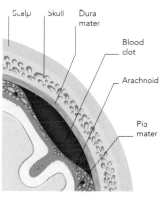

Scalp — **Skull** — **Dura mater** — **Blood clot** — **Arachnoid** — **Pia mater**

SUBDURAL HEMORRHAGE
If a vein in the dura mater is torn, bleeding (hemorrhage) leaks into the subdural space, between the dura mater and the arachnoid. The blood accumulates and forms a clot, which can compress surrounding brain tissue.

MIGRAINE

ABOUT 1 IN 10 PEOPLE HAVE MIGRAINES. THEY HAVE EPISODES OF SEVERE HEADACHE OFTEN ASSOCIATED WITH VISUAL DISTURBANCES, NAUSEA, AND VOMITING.

The underlying cause of a migraine is unknown but changes in the diameter of the blood vessels in the scalp and brain are known to occur. Current research indicates that migraines are linked to abnormal function of nerve pathways and a disturbance in the activity of brain chemicals. Triggers for a migraine attack include stress, missed meals, lack of sleep, and certain foods, such as cheese or chocolate. In many women, migraines are associated with menstruation.

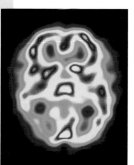

BLOOD FLOW CHANGES
A SPECT scan shows blood flow through tissues. The picture shows an area, bottom left, with a reduced blood flow during a migraine attack.

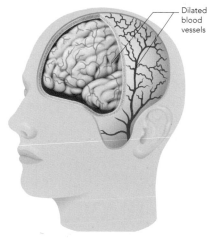

Dilated blood vessels

HEADACHE PHASE
During a migraine, severe, throbbing pain may affect half or all of the head as blood vessels in the scalp and brain widen (dilate). These vascular changes are thought to be secondary to nerve pathway abnormalities.

SYMPTOMS OF MIGRAINE

A migraine can occur with or without a preceding aura (altered perceptions and sensations). The main symptoms are sometimes preceded by what is known as a prodrome, which includes anxiety or mood changes; an altered sense of taste and smell; and either an excess or a lack of energy. People who have a migraine with aura experience a number of further symptoms before the migraine, including visual disturbances, such as blurred vision and bright flashes, and pins and needles, numbness, or a sensation of weakness on the face or on one side of the body. The main symptoms, common to both types of migraine, then develop. These symptoms include headache that is severe, throbbing, made worse by movement, and usually felt on one side of the head, over one eye, or around one temple; nausea or vomiting; and dislike of bright light or loud noises. A migraine may last for anything from a few hours to a few days.

BRAIN AND SPINAL CORD DISORDERS

STRUCTURAL, BIOCHEMICAL, OR ELECTRICAL CHANGES IN THE BRAIN AND SPINAL CORD, OR IN THE NERVES LEADING TO OR FROM THEM, MAY CAUSE DISORDERS THAT RESULT IN PARALYSIS, WEAKNESS, POOR COORDINATION, SEIZURES, OR LOSS OF SENSATION. ALTHOUGH INCREASED UNDERSTANDING OF BRAIN FUNCTION HAS GENERATED IMPROVEMENTS IN TREATMENT, SOME COMMON CONDITIONS ARE DIFFICULT TO REVERSE. ALL THAT CAN BE OFFERED TO THOSE AFFECTED IS RELIEF OF SYMPTOMS.

EPILEPSY

RECURRENT SEIZURES OR BRIEF EPISODES OF ALTERED CONSCIOUSNESS ARE CAUSED BY ABNORMAL ELECTRICAL ACTIVITY IN THE BRAIN.

Often the cause of epilepsy is unknown, but in some cases it may be due to a brain condition such as a tumor or an abscess, a head injury, stroke, or chemical imbalance. Epileptic seizures may be generalized or partial, depending on how much of the brain is affected by abnormal electrical activity. There are several types of seizure (fits). In a tonic-clonic seizure, the body stiffens before uncontrolled movements of the limbs and trunk begin, lasting for as long as several minutes. In absence (petit mal) seizures, which affect mainly children, the affected person may be briefly unaware of the outside world but does not lose consciousness. Some seizures are more limited, or partial. In a simple partial seizure, the affected person remains conscious. The head and eyes may turn to one side, and the hand, arm, and one side of the face may twitch or feel tingly. A complex partial seizure affects consciousness, and most often takes place in one of the two temporal lobes (see far right).

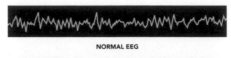

NORMAL EEG

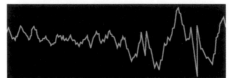

EEG DURING A SIMPLE PARTIAL SEIZURE

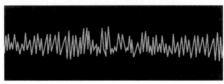

EEG DURING A GENERALIZED SEIZURE

ELECTRICAL ACTIVITY IN THE BRAIN
Normal electrical impulses in the brain show a regular pattern on an EEG recording. In a partial seizure, activity is irregular, and in a generalized seizure, the pattern is chaotic.

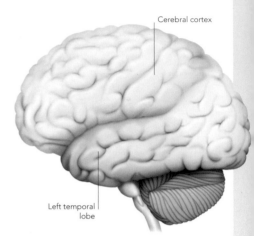

Cerebral cortex

Left temporal lobe

TEMPORAL LOBE EPILEPSY
In this condition, a seizure occurs in one of the temporal lobes. Before the seizure, the person may experience smells or sounds that others cannot detect. During the attack, there may be involuntary movements, especially chewing and sucking, and a partial loss of consciousness. The attack may also cause irrational feelings of fear or even joy.

PARKINSON'S DISEASE

DEGENERATION OF CELLS IN A PART OF THE BRAIN CALLED THE SUBSTANTIA NIGRA CAUSES SHAKING AND PROBLEMS WITH MOVEMENT THAT BECOME PROGRESSIVELY WORSE.

Normally, the cells in the substantia nigra produce a neurotransmitter called dopamine, which helps fine-tune muscle control. In Parkinson's disease, nerve cells producing dopamine degenerate, producing less and less dopamine and adversely affecting muscle control. Treatment consists of drugs that increase the amount of dopamine, act as a substitute for dopamine, or block enzymes that break dopamine down. The drugs help relieve symptoms, but none can reverse disease progress. In some cases, electrical deep-brain stimulation may be appropriate if the disease no longer responds to drugs. Stem-cell and gene therapies are also being tested.

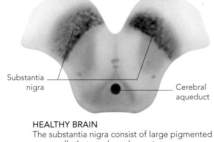

Substantia nigra

Cerebral aqueduct

HEALTHY BRAIN
The substantia nigra consist of large pigmented nerve cells that produce dopamine, a neurotransmitter (brain chemical) needed for control of movement.

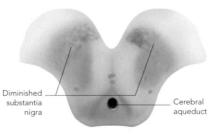

Diminished substantia nigra

Cerebral aqueduct

DISEASED BRAIN
In Parkinson's disease, the substantia nigra degenerate so that reduced amounts of dopamine are produced. As a result, problems with movement occur.

Location of substantia nigra

LOCATION OF SUBSTANTIA NIGRA
This color-enhanced MRI scan of a horizontal section through the head shows the location of the substantia nigra deep within the brain. The front of the head is at the top.

CREUTZFELDT–JAKOB DISEASE

BRAIN TISSUE IS PROGRESSIVELY DESTROYED BY AN INFECTIOUS AGENT CALLED A PRION, WHICH REPLICATES IN THE BRAIN, CAUSING BRAIN DAMAGE.

Creutzfeldt-Jakob disease (CJD) leads to a general decline in all areas of mental and physical ability and ultimately to death. Usually the source of infection is unknown, but a rare variant called vCJD is believed to be linked with eating contaminated meat from cattle with bovine spongiform encephalopathy (BSE). There is no cure for CJD but drugs can relieve some symptoms. However, the disorder is usually fatal within one year of symptoms first appearing.

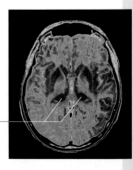

Areas of damaged brain tissue

BRAIN IN CJD
This color-enhanced MRI scan shows a brain affected by CJD. The two red areas are parts of the thalamus diseased with CJD. The thalamus relays incoming sensory information to the cerebral cortex (outer layer of brain).

MULTIPLE SCLEROSIS

PROGRESSIVE DAMAGE TO NERVES IN THE BRAIN AND SPINAL CORD CAUSES WEAKNESS AND PROBLEMS WITH SENSATION AND VISION.

Multiple sclerosis (MS) is due to immune system damage to the myelin sheaths that protect nerve fibers, so that impulses are no longer conducted normally along the nerves. The condition causes a wide range of symptoms that affect sensation, movement, body functions, and balance. For example, damage to the spinal cord nerves may affect balance. In some people, symptoms may last for days or weeks and then clear up for months or even years. In others, there is a gradual worsening of symptoms. MS cannot be cured but beta interferon drugs may help lengthen remission periods and shorten attacks. In addition, many symptoms can be relieved by drugs.

EARLY STAGE
In MS, the insulating myelin sheaths of nerve fibers are damaged. Macrophages, a type of scavenging cell, remove the damaged areas, exposing the fibers and impairing nerve conduction. In the early stages, there are only small patches of damage.

Macrophage

Myelin sheath   Nerve fiber

LATE STAGE
As MS progresses, the amount of damage to the myelin sheaths gradually increases, impairing nerve conduction further and sometimes completely preventing it. In addition, more and more nerve fibers are damaged. As damage worsens and becomes more widespread, symptoms become progressively worse.

Damaged myelin sheath

DEMENTIA

A DECLINE IN THE NUMBER OF BRAIN CELLS RESULTS IN SHRINKAGE OF BRAIN TISSUE AND CONSEQUENT DETERIORATION IN MENTAL ABILITY.

Dementia is a combination of memory loss, confusion, and general intellectual decline. The disorder mainly occurs in people over the age of 65, but young people are sometimes affected. In the early stages of dementia, a person is prone to becoming anxious or depressed due to awareness of the memory loss. As the dementia worsens, the person may become more dependent on others and may eventually need full-time care in a nursing home. Caregivers may also need support.

ALZHEIMER'S DISEASE

The most common form of dementia is Alzheimer's disease. Brain damage occurs due to the abnormal production of the protein amyloid, which builds up in the brain. No cure has been found, but drugs slow the progress of the disease in some people.

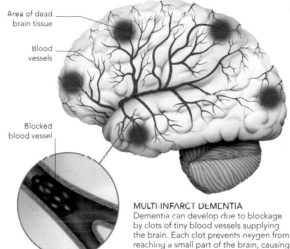

Area of dead brain tissue

Blood vessels

Blocked blood vessel

MULTI-INFARCT DEMENTIA
Dementia can develop due to blockage by clots of tiny blood vessels supplying the brain. Each clot prevents oxygen from reaching a small part of the brain, causing tissue death (infarct) in the affected part.

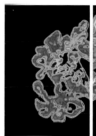

ALZHEIMER'S DISEASE   HEALTHY BRAIN

BRAIN IN ALZHEIMER'S DISEASE
This computer graphic shows a slice through the brain of a person affected by Alzheimer's disease compared to that of a healthy person. The Alzheimer's disease brain is considerably shrunken due to the degeneration and death of nerve cells. Apart from a decrease in brain volume, the surface of a brain affected by Alzheimer's disease may be more deeply folded.

SPINA BIFIDA

ABNORMAL DEVELOPMENT OF THE EMBRYO IN EARLY PREGNANCY RESULTS IN FAILURE OF THE SPINE TO CLOSE COMPLETELY.

There are three main forms of spina bifida: spina bifida occulta, meningocele, and myelomeningocele. Spina bifida occulta may require surgery to avoid serious neurological complications later in life. Meningoceles usually have a good prognosis after surgery. Myelomeningocele has effects that may include paralysis or weakness in the legs, and lack of bladder and bowel control. Children with this form are permanently disabled and may need extra support during their lives. Folic acid (folate) helps to prevent spina bifida, and some women take supplements when planning to conceive and during the first 12 weeks of pregnancy.

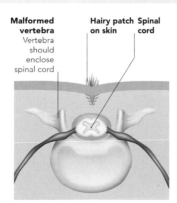

Malformed vertebra

Vertebra should enclose spinal cord

Hairy patch on skin

Spinal cord

SPINA BIFIDA OCCULTA
In the mildest form of spina bifida, one or more vertebrae are malformed. There is no damage to the spinal cord, and the external effects may be dimpling or a tuft of hair at the base of the spine, a birthmark, or a fatty lump (lipoma) on the skin.

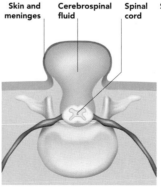

Skin and meninges

Cerebrospinal fluid

Spinal cord

MENINGOCELE
The protective coverings around the spinal cord (meninges) protrude through a malformed vertebra. They form a visible sac called a meningocele that is filled with cerebrospinal fluid. The spinal cord remains intact and the defect can be repaired.

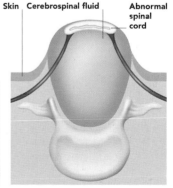

Skin

Cerebrospinal fluid

Abnormal spinal cord

MYELOMENINGOCELE
A portion of the spinal cord itself, contained within a sac of cerebrospinal fluid, protrudes through a defect in the skin, forming a large outward bulge. This is the most severe form of spina bifida.

BRAIN INFECTIONS, INJURIES, AND TUMORS

INJURIES AND DISORDERS AFFECTING THE BRAIN AND NERVOUS SYSTEM CAN RESULT IN A WIDE RANGE OF BOTH PHYSICAL AND MENTAL DISABILITIES. THE SKULL IS A CLOSED BOX, SO ANY SWELLING WITHIN THE BRAIN RAISES PRESSURE AND CAN COMPRESS STRUCTURES. THIS CAN CAUSE DAMAGE TO VITAL NERVE TISSUE, WITH SOME LOSS OF BODY CONTROL AND FUNCTION. SPINAL INJURIES CAN CAUSE HARM TO NERVE TRACTS, WHICH MAY RESULT IN SENSORY LOSS OR PARALYSIS.

BRAIN INFECTIONS

INFECTION OF BRAIN TISSUE OR OF ITS PROTECTIVE COVERINGS CAN BE CAUSED BY A WIDE VARIETY OF VIRUSES, BACTERIA, AND TROPICAL PARASITES.

Infection of the brain, or encephalitis, can be a rare complication of a viral infection, such as mumps or measles. The condition can occasionally be fatal, and babies and elderly people are most at risk.

MENINGITIS

Inflammation of the meninges (membranes) is termed meningitis and is usually caused by either viral or bacterial infection. Initially, meningitis may cause vague flulike symptoms. More pronounced symptoms may also develop, such as headache, fever, nausea and vomiting, stiff neck, and a dislike of bright light. In young children, the symptoms may be less obvious. They may include fever and other signs of being unwell such as crying, vomiting, diarrhea, reluctance to feed, and drowsiness. In the meningococcal (*Neisseria* bacterial) form, there is a distinctive reddish purple rash. If meningitis is suspected, immediate admission to the hospital is necessary. A lumbar puncture is performed to test for infection and then intravenous antibiotics are commenced. If bacterial meningitis is confirmed, treatment in intensive care is often required. It may take weeks or months to make a complete recovery from bacterial meningitis. Occasionally, problems such as memory impairment may persist. Bacterial meningitis can be fatal despite treatment. Recovery from viral meningitis usually takes up to two weeks; no specific treatment is needed.

BRAIN ABSCESS

An abscess is a collection of pus. Brain abscesses are rare and are usually caused by a bacterial infection that has spread to the brain from an infection in nearby tissues in the skull. Treatment consists of high doses of antibiotics and possibly corticosteroids to control swelling of the brain. Surgery may be needed to drain pus through a hole drilled in the skull. If given early treatment, many people with a brain abscess recover. However, some have persistent problems, such as seizures, slurred speech, or weakness of a limb.

Brain tissue
Infection of brain tissue, known as encephalitis, is often often mild, but can occasionally be life-threatening; features include headache, fever, and nausea

Dura mater

Arachnoid

Pia mater

Skull

SITES OF INFECTION
Infectious organisms can affect the brain itself, the three membranes that surround the brain, or both. Infections can reach the brain through the bloodstream but can also spread from a nearby infection (such as an ear infection) or through a skull wound. The brain can also be infected by prions (see Creutzfeldt-Jakob disease, p.114).

Meninges
Inflammation of the three membranes that cover and protect the brain, the meninges, is known as meningitis. The arachnoid and pia mater are affected more severely than the outermost layer, the dura mater.

LUMBAR PUNCTURE

A lumbar puncture, also known as a spinal tap, may be performed to look for evidence of meningitis. The procedure is carried out under local anesthesia and takes about 15 minutes. The patient lies on his or her side while a hollow needle is inserted into the spine and a sample of cerebrospinal fluid is withdrawn. The sample is analyzed in a laboratory for evidence of infection and to determine the type of infectious organism. After the procedure, the patient is advised to remain lying down and rest for an hour to prevent a severe headache.

Cerebrospinal fluid

Spinal cord

LOCATOR

Needle

Vertebra

THE PROCEDURE
A small, hollow needle is inserted between two vertebrae in the lower spine, usually the third and fourth lumbar vertebrae. The tip of the needle is pushed carefully into the space surrounding the spinal cord and a sample of cerebrospinal fluid is withdrawn into a syringe.

BRAIN ABSCESS
This MRI scan of the brain shows an abscess (blue area) due to a fungal infection in a person affected by AIDS. People with AIDS are at increased risk of developing a brain abscess.

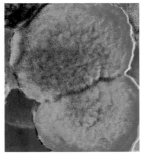

MENINGITIS BACTERIUM
This scanning electron microscope image shows the bacterium *Neisseria meningitidis*. Protection against some forms of meningitis is available through immunization programs.

TESTING A MENINGITIS RASH
In meningococcal meningitis, bacteria in the blood may cause dark red or purple spots that develop into blotches. The rash does not fade when pressed with a glass.

CEREBRAL PALSY

ABNORMALITIES OF MOVEMENT AND POSTURE ARE CAUSED BY DAMAGE TO THE IMMATURE BRAIN.

Cerebral palsy is not a specific disease but a group of disorders that result from damage to the developing brain either before or during birth or during a child's early years. Children with cerebral palsy lack normal control of limbs and posture and may also have difficulty in swallowing, speech

problems, and chronic constipation; however, intellect is often unaffected. Damage to the brain does not progress, but the disabilities it causes change as a child grows. Children with mild physical disabilities usually lead active, full, and long lives and often live independently as adults. Severely disabled children require long-term specialized support. Some, especially those with swallowing difficulties, who are more susceptible to serious chest infections, have a shorter life expectancy.

BRAIN TUMORS

CANCEROUS OR NONCANCEROUS GROWTHS CAN DEVELOP IN BRAIN TISSUE OR THE COVERINGS OF THE BRAIN.

Tumors that first develop in brain tissue or their membrane coverings are called primary and may be either cancerous or noncancerous. Secondary brain tumors (metastases) are much more common than primary tumors and are always cancerous, having developed from cells carried in the bloodstream from cancerous tumors in areas such as the breast or lungs. The general outlook for brain tumors depends on their location, size, and rate of growth. The outcome is usually better for noncancerous tumors that are slow-growing; many people with this type are cured by surgery. Most people with brain metastases do not live longer than 18 months.

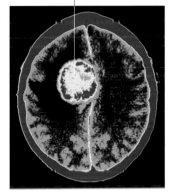
Large, noncancerous (benign) brain tumor

BRAIN TUMOR
This color-enhanced CT scan shows a meningioma, which is a noncancerous, slow-growing tumor that develops from the arachnoid, one of the membranes (meninges) that cover the brain. It may be possible to remove the tumor surgically.

HEAD INJURIES

DAMAGE TO THE SCALP, SKULL, OR BRAIN CAN VARY IN SEVERITY FROM MINOR TO LIFE-THREATENING.

Minor bumps to the head or injury to the scalp alone are not usually serious and have no long-term consequences. However, any injury that affects the brain itself is potentially extremely serious. Direct damage to the brain may occur if both the scalp and skull are penetrated.

Indirect damage occurs as a result of a blow to the head that does not damage the skull (see below). These injuries can still be serious, particularly if there is bleeding inside the skull (see Subdural hemorrhage, p.113). Anyone with a serious head injury is admitted to the hospital; treatment may consist of antibiotics, surgery, or both. About one in two people survive a serious injury, but some impairment may remain.

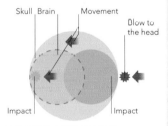

1 RAPIDLY MOVING PERSON
The skull and brain within it travel at the same speed. If movement suddenly stops, as in a fall, the brain may be injured.

2 SUDDEN DECELERATION
The brain may be injured as it smashes against the skull's hard inner surface, and sustain further injury as it rebounds.

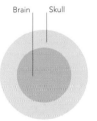

1 STATIONARY HEAD
Both the skull and the brain are motionless. If the head is suddenly struck, as in boxing, injury to the brain can occur.

2 SUDDEN ACCELERATION
The brain may be compressed against the inside of the skull and then bounces off the opposite inner surface of the skull.

PARALYSIS

LOSS OF MUSCLE FUNCTION DUE TO BRAIN OR MUSCLE DAMAGE MAY BE TEMPORARY OR PERMANENT.

Paralysis can affect anything from a small facial muscle to many of the major muscles of the body. Voluntary muscle activity as well as automatic functions, such as breathing, may be affected, and there may be loss of sensation. Paralysis results from damage to motor areas of the brain or nerve pathways of the spinal cord. It can also be caused by a muscle disorder. The underlying cause of paralysis is treated if possible. Physical therapy is used to prevent joints from becoming locked and can also help retrain muscles if paralysis is temporary. Paralyzed people confined to a wheelchair may need support to avoid complications of immobility.

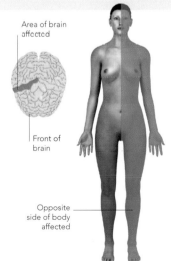
Area of brain affected
Front of brain
Opposite side of body affected

HEMIPLEGIA
Damage to the motor areas on one side of the brain can lead to paralysis of the opposite side of the body. This type of one-sided paralysis is known as hemiplegia.

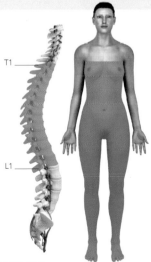
T1
L1

PARAPLEGIA
Damage to the middle or lower area of the spinal cord can cause paralysis of both legs and possibly part of the trunk. Bladder and bowel control may also be affected.

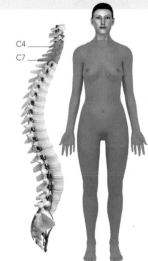
C4
C7

QUADRIPLEGIA
Damage to the spinal cord in the lower neck can cause paralysis of the whole trunk plus arms and legs. If damage is between C1 and C2 or higher, survival is unlikely.

EAR AND EYE DISORDERS

THE EARS AND EYES ARE VULNERABLE TO MANY DISORDERS, RANGING FROM DAMAGE CAUSED BY
AN EXCESS OF SOUND AND LIGHT TO THE NATURAL DEGENERATION OF THE SENSES DUE TO AGE.
HEARING AND VISION ARE MUTUALLY SUPPORTIVE, SO THAT WHEN ONE OF THEM SUFFERS REDUCED
PERFORMANCE, THE OTHER MAY BECOME MORE ACUTE AS A WAY OF COMPENSATING.

DEAFNESS

HEARING IMPAIRMENT MAY RESULT FROM DISEASE OR INJURY
OR MAY BE PRESENT FROM BIRTH; MOST PEOPLE EXPERIENCE
DETERIORATION OF HEARING WITH AGE.

There are two types of hearing loss: conductive and
sensorineural. Conductive hearing loss results from
impaired transmission of sound waves to the inner ear,
and is often temporary. In children, the most common
cause is serous otitis media (SOM; see below); in adults,
to blockage by earwax. Other causes include damage to
the eardrum or, rarely, stiffening of a bone in the middle
ear so that it cannot transmit sound. Sensorineural hearing
loss is most commonly due to deterioration of the cochlea
with age. It may also result from damage to the cochlea
by excessive noise or by Ménière's disease (see opposite).
Rarely, hearing loss is caused by an acoustic neuroma
or by certain drugs. Simple measures can be effective for
treatment of conductive deafness, such as syringing the
ear for removing earwax or antibiotics for infections. Surgery
may be required for SOM or otosclerosis. Sensorineural
deafness usually cannot be cured, but hearing aids can
help. A cochlear implant, in which electrodes are surgically
implanted in the cochlea, may help in profound deafness.

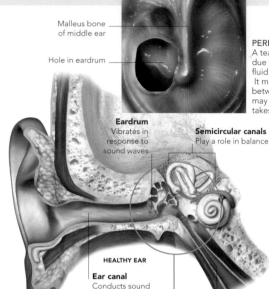

Malleus bone
of middle ear

Hole in eardrum

Eardrum
Vibrates in
response to
sound waves

Semicircular canals
Play a role in balance

HEALTHY EAR

Ear canal
Conducts sound
waves to eardrum

Bones of
middle ear
(ossicles)

**Eustachian
tube**

PERFORATED EARDRUM
A tear or hole in the eardrum may occur
due to pressure from buildup of pus or
fluid in the middle ear during an infection.
It may also occur due to unequal pressures
between the middle and outer ear, as
may happen when flying. Healing usually
takes about a month.

ACOUSTIC NEUROMA
This MRI scan shows a noncancerous,
tumor (red area) called an acoustic
neuroma. It grows around and
presses on the auditory nerve,
causing progressive hearing loss.

Auditory nerve

Cochlea

MECHANISM OF HEARING
Beyond the eardrum lies the middle ear.
It contains three tiny bones that transmit
vibrations to the cochlea in the inner ear,
which contains the receptor for hearing.
From here, messages travel along the
auditory nerve to the brain.

SEROUS OTITIS MEDIA (SOM)

In this condition, the middle ear becomes
filled with a thick, sticky, gluelike fluid,
which may impair hearing. If SOM does
not clear up with time, a small tube called
a tympanostomy tube may be surgically
inserted in the eardrum. The tube ventilates
the middle ear and allows fluid to drain
away. It usually falls out in 6–12 months,
and the hole in the eardrum closes.

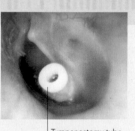

Tympanostomy tube

VERTIGO

A FALSE SENSATION OF MOVEMENT AND A SPINNING
SENSATION ARE OFTEN ASSOCIATED WITH NAUSEA
AND SOMETIMES WITH SEVERE VOMITING.

Vertigo may result from disturbance affecting the
organs of balance in the inner ear, the nerve that
connects the inner ear to the brain, or areas of the
brain concerned with balance. Rarely, it is a sign
of a serious underlying condition. Vertigo often
develops suddenly and may last from a few seconds
to several days, occurring either intermittently or
constantly. The condition can be very distressing and,
in severe cases, may make it impossible to walk or
stand. Usually, vertigo disappears on its own or
following treatment of the underlying disorder.

MOTION SICKNESS

NAUSEA AND OTHER SYMPTOMS OCCUR DURING
TRAVEL WHEN VISUAL INFORMATION CONFLICTS
WITH THAT ABOUT BALANCE.

The initial symptoms of motion sickness usually
include nausea; headache and dizziness; and
lethargy and fatigue. If the motion continues,
the initial symptoms get worse and others
develop, such as pale skin, excessive sweating,
hyperventilation, and vomiting. To avoid motion
sickness, it helps to look at the horizon or a distant
object in the direction of travel. Effective
medications for preventing or treating motion
sickness are available. To prevent symptoms,
medications should be taken before traveling.

TINNITUS

HEARING SOUNDS THAT ORIGINATE WITHIN THE EAR
ITSELF MAY INCLUDE RINGING, BUZZING, WHISTLING,
ROARING, OR HISSING NOISES.

Tinnitus may occur in brief episodes, but for
many people it is continuous. The condition is
often associated with hearing loss, and exposure
to loud noise increases the risk of developing it.
Tinnitus may occur for no apparent reason but
is often associated with certain ear disorders,
such as Ménière's disease (see opposite). Tinnitus
may improve if an underlying cause is found and
treated successfully. If it persists, a device called
a masker, worn in or behind the ear like a hearing
aid, may help by producing distracting sounds.

MÉNIÈRE'S DISEASE

SUDDEN EPISODES OF SEVERE DIZZINESS, COMBINED WITH HEARING LOSS, TINNITUS, AND A FEELING OF PRESSURE IN THE EARS, MAY OCCUR IN EPISODES OR CONTINUOUSLY.

The precise cause of Ménière's disease is unknown, but current theory suggests that it is a problem with the fluid balance regulating system in the inner ear (see right). Attacks occur suddenly and may last from a few minutes to several days, and the length of time between attacks may range from days to years. With repeated attacks, hearing often deteriorates progressively. There is no cure for the disease but medications may help relieve symptoms and reduce frequency of attacks. In cases of severe vertigo, surgical options include severing the vestibular nerve or destroying the labyrinthine structure in the ear.

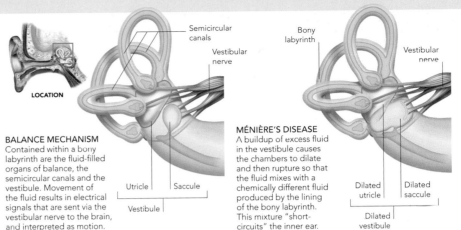

LOCATION

BALANCE MECHANISM
Contained within a bony labyrinth are the fluid-filled organs of balance, the semicircular canals and the vestibule. Movement of the fluid results in electrical signals that are sent via the vestibular nerve to the brain, and interpreted as motion.

MÉNIÈRE'S DISEASE
A buildup of excess fluid in the vestibule causes the chambers to dilate and then rupture so that the fluid mixes with a chemically different fluid produced by the lining of the bony labyrinth. This mixture "short-circuits" the inner ear.

Labels: Semicircular canals, Vestibular nerve, Bony labyrinth, Vestibular nerve, Utricle, Saccule, Vestibule, Dilated utricle, Dilated saccule, Dilated vestibule

---

VISUAL PROBLEMS

THE MOST COMMON DISORDERS OF VISION ARE PROBLEMS WITH FOCUSING, KNOWN AS REFRACTIVE ERRORS.

Problems with focusing for near vision (farsightedness or hypermetropia) or far vision (nearsightedness or myopia) result from the eyeball being either too short or too long so that light rays are focused either behind or in front of the retina rather than on it (see below). In astigmatism, vision is blurred because the cornea is irregularly curved, and the lens of the eye is unable to bring all light rays from an object into focus on the retina. Normal aging often brings on difficulty with near vision because the lens gradually loses its elasticity and cannot easily adjust its shape; this condition is called presbyopia. Refractive errors can usually be corrected by glasses or contact lenses. Surgery can also be used to correct some refractive errors (except presbyopia) permanently. The main techniques are laser-assisted in situ keratomileusis (LASIK) and photorefractive keratectomy (PRK). In LASIK, the middle layers of the cornea are reshaped by a laser, while in PRK, areas of the cornea's surface are shaved away by a laser to alter its shape.

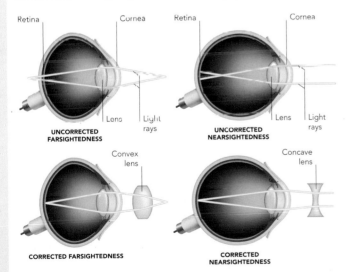

UNCORRECTED FARSIGHTEDNESS
Labels: Retina, Cornea, Lens, Light rays

UNCORRECTED NEARSIGHTEDNESS
Labels: Retina, Cornea, Lens, Light rays

CORRECTED FARSIGHTEDNESS
Label: Convex lens

CORRECTED NEARSIGHTEDNESS
Label: Concave lens

FARSIGHTEDNESS
In farsightedness, the eyeball is too short relative to the focusing power of the cornea and lens. Light rays are focused behind the retina, and the image is blurred. Convex lenses make the light rays converge (bend together) so that they are focused on the retina, correcting vision.

NEARSIGHTEDNESS
In nearsightedness, the eyeball is too long relative to the focusing power of the cornea and lens. Light rays are focused in front of the retina and the image is blurred. Concave lenses are required, which make the light rays diverge (bend apart) so that they are focused on the retina.

CAUSES OF BLINDNESS

SEVERE TO TOTAL LOSS OF VISION THAT CANNOT BE RECTIFIED BY CORRECTIVE LENSES HAS MANY DIFFERENT CAUSES.

In most developed countries, blindness occurs mainly later in life. Glaucoma is rare before age 40. Disease of the retina can result from diabetes mellitus or hypertension, both of which are more common in older people. People over 60 may be affected by macular degeneration, overlaying of the central retina by scar tissue. Cataracts are also common in elderly people.

Glaucoma

Glaucoma is abnormally high pressure inside the eye due to buildup of fluid. The pressure may permanently damage nerve fibers in the retina or the optic nerve. The condition may be acute, developing suddenly with severe pain, or chronic (see right), coming on slowly and painlessly over many years.

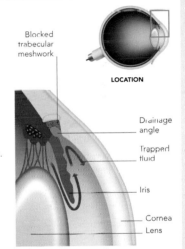

LOCATION

Labels: Blocked trabecular meshwork, Drainage angle, Trapped fluid, Iris, Cornea, Lens

CHRONIC GLAUCOMA
Fluid continually moves into and out of the eye to nourish its tissues and maintain the shape of the eye. Normally, the fluid flows out through the pupil and drains out of the trabecular meshwork within the drainage angle. In chronic glaucoma, the meshwork is blocked, and pressure builds up.

Cataracts

In a cataract, the normally transparent lens of the eye is cloudy as a result of changes in the protein fibers in the lens. The clouding affects the transmission and focusing of light entering the eye, reducing the clarity of vision. The most common cause of cataracts seems to be the general aging process, and most people over 75 have some cataract formation. Cataracts are sometimes present from birth, caused when a woman is infected with rubella during early pregnancy. Diabetes mellitus and exposure to radiation are other possible causes. Cataracts can be treated by surgically implanting an artificial replacement lens.

SEVERE CATARACT
This cataract, seen as a cloudy area behind the pupil, affects a large part of the lens. Such a cataract can cause total loss of clarity and detail of vision. However, the eye will still be able to detect light and shade.

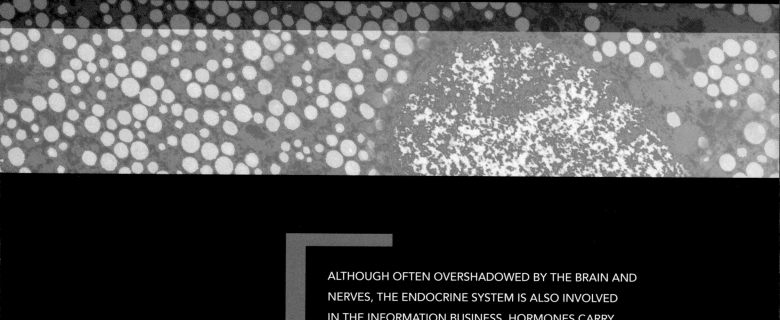

ALTHOUGH OFTEN OVERSHADOWED BY THE BRAIN AND
NERVES, THE ENDOCRINE SYSTEM IS ALSO INVOLVED
IN THE INFORMATION BUSINESS. HORMONES CARRY
ESSENTIAL MESSAGES THAT HAVE FAR-REACHING
EFFECTS. THEY CONTROL PROCESSES AT EVERY LEVEL,
FROM ENERGY UPTAKE BY A SINGLE CELL TO THE WHOLE

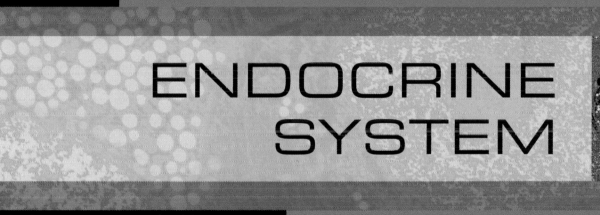

ENDOCRINE SYSTEM

ENDOCRINE ANATOMY

THE BODY'S CHEMICAL MESSENGERS (HORMONES) ARE MADE BY ENDOCRINE GLANDS. THESE GLANDS HAVE NO DUCTS; INSTEAD, THEY SECRETE THEIR HORMONES DIRECTLY INTO THE BLOOD, TO BE CARRIED IN THE BLOOD TO EVERY CELL IN THE BODY. HORMONES AFFECT CERTAIN TARGET TISSUES OR ORGANS AND REGULATE THEIR ACTIVITIES.

The endocrine system is composed of bodies of glandular tissue, such as the thyroid, as well as glands within certain organs, such as the testes, ovaries, and heart. The system uses hormones to control and coordinate body functions in much the same way as the nervous system uses tiny electrical signals. The two systems integrate in the brain and complement each other, but they tend to work at different speeds. Nerves respond within split seconds but their action soon fades; some hormones have longer-lasting effects and act over hours, weeks, and years. Hormones regulate processes such as the breakdown of chemical substances in metabolism, fluid balance and urine production, the body's growth and development, and sexual reproduction. Hormone output from a gland can be influenced by several factors, including levels of substances in the blood and input from the nervous system. Since hormones travel in the blood, each hormone reaches every body part. However, the specific molecular shape of each hormone slots only into receptors on its target tissues or organs.

Pineal gland (pineal body)
Pea-sized gland in middle of brain; makes melatonin, a hormone important in body rhythms such as the sleep–wake cycle; also influences sexual activity

Parathyroid glands
Four tiny glands embedded in back of thyroid gland; help regulate blood calcium levels

Hypothalamus
Cluster of nerve cells that serves as the main link between nerves and hormones; produces "releasing factors" (regulatory hormones) that travel to pituitary gland

Pituitary gland
Called the "master gland," controls many other endocrine glands

Thyroid gland
Controls rate of metabolism, including maintenance of body weight, rate of energy use, and heart rate; unlike other endocrine glands, it can store its hormones

Thymus gland
Produces three hormones involved in development of white blood cells called T cells, which function in the immune system

360-DEGREE VIEW

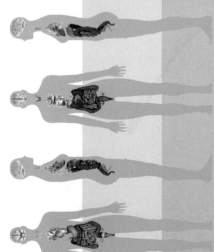

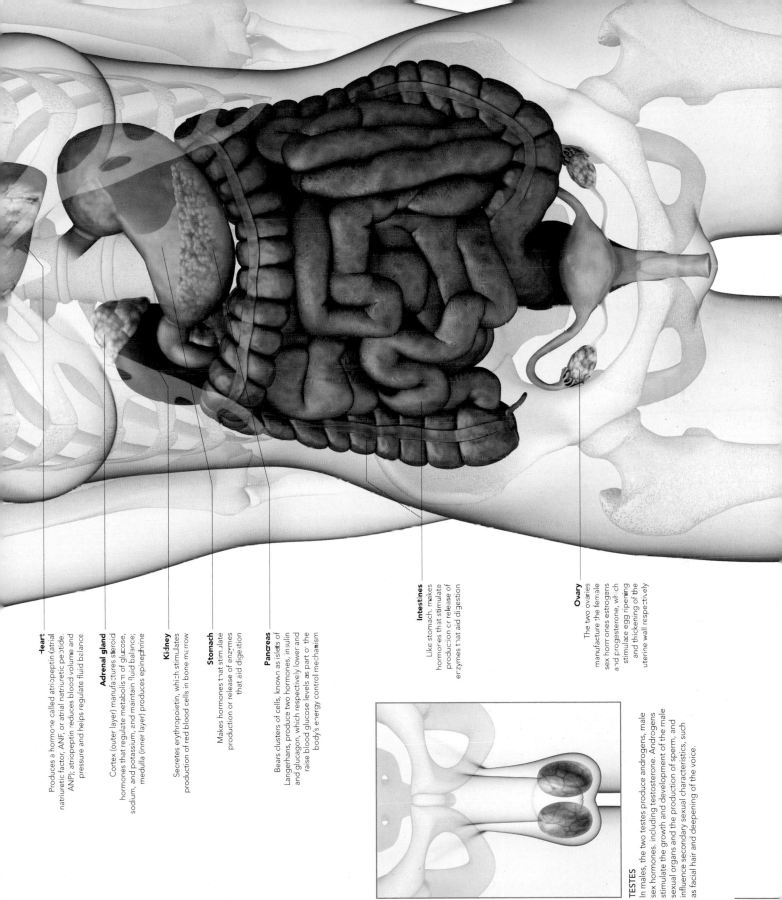

Heart
Produces a hormone called atriopeptin (atrial natriuretic factor, ANF, or atrial natriuretic peptide, ANP); atriopeptin reduces blood volume and pressure and helps regulate fluid balance

Adrenal gland
Cortex (outer layer) manufactures steroid hormones that regulate metabolism of glucose, sodium, and potassium, and maintain fluid balance; medulla (inner layer) produces epinephrine

Kidney
Secretes erythropoietin, which stimulates production of red blood cells in bone marrow

Stomach
Makes hormones that stimulate production or release of enzymes that aid digestion

Pancreas
Bears clusters of cells, known as islets of Langerhans, produce two hormones, insulin and glucagon, which respectively lower and raise blood glucose levels as part of the body's energy control mechanism

Intestines
Like stomach, makes hormones that stimulate production or release of enzymes that aid digestion

Ovary
The two ovaries manufacture the female sex hormones estrogens and progesterone, which stimulate egg ripening and thickening of the uterine wall respectively

TESTES
In males, the two testes produce androgens, male sex hormones, including testosterone. Androgens stimulate the growth and development of the male sexual organs and the production of sperm, and influence secondary sexual characteristics, such as facial hair and deepening of the voice.

HORMONE PRODUCERS

HORMONES CARRY THE CHEMICAL DATA THAT CONTROL
THE RATE AT WHICH GLANDS AND OTHER ORGANS WORK.
HORMONE-PRODUCING CELLS ARE FOUND ALL AROUND
THE BODY. MANY ARE GROUPED IN GLANDS THAT HAVE
SPECIALIZED FUNCTIONS SUCH AS THE THYROID.

MASTER GLAND

The pituitary, or hypophysis, is the most influential gland in the
endocrine system. It is actually two distinct glands in one. Its front,
or anterior lobe, also known as the adenohypophysis, makes up the
majority of its bulk. Behind is the posterior lobe, or neurohypophysis.
The anterior pituitary manufactures its eight major hormones on site
and releases them into the bloodstream. The posterior pituitary receives
its two main hormones from the hypothalamus, which lies above
it, where they are manufactured by neurosecretory cells.
Other neurosecretory cells make regulatory hormones,
which travel via capillaries to the anterior lobe and
control the release of hormones there.

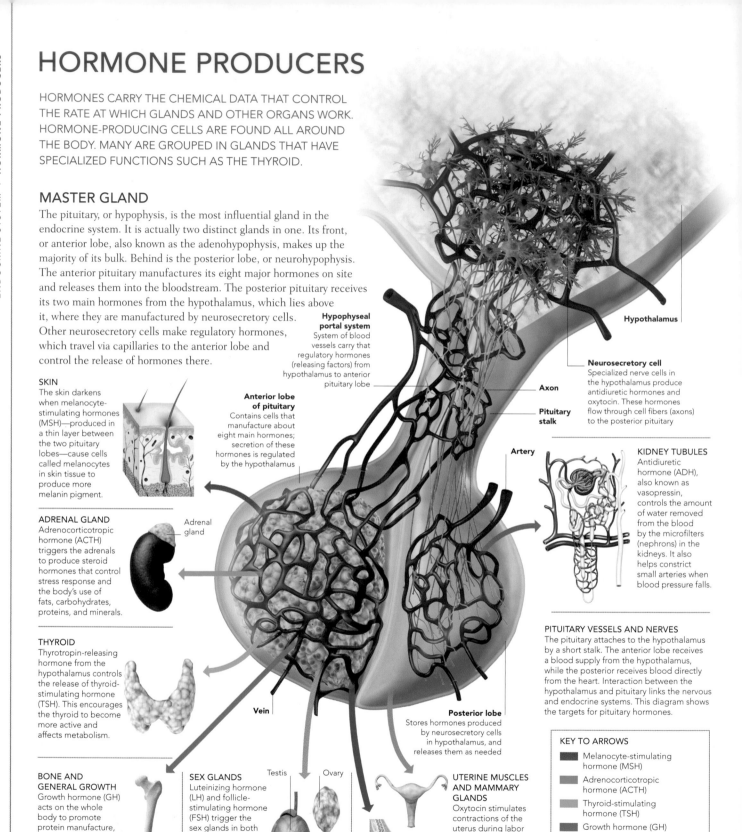

SKIN
The skin darkens
when melanocyte-
stimulating hormones
(MSH)—produced in
a thin layer between
the two pituitary
lobes—cause cells
called melanocytes
in skin tissue to
produce more
melanin pigment.

ADRENAL GLAND
Adrenocorticotropic
hormone (ACTH)
triggers the adrenals
to produce steroid
hormones that control
stress response and
the body's use of
fats, carbohydrates,
proteins, and minerals.

Adrenal
gland

THYROID
Thyrotropin-releasing
hormone from the
hypothalamus controls
the release of thyroid-
stimulating hormone
(TSH). This encourages
the thyroid to become
more active and
affects metabolism.

BONE AND
GENERAL GROWTH
Growth hormone (GH)
acts on the whole
body to promote
protein manufacture,
bone enlargement,
and building of new
tissues throughout
life, but is especially
important for growth
and development
in children.

SEX GLANDS
Luteinizing hormone
(LH) and follicle-
stimulating hormone
(FSH) trigger the
sex glands in both
males and females
to make their own
hormones, and
also to produce
ripe egg cells in
females and mature
sperm cells in males.

Testis

Ovary

**Hypophyseal
portal system**
System of blood
vessels carry that
regulatory hormones
(releasing factors) from
hypothalamus to anterior
pituitary lobe

**Anterior lobe
of pituitary**
Contains cells that
manufacture about
eight main hormones;
secretion of these
hormones is regulated
by the hypothalamus

Vein

Posterior lobe
Stores hormones produced
by neurosecretory cells
in hypothalamus, and
releases them as needed

Hypothalamus

Neurosecretory cell
Specialized nerve cells in
the hypothalamus produce
antidiuretic hormones and
oxytocin. These hormones
flow through cell fibers (axons)
to the posterior pituitary

Axon

**Pituitary
stalk**

Artery

KIDNEY TUBULES
Antidiuretic
hormone (ADH),
also known as
vasopressin,
controls the amount
of water removed
from the blood
by the microfilters
(nephrons) in the
kidneys. It also
helps constrict
small arteries when
blood pressure falls.

PITUITARY VESSELS AND NERVES
The pituitary attaches to the hypothalamus
by a short stalk. The anterior lobe receives
a blood supply from the hypothalamus,
while the posterior receives blood directly
from the heart. Interaction between the
hypothalamus and pituitary links the nervous
and endocrine systems. This diagram shows
the targets for pituitary hormones.

UTERINE MUSCLES
AND MAMMARY
GLANDS
Oxytocin stimulates
contractions of the
uterus during labor
and—together with
prolactin from the
anterior pituitary—
triggers the release
of milk from the
mammary glands in the
breasts for the baby.

KEY TO ARROWS
- Melanocyte-stimulating
  hormone (MSH)
- Adrenocorticotropic
  hormone (ACTH)
- Thyroid-stimulating
  hormone (TSH)
- Growth hormone (GH)
- Luteinizing hormone (LH) and
  follicle-stimulating hormone (FSH)
- Oxytocin
- Antidiuretic hormone (ADH)
- Prolactin

PANCREAS

The pancreas is a dual-purpose gland. It produces digestive enzymes in cells called acini, and also has an endocrine function. Within the acinar tissues are about one million cell clusters called islets of Langerhans. These contain cells that produce hormones involved in controlling glucose (blood sugar), the body's main energy source. Beta cells make the hormone insulin, which promotes glucose uptake by cells and speeds conversion of glucose into glycogen for storage in the liver. In this way, insulin lowers blood glucose levels. Another hormone, glucagon, is produced by alpha cells and has opposing actions, raising blood glucose levels. Delta cells make somatostatin, which regulates the alpha and beta cells.

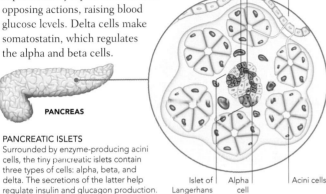

PANCREATIC ISLETS
Surrounded by enzyme-producing acini cells, the tiny pancreatic islets contain three types of cells: alpha, beta, and delta. The secretions of the latter help regulate insulin and glucagon production.

THYROID AND PARATHYROID GLANDS

The thyroid is located in the front of the neck, with the four tiny parathyroid glands embedded in its rearmost "wings." The hormones it produces have wide-ranging effects on body chemistry, including the maintenance of body weight, the rate of energy use from blood glucose, and heart rate. Unlike other glands, it can store the hormones it produces. The parathyroids produce parathormone (PTH), which increases the levels of calcium in the blood. PTH acts on bones to release their stored calcium, on the intestines to increase calcium absorption, and on the kidneys to prevent calcium loss.

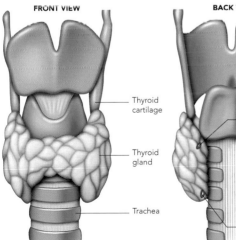

THYROID
The bow-tie-like thyroid straddles the upper windpipe (trachea). Its ball-shaped groups of follicular cells produce thyroxine (T_4) and triiodothyronine (T_3), which regulate the body's metabolism.

PARATHYROIDS
The small parathyroid glands are set into the rear corners of the thyroid's lobes, at the back of the trachea. There are usually four, but their number and exact locations vary.

ADRENAL GLANDS

The inner medulla and outer cortex of the adrenal gland each secretes different hormones. The cortical hormones are steroids (see p.126) and include glucocorticoids, such as cortisol, which affect metabolism; mineralocorticoids, such as aldosterone, which influence salt and mineral balance; and gonadocorticoids, which act on the ovaries and testes. The inner medulla functions as a separate gland. Its nerve fibers link to the sympathetic nervous system, and it makes the fight-or-flight hormones, such as epinephrine.

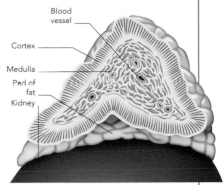

ADRENAL ANATOMY
Each adrenal gland is shaped like a low cone or pyramid located on top of the kidney, cushioned by a pad of fat. The glands consist of two parts: the cortex, which forms about nine-tenths of the gland's bulk and has three layers, and the medulla, containing nerve fibers and blood vessels.

ADRENAL HORMONES

Adrenal cortical hormones have life-sustaining actions in helping coordinate and maintain internal conditions (homeostasis), while the medullary hormones are involved in the body's response to stress.

ALDOSTERONE	Secreted by the outer zone of the cortex, this hormone inhibits the level of sodium excreted in urine and promotes potassium loss, maintaining blood volume and pressure.
CORTISOL	The middle layer of the cortex produces this hormone, which controls how the body uses fat, protein, carbohydrates, and minerals, and helps reduce inflammation.
GONADOCORTICOIDS	Produced by the inner cortex layer, these sex hormones affect sperm production in males and the distribution of body hair in females, working in conjunction with ACTH.
EPINEPHRINE AND NOREPINEPHRINE	These medullary hormones work with the sympathetic nervous system to raise heart rate and blood pressure, trigger carbohydrate metabolism, and prime the body for action.

SEX GLANDS AND HORMONES

The main sex glands are the ovaries in females and testes in males. The sex hormones they produce stimulate the production of eggs and sperm respectively, and influence the early development of the embryo into a boy or girl. After birth, the circulating levels remain low until puberty. Then, in males, the testes increase their output of androgens (male sex hormones), such as testosterone. In females, the ovaries produce more estrogens and progesterone.

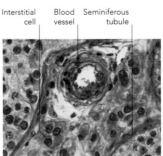

TESTOSTERONE PRODUCERS
Interstitial cells in the testis shown in pink in this microscopic image, secrete testosterone. They are found in the connective tissue between seminiferous tubules.

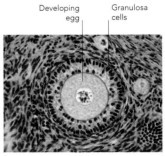

OESTROGEN PRODUCERS
This microscope picture shows a developing egg surrounded by a ring of granulosa cells. These secrete estrogens.

HORMONAL ACTION

HORMONES WORK BY ALTERING THE CHEMISTRY OF THEIR TARGET CELLS. A HORMONE DOES NOT INITIATE A CELL'S BIOCHEMICAL REACTIONS, BUT ADJUSTS THE RATE AT WHICH THEY OCCUR. DIFFERENT HORMONES ARE RELEASED ACCORDING TO DIFFERENT TRIGGER MECHANISMS.

HORMONAL TRIGGERS

The stimuli that cause an endocrine gland to release its hormone vary. In some cases, the gland responds directly to the level of a certain substance in the blood, using a feedback loop (see below). In other cases, there is an intermediate mechanism in the feedback system, such as the hypothalamus–pituitary complex. The adrenal gland has a dual trigger. Its outer part, the cortex, is controlled by circulating adrenocorticotropic hormone (ACTH), released by the pituitary on cue from the hypothalamus. The inner medulla is stimulated by nerve impulses direct from the hypothalamus.

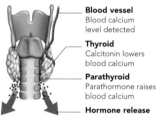

Blood vessel
Blood calcium level detected

Thyroid
Calcitonin lowers blood calcium

Parathyroid
Parathormone raises blood calcium

Hormone release

BLOOD LEVEL STIMULATION
Falling calcium levels in the blood inhibit the release of calcitonin from the thyroid and stimulate the parathyroids to release parathormone; calcium levels are raised.

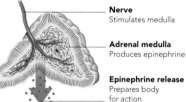

Nerve
Stimulates medulla

Adrenal medulla
Produces epinephrine

Epinephrine release
Prepares body for action

DIRECT INNERVATION
The adrenal medulla receives nerve fibers (is innervated) from the hypothalamus via the sympathetic nervous system.

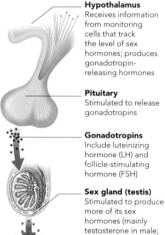

Hypothalamus
Receives information from monitoring cells that track the level of sex hormones; produces gonadotropin-releasing hormones

Pituitary
Stimulated to release gonadotropins

Gonadotropins
Include luteinizing hormone (LH) and follicle-stimulating hormone (FSH)

Sex gland (testis)
Stimulated to produce more of its sex hormones (mainly testosterone in male; ovaries produce mostly estrogen in female)

HYPOTHALMIC–PITUITARY CONTROL
As sex hormone levels fall, gonadotropin-releasing hormones (GnRH) are sent from the hypothalamus to the pituitary, which in turn releases gonadotropic hormones. These increase sex gland activity.

HORMONE CONTROL MECHANISMS

Chemically, there are two main types of hormones, those consisting of protein and amine molecules, and those made of steroid molecules. These two groups work in a similar way overall. They act biochemically to alter the rate of production of a certain substance, usually by either increasing or decreasing production of the enzyme that speeds up the manufacture of that substance. At the cellular level, the two hormone groups have different mechanisms of action. Protein and amine hormones have their effect on fixed receptor sites at the cell's surface, while steroid hormones act on mobile receptors inside the cell.

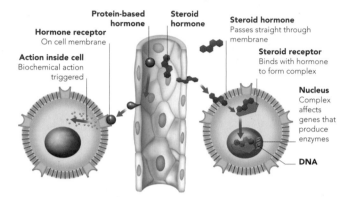

Protein-based hormone

Hormone receptor
On cell membrane

Action inside cell
Biochemical action triggered

Steroid hormone

Steroid hormone
Passes straight through membrane

Steroid receptor
Binds with hormone to form complex

Nucleus
Complex affects genes that produce enzymes

DNA

PROTEIN-BASED HORMONES
Most protein-derived hormones are water soluble and cannot pass through the fatty cell membrane. They bind to receptor sites on the membrane, activating an enzyme that controls the cell's biochemical action.

STEROID-BASED HORMONES
Steroids are fat soluble and pass through the cell membrane into the cytoplasm. The hormone binds to a receptor and enters the nucleus. It triggers genes to produce enzymes that prompt biochemical action.

FEEDBACK MECHANISMS

The level of hormones within the blood is controlled by feedback mechanisms, or loops. These mechanisms work much as a thermostat controls a central heating system. The amount of a particular hormone circulating or being secreted into the bloodstream is detected and passed on to a control unit. For many hormones, the control unit is the hypothalamus–pituitary complex in the brain, as in the case of the thyroid hormones below. If the level of a particular hormone increases beyond a normal level, the control unit responds by reducing hormone production. Likewise, if the hormone level decreases, the control unit acts again, stimulating production to raise the amount of hormone to the required level.

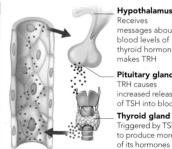

Hypothalamus
Receives messages about blood levels of thyroid hormones, makes TRH

Pituitary gland
TRH causes increased release of TSH into blood

Thyroid gland
Triggered by TSH to produce more of its hormones

INCREASING LEVELS
For low levels of thyroid hormones, the hypothalamus makes thyrotropin-releasing hormone (TRH). This triggers the pituitary to secrete thyroid-stimulating hormone (TSH).

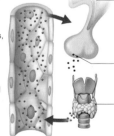

Hypothalamus
Detects rising blood levels of thyroid hormones; produces less TRH for pituitary

Pituitary gland
Releases less TSH into blood

Thyroid gland
Reduced hormone production

DECREASING LEVELS
High thyroid hormone levels prompt negative feedback, so the hypothalamus produces less TRH. This reduces TSH levels and the thyroid produces fewer hormones.

PINEAL GLAND TRIGGER

The pea-size pineal gland is near the center of the brain, just behind the thalamus. It is closely involved in the body's sleep–wake cycle and diurnal (24-hour) rhythms, and is triggered by darkness. Pineal activity is inhibited by light, which is detected by the eye's retina, and sent by a series of nerve connections to the gland (see pp.26–27). Darkness removes this inhibition, and the pineal releases its sleep hormone, melatonin.

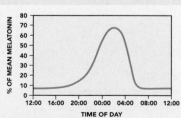

TIME OF DAY

MELATONIN LEVELS
The amount of circulating melatonin rises at night or in dark conditions, creating a daily rhythm of rising and falling hormone levels.

HORMONAL DISORDERS

SOME HORMONES HAVE WIDESPREAD EFFECTS, SO HORMONAL DISORDERS CAUSE EXTENSIVE PROBLEMS AROUND THE BODY. "HYPER-" IMPLIES AN EXCESS OF HORMONE, MAKING ITS TARGETS TOO ACTIVE. "HYPO-" IMPLIES REDUCED HORMONE ACTION. DISORDERS ARE OFTEN DUE TO DAMAGE TO A GLAND, WHICH MAY BE THE RESULT OF AN AUTOIMMUNE CONDITION OR DAMAGE TO THE BLOOD SUPPLY.

PITUITARY TUMORS

THE PITUITARY CONTROLS MANY OTHER ENDOCRINE GLANDS, AS WELL AS MAKING ITS OWN HORMONES, SO ITS DISORDERS CAN HAVE WIDE-RANGING EFFECTS.

The central role of the pituitary in the endocrine system is reflected in the problems caused by a pituitary tumor, which may grow in any part of the gland; those in the anterior lobe are more likely to be benign (noncancerous). One result may be excess growth hormone, which causes enlargement of certain bones, such as those in the face, hands and feet, and of some tissues, such as the tongue, as well as the appearance of coarse body hair and deepening of the voice. This condition is known as acromegaly. Some tumors cause excessive prolactin secretion or overstimulate the adrenal cortex (see right, far right).

PROLACTINOMAS

About 4 in 10 pituitary tumors are prolactinomas—slow-growing, noncancerous tumors that cause the anterior lobe to secrete excessive prolactin. Normally this hormone promotes breast development and milk production in pregnancy. Symptoms include irregular menstrual periods and lowered fertility in women; breast enlargement and erectile dysfunction in men; and fluid leakage from the nipples along with reduced sexual desire. In most cases, drug medication helps shrink the tumor and reduce prolactin output; otherwise, surgery or radiation therapy may be necessary

PITUITARY TUMOR
An enlarging tumor may press on the optic nerves that pass just above it, causing headache and visual disturbances, such as losing part of the visual field.

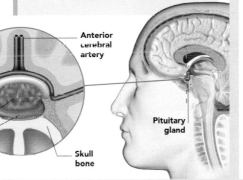

Anterior cerebral artery

Compressed optic nerve

Pituitary tumor
Tumor presses on optic nerve

Pituitary gland
May fail to function normally

Pituitary gland

Skull bone

CUSHING'S SYNDROME

THIS CHARACTERISTIC GROUP OF SYMPTOMS IS DUE TO OVERACTIVITY OF THE CORTICOSTEROID HORMONES PRODUCED BY THE ADRENALS.

Corticosteroids help regulate metabolic rate, salt and water balance, and blood pressure. The effects of this syndrome are linked to disrupted regulation: rounded, reddened face, weight gain, irregular or absent menstrual periods, more noticeable body hair, muscle weakness, and depression. The main cause is long-term oral corticosteroid drug treatment, which enhances the effects of the adrenals' natural corticosteroids. Less common is an adrenal tumor elevating corticosteroid production, or a pituitary tumor overstimulating the adrenal.

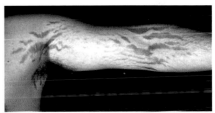

STRETCH MARKS
Characteristic of Cushing syndrome are easily bruised skin and reddish or red-purple stretch marks, especially on the abdomen and also the thighs and arms.

HYPERTHYROIDISM

THYROID HORMONES AFFECT THE RATE OF METABOLISM AND ENERGY USE. AN EXCESS MAKES THE BODY "SPEED UP."

Three-quarters of overstimulated thyroid cases are due to Grave's disease, an autoimmune disorder in which antibodies stimulate the thyroid, causing excessive hormone production. It is one of the most common hormonal disorders, especially in women aged 20-50. A less common cause is small lumps (nodules) in the gland. Raised hormone levels push up metabolic rate, with weight loss due to increased energy usage, rapid irregular heartbeat, trembling, sweating, anxiety, insomnia, weakness, and more bowel movements; the enlarged thyroid may show as a swelling in the neck (goiter). Drugs usually control the condition.

GRAVE'S DISEASE
Hyperthyroidism due to Grave's disease can cause bulging eyes, giving a staring appearance and possibly blurred vision.

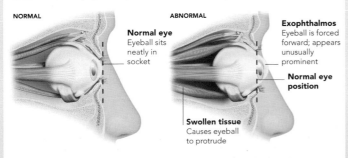

NORMAL

Normal eye
Eyeball sits neatly in socket

ABNORMAL

Exophthalmos
Eyeball is forced forward; appears unusually prominent

Normal eye position

Swollen tissue
Causes eyeball to protrude

HYPOTHRYOIDISM

THIS CONDITION INVOLVES DECREASED OUTPUT OF THYROID HORMONES, SO THE BODY GRADUALLY SLOWS DOWN.

In hypothyroidism, the thyroid hormones, tri-iodothyronine and thyroxine, are underproduced. Since these govern the speed of many metabolic processes, their lack leads to a slowing down of body functions. Symptoms include fatigue, weight gain, slow bowel activity and constipation, swollen face, puffy eyes, thickened skin, thinned hair, hoarse voice, and inability to cope with cold. The usual cause is inflammation of the thyroid gland due to an autoimmune condition called Hashimoto's thyroiditis, in which antibodies mistakenly damage the gland. Hashimoto's thyroiditis runs in families and is more common in older women. The thyroid gland may swell considerably as a lump, or goiter, in the neck. A less common cause, which affects women in underdeveloped regions, is a lack of the mineral iodine (which is needed to make the thyroid hormones) in the diet. A rarer possibility is damage to the pituitary gland by a tumor. Treatment of all cases of hypothyroidism is with synthetic thyroid hormones.

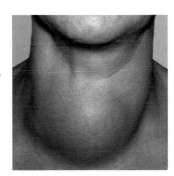

GOITRE
A swollen thyroid (goiter) may be due to thyroiditis, hyper- or hypothyroidism, thyroid nodules, or cancer of the thyroid gland.

DIABETES MELLITUS

The main energy source for body cells is glucose, which they absorb from the blood with the help of a hormone called insulin. In diabetes mellitus, this process does not work properly, so cells cannot take up enough glucose and too much remains in the blood. There are two main types of diabetes mellitus, type 1 and type 2; a third form, gestational diabetes, develops in pregnancy.

BLOOD SUGAR REGULATION

THE BODY NEEDS TO REGULATE BLOOD GLUCOSE LEVELS AT ALL TIMES SO THAT CELLS RECEIVE ENOUGH ENERGY TO MEET THEIR NEEDS EXACTLY.

During digestion, the body breaks down nutrients from food and drinks to make substances that cells use to fuel and repair themselves. The main source of fuel is glucose (blood sugar), which is carried in the bloodstream to cells. Any excess is stored in the liver, muscle cells, and fat cells, to be released later if needed. The body has to adjust blood glucose levels to keep them steady. If the levels fall too low, the cells will not have enough energy, but excess blood glucose can cause autoimmune disease and pancreatitis. Regulation is carried out by two groups of hormone-secreting cells in the pancreas, in structures called the islets of Langerhans. Beta cells secrete insulin, which corrects high levels, and alpha cells secrete glucagon, which increases blood sugar if levels are low.

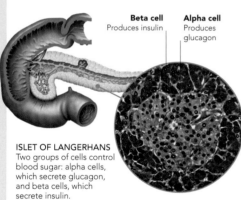

Beta cell
Produces insulin

Alpha cell
Produces glucagon

ISLET OF LANGERHANS
Two groups of cells control blood sugar: alpha cells, which secrete glucagon, and beta cells, which secrete insulin.

HIGH BLOOD SUGAR
After every meal, blood glucose levels increase. Excess glucose stimulates the beta cells in the pancreas to release insulin, which enables the surplus glucose to be stored in the form of glycogen and fatty acids. As a result, blood levels return to normal.

GLUCOSE IN BLOODSTREAM

BETA CELLS

PANCREAS

Insulin released
Beta cells in pancreas release insulin, stimulating body to store glucose.

Glucose stored in liver
Liver converts glucose to glycogen for storage, ready for quick release when needed.

Glucose stored in muscle
Muscle cells are stimulated to take up glucose and convert it to glycogen for storage.

Glucose stored as fatty acids
If glycogen stores are full, excess glucose is converted into fatty acids for storage.

Blood sugar stabilized

LOW BLOOD SUGAR
If the body is not fed for several hours, blood glucose levels drop. This decrease stimulates alpha cells in the pancreas to secrete glucagon, which enables the body to release glucose from its stores. Blood glucose levels then return to normal.

GLUCOSE IN BLOODSTREAM

ALPHA CELLS

PANCREAS

Glucagon released
Alpha cells in pancreas release glucagon, causing release of stored glucose.

Liver releases glucose
Liver breaks down its stored glycogen to form glucose, which is then released into bloodstream.

Blood sugar stabilized

TYPE 1 DIABETES

THIS FORM OF DIABETES OCCURS WHEN BETA CELLS IN THE PANCREAS ARE DESTROYED, SO THE PANCREAS PRODUCES TOO LITTLE INSULIN OR NONE AT ALL.

Type 1 diabetes mellitus is an autoimmune disorder (see pp.186–87). It occurs when the immune system misidentifies the beta cells as foreign and destroys them. The cause is unknown, but the disease may be triggered by a viral infection or by inflammation in the pancreas. It usually develops rapidly in childhood or adolescence. Symptoms include thirst, dry mouth, hunger, frequent urination, fatigue, blurred vision, and weight loss. If untreated, the disorder can cause ketoacidosis, in which toxic chemicals called ketones build up in the blood, leading to a coma. There can also be long-term complications (see Type 2 diabetes, opposite). Treatment involves frequent insulin injections or an insulin pump combined with bolus doses. There is no cure; a kidney and pancreas transplant can relieve the disorder, but drugs are needed for life so the body will not reject the organs.

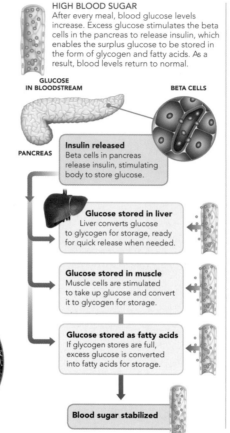

Beta cells
Insulin-producing cells

Insulin
Insulin secreted into capillaries

NORMAL BETA-CELL FUNCTION
As food and drinks are digested, the presence of glucose, amino acids, and fatty acids in the intestine stimulates beta cells to release insulin into the bloodstream via tiny blood vessels called capillaries, which run through the islets of Langerhans.

Damaged beta cells
Insulin-producing cells destroyed

Capillary
No insulin is secreted into capillaries

DAMAGED BETA CELLS
If the beta cells are damaged, they cannot release insulin. As a result, body cells cannot take up glucose, and blood glucose levels rise too high. The lack of insulin allows the alpha cells to produce more glucagon, which raises blood glucose levels still further.

INSULIN THERAPY

Injections of insulin are given to make up for the insulin that the body fails to produce. Treatment follows natural patterns of insulin production. Short-acting insulins are given before meals, creating high levels to cope with the glucose that comes into the body. Longer-acting insulins are taken once or twice a day to maintain a constant background level of the hormone.

TYPE 2 DIABETES

THE MOST COMMON FORM OF DIABETES MELLITUS, TYPE 2 DIABETES DEVELOPS WHEN BODY CELLS BECOME RESISTANT TO THE EFFECTS OF INSULIN.

In type 2 diabetes, the pancreas secretes insulin, but the body cells are unable to respond to it. The causes are complex, including genetic predisposition and lifestyle factors. This form of diabetes is often associated with obesity and is a growing problem in affluent societies. The disorder develops slowly. There may be initial symptoms such as thirst, fatigue, and frequent urination, but in some cases, the diabetes goes unnoticed for several years. As a result, complications may arise. Persistent high glucose levels can cause damage to small blood vessels around the body. People with type 2 diabetes are also more prone to high cholesterol levels, atherosclerosis (see p.140), and high blood pressure. The condition can be controlled with a healthy diet, regular exercise, and daily monitoring of blood glucose. However, in some cases, drugs are needed to boost insulin production or to help the cells absorb glucose.

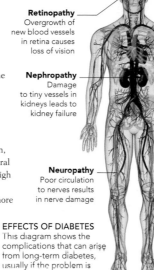

Retinopathy
Overgrowth of new blood vessels in retina causes loss of vision

Nephropathy
Damage to tiny vessels in kidneys leads to kidney failure

Neuropathy
Poor circulation to nerves results in nerve damage

Coronary artery disease
Caused by atherosclerosis; is more likely, and develops at an earlier age, in people with diabetes

Small vessel disease
Vessel walls thicken and restrict oxygen supply to tissues

Foot problems
Poor circulation and loss of feeling leads to skin ulcers and gangrene

EFFECTS OF DIABETES
This diagram shows the complications that can arise from long-term diabetes, usually if the problem is poorly controlled.

GESTATIONAL DIABETES

This form of diabetes mellitus develops in about 1 in 20 women during pregnancy; it is more common in overweight women, those over the age of 30, and those with a family history of diabetes. Some of the hormones that the placenta produces during pregnancy have an anti-insulin effect. If the body cannot produce enough insulin to counter this effect, blood glucose levels rise too high and gestational diabetes develops. The symptoms include fatigue, thirst, increased urination, and possibly yeast infections or bladder infections. If the diabetes is not controlled, the fetus may grow too large and birth may be difficult. The condition is diagnosed with blood and urine tests to detect glucose. It can be controlled with a low-sugar diet and usually disappears after the birth. However, some women who have had gestational diabetes go on to develop type 2 diabetes a few years later.

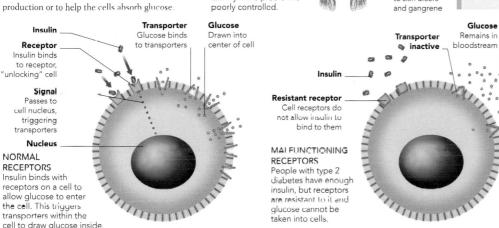

Insulin

Receptor
Insulin binds to receptor, "unlocking" cell

Signal
Passes to cell nucleus, triggering transporters

Nucleus

Transporter
Glucose binds to transporters

Glucose
Drawn into center of cell

NORMAL RECEPTORS
Insulin binds with receptors on a cell to allow glucose to enter the cell. This triggers transporters within the cell to draw glucose inside.

Insulin

Resistant receptor
Cell receptors do not allow insulin to bind to them

Transporter inactive

Glucose
Remains in bloodstream

MALFUNCTIONING RECEPTORS
People with type 2 diabetes have enough insulin, but receptors are resistant to it and glucose cannot be taken into cells.

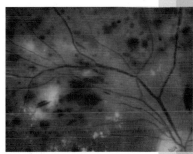

DIABETIC RETINOPATHY
This image shows damaged blood vessels of the retina (diabetic retinopathy). The blotches are aneurysms and hemorrhages, where blood vessels have leaked.

OBESITY

An excessive level of body fat is called obesity. The condition is usually caused by overeating and lack of exercise. Obesity is a major problem in rich countries but, is increasingly common across the world. Obese people are at higher risk of developing serious disorders. Major threats include coronary artery disease (see p.140) and stroke (see p.112), due to the buildup of fatty deposits in arteries, and type 2 diabetes (see above); these are particular risks for people with excess fat around the abdomen. Other problems include certain cancers, such as breast cancer and colon cancer. Excess weight puts strain on muscles and joints, and fat around the face and neck can interfere with breathing during sleep.

One measure of obesity is the body mass index (see right), by which people are defined as obese if they are more than 20 percent above the maximum healthy weight for their body size. However, this index does not take account of skeletal size and muscle mass; some muscular people may actually be classed as "overweight." A more helpful guide is waist measurement. A figure above 40in (102cm) for men or 35in (89cm) for women may indicate excess abdominal fat.

EXCESS BODY FAT
Body fat may be laid down just under the skin (subcutaneous fat) and can also collect in the abdominal cavity (central or visceral fat). Fat distribution partly depends on sex; males are more likely to have excess fat concentrated around and inside the abdomen, whereas females tend to accumulate surplus fat on the hips, thighs, and buttocks.

Central or visceral fat
Collects in abdominal cavity and around organs

Spinal column

Back muscles

Subcutaneous fat
Accumulates in layer beneath skin

SECTION THROUGH BODY

Overweight
BMI over 25

Underweight
BMI less than 18.5

Ideal range
BMI between 18.5 and 25

WEIGHT IN POUNDS: 0, 50, 100, 150, 200
HEIGHT IN INCHES: 55, 59, 63, 67, 71, 75, 79

BODY MASS INDEX
The body mass index (BMI) is used to determine whether a person is a healthy weight for his or her size. BMI is usually expressed as a number. The ideal range is between 18.5 and 25 (shown here by the red band). Anyone over 25 is overweight, and anyone over 30 is obese. A person with a BMI under 18.5 is underweight.

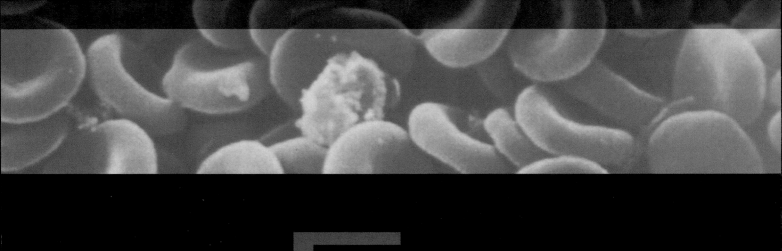

THROBBING HEART, PULSING VESSELS, BLOOD LEAKING FROM A WOUND—THE CARDIOVASCULAR SYSTEM IMPACTS DEEPLY ON OUR CONSCIOUSNESS. EVERY PART OF THE BODY RELIES ON A STEADY FLOW OF LIFE-GIVING BLOOD. THAT MOST VITAL OF PUMPS, THE HEART, IS MOSTLY MUSCLE AND IF MALTREATED, IT CAN WEAKEN AND WASTE, COMPROMISING ITS OWN BLOOD SUPPLY. DISORDERS OF THE HEART AND CIRCULATION ARE GENERALLY THOSE CAUSED BY ABUSE AND EXCESS: SMOKING TOBACCO, TOO MUCH FOOD LEADING TO OBESITY, AND TOO LITTLE EXERCISE GIVING THE WHOLE SYSTEM A LACK OF PURPOSE IN LIFE.

CARDIOVASCULAR SYSTEM

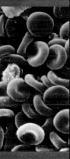

CARDIOVASCULAR ANATOMY

THE CIRCULATORY SYSTEM (OR CARDIOVASCULAR SYSTEM) IS RESPONSIBLE FOR DELIVERING OXYGEN AND OTHER NUTRIENTS TO VIRTUALLY ALL BODY CELLS AND REMOVING CARBON DIOXIDE AND OTHER WASTE PRODUCTS FROM THEM. LIKE THE NERVOUS AND LYMPHATIC SYSTEMS, THIS COMPLEX NETWORK EXTENDS INTO EVERY CREVICE OF THE BODY.

The circulatory system is composed of the heart, blood vessels, and blood. Although the heart is linked to emotions and virtues, such as love and courage, it is actually a muscular pump. Its regular contractions send blood into tough, elastic tubes called arteries, which branch into smaller vessels and convey oxygen-rich blood through the body. The arteries eventually divide into tiny capillaries, which have such thin walls that oxygen, nutrients, minerals, and other substances pass through to surrounding cells and tissues. Waste substances flow from the tissues and cells into the blood for disposal. The capillaries join and enlarge to create tubes that eventually become veins, which take blood back to the heart. Vessels carrying oxygenated blood (usually arteries) are shown in red and those carrying deoxygenated blood (usually veins) are blue. The intricate network has a length of some 90,000 miles (150,000 km)—equivalent to almost four times around the Earth.

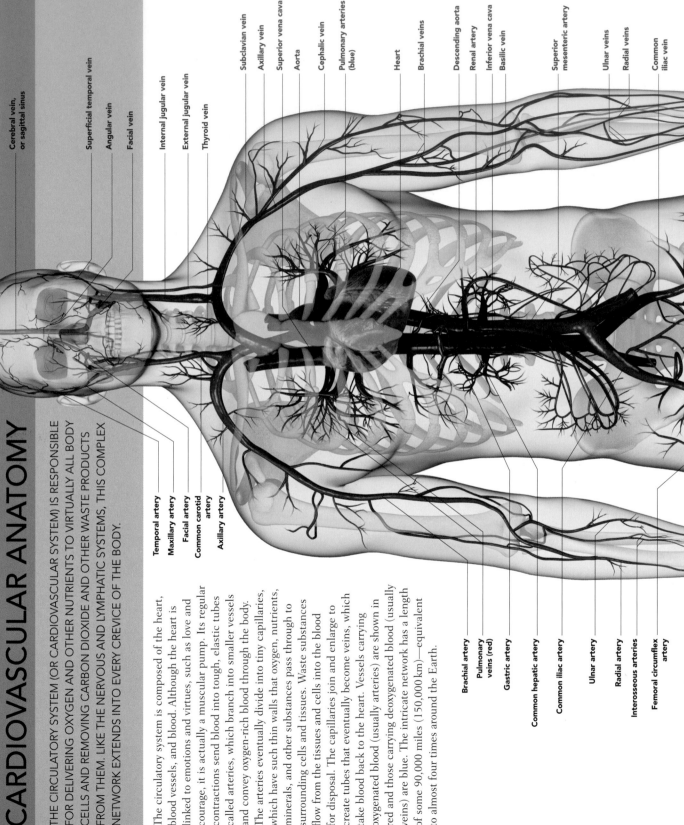

Cerebral vein, or sagittal sinus

Superficial temporal vein

Angular vein

Facial vein

Internal jugular vein

External jugular vein

Thyroid vein

Subclavian vein

Axillary vein

Superior vena cava

Aorta

Cephalic vein

Pulmonary arteries (blue)

Heart

Brachial veins

Descending aorta

Renal artery

Inferior vena cava

Basilic vein

Superior mesenteric artery

Ulnar veins

Radial veins

Common iliac vein

Temporal artery

Maxillary artery

Facial artery

Common carotid artery

Axillary artery

Brachial artery

Pulmonary veins (red)

Gastric artery

Common hepatic artery

Common iliac artery

Ulnar artery

Radial artery

Interosseous arteries

Femoral circumflex artery

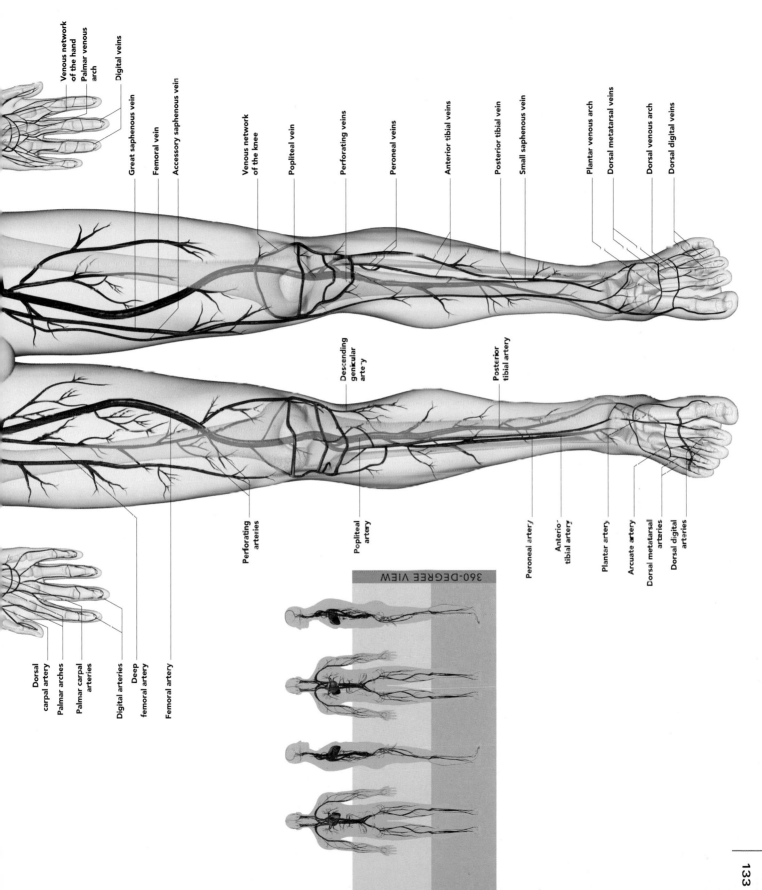

Venous network of the hand
Palmar venous arch
Digital veins

Great saphenous vein
Femoral vein
Accessory saphenous vein

Venous network of the knee
Popliteal vein
Perforating veins
Peroneal veins
Anterior tibial veins
Posterior tibial vein
Small saphenous vein
Plantar venous arch
Dorsal metatarsal veins
Dorsal venous arch
Dorsal digital veins

Descending genicular artery
Posterior tibial artery

Perforating arteries
Popliteal artery

Peroneal artery
Anterior tibial artery
Plantar artery
Arcuate artery
Dorsal metatarsal arteries
Dorsal digital arteries

Dorsal carpal artery
Palmar arches
Palmar carpal arteries
Digital arteries
Deep femoral artery
Femoral artery

360-DEGREE VIEW

133

BLOOD AND BLOOD VESSELS

BLOOD IS A COLLECTION OF SPECIALIZED CELLS SUSPENDED IN A STRAW-COLORED LIQUID CALLED PLASMA. BLOOD DELIVERS OXYGEN AND NUTRIENTS TO BODY CELLS, COLLECTS WASTE, DISTRIBUTES HORMONES, SPREADS HEAT AROUND THE BODY TO CONTROL TEMPERATURE, AND PLAYS A PART IN FIGHTING INFECTION AND HEALING INJURIES.

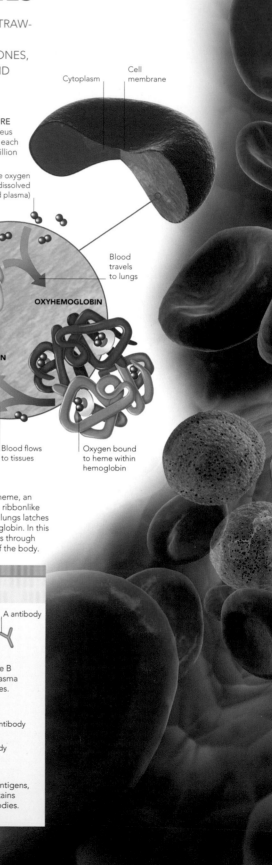

Cytoplasm   Cell membrane

WHAT IS BLOOD?

Blood forms about one-twelfth of the body weight of an adult, amounting to about 11 pints (5 liters) in volume. Roughly 50–55 percent of blood is plasma, the liquid-only portion in which cellular components are distributed. Plasma is 90 percent water containing dissolved substances such as glucose (blood sugar), hormones, enzymes, and also waste products such as urea and lactic acid. Plasma also contains proteins such as albumins, fibrinogen (important in clotting), and globular proteins or globulins. Alpha and beta globulins help transport lipids, which are fatty substances such as cholesterol. Gamma globulins are mostly the disease-fighting substances that are known as antibodies. The remaining 45–50 percent of blood is made up of three types of specialized cells. Red cells or erythrocytes carry oxygen; various white cells, known as leucocytes, are part of the defense system; and cellular fragments (platelets or thrombocytes) are involved in the process of clotting.

PARTS OF BLOOD
Blood is made up of a liquid portion (plasma), red blood cells, and a small band of platelets and white blood cells.

Plasma (about 50–55%)

White blood cells and platelets (1–2%)

Red blood cells (about 45–50%)

RED BLOOD CELL STRUCTURE
A biconcave disk with no nucleus or discernible inner structure, each red blood cell contains 300 million hemoglobin molecules.

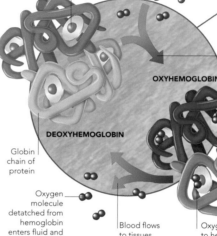

Heme molecule, including iron atom

Free oxygen molecule (dissolved in blood plasma)

Blood travels to lungs

OXYHEMOGLOBIN

DEOXYHEMOGLOBIN

Globin chain of protein

Oxygen molecule detached from hemoglobin enters fluid and body cells

Blood flows to tissues

Oxygen bound to heme within hemoglobin

ROLE OF HEMOGLOBIN
Hemoglobin is composed of heme, an iron-rich pigment, and globin, ribbonlike protein chains. Oxygen in the lungs latches onto heme to make oxyhemoglobin. In this conjoined form, oxygen travels through the bloodstream to all parts of the body.

BLOOD GROUPS

Every individual falls into one of four blood groups, which are determined by markers on red blood cells known as antigens (agglutinogens). Both of the antigens, A and B, may be present (AB), neither A nor B (O), or just one of them (A or B), and blood groups are named correspondingly. Plasma contains different antibodies (isohemagglutinins). For example, a person with blood group A has plasma containing B antibodies. If mixed with type B blood (with A antibodies in its plasma), A antibodies clump (or agglutinate) with A antigens. Therefore blood types must be matched for blood transfusions.

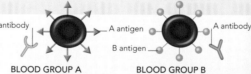

B antibody

A antigen

B antigen

A antigen

A antibody

BLOOD GROUP A
Red blood cells have A antigens with B antibodies contained in the plasma.

BLOOD GROUP B
Red blood cells have B antigens and the plasma contains A antibodies.

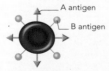

A antigen

B antigen

A antibody

B antibody

BLOOD GROUP AB
Red blood cells have A and B antigens, with neither A nor B in plasma.

BLOOD GROUP O
This lacks A and B antigens, but the plasma contains both A and B antibodies.

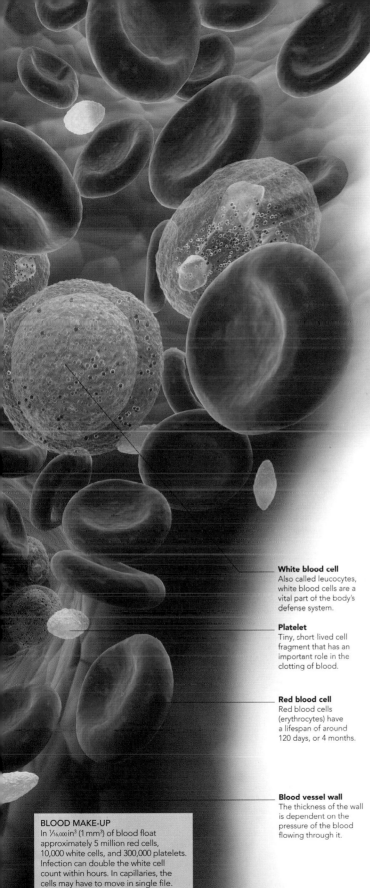

ARTERIES

Arteries carry blood away from the heart toward organs and tissues. Apart from the pulmonary arteries, all arteries carry oxygenated blood. Their thick walls and muscular and elastic layers can withstand the high pressure that occurs when the heart contracts. An artery narrows when the heart relaxes, helping push blood onward. The largest artery is the aorta, with a diameter of 1 in (25 mm); it conveys blood from the heart at up to 16 in (40 cm) per second. Most other arteries have a diameter of ⅙–¼ in (4–7 mm) and walls ¹⁄₂₅ in (1 mm) thick.

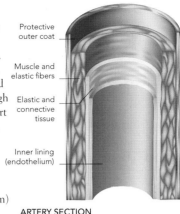

Protective outer coat

Muscle and elastic fibers

Elastic and connective tissue

Inner lining (endothelium)

ARTERY SECTION
Four distinct layers are found in an artery, with the blood-carrying space, called the lumen, in the center.

VEINS

A vein is more flexible than an artery and its walls are considerably thinner. The blood inside a vein is under relatively low pressure and, as a result, it flows slowly and smoothly. Many larger veins, particularly the long veins in the legs, contain valves that are formed from pouchlike pockets of single-cell lining tissue (endothelium). These prevent blood from flowing back down the legs, a job helped by muscles around the veins that contract during movement. The two main veins returning blood from the upper and lower parts of the body are called the superior and inferior venae cavae.

Outer layer

Inner lining

Valve cusp, or leaflet

Muscle layer

VEIN SECTION
The muscle layer of a vein is thin and enclosed by two layers; the innermost layer of some veins has valves at regular intervals.

White blood cell
Also called leucocytes, white blood cells are a vital part of the body's defense system.

Platelet
Tiny, short lived cell fragment that has an important role in the clotting of blood.

Red blood cell
Red blood cells (erythrocytes) have a lifespan of around 120 days, or 4 months.

Blood vessel wall
The thickness of the wall is dependent on the pressure of the blood flowing through it.

BLOOD MAKE-UP
In ¹⁄₁₆,₀₀₀ in³ (1 mm³) of blood float approximately 5 million red cells, 10,000 white cells, and 300,000 platelets. Infection can double the white cell count within hours. In capillaries, the cells may have to move in single file.

CAPILLARIES

The smallest and most numerous of the blood vessels, capillaries convey blood between arteries and veins. A typical capillary is ¹⁄₂₅ in (1 mm) or less in length, about ¹⁄₂,₅₀₀ in (0.01 mm) in diameter, and only slightly wider than a red blood cell, which is ¹⁄₃,₅₀₀ in (0.007 mm) across. Many capillaries enter tissue to form a capillary bed—the area where oxygen and other nutrients are released, and where waste matter passes into the blood. At any moment, only 5 percent of the body's blood is traveling in capillaries, with 20 percent in arteries, and 75 percent in veins.

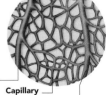

CAPILLARY BED
Capillaries connect small arteries (arterioles) to veins (venules).

Arteriole
Carries red blood rich in oxygen and nutrients

Capillary

Venule
Contains dark reddish blue blood low in oxygen

Capillary wall
Made up of a single layer of curved cells

Cell nucleus

CAPILLARY SECTION
The thinness of the capillary wall allows smooth, effortless movement of substances between surrounding tissues.

HEART STRUCTURE

THE HEART IS A POWERFUL ORGAN ABOUT THE SIZE OF A CLENCHED FIST. LOCATED JUST TO THE LEFT OF CENTER BETWEEN THE LUNGS, IT OPERATES AS TWO COORDINATED PUMPS THAT SEND BLOOD AROUND THE BODY.

THE HEART'S BLOOD SUPPLY

The muscular wall, or myocardium, of the heart is constantly active and needs a generous supply of oxygen and energy from blood. To provide this, the heart muscle has its own network of blood vessels known as the coronary arteries. These two arteries—the right and the left—branch from the main artery, the aorta, just after it leaves the heart, divide over the heart's surface, and send smaller blood vessels into the heart muscle. The pattern of the coronary veins, which collect wastes from the muscle tissue, is similar.

Most of the blood in these veins is collected by the coronary sinus, a large vein at the back of the heart that empties into the right atrium.

Right coronary artery
Aorta
Left coronary artery
Coronary vein
Main branch of left coronary artery
Coronary sinus
Small connecting blood vessels

CORONARY VESSELS
There are many connecting vessels between the coronary arteries. If an artery becomes blocked, these can provide an alternative route for the blood flow.

DOUBLE CIRCULATION

In the pulmonary circulation, the right side of the heart pumps blood to the lungs to be oxygenated and then back to the left side of the heart. In the systemic circulation, the left side of the heart pumps oxygen-rich blood to all the body's tissues and oxygen and depleted blood is returned to the right side of the heart.

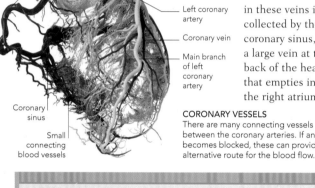

Vessels in upper body

Aorta
Carries oxygen-rich blood from the heart to the tissues

Pulmonary veins
Take oxygen-rich blood from the lungs to the heart

Network of vessels in right lung
Gas exchange takes place in the lungs' capillary network; oxygen passes into the blood, and carbon dioxide passes out

Network of vessels in left lung

Superior vena cava
Collects oxygen-depleted blood from the upper parts of the body and arms

Pulmonary artery
Carries oxygen-depleted blood from the heart to the lungs

Inferior vena cava
Collects oxygen-depleted blood from the lower parts of the body and legs

Portal vein
Conveys nutrient-rich blood from intestines to liver

Blood vessels in liver
Vessels in lower body
Blood vessels in digestive system

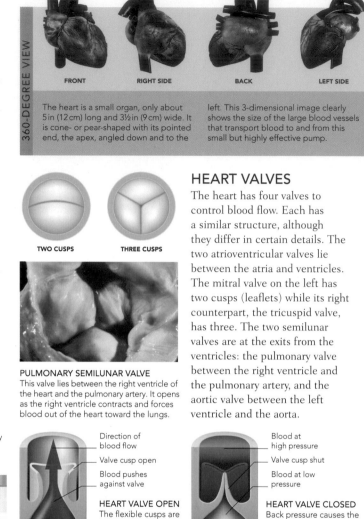

FRONT   RIGHT SIDE   BACK   LEFT SIDE

The heart is a small organ, only about 5 in (12 cm) long and 3½ in (9 cm) wide. It is cone- or pear-shaped with its pointed end, the apex, angled down and to the left. This 3-dimensional image clearly shows the size of the large blood vessels that transport blood to and from this small but highly effective pump.

TWO CUSPS   THREE CUSPS

PULMONARY SEMILUNAR VALVE
This valve lies between the right ventricle of the heart and the pulmonary artery. It opens as the right ventricle contracts and forces blood out of the heart toward the lungs.

HEART VALVES

The heart has four valves to control blood flow. Each has a similar structure, although they differ in certain details. The two atrioventricular valves lie between the atria and ventricles. The mitral valve on the left has two cusps (leaflets) while its right counterpart, the tricuspid valve, has three. The two semilunar valves are at the exits from the ventricles: the pulmonary valve between the right ventricle and the pulmonary artery, and the aortic valve between the left ventricle and the aorta.

Direction of blood flow
Valve cusp open
Blood pushes against valve

Blood at high pressure
Valve cusp shut
Blood at low pressure

HEART VALVE OPEN
The flexible cusps are forced apart by the pressure of blood as the heart contracts.

HEART VALVE CLOSED
Back pressure causes the cusps to close and seal at the edges, preventing reverse blood flow.

CARDIAC SKELETON

A set of four fibrous, cufflike rings known as the cardiac skeleton is built into the upper heart. The rings provide rigid points of attachment for the four heart valves and for the various sections of heart muscle. The wraparound arrangement of the muscle fibers in ventricle walls and the timing of their contractions enable the ventricles to squirt blood from the apex (lower pointed end) upward, and out through the pulmonary and aortic valves, rather than squeezing blood down so that it pools into the apex region.

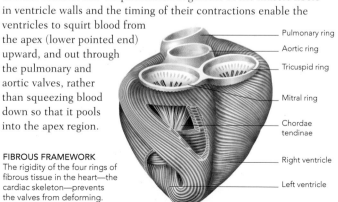

Pulmonary ring
Aortic ring
Tricuspid ring
Mitral ring
Chordae tendinae
Right ventricle
Left ventricle

FIBROUS FRAMEWORK
The rigidity of the four rings of fibrous tissue in the heart—the cardiac skeleton—prevents the valves from deforming.

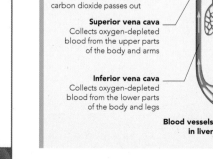

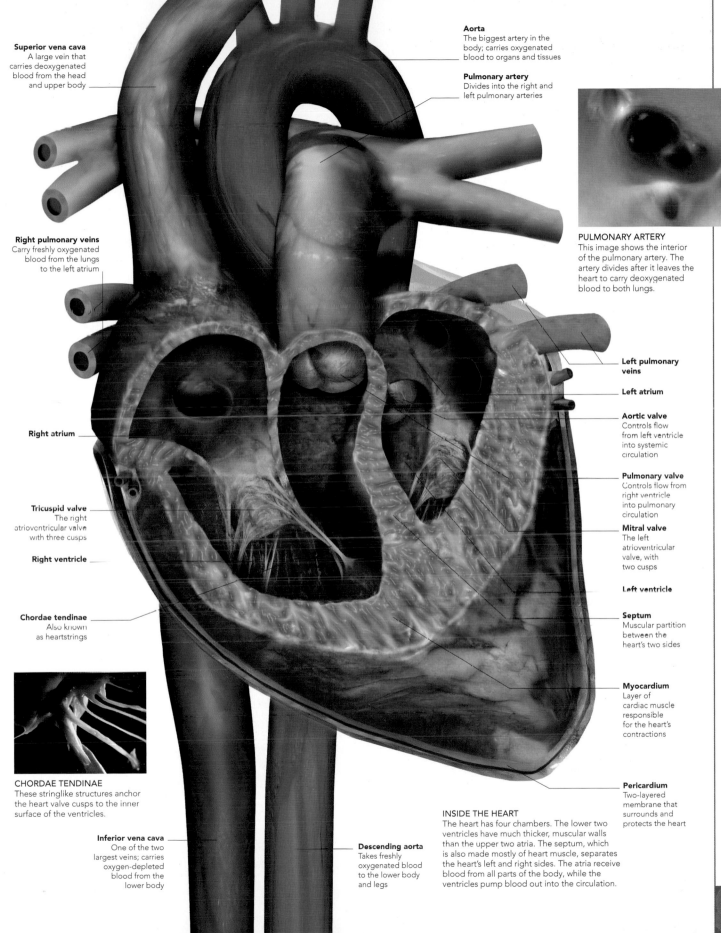

Superior vena cava
A large vein that carries deoxygenated blood from the head and upper body

Aorta
The biggest artery in the body; carries oxygenated blood to organs and tissues

Pulmonary artery
Divides into the right and left pulmonary arteries

Right pulmonary veins
Carry freshly oxygenated blood from the lungs to the left atrium

PULMONARY ARTERY
This image shows the interior of the pulmonary artery. The artery divides after it leaves the heart to carry deoxygenated blood to both lungs.

Left pulmonary veins

Left atrium

Aortic valve
Controls flow from left ventricle into systemic circulation

Right atrium

Pulmonary valve
Controls flow from right ventricle into pulmonary circulation

Mitral valve
The left atrioventricular valve, with two cusps

Tricuspid valve
The right atrioventricular valve with three cusps

Right ventricle

Left ventricle

Septum
Muscular partition between the heart's two sides

Chordae tendinae
Also known as heartstrings

Myocardium
Layer of cardiac muscle responsible for the heart's contractions

CHORDAE TENDINAE
These stringlike structures anchor the heart valve cusps to the inner surface of the ventricles.

Pericardium
Two-layered membrane that surrounds and protects the heart

INSIDE THE HEART
The heart has four chambers. The lower two ventricles have much thicker, muscular walls than the upper two atria. The septum, which is also made mostly of heart muscle, separates the heart's left and right sides. The atria receive blood from all parts of the body, while the ventricles pump blood out into the circulation.

Inferior vena cava
One of the two largest veins; carries oxygen-depleted blood from the lower body

Descending aorta
Takes freshly oxygenated blood to the lower body and legs

HOW THE HEART BEATS

THE HEART IS A DYNAMIC, UNTIRING, PRECISELY ADJUSTABLE DOUBLE-PUMP THAT FORCES BLOOD AROUND THE BODY'S IMMENSE NETWORK OF BLOOD VESSELS—PERHAPS MORE THAN THREE BILLION TIMES DURING A LIFETIME.

The heart's power comes from its two lower chambers (ventricles), which have thick muscular walls that contract to squeeze blood out into the arteries. The upper chambers (atria) have thinner walls and function partly as passive reservoirs for blood oozing in from the main veins. Each heartbeat has two main phases:

in the first phase (diastole), the heart relaxes and refills with blood; during the second stage (systole), it contracts, forcing the blood out. The whole cycle takes, on average, less than a second. During vigorous activity or stress, both the beating rate and the volume of blood pumped out of the heart increase greatly.

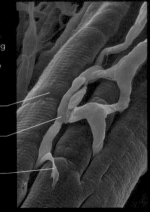

CONDUCTING FIBERS
The heart's conducting fibers are specialized long, thin cardiac muscle cells known as conducting myofibers or Purkinje fibers. These cells convey electrical impulses through the heart.

Cardiac muscle fiber

Capillary

Cardiac conducting myofiber

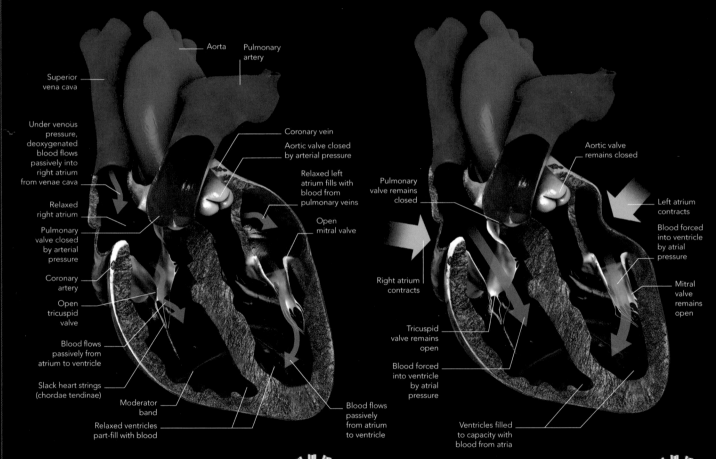

Aorta

Pulmonary artery

Superior vena cava

Under venous pressure, deoxygenated blood flows passively into right atrium from venae cava

Relaxed right atrium

Pulmonary valve closed by arterial pressure

Coronary artery

Open tricuspid valve

Blood flows passively from atrium to ventricle

Slack heart strings (chordae tendinae)

Moderator band

Relaxed ventricles part-fill with blood

Coronary vein

Aortic valve closed by arterial pressure

Relaxed left atrium fills with blood from pulmonary veins

Open mitral valve

Blood flows passively from atrium to ventricle

Pulmonary valve remains closed

Right atrium contracts

Tricuspid valve remains open

Blood forced into ventricle by atrial pressure

Ventricles filled to capacity with blood from atria

Aortic valve remains closed

Left atrium contracts

Blood forced into ventricle by atrial pressure

Mitral valve remains open

1 RELAXATION (LATE DIASTOLE)

During this phase of the heartbeat sequence, the muscular walls of the heart relax. The atrial chambers balloon slightly as they fill with blood coming in under low pressure from the main veins. Deoxygenated blood from the body enters the right atrium, while oxygenated blood from the lungs enters the left atrium. Some of the blood in the atria flows down into the ventricles. By the end of this phase, the ventricles are filled to about 80 percent of capacity.

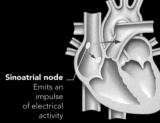

Sinoatrial node
Emits an impulse of electrical activity

PACEMAKER (SINOATRIAL NODE)
The sinoatrial node is inactive during most of diastole. As systole approaches, it begins to send out a wave of electrical impulses which will coordinate the heartbeat.

2 CONTRACTION OF THE ATRIA (ATRIAL SYSTOLE)

The heart's natural pacemaker, known as the sinoatrial node, is located in the upper part of the right atrium. It "fires" electrical impulses, much like those generated by nerves, which set off the contraction phase. Some impulses spread through the atrial walls and stimulate their cardiac muscle to contract. This squeezes blood inside the atria through the atrioventricular (tricuspid and mitral) valves into the ventricles, whose walls remain relaxed.

Electrical impulse
Spreads over surface of both atria, simulating them to contract

Atrioventricular node

ELECTRICAL IMPULSES SPREAD
Impulses traveling through atrial muscles make them contract within 0.1 seconds. Some signals pass faster along conducting fibers to the atrioventricular node.

CONTROL OF THE HEART RATE

Without control, the heart would beat at its natural, intrinsic rate of about 100 times per minute. However, a region known as the cardioregulatory centre in the medulla of the brainstem sends electrical impulses along nerves (especially the cranial vagus nerve) to set an average resting rate of about 70 beats per minute. During activity or stress, the sympathetic cardiac nerve signals, controlled by the hypothalamus, convey overriding signals to speed up the heart rate. The rate is also influenced by hormones such as adrenaline.

THE BRAIN'S INFLUENCE
The heart controls its own rhythm, but its rate is controlled by the central nervous system.

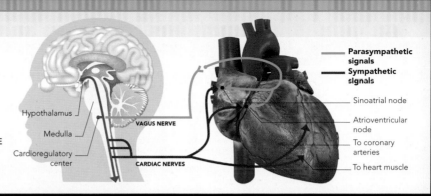

Parasympathetic signals
Sympathetic signals

Hypothalamus
Medulla
Cardioregulatory center

VAGUS NERVE
CARDIAC NERVES

Sinoatrial node
Atrioventricular node
To coronary arteries
To heart muscle

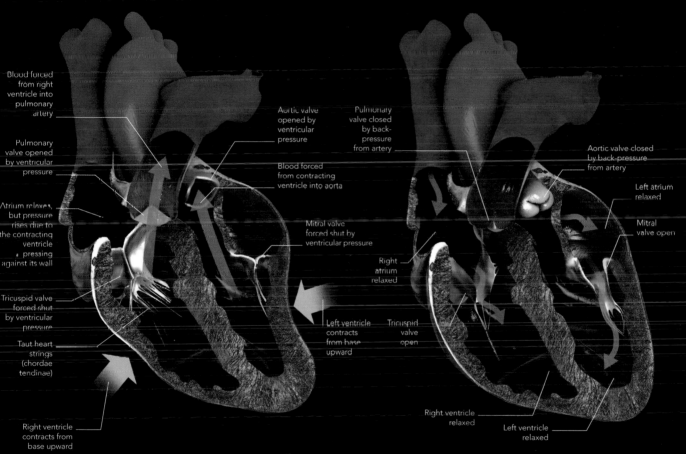

Blood forced from right ventricle into pulmonary artery

Pulmonary valve opened by ventricular pressure

Atrium relaxes, but pressure rises due to the contracting ventricle pressing against its wall

Tricuspid valve forced shut by ventricular pressure

Taut heart strings (chordae tendinae)

Right ventricle contracts from base upward

Aortic valve opened by ventricular pressure

Blood forced from contracting ventricle into aorta

Mitral valve forced shut by ventricular pressure

Pulmonary valve closed by back-pressure from artery

Aortic valve closed by back-pressure from artery

Left atrium relaxed

Mitral valve open

Right atrium relaxed

Left ventricle contracts from base upward

Tricuspid valve open

Right ventricle relaxed

Left ventricle relaxed

3 CONTRACTION OF THE VENTRICLES (VENTRICULAR SYSTOLE)

During this most active and powerful stage of the heartbeat, the thick cardiac muscle in the ventricle walls contracts, stimulated by electrical impulses relayed by the atrioventricular node. This causes a rise in ventricular pressure, which opens the aortic and pulmonary valves at the exits of the ventricles. Blood is forced out into the main arteries, making the atrioventricular valves snap shut.

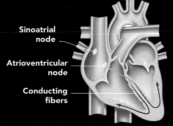

Sinoatrial node
Atrioventricular node
Conducting fibers

ATRIOVENTRICULAR SIGNALS FIRE
The atrioventricular node "fast-tracks" impulses along conducting fibers within the septum (dividing wall) to the lower ventricles and up through ventricle muscle.

4 RELAXATION (EARLY DIASTOLE)

The walls of the ventricles begin to relax, causing ventricular pressure to reduce. The pressure of the recently ejected blood in the main arteries is now high, so both the aortic and pulmonary valves close. This prevents back-flow into the ventricles. As ventricular pressure on the atrioventricular valves relaxes, the valves open. This reduces pressure in the atria, allowing blood to enter once again from the main veins.

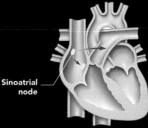

Sinoatrial node

ELECTRICAL IMPULSES FADE
Impulses spread through the ventricular walls back toward the atria within 0.2 seconds of leaving the sinoatrial node, which then fires again to continue the cycle.

CORONARY ARTERY DISEASE

THE CARDIAC MUSCLE, OR MYOCARDIUM, OF THE HEART WALL DEPENDS ON A CONSTANT FLOW OF BLOOD SUPPLIED BY THE CORONARY ARTERIES. IF THIS SUPPLY IS RESTRICTED, THEN OXYGEN AND NUTRIENTS CANNOT REACH THE MUSCLE AND THE RESULT COULD BE A FORM OF CORONARY ARTERY DISEASE (CAD). THE EXTENT OF THE SYMPTOMS OF CAD DEPENDS ON THE LOCATION, SEVERITY, AND SPEED OF ONSET OF THE RESTRICTED BLOOD SUPPLY.

ATHEROSCLEROSIS

ATHEROSCLEROSIS IS CAUSED BY THE NARROWING AND STIFFENING OF THE ARTERIES DUE TO FATTY DEPOSITS, KNOWN AS ATHEROMA, ACCUMULATING IN THEIR WALLS.

The process that leads to atherosclerosis begins with abnormally high levels of excess fats and cholesterol in the blood. These substances infiltrate the lining of arteries at sites of microscopic damage, forming deposits known as atheroma. This can happen in any of the body's arteries, including those supplying the brain with blood, when the result may be a stroke. The atheromatous deposits gradually form raised patches known as plaques. These consist of fatty cores within the arterial wall, covered by fibrous caps. The plaques narrow the space, or lumen, within the artery and this restricts the overall flow of blood to tissues beyond the site. It also causes turbulence that disrupts the smooth flow of blood and the eddies over the plaque surface make it more likely that blood will clot. The major risk factors for atherosclerosis include smoking, a diet high in saturated fats, lack of exercise, and excess weight.

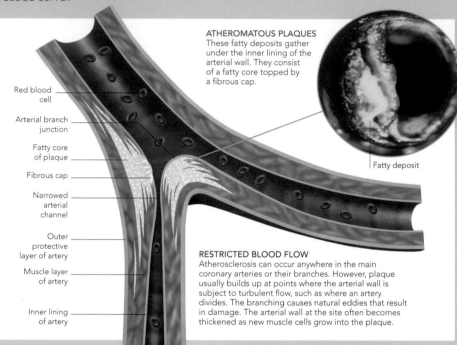

Red blood cell

Arterial branch junction

Fatty core of plaque

Fibrous cap

Narrowed arterial channel

Outer protective layer of artery

Muscle layer of artery

Inner lining of artery

ATHEROMATOUS PLAQUES
These fatty deposits gather under the inner lining of the arterial wall. They consist of a fatty core topped by a fibrous cap.

Fatty deposit

RESTRICTED BLOOD FLOW
Atherosclerosis can occur anywhere in the main coronary arteries or their branches. However, plaque usually builds up at points where the arterial wall is subject to turbulent flow, such as where an artery divides. The branching causes natural eddies that result in damage. The arterial wall at the site often becomes thickened as new muscle cells grow into the plaque.

ANGINA

ANGINA, CHEST PAINS THAT COME ON WITH EXERTION AND ARE RELIEVED BY REST, IS A SIGN THAT THE HEART MUSCLE IS NOT RECEIVING AN ADEQUATE SUPPLY OF BLOOD.

Angina is caused by a temporarily inadequate supply of blood to the heart muscle, usually because of arterial narrowing due to atherosclerosis. The pain most often occurs when the heart's workload is increased, for example with exercise, and fades with rest. Other triggers for angina are stress, cold weather, or a large meal. An angina attack typically begins with a heavy, constricting pain behind the breastbone. This can spread into the throat and jaw, and down into the arms, especially the left one. The pain usually subsides within 10–15 minutes. People with angina often take medication that relieves the pain by causing the coronary arteries to widen (dilate).

DAMAGED HEART MUSCLE
During angina, areas of heart muscle downstream from a narrowed artery suffer from lack of oxygen. After the attack, the muscle recovers.

WHY ANGINA OCCURS
Atherosclerosis of a coronary artery causes narrowing in the vessel and a reduction in blood flow. During exertion the heart beats faster and the muscle's demand for oxygen increases. However, extra blood cannot pass through the narrowed artery and the muscle "cramps."

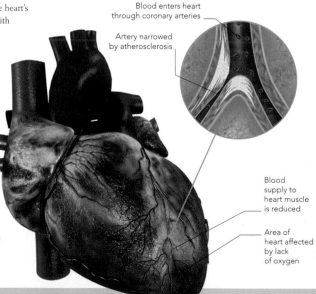

Blood enters heart through coronary arteries

Artery narrowed by atherosclerosis

Blood supply to heart muscle is reduced

Area of heart affected by lack of oxygen

ANGIOGRAPHY

The diagnostic procedure known as angiography shows the outline of blood vessels on a specialized X-ray image (called an angiogram). A fine catheter (hollow tube) is passed into an artery, usually in the leg, and then threaded up toward the heart via the aorta. A contrast medium, or radiopaque dye, is injected into the catheter and X-ray images are viewed on a monitor. These show the dye flowing through the coronary artery network and reveal any narrowing or blockage.

X-RAY IMAGE
The pattern of coronary arteries is similar in most hearts. This coronary angiogram reveals a narrowing that restricts blood flow to a region of cardiac muscle.

Narrowed coronary artery

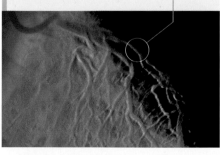

HEART ATTACK

A HEART ATTACK OCCURS WHEN AN AREA OF CARDIAC MUSCLE IS DEPRIVED OF BLOOD, AND THEREFORE OXYGEN, DUE TO A BLOCKAGE IN AN ARTERY.

A heart attack (myocardial infarction) is the result of coronary artery disease due to atherosclerosis, and the subsequent formation of a blood clot, or thrombus. Once formed, the clot can completely block blood flow to an area of heart muscle, starving it of blood and eventually causing tissue death. If possible, the blood flow must be restored to the damaged cells as quickly as possible. A heart attack usually occurs suddenly, with little or no warning. The chest pain may resemble that of angina, but it is more severe, is not necessarily brought on by exertion, and persists despite resting. A heart attack can also cause sweating, shortness of breath, nausea, and loss of consciousness.

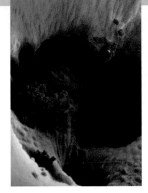

CLOTTED ARTERY
The healthy lining of a blood vessel allows blood to slip over it smoothly. Blood tends to clot where this smooth flow is disturbed by projections from the vessel wall, seen, for example, on the left of this image.

THROMBOLYTICS

The key to heart attack treatment is speed. The sooner the arterial blockage can be removed, the sooner blood flow is restored to the damaged area and it may be able to recover. Thrombolytic drugs are often introduced directly into the bloodstream after a heart attack. These help dissolve the clot that is blocking the coronary artery by increasing levels of various substances that prevent more clot formation and break down the strands of fibrin that bind the clot together. Antiplatelet drugs are usually given for some time after the heart attack because they thin the blood and prevent further clot formation.

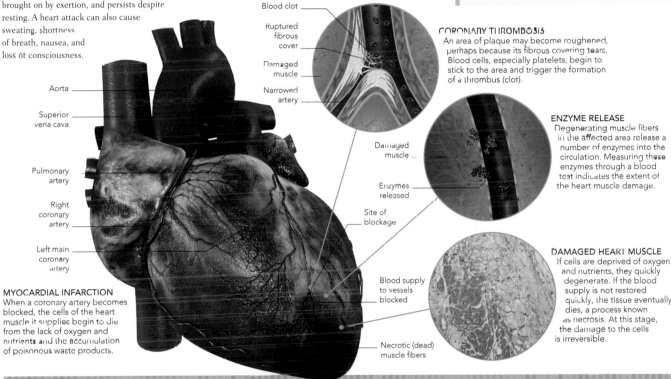

Blood clot

Ruptured fibrous cover

Damaged muscle

Narrowed artery

CORONARY THROMBOSIS
An area of plaque may become roughened, perhaps because its fibrous covering tears. Blood cells, especially platelets, begin to stick to the area and trigger the formation of a thrombus (clot).

Aorta

Superior vena cava

Pulmonary artery

Right coronary artery

Left main coronary artery

Damaged muscle

Enzymes released

Site of blockage

ENZYME RELEASE
Degenerating muscle fibers in the affected area release a number of enzymes into the circulation. Measuring these enzymes through a blood test indicates the extent of the heart muscle damage.

Blood supply to vessels blocked

MYOCARDIAL INFARCTION
When a coronary artery becomes blocked, the cells of the heart muscle it supplies begin to die from the lack of oxygen and nutrients and the accumulation of poisonous waste products.

Necrotic (dead) muscle fibers

DAMAGED HEART MUSCLE
If cells are deprived of oxygen and nutrients, they quickly degenerate. If the blood supply is not restored quickly, the tissue eventually dies, a process known as necrosis. At this stage, the damage to the cells is irreversible.

ANGIOPLASTY

This procedure is used to widen a section of coronary artery that has been narrowed or blocked by atheroma. It is often carried out to treat severe angina or after a heart attack. Angioplasty may be part of the same procedure as angiography, which visualizes the coronary arteries on an X-ray (see opposite page). Under local anesthetic, a fine catheter (hollow tube) is inserted into the femoral artery in the groin (or sometimes the arm), and passed up the aorta and into the coronary artery network. When the affected site is reached, a tiny balloon at the end of the catheter is inflated to widen the narrowed area. An expandable stainless steel mesh stent is often left permanently in place after withdrawal of the balloon catheter. This prevents the artery from narrowing again.

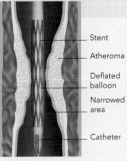

Stent

Atheroma

Deflated balloon

Narrowed area

Catheter

1 CATHETER INSERTED
The catheter is equipped with a small inflatable balloon near its end and, in this case, a metal mesh self-expanding tube called a stent.

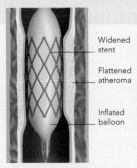

Widened stent

Flattened atheroma

Inflated balloon

2 BALLOON INFLATED
When the balloon is positioned within the narrowed area, it is inflated with gas or liquid to stretch the artery and widen the stent.

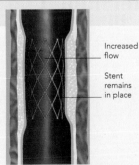

Increased flow

Stent remains in place

3 CATHETER REMOVED
The balloon is deflated and withdrawn, leaving the stent in place in its expanded form. In a few weeks, a thin layer of cells grows over the stent.

HEART MUSCLE DISORDERS

THE HEART IS COMPOSED MOSTLY OF SPECIALIZED MUSCLE, KNOWN AS THE CARDIAC MUSCLE, OR MYOCARDIUM. SOME HEART DISORDERS ARE CAUSED BY PROBLEMS WITH THIS MUSCLE OR WITH THE SACLIKE PERICARDIUM SURROUNDING THE HEART. LONG-STANDING OR SEVERE HEART MUSCLE PROBLEMS CAN LEAD TO HEART FAILURE, WHEN THE HEART'S PUMPING POWER IS REDUCED.

HEART MUSCLE DISEASE

INFLAMMATION OF THE HEART MUSCLE IS KNOWN AS MYOCARDITIS; NONINFLAMMATORY HEART MUSCLE DISEASE IS CALLED CARDIOMYOPATHY.

Many cases of myocarditis are due to infection, often with a virus such as coxsackie. The problem may go unnoticed, but, if severe, can lead to chest pain and long-term heart failure. Other causes of myocarditis include rheumatic fever, exposure to radiation or certain drugs or chemicals, or an autoimmune condition such as systemic lupus erythematosus (see p.186). Cardiomyopathy is noninflammatory heart muscle disease in which the muscle becomes weakened, damaged, and stretched. This condition takes several forms with various causes, as illustrated below.

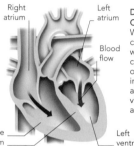

NORMAL HEART
The muscular walls of a normal heart, especially the ventricles, are substantial. They are also flexible and bend as they contract and squeeze out blood. The pumping rate and volume adjust to cope with the body's demands for oxygenated blood.

Right atrium
Left atrium
Blood flow
Right ventricle
Septum
Left ventricle

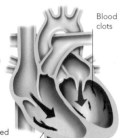

DILATED CARDIOMYOPATHY
Widening (dilation) causes the ventricle walls to thin. In some cases, blood clots form on the linings. Causes include excessive alcohol intake, viral illness, or an autoimmune disorder.

Blood clots
Dilated ventricle walls

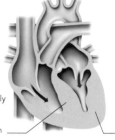

HYPERTROPHIC CARDIOMYOPATHY
This condition causes the heart muscle to thicken, especially in the left ventricle and septum, so that the heart cannot fill with blood properly. It is usually an inherited problem and a cause of sudden death in apparently healthy young people.

Thickened septum

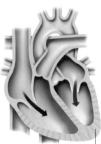

RESTRICTIVE CARDIOMYOPATHY
The walls of the ventricles become rigid, which restricts their ability to stretch when filled with blood, and also to flex on contraction to expel it. This problem is caused by scar tissue or by deposits of iron or abnormal protein.

Thickened wall of left ventricle
Rigid ventricular walls

PERICARDITIS

INFLAMMATION OF THE PERICARDIUM—THE TWO-LAYERED MEMBRANOUS SAC THAT SURROUNDS THE HEART—IS OFTEN DUE TO A VIRAL INFECTION OR A HEART ATTACK.

The most common cause of pericarditis is a viral infection that inflames the pericardium. Other causes include bacterial pneumonia, tuberculosis, the spread of a cancerous tumour to the pericardium, an autoimmune disorder such as rheumatoid arthritis, kidney failure, a heart attack, or a penetrating wound to the area. Any inflamed pericardium cannot lubricate the heart's beating motions normally, so it rubs and scrapes. Symptoms include pain in the center of the chest, which is relieved by leaning forward but worsened by a deep breath, breathlessness, or fever.

PERICARDIAL EFFUSION
The outer (fibrous) layer of the pericardium is tough and elastic. The inner (serous) membrane forms a double layer around the heart, separated by a thin film of lubricating fluid. Pericardial effusion is an excess of fluid, caused by inflammation of the serous membrane, and can interfere with the heart's pumping.

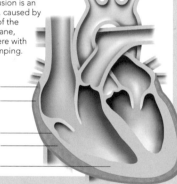

Outer fibrous layer of pericardium
Pericardial fluid
Inner serous layer of pericardium
Pericardial effusion
Heart muscle

HEART FAILURE

HEART FAILURE DEVELOPS WHEN THE HEART CAN NO LONGER PUMP BLOOD EFFECTIVELY TO THE BODY TISSUES AND THE LUNGS.

The failure of the heart to pump blood adequately leads to an accumulation of fluid in the tissues. The site of fluid accumulation is determined by which part of the heart is failing. In left-sided heart failure, the heart's left ventricle fails to pump blood out to the body as fast as it enters from the lungs. As a result, blood backs up in the pulmonary veins and lungs, causing congestion.

The pressure in the lungs causes fluid to collect there (a condition known as pulmonary edema) and oxygen is absorbed less efficiently, producing symptoms such as breathlessness, coughing, and fatigue. In right-sided heart failure, the right ventricle cannot pump blood out to the lungs as fast as it comes in from body tissues. Blood backs up in the main veins, again causing congestion. Increased venous pressure forces fluid out of the capillaries into the tissues with noticeable swelling (edema) in the ankles and lower back. Other symptoms include breathlessness, fatigue, and nausea. Heart failure has many possible causes—these include a heart attack, coronary heart disease, persistent high blood pressure, cardiomyopathy, a heart valve or rhythm disorder, or chronic obstructive pulmonary disease (see pp.160–61). Heart failure is a serious long-term condition and there is no cure. However, there is a range of drugs which can improve symptoms and life expectancy. In some cases, a pacemaker will be implanted to improve the pumping action of the heart.

Indentation

FLUID RETENTION
Fluid accumulation due to chronic heart failure makes the tissues swollen and soggy. An indentation caused by pressing tends to stay after the pressure is removed.

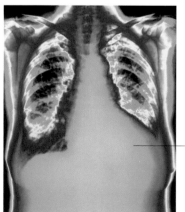

Enlarged heart due to heart failure

ENLARGED HEART
In heart failure, the heart becomes grossly enlarged over time as it struggles to pump blood through the body's circulation.

STRUCTURAL DISORDERS

STRUCTURAL HEART DISORDERS CAN AFFECT PEOPLE OF ANY AGE; CONGENITAL HEART DEFECTS ARE
PRESENT AT BIRTH, WHILE VALVE DISORDERS GENERALLY ARISE LATER IN LIFE. MEDICAL ADVANCES
ALLOW MANY DEFECTS IN THE HEART TO BE EFFECTIVELY TREATED WITH SURGICAL TECHNIQUES.
SIMILARLY, DISEASED VALVES CAN BE SURGICALLY WIDENED OR REPLACED.

CONGENITAL HEART DEFECTS

HEART DEFECTS PRESENT FROM BIRTH (CONGENITAL)
MAY BE DUE TO A FAULT IN DEVELOPMENT DURING
EARLY EMBRYO DEVELOPMENT.

Some types of congenital heart defect (CHD) run in families,
suggesting there is a genetic influence, although usually there
is no obvious cause. However, in some cases there is a link
with the mother catching an infection such as rubella during
pregnancy or being exposed to certain drugs including
alcohol. Symptoms of CHD include breathlessness (which
can affect feeding) and slow weight gain. Ultrasound scans
can help diagnose some types of CHD so that appropriate
treatment can be prepared.

HEART DEVELOPMENT
In the embryo, the heart
develops as a section of
blood vessel that thickens
its walls and begins to twist
and loop, creating atrial
and ventricular chambers.
Complex connections of
arteries and veins begin to
take shape. Many congenital
heart defects arise from a
problem during this initial
stage of development.
Here, deoxygenated
blood is blue; red is
oxygenated blood.

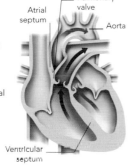

Atrial septum — Pulmonary valve — Aorta — Ventricular septum

COARCTATION OF AORTA
A short section of the aorta
is narrowed, usually at a
point where the main arteries
branch off for the head, brain,
arms, and upper body.
This results in restricted
blood flow to the lower
body and legs. The
heart works harder to
compensate, so blood
pressure in the upper
body is elevated. A
baby with this condition
is usually pale and finds
it difficult to breathe or
feed. Urgent corrective
surgery may be needed.

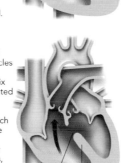

Aortic narrowing — Reduced blood flow

VENTRICULAR SEPTAL DEFECT
This is a hole in the wall
between the two ventricles
(ventricular septum),
causing the blood to mix
(purple color). Oxygenated
blood from the left
ventricle flows through
the hole so that too much
blood is pumped by the
right ventricle into the
lungs. A small hole may
close as the child grows,
but a larger one will
require surgical repair.

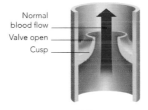

Ventricular septal defect — Septum

ATRIAL SEPTAL DEFECT
This is an abnormal
opening in the wall
(atrial septum) between
the two upper chambers
(atria). As a result,
blood shunts from
the high-pressure left
side of the heart into
the right side (purple
area). The blood flow
to the lungs increases
in consequence and less
is pumped around the
body. Both atrial and
ventricular septal defects
are common in children
with Down Syndrome.

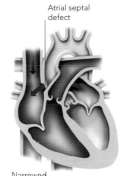

Atrial septal defect — Narrowed pulmonary valve — Displaced aorta

TETRALOGY OF FALLOT
A combination of four
structural defects: a
ventricular septal defect;
an aorta that is displaced
towards the right side so
that deoxygenated blood
can flow into it from the
right ventricle (purple
area); narrowed pulmonary
valve (pulmonary stenosis);
and a thickened right
ventricle wall. An affected
person is breathless and
has a distinctive blueish
skin color (cyanosis).

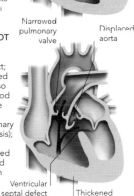

Ventricular septal defect — Thickened wall

VALVE DISORDERS

THERE ARE SEVERAL CONDITIONS THAT CAN
AFFECT THE EFFICIENT FUNCTIONING OF ANY
OF THE HEART'S FOUR VALVES.

There are two main types of valve disorder. In
stenosis, the valve outlet is too narrow and restricts
blood flow. It may be congenital or due to an
infection such as rheumatic fever. Stenosis is
also part of the aging process. In incompetence,
the valve does not close fully, allowing backflow
of blood. This problem can occur as a result
of a heart attack or an infection of the valve.

MITRAL VALVE
This image of a
healthy human
heart valve shows
the heart strings
(chordae tendinae)
and cusps. The
mitral valve lies
between the left
atrium and the
left ventricle.

Cusp — Chordae tendinae

Normal blood flow — Valve open — Cusp

NORMAL VALVE OPEN
As a heart chamber contracts, the high
pressure pushes against the cusps of
the valve, forcing it open and allowing
blood to flow past.

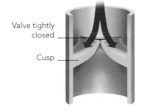

Valve tightly closed — Cusp

NORMAL VALVE CLOSED
The pressure on the other side of
the valve increases and the valve
cusps snap shut so that blood
cannot flow backward.

Restricted blood flow — Valve partially open — Abnormal cusp

STENOSIS
The valve tissue is stiffened and
cannot open fully. Blood passing
through it is restricted so the heart
beats harder to maintain flow.

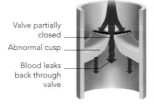

Valve partially closed — Abnormal cusp — Blood leaks back through valve

INCOMPETENCE
The valve cusps do not close
properly and allow blood to leak
backward. As a result, the heart has
to work harder to circulate blood.

HEART MURMURS

UNUSUAL HEART SOUNDS PRODUCED
BY TURBULENT BLOOD FLOW MAY BE
DUE TO A HEART VALVE DEFECT.

The "lub-dub" sound of the heartbeat is
made by healthy valves snapping shut.
Some types of unusual sounds are known
as "murmurs" and may indicate an
abnormality. However, many murmurs,
particularly in children, do not indicate
valve abnormalities.

ABNORMAL FLOW
Murmurs can be produced by turbulent
flow as blood rushes around the cusps of
a stenosed valve or leaks back through
an incompetent valve
and collides with
oncoming
blood.

Pulmonary valve stenosis — Mitral valve incompetence

CIRCULATORY AND HEART RATE DISORDERS

A CONSTANT AND ADEQUATE BLOOD SUPPLY IS ESSENTIAL FOR HEALTHY TISSUES. SHOULD A
BLOCKAGE OCCUR IN A BLOOD VESSEL, THE TISSUES BEYOND IT MAY BE STARVED OF OXYGEN,
CAUSING TISSUE DAMAGE OR, IN MORE SEVERE CASES, TISSUE DEATH. THE HEART MAY ALSO BE
AFFECTED IF THE ELECTRICAL SYSTEM THAT MAINTAINS HEART RATE AND RHYTHM IS DISTURBED.

EMBOLISM

AN EMBOLUS—A FRAGMENT OF MATERIAL THAT
BREAKS AWAY FROM ITS ORIGINAL SITE—CAN CAUSE
THE PARTIAL OR TOTAL BLOCKAGE OF A BLOOD VESSEL.

Most emboli are fragments of a blood clot (thrombus),
or even a whole clot, that has detached from its original
site and traveled in the bloodstream to lodge in a blood
vessel. An embolus may also be made of fatty material
from an atheromatous plaque (see p.140) in an arterial
wall, crystals of cholesterol, fatty bone marrow that has
entered the circulation following a bone fracture, or an
air bubble or amniotic fluid. In a pulmonary embolism, a
clot originating elsewhere in the body travels to the lungs
in veins. Clots that form in the heart or arteries can block
circulation anywhere in the body. An embolus is most
likely to block a blood vessel where it narrows or branches,
depriving tissues beyond it of vital oxygen. Symptoms
depend on the site affected; for example, an embolus
blocking an artery supplying the brain may lead to a stroke.
If the embolus is a fragment of a clot, it can be treated
with thrombolytic, or "clot-busting," drugs.

Inferior
vena cava

Path of
embolus

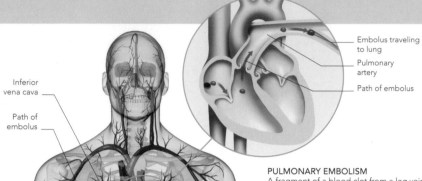

Embolus traveling
to lung
Pulmonary
artery
Path of embolus

PULMONARY EMBOLISM
A fragment of a blood clot from a leg vein
may travel through the venous system to
the heart's right side, then out along the
pulmonary arteries to a lung. It may lodge
here, depriving lung tissue of vital oxygen
and reducing oxygen uptake by the
pulmonary circulation.

Thrombotic embolus
A fragment (embolus) composed of
blood clot (thrombotic) material; may
arise anywhere in body, but veins of
the legs and pelvis are common sites

THROMBOSIS

THE PARTIAL OR TOTAL BLOCKAGE OF AN ARTERY, VEIN,
OR EVEN THE HEART CAN OCCUR WHEN A BLOOD CLOT
(THROMBUS) FORMS DUE TO A CIRCULATORY PROBLEM.

Thrombosis is most likely to occur where the normal
smooth flow of blood is disrupted and either slows down
or becomes turbulent. This disruption may be caused by
plaques of fatty atheromatous tissue in the walls of an artery
or by inflammation of the blood vessel. The clot eventually
narrows or blocks the passage for blood so that tissues
downstream are deprived of oxygen and nutrients. The
effects depend on the site of the thrombosis.

THROMBUS FORMATION
Thrombosis can occur in arteries and veins, but
commonly happens at a site of atherosclerosis in
an artery wall, which disrupts normal blood flow.

Fibrin
strands

Thrombus
blocking artery;
thrombi can also
form in veins

DEEP VEIN THROMBOSIS

Blood that flows slowly is more likely to thrombose,
or clot. This can happen in the deep veins of the
legs and lower body, which rely to some extent on
contracting muscles to assist blood flow. Deep vein
thrombosis (DVT) tends to occur during periods of
immobility, particularly during long trips, when
muscles are relaxed and blood pools in the veins.
To help prevent clots from forming, keep moving
and drink nonalcoholic beverages. Symptoms of
DVT include tenderness, pain, and swelling in the
leg, and visibly engorged veins. Treatment with
anticoagulant drugs reduces the risk that part of the
clot will break off and travel to the lung (see above).

X-RAY DIAGNOSIS
A deep-vein thrombosis in the calf is
revealed here by injecting a radiopaque
dye into the circulation and taking an X-ray.

Visible
clot

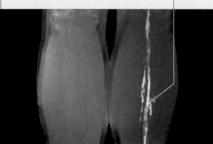

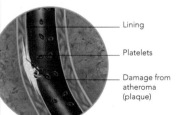

Lining

Platelets

Damage from
atheroma
(plaque)

1 INTERNAL DAMAGE
When an artery lining is damaged by
rupture of a plaque, platelets in the area
clump together and release chemicals that
begin the clotting or coagulating process.

2 CLOT FORMATION
The chemicals help convert
fibrinogen into insoluble fibrin strands.
These trap platelets and other blood
cells, and clot formation escalates.

ANEURYSM

ABNORMAL SWELLING OF A WEAKENED ARTERIAL WALL MAKES THE WALL BULGE OUT LIKE A BALLOON.

This defect in an arterial wall may be due to disease or injury, or it can be congenital. Although aneurysms may occur in arteries anywhere in the body, they most often affect the main artery from the heart, the aorta. Most aortic aneurysms occur in the abdominal section below the kidneys, rather than in the chest, and this type of aneurysm tends to run in families. Small aortic aneurysms are usually symptomless, although large ones may cause localized pain. Aneurysms may be treated by surgery, the aim of which is to repair the artery before the aneurysm dissects, or ruptures (see right). Berry aneurysms occur in the small arteries at the base of the brain. There may be one or several of them, and they are thought to be present from birth. If a berry aneurysm ruptures, it causes a subarachnoid hemorrhage (see p.112) and results in an intensely painful headache.

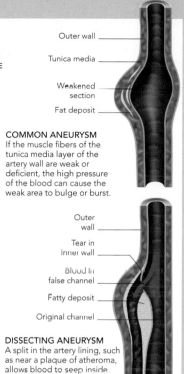

Outer wall
Tunica media
Weakened section
Fat deposit

COMMON ANEURYSM
If the muscle fibers of the tunica media layer of the artery wall are weak or deficient, the high pressure of the blood can cause the weak area to bulge or burst.

Outer wall
Tear in inner wall
Blood in false channel
Fatty deposit
Original channel

DISSECTING ANEURYSM
A split in the artery lining, such as near a plaque of atheroma, allows blood to seep inside. The artery swells and its walls thin and may burst.

HYPERTENSION

PERSISTENT, HIGHER-THAN-NORMAL BLOOD PRESSURE CAN DAMAGE INTERNAL ORGANS IF UNTREATED.

Normally, blood is under pressure as the heart forces it around the circulation. In hypertension, this pressure is above normal limits. There are no symptoms at first, but despite this, over time it increases the risk of many serious disorders, such as stroke, heart disease, and kidney failure. Contributing factors to hypertension include certain genetic influences and diet and lifestyle factors, such as being overweight, drinking excessive amounts of alcohol, smoking, and having a high-salt diet. It is most common in middle-aged and elderly people. A stressful lifestyle may aggravate the condition. Hypertension cannot be cured, but it can be controlled. A change of diet and lifestyle may be all that is necessary, but more severe cases may be treated with antihypertensive drugs.

BLOOD PRESSURE GRAPH
Normal blood pressure varies according to activity levels. This graph shows that during sleep, both the systolic and diastolic pressures (see pp.138–39) are much lower.

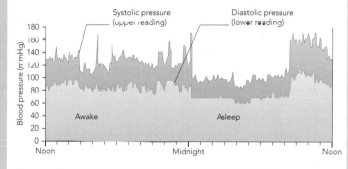

ARRHYTHMIA

AN ABNORMAL HEART RATE OR RHYTHM IS CAUSED BY A DISTURBANCE IN THE ELECTRICAL SYSTEM THAT CONTROLS THE WAY HEART MUSCLE CONTRACTS.

An arrhythmia is a heart rate that is unusually slow or fast, or erratic. A normal heartbeat is initiated by specialized cells in the natural "pacemaker," the sinoatrial (SA) node, at the top of the right atrium. They send electrical signals resembling nerve impulses out through the atrial muscle tissue, stimulating it to contract. These signals are relayed by the atrioventricular (AV) node along nervelike fibers through the septum (central dividing wall) and into the thick muscle tissue of the ventricle walls. A fault in the system can lead to the arrhythmias described here.

SINUS TACHYCARDIA
In sinus tachycardia there is a regular but rapid heart rate, usually more than 100 beats per minute. This can occur during fever, exercise, great stress, or as a response to stimulants, such as caffeine.

SA node
AV node

Very fast heartbeat
Atrium
Ventricle

ATRIAL FIBRILLATION
Fibrillations are extremely rapid, disordered, weak contractions with a rate as high as 500 per minute. A blockage at the AV node can make them happen in the atria, which can also make the ventricles beat faster, at up to 160 beats per minute.

Variable blockage at AV node

Irregular impulses through atria

Extremely rapid and irregular heartbeat

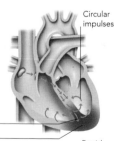

BUNDLE-BRANCH BLOCK
Damage to a branch of the nervelike fiber bundles that carry the electrical signals impedes their flow. Some signals may "leak" across from the other, healthy side. If both right and left bundles are affected, the heart rate slows to a very low rate.

Blockage

Impulses from healthy side
Double peak due to uncoordinated ventricles
Slow heartbeat

VENTRICULAR TACHYCARDIA
Very fast contractions of the ventricles may be caused by damaged heart muscle, for example due to heart disease or a heart attack. Electrical impulses have difficulty passing through the scarred heart muscle so they recirculate.

Circular impulses

Slowed conduction through damaged area
Damaged heart muscle
Rapid heartbeat

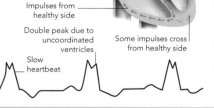

TREATMENTS

Often, heart arrhythmias can be treated with drugs. Another solution is to implant an artificial pacemaker into the chest wall. The pacemaker is connected to the heart by wires and takes over the role of supplying electrical signals to the heart muscle. In some cases, cardioversion (sometimes known as defibrillation) is also possible using an implant. A cardiac defibrillator (ICD), about the size of a thumb, can be implanted just below the collarbone. It monitors heart rate and can detect life-threatening arrhythmias. The ICD reacts by shocking the heart with a jolt of electricity, which restores its normal rhythm.

OXYGEN IS VITAL FOR LIFE. THE RESPIRATORY SYSTEM
TRANSFERS OXYGEN FROM AIR TO BLOOD SO THAT THE
CARDIOVASCULAR SYSTEM CAN DISTRIBUTE IT, WHILE THE
MUSCULAR AND SKELETAL SYSTEMS DRIVE THE MOVEMENTS
OF BREATHING. THE AIR IS OFTEN CONTAMINATED WITH
DUST PARTICLES, HARMFUL MICROBES, ALLERGENS, AND
HAZARDOUS, IRRITANT, AND CANCER-CAUSING CHEMICALS;
SMOKERS FURTHER BOOST THESE LAST THREE CATEGORIES.
ALL OF THESE ITEMS CAN DAMAGE THE SYSTEM'S DELICATE
PARTS, MAKING RESPIRATORY DISORDERS AMONG THE MOST

RESPIRATORY SYSTEM

RESPIRATORY ANATOMY

THE RESPIRATORY SYSTEM, IN CLOSE CONJUNCTION WITH THE CIRCULATORY SYSTEM, IS RESPONSIBLE FOR SUPPLYING ALL BODY CELLS WITH ESSENTIAL OXYGEN AND REMOVING POTENTIALLY HARMFUL CARBON DIOXIDE FROM THE BODY. THE MOUTH AND NOSE CHANNEL AIR FROM OUTSIDE THE BODY THROUGH A SYSTEM OF TUBES OF DIMINISHING SIZE THAT EVENTUALLY REACH THE TWO LUNGS SITUATED ON EITHER SIDE OF THE HEART WITHIN THE CHEST CAVITY.

Air enters the body mainly through the nostrils (but sometimes through the mouth). The nostrils lead into the nasal cavity, which opens up within the skull and joins with the pharynx (part of the throat) toward the rear. The pharynx is a short funnel-shaped tube that extends partway down the neck. The first part of the pharynx conveys only air, but lower down, food and liquids also travel through. The larynx, home to the vocal cords, joins the pharynx to the windpipe (trachea). A loose flap of cartilage, the epiglottis, lies just above the larynx and blocks it off during swallowing to prevent food and liquids from entering the trachea. The trachea splits into two airways called primary bronchi, one of which enters the right lung and the other the left lung. Each bronchus divides further into secondary and tertiary bronchi, and eventually into tiny bronchioles. This continuous branching is referred to as the bronchial tree. Deep within the paired, cone-shaped lungs, exchange of gases takes place.

360-DEGREE VIEW

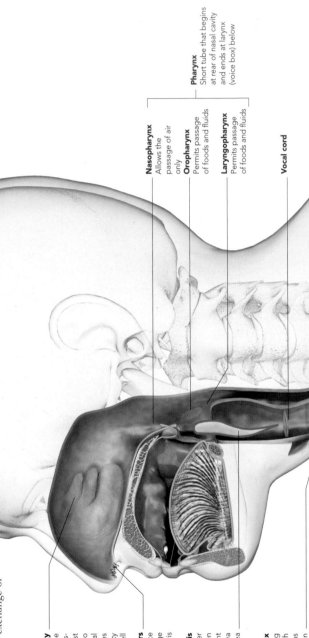

Nasopharynx
Allows the passage of air only

Oropharynx
Permits passage of foods and fluids

Pharynx
Short tube that begins at rear of nasal cavity and ends at larynx (voice box) below

Laryngopharynx
Permits passage of foods and fluids

Vocal cord

Nasal cavity
Main route for air to and from the lungs; lined with a sticky, mucus-covered membrane that traps dust particles and germs; divided in two by central plate of cartilage (nasal septum); fuzzy-looking patches (olfactory epithelia) in roof of cavity are the sensory organs of smell

Nose hairs
Situated inside entrance of nostrils; help filter large particles of dust and debris

Epiglottis
Cartilage flap that tilts over entrance to larynx when swallowing, to prevent food, drinks, and saliva from entering trachea

Larynx
Short, cartilaginous tube joining pharynx with trachea; together with vocal cords within the larynx, it has a vital role in speech production

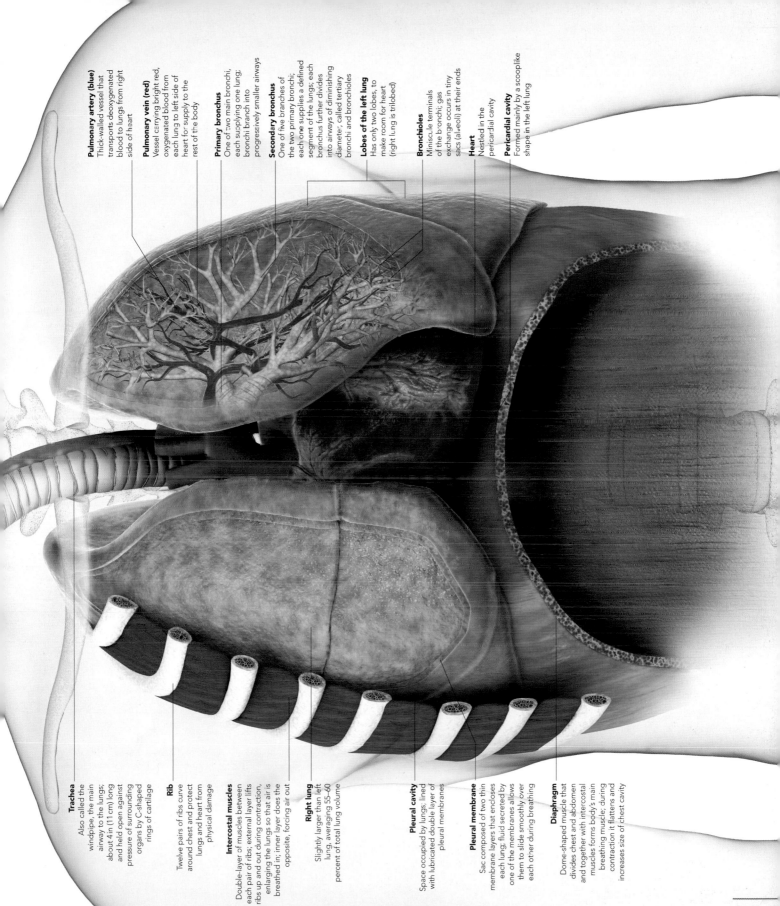

Trachea
Also called the windpipe, the main airway to the lungs; about 4 in (11 cm) long and held open against pressure of surrounding organs by C-shaped rings of cartilage

Rib
Twelve pairs of ribs curve around chest and protect lungs and heart from physical damage

Intercostal muscles
Double-layer of muscles between each pair of ribs; external layer lifts ribs up and out during contraction, enlarging the lungs so that air is breathed in; inner layer does the opposite, forcing air out

Right lung
Slightly larger than left lung, averaging 55–60 percent of total lung volume

Pleural cavity
Space occupied by lungs; lined with lubricated double layer of pleural membranes

Pleural membrane
Sac composed of two thin membrane layers that encloses each lung; fluid secreted by one of the membranes allows them to slide smoothly over each other during breathing

Diaphragm
Dome-shaped muscle that divides chest and abdomen and together with intercostal muscles forms body's main breathing muscle; during contraction it flattens and increases size of chest cavity

Pulmonary artery (blue)
Thick-walled vessel that transports deoxygenated blood to lungs from right side of heart

Pulmonary vein (red)
Vessel carrying bright red, oxygenated blood from each lung to left side of heart for supply to the rest of the body

Primary bronchus
One of two main bronchi, each supplying one lung; bronchi branch into progressively smaller airways

Secondary bronchus
One of five branches of the two primary bronchi; each one supplies a defined segment of the lungs; each bronchus further divides into airways of diminishing diameter, called tertiary bronchi and bronchioles

Lobes of the left lung
Has only two lobes, to make room for heart (right lung is trilobed)

Bronchioles
Miniscule terminals of the bronchi; gas exchange occurs in tiny sacs (alveoli) at their ends

Heart
Nestled in the pericardial cavity

Pericardial cavity
Formed mainly by a scooplike shape in the left lung

LUNGS

THE TWO SPONGELIKE LUNGS FILL MOST OF THE CHEST CAVITY AND ARE PROTECTED BY THE FLEXIBLE RIB CAGE. TOGETHER, THEY FORM ONE OF THE BODY'S LARGEST ORGANS. THEIR ESSENTIAL FUNCTION IS GAS EXCHANGE—TAKING IN VITAL OXYGEN FROM THE AIR AND EXPELLING WASTE CARBON DIOXIDE TO THE AIR.

LUNG STRUCTURE

Air enters the lungs from the trachea, which branches at its base into two main airways, the primary bronchi. Each primary bronchus enters its lung at a site called the hilum, which is also where the main blood vessels pass in and out of the lung. The primary bronchus divides into secondary bronchi, and these subdivide into tertiary bronchi, all the time decreasing in diameter. Many subsequent divisions form the narrowest airways: the terminal and then respiratory bronchioles, which distribute air to the alveoli. This intricate network of air passages resembles an inverted tree, with the trachea as the trunk, and is known as the bronchial tree. There are corresponding trees for the pulmonary arteries and arterioles, bringing low-oxygen blood from the heart's right side, and the pulmonary venules and veins, returning high-oxygen blood to the heart's left side.

CHEST SECTION
This CT scan shows a horizontal slice through the chest. The heart nestles in the left side of the chest cavity.

Heart

Area occupied by right lung

Vertebra

Rib

Area occupied by left lung

Descending aorta

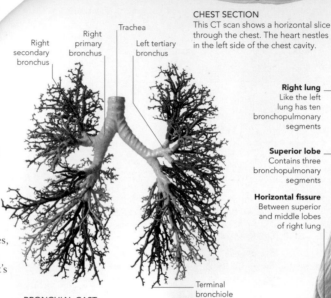

Right secondary bronchus

Right primary bronchus

Trachea

Left tertiary bronchus

Right primary bronchus

Terminal bronchiole

Right lung
Like the left lung has ten bronchopulmonary segments

Superior lobe
Contains three bronchopulmonary segments

Horizontal fissure
Between superior and middle lobes of right lung

BRONCHIAL CAST
By filling the airways of a lung with a resin that hardens, a cast such as this can be made of the bronchial tree. Each color indicates an individual bronchopulmonary segment aerated by a tertiary, or segmental, bronchus.

KEEPING CLEAN
The airway linings have millions of cilia (microhairs). These beat with a wavelike motion to propel mucus, microbes, and dust up the trachea, to be coughed up.

ALVEOLI

The lungs' microscopic air sacs, alveoli, are elastic, thin-walled structures arranged in clumps at the ends of respiratory bronchioles. They resemble bunches of grapes, although the alveoli are partly merged with each other. White blood cells known as macrophages are always present on their inner surfaces, where they ingest and destroy airborne irritants such as bacteria, chemicals, and dust. Around the alveoli are networks of capillaries. Oxygen passes from the air in the alveoli into the blood by diffusion through the alveolar and capillary walls (see p.152). Carbon dioxide diffuses from blood into the alveoli. More than 300 million alveoli in both lungs provide a huge surface area for gas exchange, about 40 times greater than the body's outer surface.

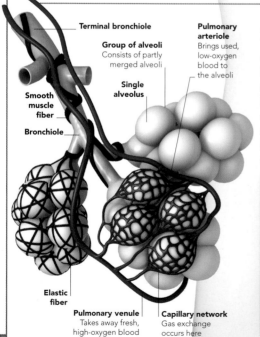

Terminal bronchiole

Group of alveoli
Consists of partly merged alveoli

Single alveolus

Smooth muscle fiber

Bronchiole

Pulmonary arteriole
Brings used, low-oxygen blood to the alveoli

Elastic fiber

Pulmonary venule
Takes away fresh, high-oxygen blood

Capillary network
Gas exchange occurs here

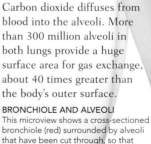

BRONCHIOLE AND ALVEOLI
This microview shows a cross-sectioned bronchiole (red) surrounded by alveoli that have been cut through, so that they resemble air bubbles in a sponge.

Inferior lobe
Contains five bronchopulmonary segments

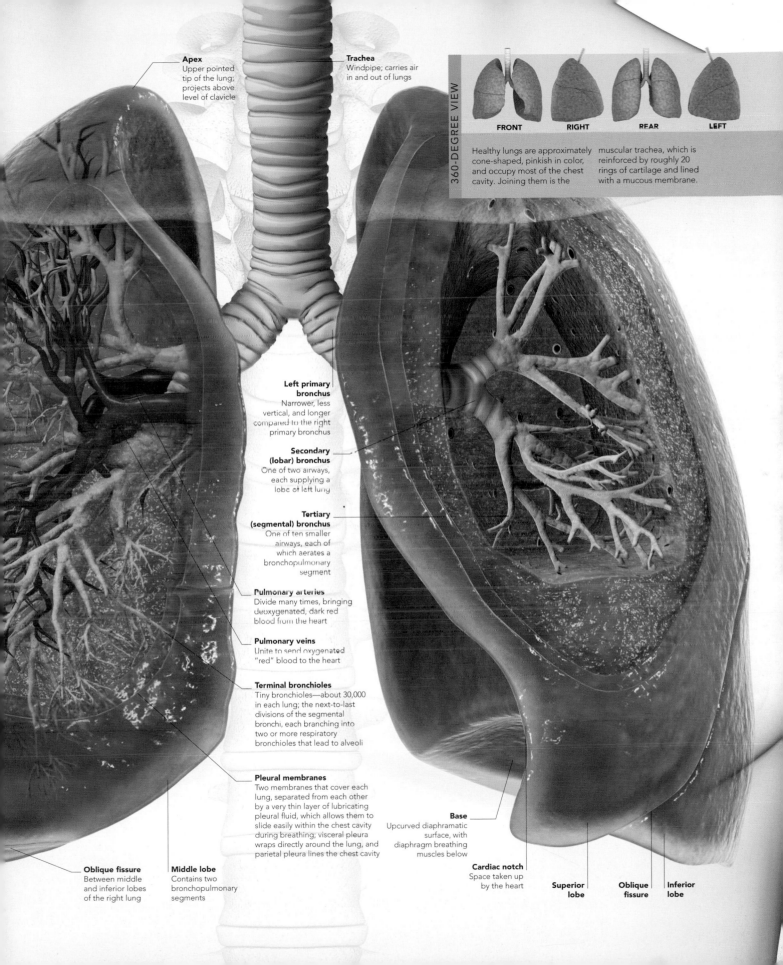

Apex
Upper pointed tip of the lung; projects above level of clavicle

Trachea
Windpipe; carries air in and out of lungs

360-DEGREE VIEW

FRONT   RIGHT   REAR   LEFT

Healthy lungs are approximately cone-shaped, pinkish in color, and occupy most of the chest cavity. Joining them is the muscular trachea, which is reinforced by roughly 20 rings of cartilage and lined with a mucous membrane.

Left primary bronchus
Narrower, less vertical, and longer compared to the right primary bronchus

Secondary (lobar) bronchus
One of two airways, each supplying a lobe of left lung

Tertiary (segmental) bronchus
One of ten smaller airways, each of which aerates a bronchopulmonary segment

Pulmonary arteries
Divide many times, bringing deoxygenated, dark red blood from the heart

Pulmonary veins
Unite to send oxygenated "red" blood to the heart

Terminal bronchioles
Tiny bronchioles—about 30,000 in each lung; the next-to-last divisions of the segmental bronchi, each branching into two or more respiratory bronchioles that lead to alveoli

Pleural membranes
Two membranes that cover each lung, separated from each other by a very thin layer of lubricating pleural fluid, which allows them to slide easily within the chest cavity during breathing; visceral pleura wraps directly around the lung, and parietal pleura lines the chest cavity

Base
Upcurved diaphramatic surface, with diaphragm breathing muscles below

Cardiac notch
Space taken up by the heart

Oblique fissure
Between middle and inferior lobes of the right lung

Middle lobe
Contains two bronchopulmonary segments

Superior lobe

Oblique fissure

Inferior lobe

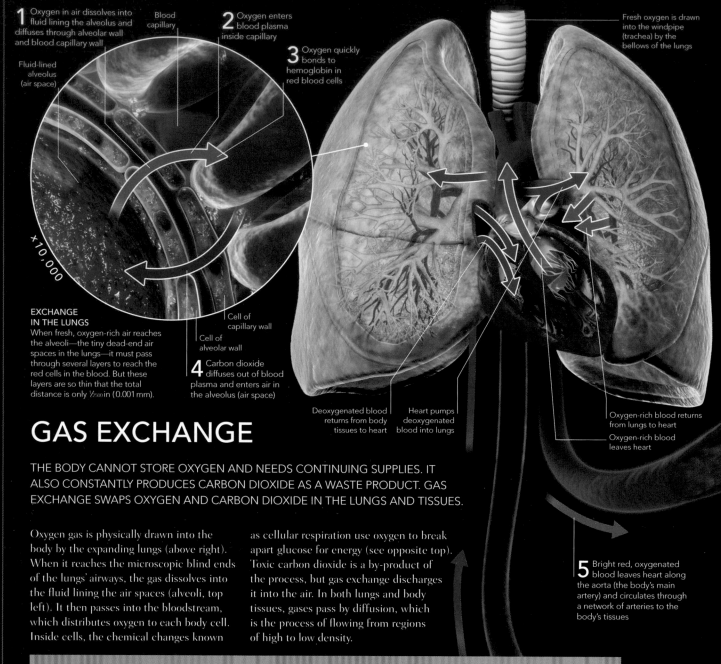

1 Oxygen in air dissolves into fluid lining the alveolus and diffuses through alveolar wall and blood capillary wall

Blood capillary

2 Oxygen enters blood plasma inside capillary

3 Oxygen quickly bonds to hemoglobin in red blood cells

Fresh oxygen is drawn into the windpipe (trachea) by the bellows of the lungs

Fluid-lined alveolus (air space)

×10,000

EXCHANGE IN THE LUNGS

When fresh, oxygen-rich air reaches the alveoli—the tiny dead-end air spaces in the lungs—it must pass through several layers to reach the red cells in the blood. But these layers are so thin that the total distance is only ½₅₀₀ in (0.001 mm).

Cell of capillary wall

Cell of alveolar wall

4 Carbon dioxide diffuses out of blood plasma and enters air in the alveolus (air space)

Deoxygenated blood returns from body tissues to heart

Heart pumps deoxygenated blood into lungs

Oxygen-rich blood returns from lungs to heart

Oxygen-rich blood leaves heart

GAS EXCHANGE

THE BODY CANNOT STORE OXYGEN AND NEEDS CONTINUING SUPPLIES. IT ALSO CONSTANTLY PRODUCES CARBON DIOXIDE AS A WASTE PRODUCT. GAS EXCHANGE SWAPS OXYGEN AND CARBON DIOXIDE IN THE LUNGS AND TISSUES.

Oxygen gas is physically drawn into the body by the expanding lungs (above right). When it reaches the microscopic blind ends of the lungs' airways, the gas dissolves into the fluid lining the air spaces (alveoli, top left). It then passes into the bloodstream, which distributes oxygen to each body cell. Inside cells, the chemical changes known as cellular respiration use oxygen to break apart glucose for energy (see opposite top). Toxic carbon dioxide is a by-product of the process, but gas exchange discharges it into the air. In both lungs and body tissues, gases pass by diffusion, which is the process of flowing from regions of high to low density.

5 Bright red, oxygenated blood leaves heart along the aorta (the body's main artery) and circulates through a network of arteries to the body's tissues

Lower vena cava (one of the body's two main veins) returns deoxygenated blood from lower body to heart

SUPPORTING THE ALVEOLI

Alveoli are only ¹⁄₁₂₅ in (0.2 mm) across when fully inflated. They should collapse inward like deflated balloons due to powerful surface tension in their fluid lining. Their collapse is prevented by surfactant, a natural substance with detergent-like properties. It is produced by alveolar cells and consists mainly of fatty substances, such as cholesterol and phospholipids, and proteins. In addition to keeping alveoli inflated, surfactant plays a role in disabling bacteria, preventing certain lung infections.

Alveolar wall

Air

Fluid layer

ALVEOLUS (AIR SPACE)

Cohesive force

Forces collapsing alveolar wall

Fluid molecule

WITHOUT SURFACTANT
Molecules in the watery fluid lining attract and cohere to each other, making the alveolar wall pull inward and collapse.

Stable alveolar wall

Surfactant molecules

Weakened forces between fluid molecules

WITH SURFACTANT
Molecules of surfactant flow between the fluid molecules and reduce their cohesive forces, allowing the alveoli to stay inflated.

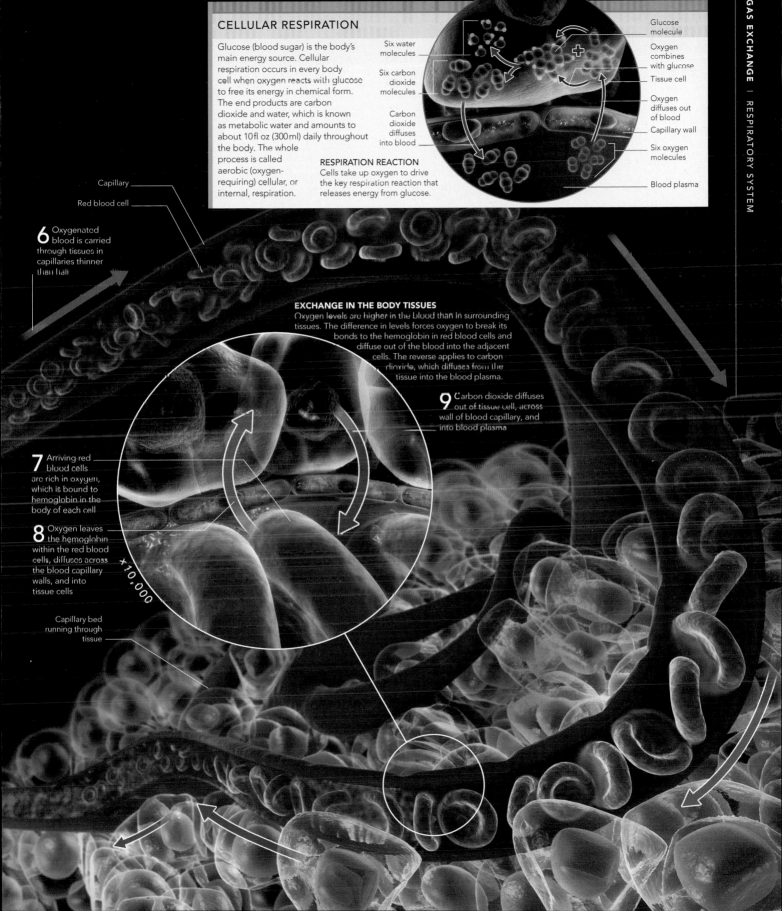

CELLULAR RESPIRATION

Glucose (blood sugar) is the body's main energy source. Cellular respiration occurs in every body cell when oxygen reacts with glucose to free its energy in chemical form. The end products are carbon dioxide and water, which is known as metabolic water and amounts to about 10 fl oz (300 ml) daily throughout the body. The whole process is called aerobic (oxygen-requiring) cellular, or internal, respiration.

RESPIRATION REACTION
Cells take up oxygen to drive the key respiration reaction that releases energy from glucose.

Six water molecules

Six carbon dioxide molecules

Carbon dioxide diffuses into blood

Glucose molecule

Oxygen combines with glucose

Tissue cell

Oxygen diffuses out of blood

Capillary wall

Six oxygen molecules

Blood plasma

Capillary

Red blood cell

6 Oxygenated blood is carried through tissues in capillaries thinner than hair

EXCHANGE IN THE BODY TISSUES
Oxygen levels are higher in the blood than in surrounding tissues. The difference in levels forces oxygen to break its bonds to the hemoglobin in red blood cells and diffuse out of the blood into the adjacent cells. The reverse applies to carbon dioxide, which diffuses from the tissue into the blood plasma.

9 Carbon dioxide diffuses out of tissue cell, across wall of blood capillary, and into blood plasma

7 Arriving red blood cells are rich in oxygen, which is bound to hemoglobin in the body of each cell

8 Oxygen leaves the hemoglobin within the red blood cells, diffuses across the blood capillary walls, and into tissue cells

×10,000

Capillary bed running through tissue

BREATHING AND VOCALIZATION

THE MOVEMENTS OF BREATHING, ALSO KNOWN AS BODILY RESPIRATION, BRING FRESH AIR CONTAINING OXYGEN DEEP INTO THE LUNGS AND THEN REMOVE STALE AIR CONTAINING THE WASTE PRODUCT CARBON DIOXIDE.

BREATHING

The physical movement of air into and out of the lungs is generated by differences in pressure within the lungs compared to the surrounding atmospheric pressure. The pressure differences are produced by forcefully expanding the chest and lungs by muscular action, and then passively allowing them to return to their former size. The rate and depth of breathing can be consciously modified. However, the underlying need to breathe is controlled by areas within the brain stem, where responses to regulate the breathing muscles (of which we are usually not aware) occur according to the levels of carbon dioxide and oxygen in the blood.

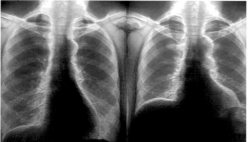

DIAPHRAGM MOVEMENT
The abdominal contents (dark area at the bottom of this X-ray) are flattened by the diaphragm muscle during inhalation (left) and then rise up during exhalation (right).

INHALATION
The chief muscles used in respiration at rest are the diaphragm at the base of the chest and the external intercostals between the ribs. For forceful inhalation, additional muscles assist in moving the ribs and sternum to expand the chest further, and to stretch the lungs even more.

Lung
Expands as diaphragm pulls down and ribs move up and out

Diaphragm
Contracts and becomes flatter to stretch lungs downward

Sternocleidomastoid
Pulls collarbone (clavicle) and sternum up to enlarge upper chest cavity

Scalenes
Three scalene muscles help elevate the uppermost two ribs

Pectoralis minor
Pulls up the third, fourth, and fifth ribs

External intercostals
Narrow the gaps between ribs, making them swing up and out

Ribs
Tilt up and out to expand chest

VOLUME AND PRESSURE

Breathing alters the volume of the chest (thoracic cavity). The lungs "suck" onto the inner chest wall, so that as the cavity expands, they also become larger. The main expanding forces are provided by the diaphragm and intercostal muscles. At rest, the diaphragm carries out most of the work, as 17 fl oz (0.5 liters) of air—the tidal volume—shifts in and out with each breath (12 to 17 times every minute). Rate and volume increase automatically if the body needs more oxygen, such as during exercise. Then forced inspiration can suck in an extra 70 fl oz (2 liters), and forced expiration expels almost as much, leading to a total air shift, or vital capacity, of more than 150 fl oz (4.5 liters) in a large, healthy adult. The breathing rate can triple, producing a total air exchange more than 20 times greater than at rest.

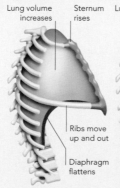

Lung volume increases
Sternum rises
Ribs move up and out
Diaphragm flattens

INHALING
The diaphragm contracts to become less domelike, while the ribs swing upward and outward with a "bucket handle" action to raise the sternum.

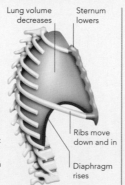

Lung volume decreases
Sternum lowers
Ribs move down and in
Diaphragm rises

EXHALING
The diaphragm relaxes, and the elastic, stretched lungs recoil to become smaller again, allowing the sternum and ribs to move down and inward.

NEGATIVE PRESSURE
As lung volume increases, air pressure within decreases. Atmospheric pressure outside the body is now higher and air is drawn down the airways and into the lungs—in effect, air is "sucked" in.

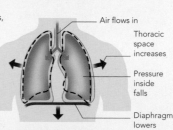

Air flows in
Thoracic space increases
Pressure inside falls
Diaphragm lowers

POSITIVE PRESSURE
As the lung volume diminishes when exhaling, the air is compressed, raising its pressure within the lungs. So the air is pushed back along the airways, and out of the nose and mouth.

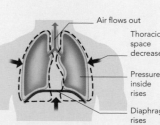

Air flows out
Thoracic space decreases
Pressure inside rises
Diaphragm rises

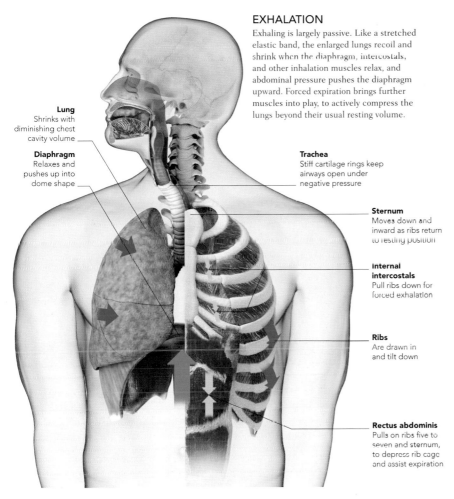

EXHALATION

Exhaling is largely passive. Like a stretched elastic band, the enlarged lungs recoil and shrink when the diaphragm, intercostals, and other inhalation muscles relax, and abdominal pressure pushes the diaphragm upward. Forced expiration brings further muscles into play, to actively compress the lungs beyond their usual resting volume.

Lung
Shrinks with diminishing chest cavity volume

Diaphragm
Relaxes and pushes up into dome shape

Trachea
Stiff cartilage rings keep airways open under negative pressure

Sternum
Moves down and inward as ribs return to resting position

Internal intercostals
Pull ribs down for forced exhalation

Ribs
Are drawn in and tilt down

Rectus abdominis
Pulls on ribs five to seven and sternum, to depress rib cage and assist expiration

VOCALIZATION

The vocal cords (vocal folds) are paired bands of fibrous tissue near the base of the larynx. In normal breathing there is a V-shaped gap between them, called the glottis. Sound is produced when the cords close together, tighten by muscle action, and vibrate as air from the lungs passes between them. The greater the tension in the cords, the higher the pitch (frequency). Above are the false vocal cords (vestibular folds). These do not produce sound but help close off the larynx when swallowing.

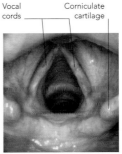

Vocal cords   Corniculate cartilage

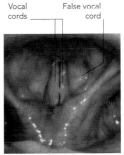

Vocal cords   False vocal cord

CORDS APART
A laryngoscope view shows the vocal cords angled apart during normal breathing, when air passes through the gap between them.

CORDS ADJACENT
Laryngeal muscles swing the arytenoid cartilages, to which the vocal cords are attached, and bring them together.

RESPIRATORY REFLEXES

The two important respiratory reflexes are coughing and sneezing. Both serve to blow out excess mucus, dust, irritants, and obstructions—coughing from the lower pharynx, larynx, trachea, and lung airways, and sneezing from the nasal chambers and nasopharynx. In both cases, a deep inhalation is followed by sudden contraction of the muscles involved in forceful exhalation (see above left). For a cough, the lower pharynx, epiglottis, and larynx close so that air pressure builds up in the lungs, and is released explosively, rattling the vocal cords. In a sneeze, the tongue closes off the mouth, to force air up and out through the nose.

MUCUS SPRAY
Both coughs and sneezes propel a spray of tiny mucus droplets from the respiratory airways for distances of up to 10ft (3m). This image shows spray from a cough.

THE LARYNX

The larynx is sited between the pharynx and the trachea. It has a framework of nine cartilages: the paired arytenoids, cuneiforms, and corniculates, and the unpaired epiglottic, thyroid, and cricoid. The thyroid cartilage forms a prominent mound under the skin of the neck, called the "Adam's apple," which is larger and more pronounced in adult males. The cartilages are held in position by numerous muscles and ligaments, and the larynx is also associated with the hyoid bone just above it, which anchors some of these muscles.

INTERNAL STRUCTURE

The larynx forms a hollow chamber through which air flows silently during normal breathing, and which can tilt its cartilages to bring the vocal cords together for speech.

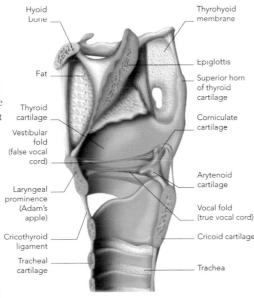

Hyoid bone

Fat

Thyroid cartilage

Vestibular fold (false vocal cord)

Laryngeal prominence (Adam's apple)

Cricothyroid ligament

Tracheal cartilage

Thyrohyoid membrane

Epiglottis

Superior horn of thyroid cartilage

Corniculate cartilage

Arytenoid cartilage

Vocal fold (true vocal cord)

Cricoid cartilage

Trachea

RESPIRATORY DISORDERS

MILLIONS OF MICROBES FLOAT IN EVEN THE CLEANEST OF AIR AND EACH BREATH BRINGS THOSE PARTICLES INTO THE RESPIRATORY TRACT. DESPITE DEFENSE SYSTEMS SUCH AS MUCUS AND CILIA, THESE MICROBES HEIGHTEN THE RISK OF A RESPIRATORY INFECTION. IF THE NOSE, THROAT, OR LARYNX IS INVOLVED, THIS IS KNOWN AS AN UPPER RESPIRATORY TRACT INFECTION (URTI).

COMMON COLD

THIS VIRAL INFECTION IS VERY COMMON, AFFECTING SOME PEOPLE EVERY TWO OR THREE YEARS, BUT OTHERS TWO OR THREE TIMES A YEAR, ESPECIALLY IN CHILDHOOD.

The common cold is one of the most frequently experienced illnesses but also generally one of the less serious. At least 200 different and highly contagious types of virus can cause the problem. They spread in fluid that floats through air, in tiny droplets of mucus coughed or sneezed out by people with colds, and also in films of moisture transferred from person to person by close contact, such as shaking hands, or via shared objects, such as cups. Symptoms involve frequent sneezing, a runny nose, which at first runs with a clear, thin discharge that may later become thicker and greenish yellow, a headache, slightly raised temperature, and perhaps an accompanying sore throat, cough, and sore, reddened eyes. Antibiotic drugs are ineffective since they do not work against viruses. Cold viruses change (mutate) their surface coatings so rapidly that even if antiviral drugs could be made to tackle existing strains, they would be ineffective against the new ones. Most cold remedies, such as decongestants or inhalants to relieve nasal stuffiness, treat the symptoms while the body's immune system attacks the invading microbes.

SPREADING INFECTION
Coughs and sneezes spread diseases—especially common cold viruses, which can be sprayed more than 9½ ft (3 m) in mucous droplets.

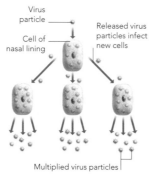

1 VIRUS INVADES CELLS
Virus particles in air land on and invade the cells lining the nose and throat. They rapidly replicate, killing their host cells.

Virus particle
Cell of nasal lining
Released virus particles infect new cells
Multiplied virus particles

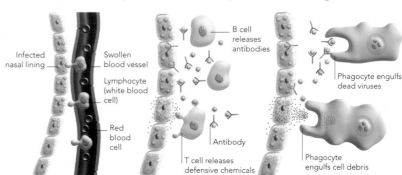

2 WHITE CELLS ARRIVE
Defensive white blood cells squeeze from capillaries toward the infected lining cells, which are creating thin mucus.

Infected nasal lining
Swollen blood vessel
Lymphocyte (white blood cell)
Red blood cell

3 ANTIBODY PRODUCTION
White blood cells called B cells produce antibodies, which immobilize the virus; other white blood cells destroy infected cells.

B cell releases antibodies
Antibody
T cell releases defensive chemicals

4 CLEARING UP
Other white blood cells called phagocytes engulf virus particles, damaged nasal lining cells, and other debris. The cold subsides.

Phagocyte engulfs dead viruses
Phagocyte engulfs cell debris

INFLUENZA

USUALLY CALLED FLU, THIS VIRAL INFECTION CAUSES SYMPTOMS INCLUDING FEVER, CHILLS, SNEEZING, SORE THROAT, HEADACHE, MUSCLE ACHES, AND FATIGUE.

Influenza is primarily an upper respiratory tract infection, but it also has body-wide symptoms: raised temperature, sensations of being hot and sweaty and then cold with shivers, muscle aches, and exhaustion. Even after the main infection has cleared up there may be lingering depression and fatigue. The influenza viruses are coded A, B, and C and are very contagious. Influenza A tends to produce regular outbreaks and can also affect domestic animals such as pigs, horses, and fowl. Influenza B usually causes more sporadic outbreaks in places where many people gather and interact. Influenza C is less likely to produce serious symptoms. The type A virus is most likely to change or mutate. People at risk of complications, such as those with existing illnesses, can be vaccinated before the main risk time of the winter season. Because the virus can mutate, new vaccines are prepared annually. Complications include respiratory tract infections, such as pneumonia and acute bronchitis. Influenza can be life-threatening to very young and elderly people; some epidemics kill people of all ages.

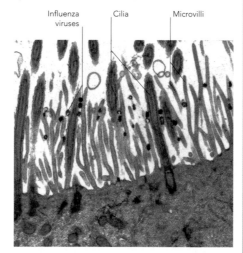

Influenza viruses
Cilia
Microvilli

VIRAL INVASION
Influenza viruses (blue) attach themselves to hairlike microvilli and cilia on the surface of cells that line the upper respiratory tract. They then enter the cells and start to proliferate, eventually causing the cells to die and the symptoms of influenza to become apparent.

BIRD FLU

One type of influenza A virus (technically from a group of viruses known as the Orthomyxoviridae) had its origins in birds, causing the types of illness generally known as avian, or bird, flu. It has relatively recently "crossed over" into mammals, including humans. Virus subtype H5N1 infects several types of birds, including chickens. This strain can infect humans, causing a serious form of influenza with respiratory complications, which is fatal in up to half of the affected people. However, it is only contracted after close contact with infected birds. There is no clear evidence of human-to-human spread as in the other types.

H5N1 VIRUS
A transmission electron micrograph of H5N1. Within the lipid envelope (green) are the proteins hemagglutinin (H) and neuraminidase (N).

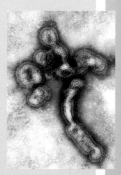

UPPER AIRWAY INFECTION

MANY BACTERIA AND VIRUSES CAUSE UPPER RESPIRATORY TRACT INFECTIONS (URTI), WITH THE NAME OF THE INFECTION DEPENDING ON THE PART MOST AFFECTED.

The upper airway is exposed to a continual intake of microbes with each breath. Harmful microbes may manage to break through the mucous lining and other defenses at various places, and set up an infection zone there. Apart from the nasal chambers, which suffer most during a common cold, other sites at risk include the sinuses; the pharynx, or throat; and the larynx, or voicebox. The sinuses are air-filled cavities that branch from the nasal airways into the facial skull bones. There are also lumps of lymphoid tissue in the upper airways, which may swell markedly with infection. They include the pharyngeal tonsils, or adenoids, in the upper pharynx (nasopharynx), at the rear of the nasal chamber, and the palatine tonsils, or "the tonsils," on either side of the mid-pharynx, near the rear of the soft palate. Each area of infection causes specific symptoms. Inflammation and soreness in various combinations of the pharynx, palatine tonsils, and larynx are often known by the general name of "sore throat." The usual causes are viruses, which may be associated with infection spreading from a common cold. The adenoids and tonsils tend to be larger during childhood because this age group catches more infectious diseases while their immune system is still in the process of developing.

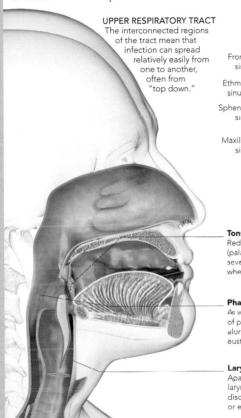

UPPER RESPIRATORY TRACT
The interconnected regions of the tract mean that infection can spread relatively easily from one to another, often from "top down."

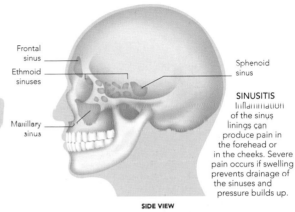

Frontal sinus
Ethmoid sinuses
Sphenoid sinus
Maxillary sinus

FRONT VIEW

Frontal sinus
Ethmoid sinuses
Maxillary sinus
Sphenoid sinus

SIDE VIEW

SINUSITIS
Inflammation of the sinus linings can produce pain in the forehead or in the cheeks. Severe pain occurs if swelling prevents drainage of the sinuses and pressure builds up.

Tonsillitis
Red, inflamed, swollen tonsils (palatine tonsils) can cause a severe sore throat and pain when swallowing.

Pharyngitis
As with other URTIs, the pain of pharyngitis may spread along the airway called the eustachian tube to the ear.

Laryngitis
Apart from a sore throat, laryngitis may produce discomfort when speaking, or even total voice loss.

Infected tonsils

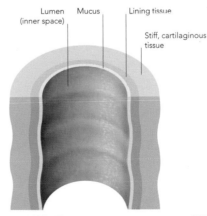

TONSILLITIS
A view into the throat shows the tonsils on either side as enlarged, reddened, and inflamed, or "angry." The white coating is commonly associated with this infection.

Infected larynx

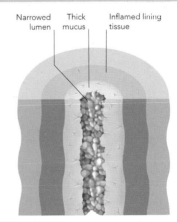

LARYNGITIS
The vocal cords (vocal folds) and laryngeal tissues are swollen and sore. The swelling prevents the cords from vibrating and the voice becomes husky or is lost entirely.

ACUTE BRONCHITIS

BRONCHITIS IS INFLAMMATION OF THE BRONCHI, WHICH ARE THE LARGER AIRWAYS THAT BRANCH FROM THE BASE OF THE TRACHEA, OR WINDPIPE, INTO THE LUNGS.

Acute bronchitis develops suddenly, within 24 to 48 hours. Its symptoms include a persistent and irritating cough yielding clear sputum (phlegm), a tight chest, wheezing and perhaps breathlessness, pain with the cough, and often a slightly raised temperature. This disorder may be a complication of an infection elsewhere in the upper respiratory tract, for example tonsillitis. Usually only the larger and medium-sized bronchi are affected, and they become inflamed and narrowed. Healthy adults usually manage to shake off the infection after a few days, without any need for medical intervention. However, in older people, or those with other respiratory problems, the condition may spread deeper into the lungs, causing a secondary infection, such as bacterial pneumonia.

Lumen (inner space)
Mucus
Lining tissue
Stiff, cartilaginous tissue

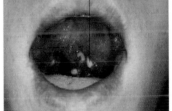

Narrowed lumen
Thick mucus
Inflamed lining tissue

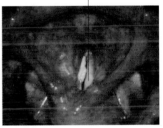

NORMAL BRONCHUS
The airway lining secretes a thin but adequate layer of its protective mucus. This leaves a wide passageway, or lumen, for air to flow in and out of the lung tissues.

INFLAMED BRONCHUS
The lining tissue swells and produces excessive mucus, some of which will be coughed up. Any mucus remaining increases the risk of infection spreading deeper into the lungs.

PNEUMONIA

INFLAMMATION OF THE LUNG'S MICROSCOPIC AIR SACS, THE ALVEOLI, AND THE SMALLEST AIRWAYS, THE BRONCHIOLES, IS KNOWN AS PNEUMONIA.

Pneumonia can develop in different areas of the lung. Lobar pneumonia affects one lobe (large division) of the lung. Bronchopneumonia affects patches of tissue in one or both lungs. The usual cause is bacterial infection, commonly *Streptococcus pneumoniae*. Pneumonia can be triggered as a secondary problem by an upper respiratory tract viral infection, such as a common cold. Pneumonia can also be caused by a range of other bacteria, as well as viruses such as those for influenza and chickenpox, and more rarely, other microorganisms such as protists (protozoa) and fungi. The main symptoms are a cough that brings up blood-stained sputum (phlegm), breathlessness, chest pain, and a high fever with confusion. If the cause of the infection is bacterial, treatment is with antibiotics.

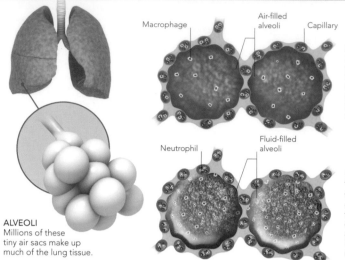

ALVEOLI
Millions of these tiny air sacs make up much of the lung tissue.

HEALTHY ALVEOLI
Macrophages, a type of white blood cell, scavenge in healthy alveoli. They ingest particles of dust and other inert, inhaled irritants but respond slowly to bacteria.

INFLAMED ALVEOLI
The infective process triggers capillary wall changes, and other types of white blood cell, including neutrophils, arrive to attack the bacteria. Fluid accumulates, which reduces oxygen absorption.

LEGIONNAIRES' DISEASE

THIS FORM OF PNEUMONIA-LIKE LUNG INFECTION IS DUE TO THE BACTERIUM *LEGIONELLA PNEUMOPHILA*.

Legionnaire's disease was first described in 1976 after an outbreak of a severe pneumonia-like illness among war veterans at an American Legion convention. It affects men more often than women. Symptoms resemble those of other pneumonias, in particular the respiratory symptoms, but in addition affected people may have diarrhea, abdominal pain, or jaundice. It occurs most often in middle-aged and older people, and the disease may become very serious or even fatal in people with a weakened immune system.

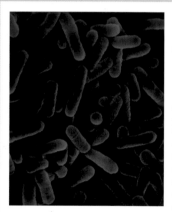

LEGIONELLA BACTERIA
Rod-shaped *Legionella* bacteria are present in most water supplies. They multiply rapidly in water-cooled air-conditioning systems and in plumbing where water can stagnate.

PLEURAL EFFUSION

EXCESS FLUID IN THE TWO-LAYERED MEMBRANE SURROUNDING THE LUNG IS KNOWN AS PLEURAL EFFUSION.

The two layers of membrane, or pleura, are lubricated by a small amount of fluid and allow the lungs to expand and contract smoothly within the chest wall. Infections such as pneumonia and tuberculosis, heart failure, and some cancers can lead to an accumulation of fluid between the pleura, up to 6 pints (3 liters) in volume, which presses on the lungs, causing breathlessness and chest pain. Treatment may initially involve removal of the fluid using a hollow needle or by a tube (chest drain) inserted through the chest wall.

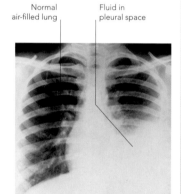

PLEURAL EFFUSION
In this X-ray, the white area over the lower left lung (on the right of the image) is a pleural effusion, which partly obscures the lung tissues, seen as dark areas.

TUBERCULOSIS (TB)

THIS INFECTIOUS DISEASE, MAINLY AFFECTING THE LUNG TISSUE, IS CAUSED BY THE BACTERIUM *MYCOBACTERIUM TUBERCULOSIS*.

Many people harbor the TB microbe, but it causes the disease in only a small proportion, usually if an individual's immunity or resistance is lowered. Symptoms include fever and a persistent cough, loss of appetite, and general weakness. Some new cases are caused by lowered immunity due to the condition HIV/AIDS. Overall, however, vaccination and oral antibiotics have been very successful in preventing and treating TB, and since the year 2000 the numbers of new cases have shown a slow decline.

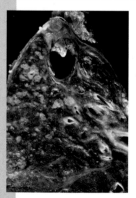

DAMAGED TISSUE
In advanced TB, the lung tissue becomes riddled with tubercles, which are small, firm lumps formed to seal off the centers of infection.

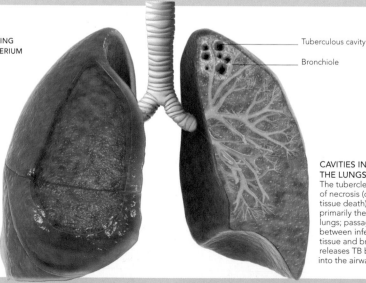

CAVITIES IN THE LUNGS
The tubercle areas of necrosis (cell and tissue death) affect primarily the upper lungs; passage of air between infected tissue and bronchi releases TB bacteria into the airways.

Tuberculous cavity

Bronchiole

PNEUMOTHORAX

PNEUMOTHORAX OCCURS WHEN ONE OR BOTH OF
THE PLEURAL MEMBRANES IS BREACHED AND AIR ENTERS
THE PLEURAL SPACE, CAUSING THE LUNG TO COLLAPSE.

The pleural membranes are separated by a very
thin layer of pleural fluid that lubricates their
movements. The balance of pressures between the
chest wall, pleural layers, and lung tissue makes
the lungs "suck" onto the inside of the chest wall.
In a pneumothorax, air is allowed into the pleural
space. The pressure balance changes, and the lung
collapses. This leads to chest tightness, pain, and
breathlessness. If more air enters the space but
cannot escape (tension pneumothorax), the pressure
around the lung compresses it even further, which
can be life-threatening. A spontaneous pneumothorax
may be due to rupture of an abnormally enlarged alveolus
on the lung surface, or to a lung condition such as asthma,
or by trauma such as rib fracture and chest wounds.

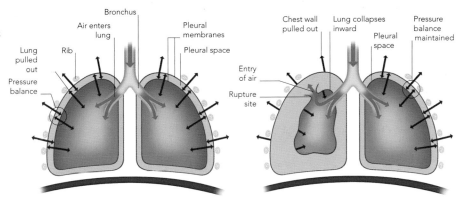

NORMAL BREATHING
The lungs inflate by being pulled out as
they "suck" onto the chest wall. Pressure is
maintained within the fluid-filled pleural space.

COLLAPSED (RIGHT) LUNG
Air from the right lung enters the surrounding
pleural space and changes the pressure balance.
The lung shrinks away from the chest wall

ASTHMA

ASTHMA IS AN INFLAMMATORY LUNG DISEASE THAT
CAUSES RECURRENT ATTACKS OF BREATHLESSNESS AND
WHEEZING, DUE TO NARROWED AIRWAYS IN THE LUNGS.

Asthma is one of the most common and most variable
of lung conditions, affecting as many as one in four
children in some regions. Some people have the
occasional slight episode, others are prone to severe
breathlessness that can threaten life; and some have
attacks that are variable and unpredictable from one
day to the next. The muscle in the walls of the airways
contracts spasmodically, causing the airways to be
constricted and bringing on an attack of breathlessness.
The narrowing is worsened by the secretion of excess
mucus. Most cases develop in childhood and may
be linked to allergy-based problems such as eczema,
with both having an inherited component. In many
children, the trigger for an attack is an allergic reaction
to a foreign substance, or allergen, which can include
tiny inhaled particles such as pollen, mold from the
droppings of house-dust mites, and particles from animal
hair or feathers. Other cases are due to food or drink
allergies, certain drugs, anxiety, stress, respiratory
infection, and vigorous activity in cold weather.

AFFECTED AIRWAYS
Asthma tends to affect not the larger
bronchi but the more slender airways
(red). These are the tertiary bronchi and
the bronchioles that lead to the alveoli.

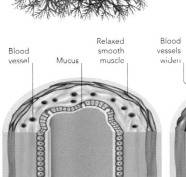

Secondary bronchus

Tertiary bronchus

Terminal
bronchiole

Primary
bronchus

PEAK FLOW METER
Asthma's severity can be
monitored by blowing into
a peak flow meter, which
measures the rate of air flow.

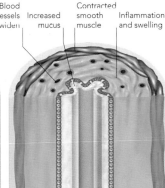

Blood
vessel

Mucus

Relaxed
smooth
muscle

Blood
vessels
widen

Increased
mucus

Contracted
smooth
muscle

Inflammation
and swelling

ASTHMA TREATMENT

There are two main approaches to treatment, which
are usually combined. Corticosteroid drugs (known
as preventers) suppress the inflammatory reaction
and should be taken regularly as prophylactics.
Bronchodilator drugs (known as relievers) are
used for quick relief to treat early symptoms
of an attack; they work rapidly but last only
a few hours. Reduced exposure to allergens
may minimize the frequency and severity
of asthma attacks.

INHALER
Inhaling a spray of antiasthma medication
gets the drug directly to the site of the
problem in the lungs' small airways.

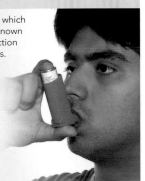

HEALTHY AIRWAY
A normal bronchiole has relaxed smooth
muscle in its walls and an adequate but
thin coating of protective mucus covering
the lining. The passageway for air, or
lumen, is wide enough for sufficient
oxygen-containing air to reach the alveoli.

ASTHMATIC AIRWAY
During an asthma attack, the smooth
muscle contracts. Inflammation due to an
allergic response causes the blood vessels
to widen and the tissues in the airway wall
to swell. The mucus layer also thickens.
This results in narrowing of the lumen.

CHRONIC OBSTRUCTIVE PULMONARY DISEASE

Chronic obstructive pulmonary disease (COPD) consists primarily of chronic bronchitis and emphysema, two conditions that usually occur together in the same person. It is a long-term disorder in which there is progressive damage to lung tissue with increasing shortness of breath. Air flow into and out of the lungs is restricted and the lungs' ability to take in oxygen for the normal body processes diminishes. By far the most important contributory factor for COPD is smoking tobacco.

CHRONIC BRONCHITIS

CHRONIC INFLAMMATION OF THE LUNGS' AIRWAYS IS USUALLY CAUSED BY SMOKING. RARELY, RECURRENT ACUTE INFECTIONS LEAD TO CHRONIC BRONCHITIS.

In chronic bronchitis the main airways leading to the lungs, the bronchi, become inflamed, congested, and narrowed due to irritation caused by tobacco smoke, frequent infections, or prolonged exposure to pollutants. The inflamed airways begin to produce too much mucus (sputum), resulting in a typical cough that at first is troublesome mostly in damp, cold months but then persists throughout the year. Symptoms such as hoarseness, wheezing, and breathlessness also develop. Eventually a person becomes short of breath even at rest. If a secondary respiratory infection develops, the sputum may change appearance from clear or white to yellow or green.

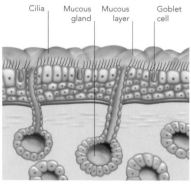

NORMAL AIRWAY LINING
Glands produce mucus that traps inhaled dust and germs. Tiny surface hairs (cilia) propel the mucus up into the throat, where it is coughed up or swallowed.

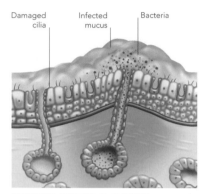

AIRWAY IN CHRONIC BRONCHITIS
Inhaled irritants cause glands to produce more mucus. Damaged cilia cannot propel mucus along, so it becomes a bacterial breeding ground.

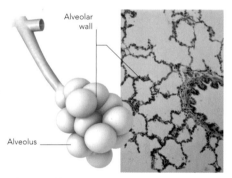

HEALTHY TISSUE
The alveoli are grouped, like grapes, and each sac is partly separate from the others. The walls can stretch because they are thin and elastic.

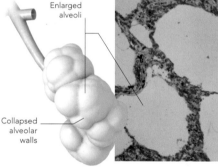

DAMAGED TISSUE
Smoke or other pollutants stimulate chemicals that cause the alveolar walls to break down and merge, reducing the area for gas exchange.

EMPHYSEMA

IN EMPHYSEMA, THE AIR SACS (ALVEOLI) BECOME OVERSTRETCHED. THEY ALSO RUPTURE AND MERGE SO THAT THEIR OXYGEN-ABSORBING SURFACES ARE REDUCED.

The alveoli not only lose their functional gas exchange area, but air also becomes trapped inside them due to their decreased wall elasticity. As a result, the lungs overinflate, the volume of air moving in and out of the lungs is reduced, and less oxygen is absorbed into the bloodstream. Most people affected by emphysema are long-term heavy smokers, although a rare inherited condition called alpha1-antitrypsin deficiency can also cause the condition. Although the damage caused by emphysema is usually irreversible, giving up smoking can sometimes slow down the progression of the disease and allow the cilia (see above) to recover.

OCCUPATIONAL DISEASES

ASBESTOSIS, SILICOSIS, AND PNEUMOCONIOSIS ARE DUE TO INHALING PARTICLES THAT IRRITATE AND INFLAME THE LUNG TISSUE, LEADING TO FIBROSIS.

The people most at risk from occupational lung diseases, such as those listed above, are those whose work exposes them to harmful particles over many years, for example, miners, quarry workers, and stone masons. In occupational lung disease, there is gradual thickening (fibrosis) of the lung tissue, which eventually leads to irreversible scarring. Symptoms such as breathlessness and a cough may develop only slowly, but worsen for years after the exposure has ceased. In developed countries, the diseases are becoming less common because many workers wear protective clothing and masks in risky environments, but regulations are often lax in developing countries.

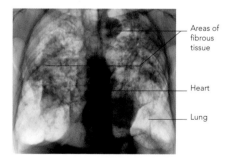

SILICOSIS
The orange patches on the lungs in this chest X-ray are areas of fibrosis caused by silicosis. Inhaled silica particles are ingested by scavenging white blood cells (macrophages). These burst, releasing the silica and other chemicals, which damage the lung tissue.

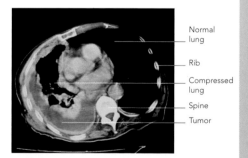

ASBESTOSIS
Asbestos is a substance that can cause serious lung damage if inhaled. In some cases of asbestosis, a form of lung cancer develops. This CT scan shows a malignant tumor, or mesothelioma, on the pleura, the thin membrane that surrounds the lung.

LUNG CANCER

A MALIGNANT TUMOR IN THE LUNG, LUNG CANCER IS THE MOST COMMON CANCER WORLDWIDE WITH OVER A MILLION NEW CASES DIAGNOSED EACH YEAR.

The most common cause of lung cancer—responsible for almost 9 in 10 cases—is tobacco smoke. In the past, lung cancer was far more common in men than women, because more men than women smoked. However, it is now less so, and passive smoking is also leading to an increase among non-smokers. The disease is also becoming increasingly common in developing countries with the spread of tobacco smoking and growing urban populations. Many inhaled irritants trigger the growth of abnormal cells in the lungs, but cigarette smoke contains thousands of known carcinogenic (cancer-causing) substances. In rare cases, lung cancer is caused by asbestos, toxic chemicals, or the radioactive gas radon.

Symptoms of lung cancer

A persistent cough is usually the earliest symptom. Because most people who develop lung cancer are smokers, this is often dismissed as a "smoker's cough." Other symptoms include coughing up blood, wheezing, weight loss, persistent hoarseness, and chest pain. If tests confirm the presence of lung cancer, a lobectomy (removal of a lung lobe) or pneumonectomy (removal of a whole lung) may be performed. This is usually advised only if the tumor is small and has not spread. Chemotherapy and radiotherapy may be given, alone or in combination.

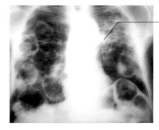

Tumor growing on hilum

TUMOR
Several tumors (white patches) are visible in the lungs. One tumor is growing on the hilum, where the main airway enters the lung.

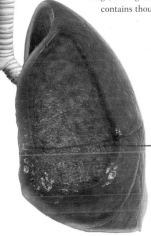

SPREADING CANCER CELLS
Tiny airborne carcinogenic particles lodge in the airways and contribute to the development of cancerous cells. Some of these cells may break away and travel in the blood or lymph to trigger secondary tumors.

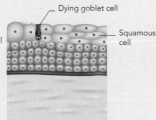

White blood cell

Carcinogens | Alveolus | Capillary

THE SPREAD OF LUNG CANCER
Lung cancer can spread (metastasize) to other parts of the body. Metastases in bones can cause pain and fractures; in the brain, headaches and contusion; and in the liver, weight loss and jaundice.

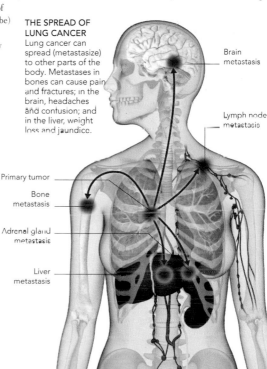

Brain metastasis

Lymph node metastasis

Primary tumor

Bone metastasis

Adrenal gland metastasis

Liver metastasis

SMOKING AND LUNG CANCER

Tobacco smoke is a complex mixture of more than 3,000 different substances, including the addictive stimulant nicotine, benzene, ammonia, hydrogen cyanide, carbon monoxide, and tar. The burning tar elements in the smoke are known to be strongly cancer-causing (carcinogenic). The risk of developing lung cancer increases with the number of cigarettes smoked per day, their tar content, the number of years that a person has smoked, and the depth of inhalation into the lungs. Another risk factor is regular exposure to other people's cigarette smoke, which is known as passive smoking.

SMOKER'S LUNG
Tar is just one of thousands of chemicals in tobacco smoke. In smokers, healthy lung tissue becomes dotted with deposits of tar, visible here to the naked eye.

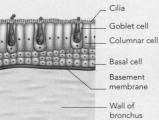

Cilia
Goblet cell
Columnar cell
Basal cell
Basement membrane
Wall of bronchus

1 HEALTHY AIRWAY LINING
Columnar cells topped by tiny, hairlike cilia line healthy airways (bronchi). Basal cells constantly divide to replace damaged columnar cells.

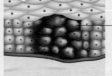

Dying goblet cell
Squamous cell

2 INITIAL DAMAGE
Over time, columnar cells damaged by smoking become squamous cells, which gradually lose their cilia. The mucus-secreting goblet cells die.

Basal cells become cancerous

3 CANCER BEGINS
To replace the damaged cells, basal cells start to multiply at an increased rate. Some of these new basal cells develop into cancerous cells.

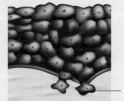

Multiplying cancer cells break through basement membrane

4 CANCER SPREADS
The cancerous cells replace healthy cells. If these cells break through the basement membrane, they can enter blood vessels to travel elsewhere.

FEW BODY PARTS RENEW AS RAPIDLY AS THE SKIN. EVERY MONTH THE OUTER LAYER OF EPIDERMIS IS COMPLETELY REPLACED, AT A RATE OF 30,000 FLAKELIKE DEAD CELLS EVERY MINUTE. THE HAIR AND NAILS ARE LIKEWISE SELF-REINSTATING AND SELF-REPAIRING. SKIN REFLECTS ASPECTS OF GENERAL HEALTH, ESPECIALLY DIET AND LIFESTYLE. ITS EXPOSED AND DYNAMIC NATURE CAN BRING PROBLEMS SUCH AS RASHES, LESIONS, SORES, AND ECZEMA. SKIN GROWTHS MAY FOLLOW EXPOSURE TO HARMFUL CHEMICALS OR CANCER-TRIGGERING

SKIN, HAIR,
AND NAILS

SKIN, HAIR, AND NAIL STRUCTURE

TOGETHER, SKIN, HAIR, AND NAILS ARE KNOWN AS THE INTEGUMENTARY SYSTEM. THE SKIN IS ONE OF THE LARGEST ORGANS IN THE BODY, WEIGHING 6–9 LB (3–4 KG) AND WITH A SURFACE AREA OF ALMOST 21 SQ FT (2 M²). IT IS A COMPLEX ORGAN FORMED OF TWO MAIN LAYERS, WHICH CONTAIN MANY DIFFERENT TYPES OF CELLS, SOME OF WHICH PRODUCE HAIR AND NAIL TISSUE.

SKIN STRUCTURE

The skin is not simply a thin, waterproof covering for the human body, but is a complex organ consisting of a number of specialized cells. Its thickness varies from about ¹⁄₅₀ in (0.5 mm) on delicate areas such as the eyelids, to ¹⁄₅ in (5 mm) or more on areas of wear and tear, such as the soles of the feet. Skin has two main structural layers. The outer epidermis is chiefly protective, and the underlying dermis contains many different tissues with varied functions. The dermis contains thousands of microsensors that enable the sense of touch, as well as sweat glands and adjustable blood vessels that contribute to body temperature regulation. Under the dermis is a layer, sometimes regarded as part of the skin, called subcutaneous fat. It acts as a buffer and provides extra thermal insulation against extreme heat and cold.

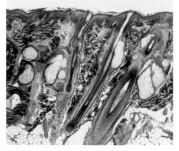

SKIN SECTION
This micrograph shows three hair follicles and globules of sebum in the dermis (blue) with the thin epidermis (pink) on top.

SKIN RENEWAL

The outer epidermis continually renews and replaces itself by cell division. The basal layer consists of boxlike cells that multiply quickly and gradually move up to the surface, pushed by new cells from below. As the cells travel upward, they develop tiny spines or prickles that bind them together tightly. They then begin to flatten and fill with a waterproofing protein known as keratin. Finally, the cells die and reach the surface fully keratinized, resembling untidy, scalelike, interlocking tiles on a roof. As they flake away with daily wear and tear, more cells arrive from below to replace them. The journey from epidermal base to surface takes about four weeks, and a typical person sheds more than 1 lb (0.5 kg) of skin every year.

EPIDERMAL LAYERS
The procession of skin cells from base to surface creates four layers (five in areas of great friction, such as the palms and soles) in the epidermis. As they move upward, the cytoplasm and nucleus in each cell is replaced by keratin.

Surface layer cell
Dead, flattened cell completely filled with keratin

Granular cell
A cell containing granules of the protein keratin

Prickle cell
A many-sided cell that binds closely with its neighbors

Basal cell
A specialized cell that multiplies continually

SKIN STRUCTURE
A patch of skin about the size of a fingernail contains 5 million microscopic cells of at least a dozen main kinds, 100 sweat glands and their pores, 1,000 touch sensors, 100-plus hairs with their sebaceous glands, up to 3⅓ ft (1 m) of tiny blood vessels, and about 1⅔ ft (0.5 m) of nerve fibers.

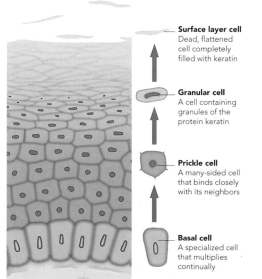

Hair shaft
Part of hair that projects above skin surface

Epidermal surface
Cornified layer of flat, dead, flakelike skin cells

Basal epidermal layer
Layer in which fast cell division renews epidermis above

Touch sensor
Specialized nerve ending at edge of epidermis; other touch sensor types lie at greater depths in dermis

Erector pili muscle
Tiny muscle that pulls hair up when body is cold

Hair bulb
Lowest part of the hair, where growth occurs

Hair follicle
Pouch of epidermis at root of hair

Sebaceous gland
Produces sebum that protects hair and lubricates skin

SKIN REPAIR

Owing to its location, skin suffers more physical damage than any other body organ. However, it has fast-acting repair mechanisms for mending small wounds. If the skin surface is breached, contents leak from damaged cells and stimulate the repair process. Platelets in the blood and the blood-clotting protein fibrinogen work together to form a meshwork of fibers that traps red cells as the beginning of a clot. Meanwhile, tissue-forming fibroblast cells collect in the area, as do white cells called neutrophils, which ingest cell debris and foreign matter such as dirt and germs. The clot gradually hardens and expels fluid to become a scab, as the tissues heal beneath.

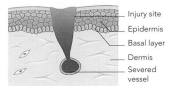

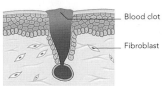

1 INJURY
The wound breaks open cells and releases their contents. These components attract various defense and repair cells.

Injury site
Epidermis
Basal layer
Dermis
Severed vessel

2 CLOTTING
Blood seeps from the vessel and forms a clot. Fibroblasts multiply and migrate to the damaged area.

Blood clot
Fibroblast

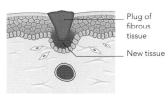

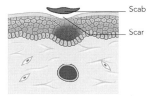

3 PLUGGING
Fibroblasts produce a plug of fibrous tissue within the clot, which contracts and shrinks. New tissue begins to form beneath.

Plug of fibrous tissue
New tissue

4 SCABBING
The plug hardens and dries into a scab, which eventually detaches. A scar may remain but usually fades with time.

Scab
Scar

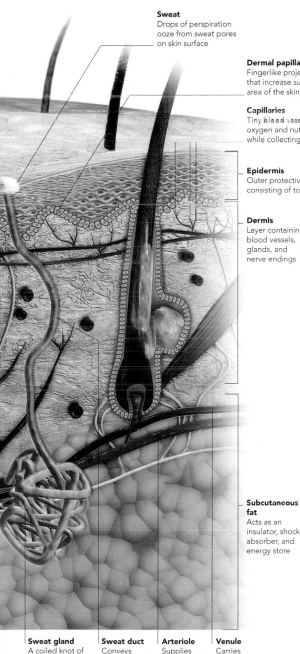

Sweat
Drops of perspiration ooze from sweat pores on skin surface

Dermal papilla
Fingerlike projections that increase surface area of the skin

Capillaries
Tiny blood vessels that supply oxygen and nutrients to tissues, while collecting waste

Epidermis
Outer protective layer consisting of tough, flat cells

Dermis
Layer containing blood vessels, glands, and nerve endings

Subcutaneous fat
Acts as an insulator, shock absorber, and energy store

Sweat gland
A coiled knot of tubes secreting watery sweat

Sweat duct
Conveys sweat to skin surface

Arteriole
Supplies oxygenated blood

Venule
Carries away waste

HAIR GROWTH

Hairs are rods of dead, flattened cells filled with keratin and have a mainly protective role. The hair's root, or bulb, is buried in a pit, the follicle. As extra cells add to the root, the hair lengthens from its base. Different kinds of hairs grow at varying rates, with scalp hairs lengthening about $\frac{1}{100}$ in (0.3 mm) each day. However, hair does not grow continuously. After three to four years, the follicle goes into a rest phase and the hair may detach at its base. Three to six months later, the follicle activates again and begins to produce a new hair.

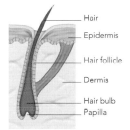

Hair
Epidermis
Hair follicle
Dermis
Hair bulb
Papilla

ACTIVELY GROWING
New cells created at the root get pushed up so the hair gets longer.

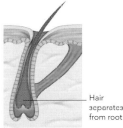

Hair separates from root

INACTIVE PHASE
Activity in the follicle stops and the hair stops growing.

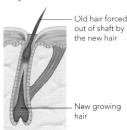

Old hair forced out of shaft by the new hair
New growing hair

NEW GROWTH
The follicle root reactivates and starts to produce a new hair as the old one falls out.

NAIL STRUCTURE

Fingernails and toenails are hard plates made of a tough protein called keratin. Growth takes place under a fold of flesh (cuticle) at the nail base. An area called the nail matrix adds keratinized cells to the nail root, and the whole nail is continuously pushed forward along the nail bed toward its free edge. Most nails grow about $\frac{1}{50}$ in (0.5 mm) each week, with fingernails lengthening faster than toenails.

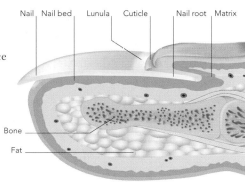

Nail | Nail bed | Lunula | Cuticle | Nail root | Matrix
Bone
Fat

CROSS-SECTION THROUGH NAIL AND FINGER

SKIN AND EPITHELIAL TISSUES

SKIN PLAYS A VITAL ROLE IN ENCLOSING AND PROTECTING THE DELICATE
UNDERLYING TISSUES BUT IT IS ALSO IMPORTANT IN PROVIDING THE SENSE
OF TOUCH. AS AN OUTER LAYER, SKIN IS A SPECIALIZED TYPE OF EPITHELIUM.
EPITHELIAL TISSUES ARE WIDESPREAD IN THE BODY, PROVIDING COVERINGS
AND LININGS FOR ALMOST ALL MAJOR BODY PARTS AND ORGANS.

COMPLEXITIES OF TOUCH

The sense of touch is based in the lower of the two skin layers,
the dermis. Touch operates by means of microsensors, the
endings of tiny nerve cells that act as receptors for various
kinds of physical change, from the lightest contact to heavy,
painful pressure. There is a wide array of microsensors, the
number and density of which vary from one location to another
on the body. On average, a skin patch approximately the size
of a fingernail contains about 1,000 receptors of various kinds.
However, the skin on the fingertips has more than 3,000
receptors that detect light touch for precise feeling. There
are also receptor fibers wrapped around the bases of hairs,
in their follicles (pits) within the dermis. Different types of
receptor respond more readily to certain types of stimulation,
but almost all respond to most stimuli. It is thought that the brain runs through
what look like random incoming nerve signals but recognizes, then picks out,
repeating patterns to determine if an object touched is hard or soft, hot or cold,
rough or smooth, wet or dry, static or moving.

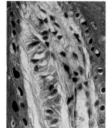

LIGHT-TOUCH SENSOR
This microscope view
shows a Meissner's
corpuscle (green) in a
fingertip. It is important for
light discriminatory touch.

TYPES OF SENSOR

Each type of microsensor is set at a particular depth
in the dermis that best suits its function. The largest
receptors, Pacinian corpuscles, are located at the
deepest level, near the base of the dermis. Sensors for
light touch are located near or just in the epidermal layer.

Free nerve endings
Branching, usually unsheathed sensors of
temperature, light touch, pressure, and
pain. They are found all over the body and
in all types of connective tissue.

Meissner's corpuscle
Encapsulated nerve ending in the skin's
upper dermis, especially on the palms,
soles, lips, eyelids, external genitals,
and nipples. Respond to light pressure.

Merkels's disk
Naked (unencapsulated) receptors, usually
in the upper dermis or lower epidermis,
especially in non-hairy areas. They sense
faint touch and light pressure.

Ruffini corpuscle
Encapsulated receptor in the skin and
deeper tissue that reacts to continuous
touch and pressure. In joint capsules, it
responds to rotational movement.

Pacinian corpuscle
Large, covered receptor located deep in
the dermis, as well as in the bladder wall,
and near joints and muscles. It senses
stronger, more sustained pressure.

**Superficial nerve
ending**
Penetrates the epidermis;
occur everywhere in the
skin and include free
nerve endings

Meissner's corpuscle
Upper dermal nerve
ending; mostly located
just below the base of
the epidermis

Merkels's disk receptor
Junction nerve ending;
sited just above or below
the boundary between
epidermis and dermis

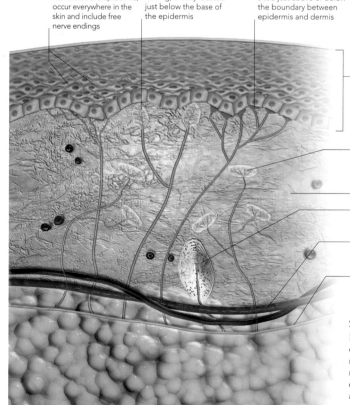

Epidermis
Layer of constantly renewing
cells; multiply at base; harden
and die as they move outward

Ruffini corpuscle
Mid-dermal nerve ending; mostly
scattered through the middle or
lower layers of the dermis

Dermis
Mix of collagen, elastin, and
other connective tissue; houses
most of the touch receptors

Pacinian corpuscle
Located deep in the dermis

Blood vessel
Brings nourishment to skin
layers and touch receptors

Nerve fiber
Receptors' nerve fibers gather
into bundles; convey signals
to the main nerves

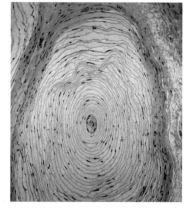

DEEP-PRESSURE SENSOR
Pacinian corpuscles have a multilayered
structure and are the largest of the skin
receptors, in some areas being more
than ½sin (1 mm) long.

SKIN MICRORECEPTORS

Deformation of the layers within a receptor, and expansion
or contraction due to temperature changes, will generate
nerve impulses. The impulses travel along the receptor's
nerve fiber and join with bundles of other fibers in the deep
dermis or below. Most receptors "fire" nerve signals infrequently
and irregularly when not stimulated, increasing their firing
rate as the skin is touched.

TEMPERATURE REGULATION

One of the skin's functions is to contribute to thermoregulation (the maintenance of a constant body temperature). It does this in three main ways: widening and narrowing of blood vessels, sweating, and hair adjustment. If the body becomes hot, blood vessels in the dermis widen (vasodilate) to allow extra blood flow so that more warmth can be lost from the surface. The skin may look flushed, and sweat oozes from sweat glands and evaporates, drawing away body heat. If the body is cold, the peripheral blood vessels narrow (vasoconstrict) to minimize heat loss, and sweating is reduced. Tiny body hairs are pulled upright by the erector pili muscles to trap air as an insulating layer (see p.32).

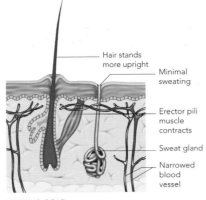

Hair stands more upright
Minimal sweating
Erector pili muscle contracts
Sweat gland
Narrowed blood vessel

FEELING COLD
Tiny body hairs, raised by contraction of the erector pili muscles, create small mounds known as goose pimples at their bases. The peripheral blood vessels constrict, and sweat glands reduce their activity.

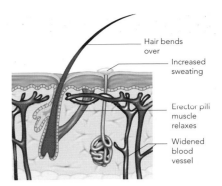

Hair bends over
Increased sweating
Erector pili muscle relaxes
Widened blood vessel

FEELING HOT
Tiny body hairs lie flatter as the erector pili muscles relax, and the small mounds at their bases disappear. Dermal blood vessels dilate, increasing blood flow, and the sweat glands raise their output of sweat.

EPITHELIUM

Epithelial tissue, also called epithelium, is an important structural element that acts as a lining or covering for other body tissues. Epithelium can be classified according to the shape and layout pattern of individual cells (see below), and also the arrangement those cells into one or more layers. Most epithelial tissues form membranes and are specialized for protection, absorption, or secretion. They do not contain blood vessels, and their cells are usually anchored to, and stabilized by, a basement membrane. There may be other cell types present, such as goblet cells that secrete blobs of mucus for release onto the surface.

PSEUDOSTRATIFIED EPITHELIUM

This type of columnar epithelium appears to be arranged in vertical layers. However, it actually consists of a single layer of cells of varying shapes and heights. The nuclei (control centers) of the different cell types are also at different levels, creating a layered (stratified) effect. Taller cells may be specialized into mucus-making goblet cells or ciliated cells that trap foreign particles. This type of epithelium occurs in the airway linings, and the excretory and male reproductive passages and ducts.

Cilia
Surface of goblet cell

TRACHEAL LINING
The electron micrograph shows cilia (green strands) projecting from the epithelial cells of the throat (trachea). Mucus-secreting goblet cells between the cilia possess tiny microvilli (yellow-brown).

TRANSITIONAL EPITHELIUM

This epithelial tissue is similar to layered (stratified) epithelium but has the ability to stretch without tearing. There are usually columnar cells in the basal layer, which become gradually rounder toward the upper surface. As these layers stretch, the cells flatten, or become more squamous. Transitional epithelium is well suited to the urinary system, where it lines areas within the kidney, ureters, bladder, and urethra. It allows these tubes to bulge as urine flows through at pressure. The epithelium also secretes mucus that protects it from acidic urine.

TYPES OF EPITHELIAL CELL

The cells that make up the epithelial layers are usually classified according to their shape. Since most epithelia, as a consequence of their locations, are subject to friction, compression, and similar physical wear, they divide rapidly to replace themselves.

Squamous
Platelike or flattened cells, wider than deep, resembling paving slabs or random paving; flattened nuclei.

Features: Cells allow selective diffusion, or permeability, allowing certain substances to pass, owing to thinness of layer.

Cuboidal
Cube- or box-shaped cells, occasionally hexagonal or polygonal; nucleus usually in cell center.

Features: Substances absorbed from one side of the layer can be altered as they pass through the cytoplasm of the cuboidal cells, before leaving.

Columnar
Tall, slim cells, often square, rectangular, or polygonal; large, oval nucleus near cell base.

Features: Cells protect and separate other tissues; may be topped with cilia for movement of fluid outside the cell or microvilli for absorption.

Glandular
Epithelial cells modified for secretion, usually cuboidal or columnar with secretory granules or vacuoles.

Features: Layers of these cells may be infolded to form pits, pockets, grooves, or ducts, as in sweat glands.

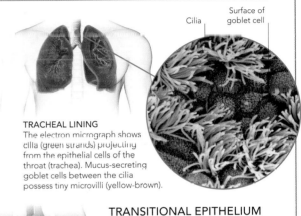

BLADDER LINING
The electron micrograph shows the tightly packed epithelial cells of the bladder lining. They are soft and pliable, enabling them to stretch as the bladder fills with urine.

Rounded epithelial cell

SIMPLE AND LAYERED EPITHELIUM

Simple epithelium is composed of a single layer of cells. It is often found in areas where substances need to pass through easily, a single-cell thickness offering minimal resistance. Layered (stratified) epithelium has two or more layers and is better for protection. Some complex epithelia have more than five layers, but two or three is more usual. The cells may be different shapes in the different layers.

EPITHELIUM IN THE EYE
The eye contains two types of epithelium: simple epithelium in the pigmented layer of the retina, and stratified squamous epithelium in the domed front "window" of the cornea.

CORNEA STRUCTURE
The epithelium covering the cornea is transparent and about five layers thick. It permits light rays to enter the eye.

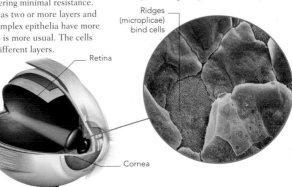

Ridges (microplicae) bind cells
Retina
Cornea

SKIN AND HAIR DEFENSIVE FUNCTIONS

Skin is the body's first line of defense against potential harm. As such, it is well equipped to prevent physical damage due to its supple, cushioned qualities. The epidermal cells that form the skin's outermost layers are tightly knit together, but allow a certain amount of pliability. The cells are almost entirely full of the tough protein keratin, which resists attack by many kinds of chemicals. The natural secretion of sebum from the millions of sebaceous glands, each associated with a hair follicle, is slightly oily at body temperature and spreads easily. It furnishes the skin with partially water-repellent and antibiotic qualities, inhibiting the growth of certain microorganisms, and prevents hairs from becoming too brittle.

SCALP HAIR
Head hairs help keep rainwater from the scalp, absorb or deflect some of the energy in knocks and blows, and shield the head from extremes of temperature.

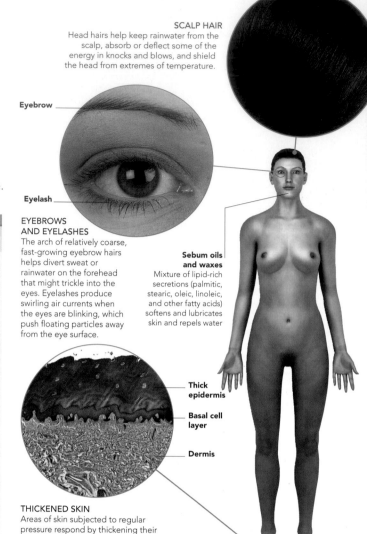

Eyebrow

Eyelash

EYEBROWS AND EYELASHES
The arch of relatively coarse, fast-growing eyebrow hairs helps divert sweat or rainwater on the forehead that might trickle into the eyes. Eyelashes produce swirling air currents when the eyes are blinking, which push floating particles away from the eye surface.

Sebum oils and waxes
Mixture of lipid-rich secretions (palmitic, stearic, oleic, linoleic, and other fatty acids) softens and lubricates skin and repels water

Thick epidermis

Basal cell layer

Dermis

THICKENED SKIN
Areas of skin subjected to regular pressure respond by thickening their epidermis for greater protection and buffering, as in this magnified image of skin from the foot.

Toenails
Made of almost solid keratin

ULTRAVIOLET DEFENSES

The Sun's rays include a spectrum of color wavelengths, including infrared or IR rays (the warming component) and ultraviolet, UV, rays. Both UV-A and UV-B wavelengths are invisible to human eyes, but exposure to the latter, in particular, is linked to forms of skin cancer (see opposite). Skin's self-defense is its dark coloring substance, or pigment, melanin. This forms a screen in the upper epidermis that shields the actively multiplying cells in the base of the epidermis.

MELANIN PRODUCTION
Melanocytes are melanin-producing cells in the base of the epidermis. They make parcels of melanin granules,melanosomes, which pass on into surrounding cells.

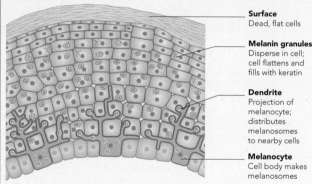

Surface
Dead, flat cells

Melanin granules
Disperse in cell; cell flattens and fills with keratin

Dendrite
Projection of melanocyte; distributes melanosomes to nearby cells

Melanocyte
Cell body makes melanosomes

SKIN PIGMENTATION

Skin color depends on the type and quantity of two main melanin pigments, reddish pheomelanin and brown-black eumelanin, in the epidermis, and on the way the pigment granules are distributed. Each melanocyte has fingerlike dendrites that contact 30–40 surrounding cells (basal keratinocytes). The melanocyte produces its pigments as granules within membrane-bound organelles called melanosomes. These move along the dendrites and are "nipped off" into the cells. Darker skin has larger melanocytes with more melanosomes, which break down to give even pigmentation through skin cells. Lighter skin has smaller melanocytes and grouped melanosomes. Exposure to UV light stimulates melanocytes so that the skin becomes darker, or tanned.

COLOR VARIATION
Darker skin tends to have larger melanin-making cells that produce more, larger, denser melanosomes, in comparison to lighter skin. The former release their pigment granules while the latter tend to stay as clumps.

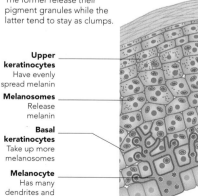

DARK **INTERMEDIATE** **LIGHT**

Upper keratinocytes
Have evenly spread melanin

Melanosomes
Release melanin

Basal keratinocytes
Take up more melanosomes

Melanocyte
Has many dendrites and is active

Surface
Tilelike cells

Upper keratinocytes
Contain little spread-out melanin

Melanosomes
Stay intact

Basal keratinocytes
Take up fewer, lighter melanosomes

Melanocyte
Has few dendrites; not very active

SKIN INJURY AND DISORDERS

SKIN CONTAINS SOME OF THE FASTEST-MULTIPLYING CELLS IN THE BODY. SEVERAL OF ITS DISORDERS RESULT FROM PROBLEMS IN THIS SELF-RENEWAL SYSTEM, INCLUDING VARIOUS GROWTHS AND TUMORS. AS THE BODY'S FIRST LINE OF DEFENSE, SKIN IS SUSCEPTIBLE TO INJURY, ALLERGIC REACTION IN THE FORM OF RASHES, AND INFECTION BY BACTERIA, FUNGI, AND OTHER MICROORGANISMS.

SKIN CANCERS

SEVERAL TYPES OF MALIGNANCIES, OF VARYING SEVERITY, AFFECT SKIN. MOST ARE ASSOCIATED WITH PROLONGED EXPOSURE TO THE HARMFUL RADIATION IN SUNLIGHT.

Basal cell carcinoma usually develops slowly and is unlikely to spread to other body parts (metastasize). Typically it starts as a small, smooth, painless lump, pink or brownish gray, with a pearly or waxlike border. As it widens it may form a central depression with rolled edges. Squamous cell carcinoma can be due to prolonged exposure to ultraviolet (UV) rays or carcinogens such as tar and oil chemicals. It begins as a red or red-brown lump with an irregular edge that is hard and painless, and may then weep and become ulcerlike. Malignant melanoma can develop from an existing mole, or as a fast-growing, dark-colored, asymmetrical spot. Features include increasing size, an irregular border, itching, bleeding, and crusting. All of these types of skin cancer require prompt medical attention.

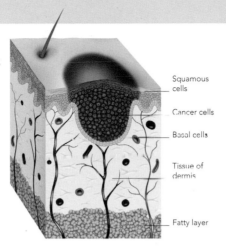

BASAL CELL CARCINOMA
Fast dividing cells in the base of the epidermis are damaged by UV exposure and begin to multiply out of control, forming a mound of flattened, or squamous, cells. The growth remains localized within the epidermis.

Squamous cells
Cancer cells
Basal cells
Tissue of dermis
Fatty layer

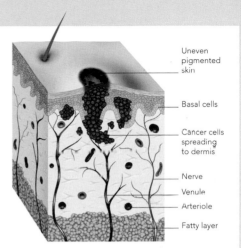

MALIGNANT MELANOMA
Radiation damage to pigment-producing cells, called melanocytes, causes them to proliferate uncontrollably. A dark, irregular mass forms, while some cancer cells break into the dermis and may travel in the blood to other sites.

Uneven pigmented skin
Basal cells
Cancer cells spreading to dermis
Nerve
Venule
Arteriole
Fatty layer

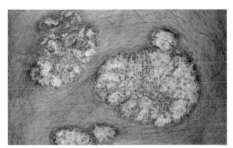

PSORIASIS
There are several types of psoriasis, mostly characterized by intermittently itchy patches of red, thickened, scaly skin, as dead epidermal cells accumulate. Common sites are the knees, elbows, lower back, scalp, and behind the ears.

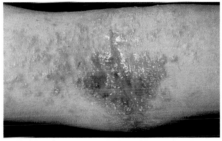

ECZEMA
A typical eczema rash is red, inflamed, and itchy, with small fluid-filled blisters or episodes of dry, scaly, thickened, and cracked skin. Common sites are the hands and creased areas of skin, such as the wrists, elbows, and knees.

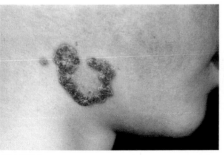

IMPETIGO
This bacterial infection is common on the face, most often around the nose and mouth. The skin reddens and develops fluid-filled blisters, which burst. This is followed by redness, weeping, and crusting that may itch.

VITILIGO
Depigmented patches of skin develop over months or years, especially on the face and hands, and usually before the age of 20. The areas are more distinct in people with dark skin. They do not carry any medical risks.

RASHES

MOST RASHES ARE AREAS OF SKIN INFLAMMATION. SOME ARE CONDITIONS OF THE SKIN ITSELF, OTHERS ARE PART OF GENERALIZED DISORDERS AFFECTING INTERNAL ORGANS.

Some skin rashes may be localized while others are more widespread. Localized rashes can occur on parts of the body exposed to, for example, sunlight, friction, or an irritant chemical. Some of these rashes have a strong inherited component, but often the actual cause or mechanism of rash formation is not clear. The condition may not pose a threat to survival, but it can be unsightly, affect quality of life, and require long-term control with self-help preventive measures and ongoing medications. Psoriasis is a common and widespread condition in which a patchy rash flares up now and again. Episodes may be triggered by infection, injury, stress, or as a side effect of drug treatment. Eczema (a term that is mostly interchangeable with dermatitis) is one of the most common skin conditions, especially in babies and children, although many people grow out of the condition. It is often linked to allergic tendencies such as asthma and perennial, or seasonal, rhinitis (hay fever), and may flare up during adolescence and adulthood. Impetigo is a blistering of the skin caused by bacterial infection, typically through a surface breach such as a cut, cold sore (*Herpes simplex* virus), or scratched, weeping eczema. Vitiligo has an autoimmune basis, where the body makes antibodies that attack the skin's pigment-making cells, melanocytes. It occurs in patchy areas, often symmetrically, over the body; in about one-third of cases, the pigmentation spontaneously returns.

SKIN MARKS AND BLEMISHES

Marks, swellings, and blemishes on the skin include small, pus-filled spots known as pustules, larger ones called boils, and acne, which usually occurs in the teenage years. Other marks or enlargements may be caused by a local increase in cell numbers, as in warts and moles. Swellings may also be due to different types of cyst. Some blemishes are caused by external factors, such as pressure and exposure to sunlight, or they can result from viral infection.

ACNE

A RASH OF SPOTS THAT APPEAR, USUALLY ON THE FACE, DUE TO BLOCKAGE AND INFLAMMATION OF GLANDS IN THE SKIN.

In acne vulgaris, the sebaceous glands produce an excessive amount of the oily-waxy secretion, sebum. This reacts on contact with air and forms a plug in the skin pore, which may be dark with pigmentation (not dirt), as a blackhead or comedone, or pale, as a whitehead. A combination of trapped sebum, dead cells, and bacterial infection inflame the area, causing a pustule. Acne is a common problem at puberty due to hormone surges.

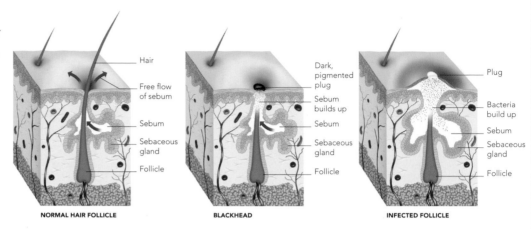

NORMAL HAIR FOLLICLE

Hair
Free flow of sebum
Sebum
Sebaceous gland
Follicle

BLACKHEAD

Dark, pigmented plug
Sebum builds up
Sebum
Sebaceous gland
Follicle

INFECTED FOLLICLE

Plug
Bacteria build up
Sebum
Sebaceous gland
Follicle

MOLE

A MOLE IS A FLAT OR RAISED MARK THAT VARIES IN SHAPE, COLOR, AND TEXTURE, AND MAY BE SINGLE OR NUMEROUS.

A mole, or nevus, is a localized overproduction and aggregation of the skin's pigment cells (melanocytes), with increased amounts of melanin pigment. Moles are very common; most adults have 10–20 moles by the age of 30 years. They can occur almost anywhere on the body and are variable in size, but usually less than ⅖ in (1 cm) across. Rarely, moles become malignant, and any change in size or appearance, itching, or bleeding, should be discussed with a doctor.

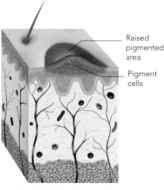

Raised pigmented area
Pigment cells

CROSS SECTION OF A MOLE
Although raised to the exterior, the area of pigmentation in this mole does not extend to cells beneath the epidermis.

CYST

A HARMLESS, SACLIKE SWELLING UNDER THE SKIN THAT CONTAINS FLUID OR SEMISOLID MATERIALS IS CALLED A CYST.

The most common type of cyst is a sebaceous cyst, or wen, that forms in a hair follicle. A cyst contains sebaceous secretions and dead cells, which are restrained in a strong baglike capsule. Its surface mound is usually smooth, and some cysts have a paler or darker central region. Common sites include the scalp, face, trunk, and genitals, although they can occur just about anywhere on the body. Treatment may be needed if the cyst becomes enlarged, unsightly, painful, or infected.

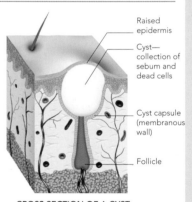

Raised epidermis
Cyst— collection of sebum and dead cells
Cyst capsule (membranous wall)
Follicle

CROSS SECTION OF A CYST
The epidermis is stretched and raised into a domelike lump where this sebaceous cyst protrudes from the dermis.

BOIL

FOUND ON THE SKIN, A BOIL IS A RED, INFLAMED, PUS-FILLED, TENDER AREA CAUSED BY A BACTERIAL INFECTION.

A boil is a collection of pus inside a hair follicle or a sebaceous gland; it may extend to both. Usually a result of bacterial infection, especially from various types of *Staphylococcus*, a boil starts as a small red lump. The bacteria multiply within the follicle or gland and the area becomes tender and swells as pus accumulates and "gathers" into a white or yellow head at the boil's center. A cluster of boils may link to form a carbuncle. Recurrent boils may signify an underlying disorder.

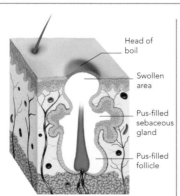

Head of boil
Swollen area
Pus-filled sebaceous gland
Pus-filled follicle

CROSS SECTION OF A BOIL
Both the hair follicle and the sebaceous gland are filled with pus causing a red and tender swelling in the overlying skin.

WART

A WART IS SMALL GROWTH CAUSED BY A VIRAL INFECTION; IT MAY BE FLAT OR RAISED AND SMOOTH OR ROUGH.

Warts are due to infection of the skin by the human papilloma virus (HPV). The virus invades the skin and causes a localized overgrowth of epidermal cells. Excess cells are pushed upward and outward to form a lump on the skin's surface. Warts may be described as common or flat and may occur in groups called crops. Warts on the sole are known as plantar warts and are flattened into the skin, causing pain. Warts often disappear of their own accord, but this may take some time.

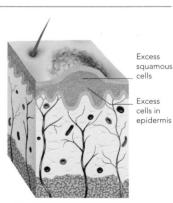

Excess squamous cells
Excess cells in epidermis

CROSS SECTION OF A WART
Overproliferation of epidermal cells causes this typical appearance of a common wart on the surface of the skin.

WOUNDS

WHEN THE SKIN'S SURFACE IS DAMAGED BY WOUNDING, HEALING CAN PROCEED ON ITS OWN OR WITH MEDICAL HELP.

Wounds may occur by accident, or as incisions during surgery. How well a wound heals depends on its extent and depth, the condition and alignment of the edges, the age and health of the affected person, and the prevention of infection. A well-closed, clean wound usually heals in a few weeks and leaves hardly a blemish. An open, jagged-edged wound takes longer to heal and may leave a puckered scar. The healing process can be aided by stitching (suturing) or adhesive closures.

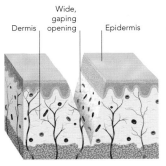

Dermis   Penetration into dermis   Epidermis

PUNCTURE WOUND
Punctures are small in area but deep in penetration. Healing is usually quick but infection is a risk, from microbes entering deep tissue—especially the tetanus bacterium (*Clostridium tetani*) found in soil.

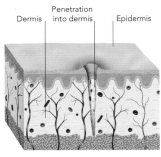

Wide, gaping opening   Dermis   Epidermis

CUT
Neat-edged cuts usually heal with only minimal scarring if they are properly cleaned, closed to promote skin-edge healing, and covered to keep out infection. Deeper cuts may need stitches (sutures) to avoid scars.

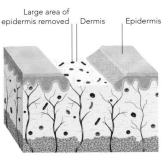

Large area of epidermis removed   Dermis   Epidermis

ABRASION
Scrapes, or abrasions, may affect a wide area of skin, damaging many nerve endings and causing considerable pain. Abrasions of the epidermis usually heal with no scar. Deeper abrasions are more likely to leave a scar.

BURNS, BRUISES, BLISTERS, AND SUNBURN

VARIOUS TYPES OF TRAUMA AND HEAT CAN HARM THE SKIN AND CAUSE PROBLEMS, SUCH AS BLISTERS FROM EXCESSIVE FRICTION OR PRESSURE, AND BURNS.

Burns may be caused by heat, electricity, radioactivity, or chemicals, and can cause extensive and, sometimes, life-threatening cell damage. A contusion, or bruise, is a discolored area of skin caused by bleeding into underlying tissues, usually due to physical impact. Bruising around the eye is called a "black eye." Local physical trauma, such as rubbing and pressure, may lead to a raised, fluid-filled area called a blister. Blisters can also be caused by heat, including the sun's ultraviolet (UV) rays. Sunburn is skin damage as a result of acute or prolonged overexposure to the sun's UV rays. The skin gets red, hot, swollen, and painful, followed by peeling. Excessive sun exposure without adequate protection can lead to skin cancer.

Red, inflamed skin

BURN
The skin is reddened (left) and the epidermis damaged in the region of a burn. If the dermis beneath is also affected, blisters soon form.

Bruise undergoes color changes

BRUISE
Leaked blood cells break down causing the bruise to change color from blue to brown–yellow. Bruising for no apparent reason requires medical advice.

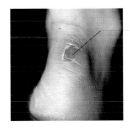

Burst blister requires covering

BLISTER
Fluid leaks from damaged vessels and collects under the skin's surface. New skin grows under the blister, and when it bursts, the old skin dries and peels away.

Dry skin peels or flakes

SUNBURN
After overexposure to the sun's UV rays, skin becomes hot, red, sore, and swollen; later it dries and peels. In severe cases, blisters may develop.

DANDRUFF AND ALOPECIA

DANDRUFF IS EXCESSIVE SHEDDING OF SKIN FLAKES FROM THE SCALP; ALOPECIA IS HAIR LOSS, SOMETIMES PERMANENT.

Dandruff is a harmless but unsightly and embarrassing condition, in which the normal shedding of skin cells from the scalp accelerates. The pale flakes show on the scalp and in the hair, and there may be itching. The condition is most common in young adults and is linked to a yeast organism, *Malassezia globosa*. Alopecia is hair loss, which can be local or general, and temporary or permanent; it is most noticeable on the scalp. Various possible causes include sensitivity to testosterone, leading to male-pattern baldness; chemotherapy, and the autoimmune disorder, alopecia areata.

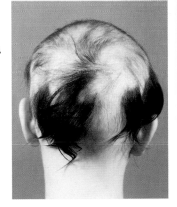

ALOPECIA AREATA
In this condition, hair is lost in patches over the scalp, leading to bald areas surrounded by shorter, broken hairs. The hair usually grows again over a few months. In rare cases, the disease can cause hair loss through the whole body and be permanent.

INGROWN NAIL

THE NAIL CURVES IN ALONG ONE OR BOTH SIDE EDGES, FORCING ITS WAY INTO THE FLESH OF A FINGER OR TOE.

An ingrowing nail penetrates into the tissue along its sides, causing inflammation, discomfort, pain, and a risk of infection. The big toe is most commonly affected, and often in young males. It may be triggered by ill-fitting footwear that presses on the nail and toe, along with incorrect trimming of the free nail edge, which should be cut straight across rather than around the curve. A toe injury is another possible cause. Poor foot hygiene increases the risk of infection and worsens the problem. Minor surgery can be carried out to remove the ingrowing part of the nail and to destroy its nail bed to prevent regrowth.

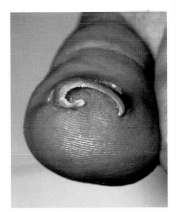

INGROWN TOENAIL
The nail of the big toe in this example clearly curves inward and cuts into the flesh. This produces redness, swelling, and broken skin, which may break and ooze pus, clear fluid, or even blood. The skin may react by overgrowing and enveloping the nail.

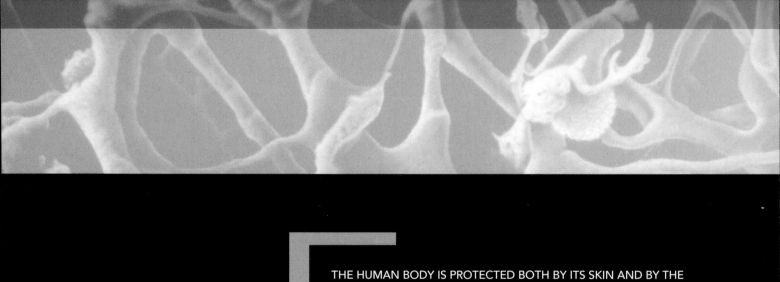

THE HUMAN BODY IS PROTECTED BOTH BY ITS SKIN AND BY THE
LYMPH AND IMMUNE SYSTEMS. EVERY DAY IT IS OPEN TO ATTACK
FROM TWO TYPES OF HOSTILITY. ONE IS EXTERNAL: THE DAILY
BATTLE AGAINST PHYSICAL HARM AND LINGERING GERMS. THE
OTHER ADVERSITIES ARE WITHIN, SUCH AS GERMS THAT HAVE
GAINED ENTRY, AND THE BODY'S OWN CELLS, WHICH CAN SET
UP DISEASES SUCH AS CANCERS. THE IMMUNE SYSTEM FIGHTS

LYMPH AND IMMUNITY

174

LYMPH AND IMMUNE SYSTEMS

SEVERAL SYSTEMS OF THE BODY HELP DEFEND IT AGAINST VARIOUS HAZARDS
—SUCH AS THE SUN'S ULTRAVIOLET RAYS, EXCESSIVE HEAT, HARMFUL CHEMICALS,
PHYSICAL DAMAGE, AND THE THREAT OF MICROORGANISMS SUCH AS BACTERIA
AND VIRUSES. HOWEVER, THE IMMUNE SYSTEM, INCORPORATING THE LYMPHATIC
SYSTEM, IS THE MAIN MEANS BY WHICH THE BODY IS PROTECTED FROM INVASION.

The lymphatic system is an integral part of the immune
system and it plays an important part in the body's
defense against disease. The active part of the system
is lymph fluid, which starts life as the interstitial fluid
that collects between cells throughout the body. It
drains into networks of tiny capillaries in tissue spaces
that unite to form larger vessels called lymph vessels.
Lymph nodes (lymph glands) are the filtering and
storage areas of the system, and they are scattered
along the routes of the lymph vessels. Unlike blood,
lymph is not pumped; instead it moves passively
as lymph vessels are compressed by contraction of
surrounding muscles during movement. Lymph fluid
enters the blood circulation via the left and right
subclavian veins. Lymphoid organs, including the
thymus and spleen, and lymphoid tissue, such as the
tonsils and Peyer's patches, complete the system. They
contain large numbers of specialized white blood cells,
particularly lymphocytes, which protect the body against
non-self material such as invading microorganisms.

Adenoids
Or pharyngeal tonsils; situated in rear
of nasal cavity; help filter incoming air
and destroy microorganisms

Tonsils
Two pairs of tonsils (palatine and lingual)
at back of mouth on either side of
pharynx and at base of tongue help
guard against inhaled microbes

Cervical (neck) nodes
Collect lymph from right
or left side of the face,
scalp, nasal chamber,
and throat

**Axillary (armpit)
nodes**
Drain lymph from arm,
breast, chest wall, and
upper abdomen

Left subclavian vein
Point at which lymph
from left and lower
body enters blood
after collecting in
thoracic duct

Thymus gland
Site of maturation
of immune-system
T cells (T lymphocytes);
T cells develop from
stem cells, which
migrate here from
bone marrow

Spleen
Largest lymph organ;
spleen acts as store
for some types of
lymphocyte and
as a major site
for filtering blood

Peyer's patch
One of a few
clusters of
lymphoid
nodules in
lower part of
small intestine;
helps protect
against
microbes
ingested
in food

Right lymphatic duct
Collects lymph from
upper-right quadrant
of body, including
right arm and right
sides of head
and chest

**Right subclavian
vein**
One of two main
exit points at which
lymph drains into
blood system

Thoracic duct
Or left lymphatic duct;
collects lymph fluid from
both legs, abdomen,
left arm, and left sides
of head and chest

Cisterna chyli
Enlarged lymph
vessel formed
from vessels that
converge from legs
and lower body;
eventually narrows
into thoracic duct

Supratrochlear node
Collects lymph from
hand and forearm

**Lumbar
lymph node**
Drains lymph
from
abdominal
organs

**External
iliac nodes**
Receive
lymph from
organs of
lower
abdomen

360-DEGREE VIEW

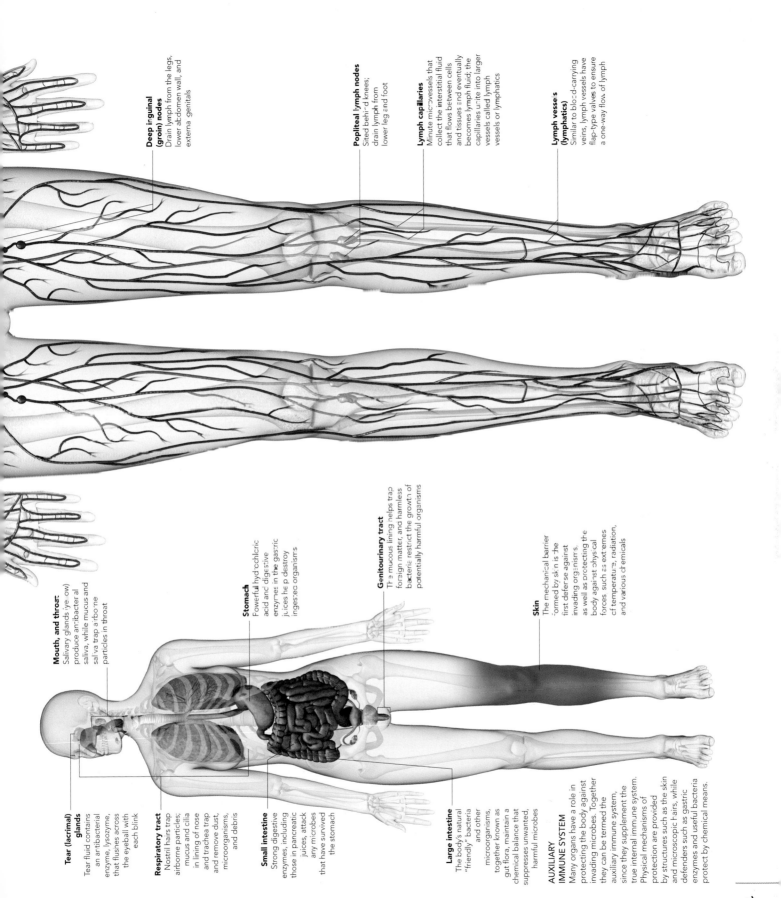

Deep inguinal (groin) nodes
Drain lymph from the legs, lower abdomen wall, and external genitals

Popliteal lymph nodes
Sited behind knees; drain lymph from lower leg and foot

Lymph capillaries
Minute microvessels that collect the interstitial fluid that flows between cells and tissues and eventually becomes lymph fluid; the capillaries unite into larger vessels called lymph vessels or lymphatics

Lymph vessels (lymphatics)
Similar to blood-carrying veins, lymph vessels have flap-type valves to ensure a one-way flow of lymph

Mouth, and throat
Salivary glands (yellow) produce antibacterial saliva, while mucus and saliva trap airborne particles in throat

Tear (lacrimal) glands
Tear fluid contains an antibacterial enzyme, lysozyme, that flushes across the eyeball with each blink

Respiratory tract
Nostril hairs trap airborne particles; mucus and cilia in lining of nose and trachea trap and remove dust, microorganisms, and debris

Stomach
Powerful hydrochloric acid and digestive enzymes in the gastric juices help destroy ingested organisms

Small intestine
Strong digestive enzymes, including those in pancreatic juices, attack any microbes that have survived the stomach

Large intestine
The body's natural "friendly" bacteria and other microorganisms, together known as gut flora, maintain a chemical balance that suppresses unwanted, harmful microbes

Genitourinary tract
The mucous lining helps trap foreign matter, and harmless bacteria restrict the growth of potentially harmful organisms

Skin
The mechanical barrier formed by skin is the first defense against invading organisms, as well as protecting the body against physical forces such as extremes of temperature, radiation, and various chemicals

AUXILIARY IMMUNE SYSTEM

Many organs have a role in protecting the body against invading microbes. Together they can be termed the auxiliary immune system, since they supplement the true internal immune system. Physical mechanisms of protection are provided by structures such as the skin and microscopic hairs, while defenders such as gastric enzymes and useful bacteria protect by chemical means.

IMMUNE SYSTEM

THE ADAPTABLE SYSTEM OF BODY DEFENSES CENTERS ON SPECIALIZED WHITE BLOOD CELLS CALLED LYMPHOCYTES, WHICH RESPOND TO INVASION BY VARIED MICROORGANISMS. THE COMPLEXITIES OF THE SYSTEM LEAD, WHEREVER POSSIBLE, TO IMMUNITY, IN WHICH, AFTER THE FIRST ATTACK, THE BODY IS PROTECTED OR RESISTANT TO FUTURE INCURSIONS BY EACH PARTICULAR TYPE OF MICROORGANISM.

LYMPH NODES

The lymph nodes (or "glands") are vital to the body's defense system because they produce and harbor immune cells (lymphocytes) that protect the body from disease. Lymph nodes are scattered throughout the body and are also concentrated in groups (see p.175). Each node is a mass of lymphatic tissue divided into compartments by partitions of connective tissue known as trabeculae. Lymph fluid from most tissues or organs flows through one or more lymph nodes, where it is filtered and cleaned, before draining into the venous bloodstream. Several smaller lymphatics (vessels) bring lymph to the node, and one larger vessel carries it away. The lymph vessels have valves to ensure a one-way flow of fluid.

INSIDE A NODE
Lymph nodes vary in diameter from ½₅ to 1 in (1 to 25 mm), although they can swell during infection or illness. Covered in a fibrous capsule, they contain sinuses, where many scavenging white blood cells, called macrophages, ingest bacteria as well as other foreign matter and debris.

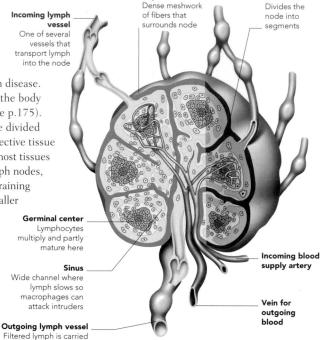

Incoming lymph vessel
One of several vessels that transport lymph into the node

Capsule
Dense meshwork of fibers that surrounds node

Trabecula
Divides the node into segments

Germinal center
Lymphocytes multiply and partly mature here

Sinus
Wide channel where lymph slows so macrophages can attack intruders

Outgoing lymph vessel
Filtered lymph is carried away by only one vessel

Incoming blood supply artery

Vein for outgoing blood

WHITE CELL TYPES

There are numerous types of white blood cell, known by the general name of leucocytes. Some grow and mature into other types. All are derived from the bone marrow.

Monocyte
Largest cell in the blood, with big and rounded, or indented, nucleus; engulfs pathogens.

Lymphocyte
Chief immune cell, with large nucleus that almost fills the cell; either B or T, depending on development.

Neutrophil
Granulocyte (having many small particles in the cytoplasm) with a multilobed nucleus; engulfs pathogens.

Basophil
Circulating granulocyte with lobed nucleus; involved in allergic reactions.

Eosinophil
Granulocyte that is important during allergic reactions; B-shaped nucleus; destroys antigen–antibody complexes.

LOCAL INFECTION

If harmful microbes enter body tissues, both the inflammatory and immune responses act swiftly to limit their spread. The infection may be confined to a naturally defined site, such as the boundary between two sets of tissues. White blood cells and invaders, living and dead, accumulate, along with fluids, toxins, and general debris. The resulting mixture is known as pus, and if it gathers in a localized area, it is an abscess. As the pus collects, it puts pressure on surrounding structures. This may cause discomfort and pain, especially if the surroundings have no flexibility, as in a tooth abscess. The pressure of a brain abscess may have serious consequences for brain functioning.

DENTAL ABSCESS
Microbes gain access through a decayed region of enamel and dentine, infect the pulp, and spread into the root, where pus collects. As pus presses on the pulp nerves, it causes the pain of toothache.

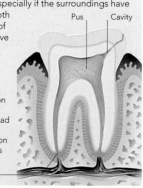

Pus   Cavity

Abscess

NONSPECIFIC RESPONSE

The immune responses involve attacks on specific microorganisms or the toxins (harmful substances) they produce, as shown opposite. Nonspecific responses react to any kind of damage, such as a physical knock, a burn, extreme cold, corrosive chemicals, and various forms of radiation, as well as living invaders, ranging from microbes to large parasites, such as worms and flukes. The main nonspecific defensive response is inflammation (see pp.160–61). Damaged tissue releases chemicals that attract white blood cells. The walls of the capillaries, the smallest blood vessels at the site, become more permeable and porous to allow the passage of white blood cells, defensive chemicals, and fluids, which accumulate as the battle proceeds. The white blood cells surround, engulf, and destroy the invading pathogens. The blood may also clot to form a barrier that not only seals the leak but also prevents further microbial penetration.

INFLAMED TISSUE
The four common signs of inflammation are redness, swelling, increased warmth, and discomfort or pain. They occur after any form of harm such as injury, irritation, or infection, in order to limit the damage and initiate repair and healing.

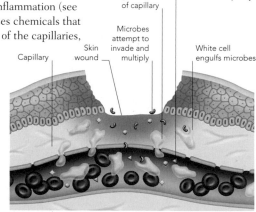

White cells squeeze out of capillary

Defense chemicals diffuse from capillary

Microbes attempt to invade and multiply

White cell engulfs microbes

Capillary

Skin wound

SPECIFIC RESPONSE

Specific responses may occur alongside nonspecific reactions such as inflammation, or follow if the infection persists. There are two main types of specific defense: cell-mediated and antibody-mediated (humoral) immunity. Both depend on the actions of lymphoctyes of two different kinds, B and T lymphocytes. B cells make protein antibodies known as gammaglobulins. These react against antigens (foreign protein substances), which differ from the body's own natural proteins. T cells multiply and attack cells infected by pathogens.

CELL-MEDIATED IMMUNITY

The cellular or cell-mediated type of immunity involves various types of T lymphocyte or T cell, named because they develop inside the thymus gland. A T cell can only recognize an antigen when it is presented by a macrophage. Once a T cell recognizes the antigen, it multiplies rapidly and its offspring differentiate into several types. Killer (cytotoxic) T cells attack and destroy infected cells, which have the antigen on their surface. Helper T cells activate both B cells, to help antibody-mediated immunity, and macrophages, to engulf debris. Suppressor (regulatory) T cells dampen down the body's immune response after the infection has been dealt with.

ANTIBODY-MEDIATED IMMUNITY

Whereas T cells attack body cells infected by microbes, B lymphocytes kill microbes in body fluids by producing chemicals called antibodies. These are generally Y or T shaped. Each type of antibody acts against a certain microorganism or "nonself" material by attaching to antigens on its surface. The presence of antigens triggers B cells to multiply. Some develop into plasma cells, which are the main antibody-producing cells. As with cell-mediated immunity, memory cells are produced, which can recognize the same antigen and initiate defense many years later.

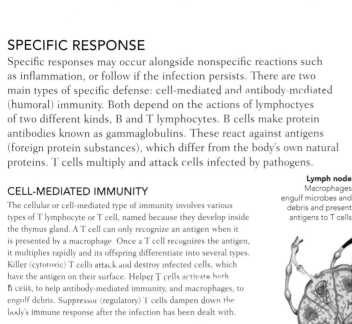

Invading microbe
Pathogenic (harmful) microorganisms such as bacteria

Antigen

Transport to lymph node
Macrophages travel in blood and lymph

Phagocytosis
Macrophage engulfs microorganisms and their antigens

Lymph node
Macrophages engulf microbes and debris and present antigens to T cells

Cell-mediated labels

Memory cell
Some T cells retain memory of the antigen for future defense

Killer T cell
Squeezes into bloodstream and travels to site of infection

Proliferation
Line, or clone, of T lymphocytes, specific to the antigen, multiplies and differentiates into helper, killer, suppressor, and memory cells

Lymphokines
Proteins toxic to microbes made by killer T cells

Macrophage
Attracted to site by lymphokines

Presentation
Macrophage presents antigens from microbe to T cell

Recognition
Preprogrammed T cell recognizes antigen (even if never encountered before)

Helper T cell
Stimulates antibody-mediated immunity by B cells

Phagocytosis
Macrophage engulfs infected cell

Antibody-mediated labels

Memory B cell
Some B cells retain memory of the antigen from a previous infection

Proliferation
One line, or clone, of B lymphocytes, specific to the antigen, multiplies and differentiates into plasma and other cells

Plasma cell
Produces antibodies specific to antigen protein

Antibodies
Float in blood and other fluids

Macrophage
Engulfs antibody–antigen complex and other debris; shares this function with eosinophils

Recognition
Preprogrammed B cell recognizes antigen (even if never encountered before)

Presentation
Macrophage presents antigens from microbe to B cell

Antibody–antigen reaction
Antibodies stick to antigen sites on microbe, forming antibody–antigen complexes (immune complexes)

COMPLEMENT SYSTEM

Circulating in the blood are over 25 proteins and related substances or factors that form the complement system. The complement (or helping and enhancing) proteins are activated by antibodies, bits of cell membrane or DNA, or other products of the battle against invading microorganisms. Once a complement reaction begins, it continues in a "cascade" fashion with one complement protein activating the next, and so on (similar to the cascade reactions of blood clotting). The complement system generally helps destroy microbes and prevent them from attacking body cells, encourages the activity of white cells such as macrophages and T cells (as shown above), widens blood vessels, and clears away the antigen–antibody complexes.

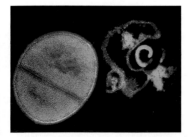

LYSED BACTERIUM
Complement causes dissolving, or lysis, of invaders such as bacterial cells by disrupting their outer membranes (cell shown right).

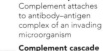
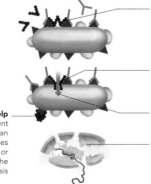

Additional help
The complement pathways can destroy microbes such as bacteria or viruses by the process of lysis

Complement protein binds to complex
Complement attaches to antibody–antigen complex of an invading microorganism

Complement cascade
Sequence of complement reactions produces more complement proteins

Membrane breached
Complement proteins break outer membrane

Swell and burst
Fluid rushes into microbe, which expands and bursts

INFLAMMATORY RESPONSE

INFLAMMATION IS THE BODY'S RAPID, GENERAL RESPONSE TO ANY KIND OF INSULT OR INJURY, SUCH AS PHYSICAL WOUNDS AND FOREIGN OBJECTS, INCLUDING INFECTING ORGANISMS, CHEMICAL TOXINS, HEAT, OR RADIATION.

Unlike the immune response, which is specific against certain invading substances, the inflammatory response is nonspecific. It is a fast, generalized reaction that passes through defined phases and involves various types of white blood cells and defensive chemicals. The four cardinal signs of inflammation, redness, swelling, heat, and pain, are known by the classical terms of rubor, tumor, calor, and dolor. The process aims to attack, break down, and remove any invading material, living or dead, and dispose of the body's own damaged cells and tissues, as well as initiate healing.

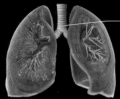

SITE OF DAMAGE
The trachea, or windpipe—the body's main airway

DEFENSIVE CELLS

Various types of white blood cells (leucocytes) become involved in inflammation, including neutrophils and monocytes. The latter are immature leaving the blood vessels and enter the tissues, but rapidly develop into, active cells called macrophages that replace neutrophils.

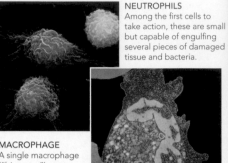

NEUTROPHILS
Among the first cells to take action, these are small but capable of engulfing several pieces of damaged tissue and bacteria.

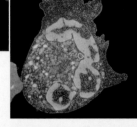

MACROPHAGE
A single macrophage ("big eater") can consume up to 100 bacteria or similar-sized items till it dies.

CAUSE OF INFLAMMATION

The respiratory system is under constant threat from tiny inhaled particles of dust and debris and attack infecting microbes. Here, the lining (epithelium) of the windpipe (trachea) mounts an inflammatory response to dust and bacteria. In reality it usually occurs alongside the specific immune response (see p.158), which targets individual foreign substances.

Red blood cell

Cells of capillary wall

Tufts of cilia
Hairlike projections borne by some cells of the tracheal lining; the cilia beat to remove protective mucus covering the cells

2 Physical damage
As the air current slows, the particles impact the tracheal lining and become trapped in the protective mucus secreted by some of the tracheal lining (epithelial) cells.

1 Causal items
Foreign particles such as microshards of fiberglass and airborne bacteria sweep into the windpipe (trachea) on the current of inhaled air

3 Physical damage
Sharp particles can fracture the epithelial cells, rupturing the delicate cell membranes.

Surface of epithelium

Foreign particle

4 Initial spread
Messenger substances, such as histamine and kinins, leak from ruptured cells, especially "mast cells" scattered through the tissue.

Histamine

Kinins

Lung airways
A network of air passages that supply the lungs

PHAGOCYTOSIS

Various kinds of white blood cells can surround, engulf, and ingest smaller items, such as bacteria and cellular debris, in a process known as phagocytosis ("cell eating"). The cell exploits its ability to change shape and move, using the intracellular components of microtubules and microfilaments (see p.38) that form its flexible, mobile internal scaffolding. The ingestion usually takes less than one second, and the consumed material is gradually broken down by enzymes and other chemicals within the cell (see p.55).

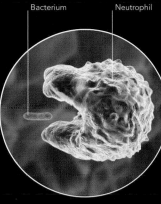

Bacterium   Neutrophil   Bacterium is digested   Digestive vesicle   Expulsion of waste products

1 ENGULFING STAGE
The white cell extends pseudopods ("false feet") toward and around the unwanted items—here a bacterium. The pseudopods merge to engulf them.

2 LYSIS STAGE
The items are trapped in phagocytic vesicles, which with enzyme-containing lysosomes, to form phagolysosomes, where lysis (breaking down) occurs.

3 EXOCYTOSIS STAGE
Harmless products of cell eating are expelled through the white blood cell's membrane, or in tiny membrane-bound exocytic vesicles to the extracellular fluid.

1 Capillaries dilate
Histamine stimulates vasodilation—the widening of blood vessels, especially capillaries. The cells forming capillary walls stretch to become thinner and narrow gaps appear between them, increasing their permeability— their tendency to allow fluids to pass through.

3 Fluid accumulation
Plasma and escaped fluids from damaged cells gather in tissue spaces, causing swelling. This presses on nerve endings, which helps cause the fourth sign of inflammation—pain.

4 Neutrophils arrive
Released chemicals attract white blood cells, such as neutrophils. Neutrophils press themselves onto the inner surface of capillaries, a stage called margination. The neutrophils squeeze between the capillary wall cells, in a process called diapedesis, leaving the blood and entering the tissues.

2 Fluid leakage
Increased blood flow produces redness and heat. Plasma (blood's liquid component, pictured as yellow) leaks into the space between the cells, carrying various proteins such as fibrinogen, which helps blood clot when the skin is broken.

5 Neutrophils enter tissues
Neutrophils are attracted to the damage the by substances released by the disrupted cells. This chemically stimulated movement is termed chemotaxis.

Particle
Remains at site of cell damage continue to release histamine and kinins (red and blue), which flow into the bloodstream

Foreign particle

Bacterium

Bronchial tree
May be affected by inflammation, or the problem may remain restricted to a patch of the trachea

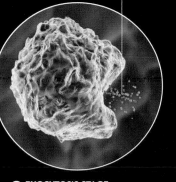

RESPONSE

Once an inflammatory response is triggered, blood flow to the damaged area is increased. The blood vessels, especially capillaries, widen and the capillary walls become thinner and more permeable, allowing plasma and fluid to leak into the space between the cells. Next, white blood cells, such as neutrophils, start to arrive. The neutrophils leave the blood and enter the tissue, drawn to the damaged area by chemicals released by the disrupted cells.

FIGHTING INFECTIONS

AN INFECTION OCCURS WHEN MICROSCOPIC ORGANISMS GAIN ENTRY INTO THE BODY, SURVIVE, MULTIPLY, AND DISRUPT NORMAL CELL FUNCTION. THE INFECTION MAY BE LOCALIZED, SUCH AS IN A PATCH OF SKIN OR IN A WOUND, OR SYSTEMIC, IN WHICH THE ORGANISMS ARE CARRIED AROUND THE BODY BY THE BLOOD AND LYMPH, INVADING MANY DIFFERENT PARTS.

VIRUS SHAPES

There are thousands of different types of virus, with shapes such as balls, boxes, polygons, sausages, golfballs, spirals, and even tiny "space rockets". Viruses are classified by their size, shape, and symmetry as well as by the groups of diseases they cause.

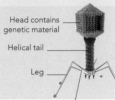

Spiral (helical)
The protein coat is corkscrew-like with the genetic material entwined. Examples include myxoviruses and paramyxoviruses.

Protein subunit (capsomer)

Genetic material

Icosahedral
Twenty equal-sided triangles connect to form a faceted container. Examples include adenoviruses and herpes viruses.

Surface protein (antigen)

Triangular face

Complex
Like a tiny rocket with "landing legs" that settle on the host cell. They only attack bacteria, so are important in health terms when they attack pathogenic bacteria in the human body. Examples include the T4 bacteriophage.

Head contains genetic material

Helical tail

Leg

VIRUSES

An important group of harmful microorganisms, or pathogenic microbes (commonly known as "germs" or "bugs"), is the viruses. Viruses are the smallest of all the microbes; millions would cover the head of a pin. Many viruses can remain inactive for long periods and survive freezing, boiling, and chemical attack, yet they can activate suddenly when they have the opportunity of invading a living cell. Viruses are obligate parasites, which means that they must have living cells, or host cells, in order to replicate themselves. The typical virus particle has a single or double strand of genetic material (nucleic acid, either DNA or RNA) surrounded by a shell-like coat of protein, the capsid, and sometimes a protective outer envelope.

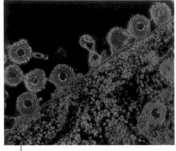

HERPES SIMPLEX VIRUS
An electron micrograph of a cluster of *Herpes simplex* virus (HSV) (orange). HSV1 is responsible for cold sores, and HSV2 for genital herpes.

LIFE CYCLE OF THE INFLUENZA VIRUS

Viruses have very few genes (typically 100–300). They are not made of cells and have no cellular "machinery" for obtaining energy or raw materials, so they cannot process nutrients or reproduce by themselves. To build copies of itself, a virus invades a host cell, and takes over the cell's own machinery. The host cell dies or functions abnormally.

1 Free virus particle
The complete virus particle, which is capable of independent survival and then infection, is known as a virion.

2 Insertion of virus
Viral surface proteins attach to specific receptor sites on the host cell's surface. After attaching itself, part or all of the virus penetrates the host cell.

3 Nucleic acid insertion
The viral RNA moves to the host cell nucleus and inserts itself into the host's nucleic acid. The result is massive replication of the viral RNA, which then moves toward the cell's surface.

4 Nucleic acid replication
The host cell makes many copies of the viral RNA molecule, using the host's raw materials and sometimes its enzymes. The viral protein subunits are also produced by the host's cellular machinery.

5 Budding virus
The nucleic acid (RNA) strands and protein-coat subunits join to form new virus particles. These form buds in the host cell membrane, using part of the membrane as their outer protective envelope.

6 Release
The buds separate as free virus particles, ready to spread and infect more cells. All eight separate segments of genetic material (RNA) must be present for the virus to successfully carry the infection further.

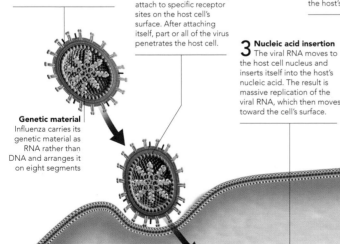

Genetic material
Influenza carries its genetic material as RNA rather than DNA and arranges it on eight segments

Virus in host cell
Virus sheds its protein coat so that RNA can enter host nucleus

Viral proteins

Infected host cell
Host cell may not die after being invaded by an enveloped virus

Viral genes are "read"
Short fragments of viral RNA (genes) are read to produce viral proteins

Replication complete
Duplicated viral RNA escapes through nuclear envelope

IMMUNIZATION

As the immune system tackles and defeats most invading microorganisms, some of the white blood cells called lymphocytes become memory cells. They retain the ability to recognize the foreign substances, antigens, on the microbe's surface. If the same microbe invades again, the memory cells stimulate a rapid immune response to destroy the microbes before they take hold. The process of becoming resistant or immune to a particular microbe as a result of infection is natural immunization. Resistance can also be developed artificially. In active immunization (right), dead or weakened (attenuated) versions of the microbe or its toxic products are injected into the body. The immune response occurs, with production of antibodies, but the illness does not develop. In passive immunization, antibodies for the virus are injected into the body.

RAPID RESPONSE
This color-enhanced microscope image shows a white blood cell (macrophage) engulfing specially weakened bacteria. They are injected into the body to simulate active immunization.

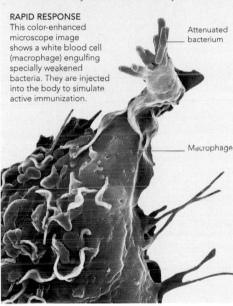

Attenuated bacterium

Macrophage

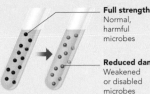

Full strength
Normal, harmful microbes

Reduced danger
Weakened or disabled microbes

1 VACCINE PRODUCTION
A vaccine contains complete or part microbes, or the toxins they make, treated so that they will stimulate the immune response without causing symptoms.

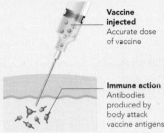

Vaccine injected
Accurate dose of vaccine

Immune action
Antibodies produced by body attack vaccine antigens

2 VACCINE DELIVERY
Introducing the vaccine into the body is called vaccination. It stimulates the immune system to produce antibodies against the antigens on the disease-carrying organisms.

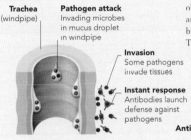

Trachea (windpipe)

Pathogen attack
Invading microbes in mucus droplet in windpipe

Invasion
Some pathogens invade tissues

Instant response
Antibodies launch defense against pathogens

3 IMMUNE RESPONSE
When the body encounters pathogens it has been vaccinated against, the memory cells are already primed and the immune system can launch an instant defense.

IMMUNIZATION AND PUBLIC HEALTH

Immunization gives individual resistance to infectious diseases, and also group protection, which is an important public health measure. For group protection to occur, a relatively high proportion of people must be immunized to minimize the chance of unprotected people coming into contact with the disease. It may eventually be eradicated. If too few people are vaccinated, the others may not only catch the infection, but also act as hosts for the microbes as they change and mutate into new strains. The existing immunization is not effective against new strains of the disease. Some individuals may be advised to abstain from immunization, due to preexisting conditions.

RUBELLA VIRUS
This microscope image shows the rubella, or German measles, virus (pink dots) on an infected cell. A combined immunization (MMR) gives lifelong protection to babies.

PASSIVE IMMUNIZATION

Active immunization (left) takes time to develop and works well in healthy people. If urgent protection is needed, or if a person's immune system is weak, passive immunization can be used. Purified antibodies against the microbe are obtained from people or animals who are immune. The antibodies provide swift resistance against the microbes, but they gradually degenerate and are not replaced. The body has no memory for making them again.

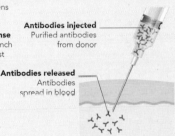

Antibodies injected
Purified antibodies from donor

Antibodies released
Antibodies spread in blood

SHORT-TERM PROTECTION
Specific antibodies are given either to treat an existing infection or to provide protection against a disease in the short term.

VIRUSES AND BACTERIA

The two main categories of harmful microorganisms are viruses and bacteria (see p.182). Viruses cannot exist independently except as inactive chemical structures, whereas bacteria have cellular machinery for obtaining energy, processing nutrients, and reproducing themselves. These latter activities make bacteria vulnerable to chemical interference, a feature exploited by antibiotics. Some bacterial infections, such as tetanus, may be prevented by immunization. Viruses are unaffected by antibiotics, but can sometimes can be treated with antiviral drugs. Immunization can also sometimes prevent viral infections.

Disk containing antibiotic

No bacterial growth

ANTIBIOTIC TREATMENT
Bacteria are painted onto a laboratory nutrient dish with disks containing different antibiotics. Lack of bacterial growth around a disk shows which antibiotics are effective.

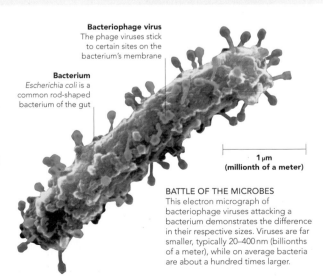

Bacteriophage virus
The phage viruses stick to certain sites on the bacterium's membrane

Bacterium
Escherichia coli is a common rod-shaped bacterium of the gut

1 μm
(millionth of a meter)

BATTLE OF THE MICROBES
This electron micrograph of bacteriophage viruses attacking a bacterium demonstrates the difference in their respective sizes. Viruses are far smaller, typically 20–400 nm (billionths of a meter), while on average bacteria are about a hundred times larger.

BACTERIA

The microorganisms known as bacteria are present almost everywhere: in soil, water, air, food, drink, and on and in our own bodies. Many types of bacteria are harmless; indeed, those present naturally in the human intestines, the "gut flora," have a beneficial effect in helping extract nutrients from food. However, hundreds of types of bacteria cause infections, ranging from mild to lethal. Bacteria are simpler than other single-cell organisms in that their genetic material (DNA) is free in the cell, rather than contained in a membrane-bound nucleus.

BACTERIAL SHAPES

There are several typical shapes for bacteria, and these, along with the way they are colored by laboratory stains, are important for classification and determining their origins and relationships. Many thousands of bacterial types are known, with more discovered each year.

Cocci
Generally spherical, may exist in clumps, chains, or pairs. Examples include *Staphylococcus* and *Streptococcus*.

Cocci in a pair ("diplococci")

Bacilli
Oval or rodlike, with or without surface hairs (pili) or whiplike flagella. Examples are *Streptobacillus* and *Clostridium*.

Pili (hairs) on surface

Spirilla
Spiral or, more accurately, helical (corkscrewlike) in shape, as open or tight coils. Examples include *Leptospira* and *Treponema*.

Open coils

STRUCTURE OF A BACTERIUM
A typical rod-shaped bacterium (bacillus) has a cell membrane enclosing cytoplasm and organelles, such as ribosomes, which are distributed in it. Unlike animal cells, it has a semirigid cell wall outside its cell membrane.

Nucleoid
Area with most of the genetic material

Ribosome
Involved in the manufacture of proteins

Cytoplasm
Complex fluid containing many dissolved substances

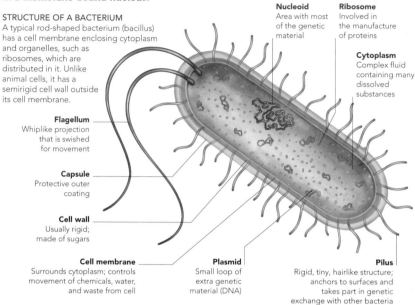

Flagellum
Whiplike projection that is swished for movement

Capsule
Protective outer coating

Cell wall
Usually rigid; made of sugars

Cell membrane
Surrounds cytoplasm; controls movement of chemicals, water, and waste from cell

Plasmid
Small loop of extra genetic material (DNA)

Pilus
Rigid, tiny, hairlike structure; anchors to surfaces and takes part in genetic exchange with other bacteria

HOW BACTERIA CAUSE DAMAGE

Disease-causing bacteria can enter the body in several ways: via the airways or digestive tract, during sexual contact, or through breaks in the skin. Once inside, some bacteria adhere to and invade body cells, such as the dysentery-causing *Shigella dysenteriae*. Others produce poisonous substances known as bacteriotoxins, or toxins. Many of these alter the biochemical reactions in body cells. The diphtheria toxin from the bacterium *Corynebacterium diphtheriae* damages heart muscle by inhibiting protein production. Some of these toxins are highly dangerous. A bucket of nerve toxin from *Clostridium botulinum* could kill everyone in the world.

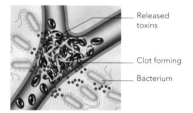

Released toxins

Clot forming

Bacterium

LEAKING VESSELS
Some bacteria release toxins that cause blood to clot in small blood vessels, depriving tissues and organs of their normal blood supply.

RESISTANCE TO ANTIBIOTICS

Many bacteria are able to develop resistance to antibiotics by changing (mutating) into new strains. Their most effective mechanism is the rapid transfer of plasmids (small looplike packages of the bacterial genetic material, DNA) between bacterial populations. The gene for antibiotic resistance crops up by accident, and the bacterium possessing it is able to pass the gene to others by the process of conjugation, or "bacterial sex," in which plasmid genetic material is donated or exchanged.

1 ROLE OF PLASMIDS
Plasmids may cause the bacterium to make enzymes against antibiotic drugs, or alter its surface receptor sites, where antibiotics bind. Then, the plasmid duplicates itself.

Drug-inactivating enzyme

Duplicated plasmid

2 PLASMID TRANSFER
Plasmid transfer takes place during a process known as conjugation. The plasmid copy is passed from the donor, through a pilus, to the recipient bacterium.

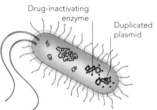

Pilus

DONOR

Plasmid transfer

RECIPIENT

3 DRUG-RESISTANT STRAINS
Recipient bacteria inherit the resistant gene. Plasmid transfer produces populations of bacteria resistant to a range of antibiotics.

Drug-inactivating enzymes

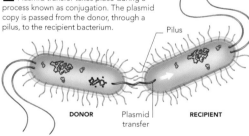

SUPERBUGS

Some bacteria pass through their life cycle in less than 20 minutes. This exceptionally fast reproductive rate, coupled with the incredible numbers of bacteria and the speed at which they pass on genetic information, gives great scope for mutation (see right). A patient taking antibiotics unwittingly provides a testing ground for the bacteria to undergo natural selection for the greatest resistance. Many strains of bacteria resistant to wide-acting, or broad-spectrum, antibiotics have appeared, touted as "superbugs." They may not be resistant to more specialized, narrow-spectrum, antibiotics, but these often have more side effects. Doctors try to reduce this risk by prescribing antibiotics only when necessary.

MRSA
Staphylococcus aureus became resistant to the antibiotic meticillin in the 1960s. It is now known as meticillin resistant *Staphylococcus aureus* (MRSA). New strains may appear in hospitals and communities.

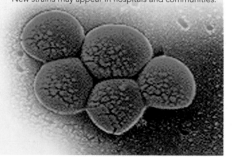

PROTISTS (PROTOZOA)

Single-celled organisms that have their genetic material inside a nucleus, unlike bacteria (opposite), are known by the general name of protists. The animal-like ones, which can move around, and obtain energy by taking in food (rather than capturing energy from sunlight), are sometimes called protozoa. There are thousands of kinds, and most live harmlessly in soil and water. However, some are parasites that cause serious diseases in humans. The protist *Plasmodium* causes the disease malaria, which affects millions of people worldwide. Single-celled parasites employ various mechanisms for evading the body's immune system. The *Leishmania* parasite, for example, which causes leishmaniasis, multiplies within white blood cells that normally destroy such microorganisms. Many protists have a flexible cell membrane and large nucleus, and may possess tail-like appendages, known as flagella, to aid movement.

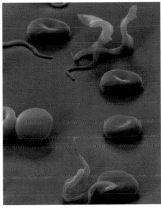

TRYPANOSOMES IN BLOOD
Trypanosomes are wormlike protists (purple), seen here with red blood cells. They cause trypanosomiasis, or sleeping sickness.

MALARIAL LIFE CYCLE

Five types of *Plasmodium* protists cause malaria. They are spread by the bite of the female *Anopheles* mosquito. Malaria produces chills and high fever, which can recur and prove fatal if not treated. Most *Plasmodium* have a similar life cycle, as shown below.

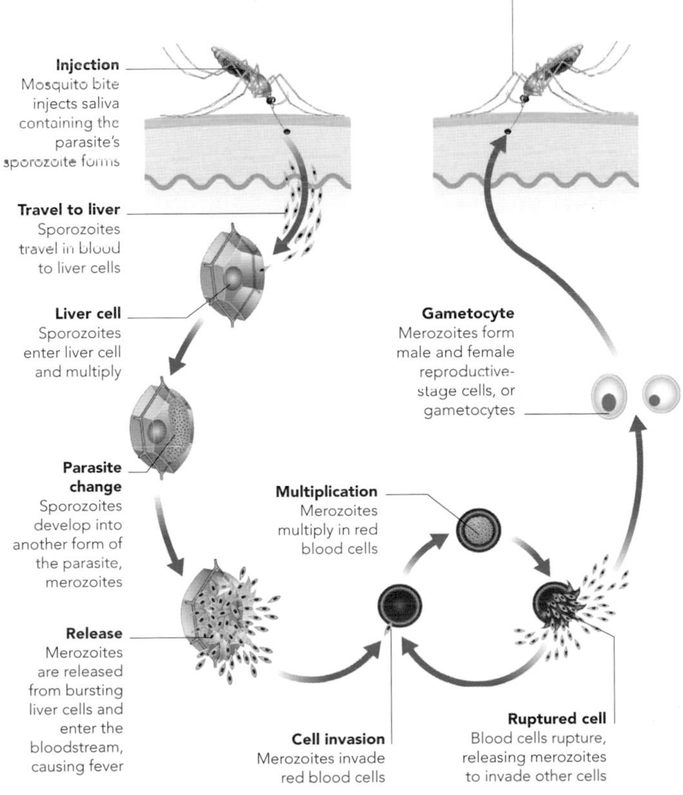

Injection
Mosquito bite injects saliva containing the parasite's sporozoite forms

Travel to liver
Sporozoites travel in blood to liver cells

Liver cell
Sporozoites enter liver cell and multiply

Parasite change
Sporozoites develop into another form of the parasite, merozoites

Release
Merozoites are released from bursting liver cells and enter the bloodstream, causing fever

Cycle continues
Gametocytes sucked up by mosquito bite mature in insect and form sporozoites; the cycle continues

Gametocyte
Merozoites form male and female reproductive-stage cells, or gametocytes

Multiplication
Merozoites multiply in red blood cells

Cell invasion
Merozoites invade red blood cells

Ruptured cell
Blood cells rupture, releasing merozoites to invade other cells

FUNGI

Fungi make up one of the great kingdoms of life and include familiar mushrooms and molds, as well as microscopic single-celled yeasts. They feed on both living and dead organic matter. Disease-causing fungi fall into two main groups: the filamentous fungi, which grow as a network of branching threads called hyphae; and the single-celled yeasts. Some types cause fairly harmless (if unsightly) superficial diseases of the skin, hair, nails, or mucous membranes, for example yeast (candidiasis). Others, such as histoplasmosis, result in potentially fatal infections of certain vital organs, for instance the lungs. Some may be linked to specific occupations such as farming or food production. Other fungal infections, for example ringworm (dermatophytosis), are more likely to affect people with damaged immune systems, such as those with HIV-AIDS.

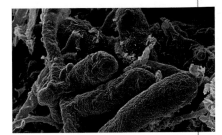

CAUSE OF ATHLETE'S FOOT
Seen here are microscopic threads of the fungus *Epidermophyton floccosum*, one cause of the white, itchy skin of athlete's foot.

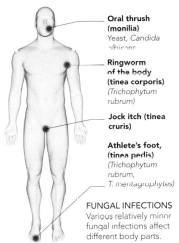

Oral thrush (monilia)
Yeast, *Candida albicans*

Ringworm of the body (tinea corporis)
(*Trichophytum rubrum*)

Jock itch (tinea cruris)

Athlete's foot, (tinea pedis)
(*Trichophytum rubrum*, *T. mentagrophytes*)

FUNGAL INFECTIONS
Various relatively minor fungal infections affect different body parts.

PARASITIC WORMS

Humans, like most other animals, can be infested with parasitic worms that derive all their nutrients from their hosts. At least 20 types of wormlike animals may live in the human body as parasites. Most spend at least part of their life cycle in the intestines. Few are members of the segmented worms group, the annelids, which includes common earthworms. Several are roundworms, or nematodes, such as ⅖in (1 cm) long hookworms *Ancylostoma duodenale*, which live in the gut. Another wormlike group is the flatworms; it includes the tapeworms, such as *Taenia*, which live in the gut, and flukes, such as *Schistosoma*, which cause schistosomiasis, or snail fever.

HOOKWORM
This microscope image shows an adult hookworm's head. The mouth contains several toothlike structures, which it uses to cling to the intestinal lining of its host.

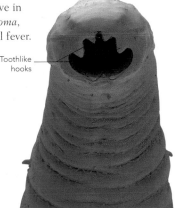

Toothlike hooks

SCHISTOSOMES
Adult flukes, such as the ⅖–⅝in (1–2 cm) long *Schistosoma* pictured above, live in blood vessels. This close-up shows some red blood cells in the fluke's mouth.

ALLERGIES

THE IMMUNE SYSTEM NORMALLY DEFENDS THE BODY AGAINST INFECTIONS, CANCER, INJURIES, AND DAMAGING SUBSTANCES SUCH AS TOXIC CHEMICALS. SOMETIMES, HOWEVER, IT OVERREACTS, ATTACKING A FOREIGN SUBSTANCE THAT IS NORMALLY HARMLESS. THIS REACTION IS AN ALLERGIC RESPONSE. SUCH RESPONSES CAN VARY FROM MILD CONDITIONS TO LIFE-THREATENING DISORDERS.

ALLERGIC RESPONSE

AN ALLERGY DEVELOPS IF THE IMMUNE SYSTEM BECOMES SENSITIZED TO A FOREIGN SUBSTANCE (AN ALLERGEN).

When first exposed to an allergen, such as pollen, nuts, or penicillin, the immune system makes antibodies to fight it. The antibodies coat the surface of mast cells, found in the skin, stomach lining, lungs, and upper airways. If the allergen enters the body again, these cells mount an allergic response.

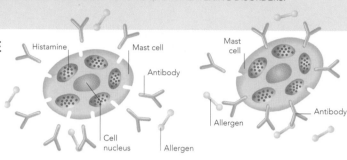

Histamine — Mast cell — Antibody — Cell nucleus — Allergen — Mast cell — Allergen — Antibody — Bursting mast cell — Histamine released — Allergen bound to antibodies

1 EXPOSURE TO ALLERGEN
Antibodies bind to the surfaces of mast cells. These cells contain histamine, which normally causes inflammation (see p.176).

2 ANTIBODIES TRIGGERED
Allergens come into contact with the antibodies. If they link two or more antibodies, they cause the cell to burst.

3 HISTAMINE RELEASED
Granules inside the mast cell release histamine as the cell bursts. Histamine causes an inflammatory response that irritates body tissues and produces the symptoms of an allergy.

ALLERGIC RHINITIS

AIRBORNE ALLERGENS THAT IRRITATE THE LINING OF THE NOSE AND THROAT CAUSE ALLERGIC RHINITIS; THIS ALLERGY MAY BE SEASONAL OR OCCUR ALL YEAR.

In allergic rhinitis, the lining of the nose and throat becomes inflamed after contact with an airborne allergen. One form is hay fever, which is brought on by pollen grains in the spring and summer. Another form, perennial rhinitis, may be caused by house dust mites, bird feathers, or animal fur or skin flakes (dander) and may occur at any time of year. Both forms can cause sneezing, a stuffy or runny nose, and itchy, watery eyes, although symptoms tend to be more severe in hay fever.

Often, the cause of rhinitis is easy to identify. If a person cannot avoid contact with the allergen, antiallergy drugs taken before or during an attack may relieve itchy eyes or a stuffy nose. Drugs can be applied directly to the inside of the nose or eyes, or taken orally.

COMMON ALLERGENS
Many people are allergic to pollen grains (above) and have hay fever as a result. The dead bodies and excrement of dust mites (left) can also cause rhinitis.

FOOD ALLERGIES

SOME ALLERGIES ARE CAUSED BY AN EXCESSIVE IMMUNE RESPONSE TO CERTAIN FOODS, MOST COMMONLY NUTS, SEAFOOD, EGGS, AND MILK.

Symptoms of food allergies may appear as soon as the food is eaten or develop over a few hours. Some affect the digestive system, causing swelling and itching in the mouth and throat, nausea and vomiting, and diarrhea. Others affect the whole body, causing skin rashes, swollen tissues (see angioedema, below), and shortness of breath. In very severe cases, food allergies provoke anaphylaxis (see below left). The only effective treatment is to avoid the problem food.

ANAPHYLAXIS

THIS RARE, BUT POTENTIALLY FATAL, ALLERGIC REACTION RESULTS FROM AN EXTREME SENSITIVITY TO AN ALLERGEN.

Anaphylaxis is a massive immune system response that involves the whole body. Widespread release of huge amounts of histamine causes a sudden fall in blood pressure (shock) and narrowing of the airways, and can be fatal unless treated immediately. Other possible symptoms include a red, itchy, lumpy rash called urticaria (hives), swelling of the face, lips, and tongue (see angioedema, right), and loss of consciousness. Triggers for anaphylaxis include foods such as nuts, drugs such as penicillin, and insect stings. A person with anaphylaxis needs

emergency medical treatment. If someone is known to be at risk of anaphylaxis, his or her doctor may prescribe syringes of epinephrine (adrenaline) that the person can self-administer as soon as an attack starts. Susceptible people should avoid trigger substances if at all possible.

Typical white lumps

URTICARIA
This itchy, red rash, often with white lumps, can result from various allergies. It can also be a symptom of anaphylaxis.

ANGIOEDEMA

SOME ALLERGIC REACTIONS CAUSE SWELLING OF BODY TISSUES, WHICH IS CALLED ANGIOEDEMA.

Swelling usually comes on suddenly, in tissues just under the skin and in mucous membranes. Angioedema often affects the face and lips. It may also occur in the mouth, tongue, and airways, interfering with breathing and swallowing. The most common triggers are foods such as nuts and seafood. Other possible triggers are antibiotic drugs and insect bites. Severe angioedema needs emergency medical treatment. For milder cases, corticosteroid or antihistamine drugs may be given to reduce the swelling.

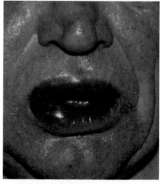

ANGIOEDEMA OF THE LIPS
Sudden, severe swelling of the soft tissues in the face, lips, or larynx is known as angioedema. It is usually caused by an allergic reaction to certain foods.

HIV–AIDS

POTENTIALLY LIFE-THREATENING INFECTION WITH THE HUMAN IMMUNODEFICIENCY VIRUS (HIV) DAMAGES THE BODY'S OWN SELF-DEFENSE SYSTEMS. IT MAY THEN LEAD TO AIDS—ACQUIRED IMMUNE DEFICIENCY SYNDROME. IN THIS LIFE-THREATENING CONDITION, THE IMMUNE SYSTEM BECOMES SO WEAK THAT EVEN NORMALLY HARMLESS MICROORGANISMS CAN CAUSE SEVERE INFECTIONS.

HIV INFECTION

HIV IS CARRIED IN BODY FLUIDS, SUCH AS BLOOD, SEMEN, VAGINAL SECRETIONS, AND BREAST MILK. IT IS PASSED ON WHEN INFECTED FLUIDS ENTER THE BODY.

The virus is most commonly transmitted by sexual intercourse. It can also be passed among drug users if they share infected needles, or from a mother to her fetus or newborn baby. Once in the bloodstream, HIV infects cells with structures called CD4 molecules on their surfaces. These cells, CD4+ cells, include white blood cells called CD4+ lymphocytes, which fight infection. The virus multiplies rapidly in CD4+ cells, destroying them in the process. The initial infection may cause a flulike illness for a few weeks, then there may be no further symptoms for years. If HIV goes untreated, the number of CD4+ lymphocytes eventually falls so low that the immune system is severely weakened, and serious disorders develop.

HIV REPLICATION
HIV is a type of virus called a retrovirus, which carries its genetic material in the form of RNA. It invades body cells and uses the cells' own processes to multiply.

1 Free HIV particle
The core (capsid) contains two strands of ribonucleic acid (RNA), each carrying a set of genes for the virus. The spikes on the surface are proteins called gp120 antigens (docking protein). They enable the virus to "dock" on the surface of CD4+ cells.

2 Binding and injection
The gp120 binds to CD4 molecules and then to coreceptors on the cell surface. The virus fuses with the cell, penetrating the surface. The capsid releases the viral RNA.

3 Reverse transcription
The virus releases an enzyme called reverse transcriptase into the cell. The enzyme copies the single strands of viral RNA as double-stranded DNA.

4 Insertion of viral DNA
The viral DNA enters the cell nucleus, where it is incorporated into the cell's DNA. The cell produces mRNA, which transmits instructions for making new proteins, including HIV proteins.

5 Creation of proteins
The mRNA enters the cell cytoplasm, where it is "read" and chains of HIV proteins and viral RNA are made. These molecules are to become the components of new HIV particles.

6 New HIV created
The HIV constituents gather at the cell wall. An immature virus forms and buds from the cell, taking some cell membrane with it. Enzymes within the virus bring about changes resulting in a mature virus particle.

Free-floating mature HIV particle (shown here in cross-section) repeats the cycle

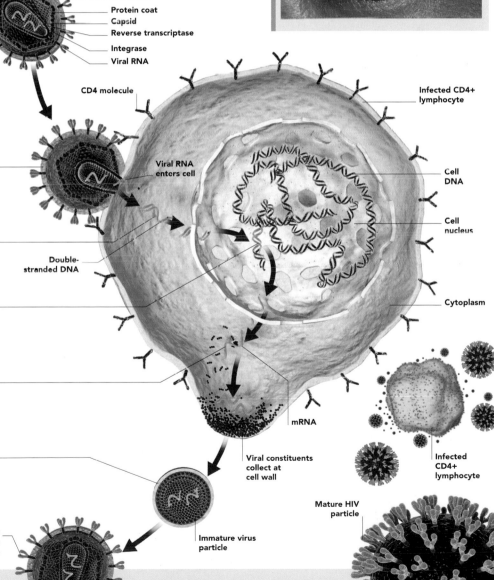

gp120 antigen (docking protein)
Membrane
Protein coat
Capsid
Reverse transcriptase
Integrase
Viral RNA

CD4 molecule
Viral RNA enters cell
Infected CD4+ lymphocyte
Cell DNA
Cell nucleus
Double-stranded DNA
Cytoplasm
mRNA
Viral constituents collect at cell wall
Infected CD4+ lymphocyte
Mature HIV particle
Immature virus particle

AIDS

HIV can be identified by specific blood or fluid tests. Being HIV positive may lead to AIDS-related illnesses, especially opportunistic infections, caused by organisms that are harmless to healthy people but dangerous to those with reduced immunity; one example is infection by *Candida albicans*, which causes yeast. People with AIDS may also develop various types of cancer, notably Kaposi's sarcoma.

KAPOSI'S SARCOMA
Kaposi's sarcoma is characterized by defined, brownish, raised nodules, seen here under the eye. They can occur anywhere in the body, including the internal organs.

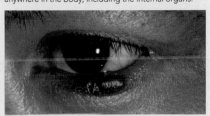

AUTOIMMUNE AND LYMPHATIC DISORDERS

THE IMMUNE SYSTEM NORMALLY PROTECTS THE BODY FROM INFECTIONS, BUT IF IT DOES NOT WORK CORRECTLY ILLNESS MAY RESULT. IN AUTOIMMUNE DISORDERS, A FAULTY IMMUNE RESPONSE IDENTIFIES THE BODY'S OWN TISSUES AS FOREIGN AND PRODUCES ANTIBODIES AGAINST THEM. THE LYMPHATIC SYSTEM DESTROYS INFECTIOUS MICROORGANISMS AND CANCEROUS CELLS, BUT THE SYSTEM ITSELF MAY SUCCUMB TO INFECTION OR CANCER.

LUPUS

A WIDE-RANGING DISORDER, LUPUS OCCURS WHEN THE IMMUNE SYSTEM ATTACKS THE CONNECTIVE TISSUES.

Systemic lupus erythematosus (SLE), or lupus, causes inflammation and swelling of the connective tissues, which hold the skin, joints, and internal organs together. Symptoms vary in severity and may flare up for a few weeks every so often. The cause is unknown, but lupus may be triggered by viral infection, stress, or exposure to sunlight. It is much more common in women and in black or Asian people, and sometimes runs in families. There is no cure; treatment aims to relieve symptoms and control the disease, but in some cases lupus can be fatal.

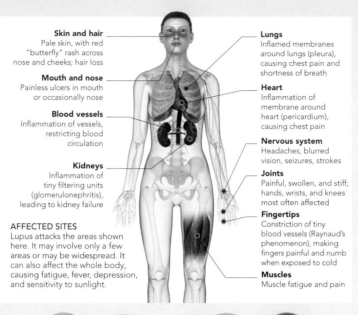

Skin and hair
Pale skin, with red "butterfly" rash across nose and cheeks; hair loss

Mouth and nose
Painless ulcers in mouth or occasionally nose

Blood vessels
Inflammation of vessels, restricting blood circulation

Kidneys
Inflammation of tiny filtering units (glomerulonephritis), leading to kidney failure

Lungs
Inflamed membranes around lungs (pleura), causing chest pain and shortness of breath

Heart
Inflammation of membrane around heart (pericardium), causing chest pain

Nervous system
Headaches, blurred vision, seizures, strokes

Joints
Painful, swollen, and stiff; hands, wrists, and knees most often affected

Fingertips
Constriction of tiny blood vessels (Raynaud's phenomenon), making fingers painful and numb when exposed to cold

Muscles
Muscle fatigue and pain

AFFECTED SITES
Lupus attacks the areas shown here. It may involve only a few areas or may be widespread. It can also affect the whole body, causing fatigue, fever, depression, and sensitivity to sunlight.

SCLERODERMA

IN THIS RARE CONDITION, ANTIBODIES DAMAGE THE SKIN, JOINT TISSUES, AND OTHER CONNECTIVE TISSUES.

Scleroderma is an autoimmune disorder in which the immune system attacks the connective tissues, which hold the body's structures together. The tissues become inflamed and thickened and may harden and contract. The skin is most often affected and may become stiff and tight. The joints, particularly in the hands, may be swollen and painful. The fingers may develop ulcers and hard patches and become painfully sensitive to cold (a condition called Raynaud's phenomenon). The cause of scleroderma is unknown. There is no cure, but treatment may relieve the symptoms and slow the disease.

PULMONARY FIBROSIS

IF ANTIBODIES ATTACK LUNG TISSUE, THEY CAUSE FIBROSIS (THICKENING AND SCARRING) OF THE AIR SACS (ALVEOLI).

In pulmonary fibrosis, an autoimmune reaction causes inflammation of the alveoli in the lungs. The alveolar walls become scarred and less able to take in oxygen. Symptoms include a dry cough and shortness of breath, which may become so severe that the person needs oxygen. The cause is unknown. There is no cure, but corticosteroid drugs may slow the rate of lung damage.

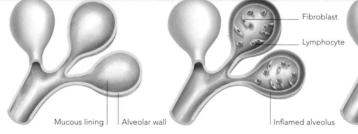

Mucous lining | Alveolar wall

Fibroblast

Lymphocyte

Inflamed alveolus

Thickened alveolar wall

Scar tissue forms

1 NORMAL ALVEOLI
The delicate walls of the alveoli, only one cell thick, easily allow oxygen from the air to pass into the blood and carbon dioxide to leave the body. The inner surfaces of the alveoli are protected by a layer of mucus.

2 INFLAMMATION
Large numbers of disease-fighting cells called lymphocytes enter the alveoli. As they break down, they secrete substances that cause inflammation. This process stimulates cells called fibroblasts to form fibrous tissue.

3 FIBROSIS
The formation of scar tissue (fibrosis) causes the alveolar walls to thicken, and restricts the flow of gases through the walls. Fibrosis gradually destroys the alveoli, and scar tissue may restrict lung expansion.

POLYARTERITIS

THIS RARE BUT SERIOUS AUTOIMMUNE DISORDER CAUSES WIDESPREAD DAMAGE TO SMALL AND MEDIUM-SIZED ARTERIES.

Polyarteritis nodosa is inflammation of artery walls due to an autoimmune reaction, which restricts the blood flow to body tissues. Symptoms include skin lesions and ulcers, abdominal pain, joint pain, and numb fingers and toes. Polyarteritis may also lead to kidney failure or a heart attack. The cause is not known. There is no cure, but corticosteroids may relieve symptoms.

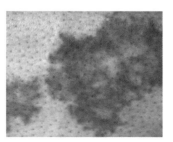

POLYARTERITIS DAMAGE
The purplish areas, seen here on the leg of a person with polyarteritis, indicate tissue starved of blood and oxygen as a result of inflamed blood vessels restricting blood flow.

SARCOIDOSIS

A DISEASE THAT MAY BE ACUTE OR CHRONIC, SARCOIDOSIS CAUSES SORES CALLED GRANULOMAS TO FORM.

Sarcoidosis is thought to be caused by an excessive immune reaction to a chemical or infection in someone with a genetic predisposition to the disease. It most often affects the lungs, causing coughing and shortness of breath but can also develop in the lymph nodes, liver, spleen, kidneys, skin, or eyes. There is no cure, but in most cases, the symptoms disappear by themselves.

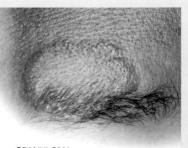

GRANULOMA
This image shows a granuloma above the eye. Granulomas are clusters of disease-fighting cells called macrophages at points where an immune response has been triggered.

ANEMIA

VARIOUS CONDITIONS, INCLUDING AN ABNORMAL IMMUNE RESPONSE, CAN CAUSE ANEMIA.

The term "anemia" is used of disorders in which hemoglobin, the pigment that gives red blood cells their color, is deficient or abnormal. Hemoglobin carries oxygen in the blood, so if it is deficient it cannot supply enough oxygen to body tissues. There are various types of anemia. Hemolytic anemia results from large-scale, rapid destruction of red blood cells (hemolysis). The autoimmune form involves an excessive immune response in which the body produces antibodies that attack red blood cells. This reaction can be triggered by another, perhaps autoimmune disorder or by drugs, such as penicillin or quinine. The most common type of anemia is caused by a lack of substances needed to make healthy red blood cells, such as iron. Another type results from inherited disorders that cause the body to produce abnormal forms of hemoglobin, such as sickle-cell disease, in which the red blood cells are distorted into a curved sickle shape. A third type, aplastic anemia, occurs when the bone marrow fails to produce enough red blood cells.

HEMOLYSIS

This color-enhanced electron microscope image shows a white blood cell called a macrophage (brown) destroying red blood cells. The loss of red blood cells in this way is called autoimmune hemolytic anemia.

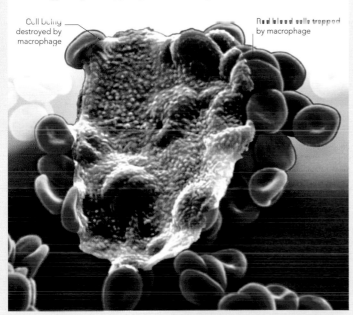

Cell being destroyed by macrophage

Red blood cells trapped by macrophage

LEUKEMIA

THERE ARE SEVERAL TYPES OF LEUKEMIA; ALL INVOLVE GROWTH OF CANCEROUS WHITE BLOOD CELLS IN BONE MARROW.

Leukemia is cancer of the white blood cells. Cancerous cells multiply in the bone marrow, where blood cells are normally made. This process reduces the production of healthy red and white blood cells and of platelets to abnormally low levels. A deficiency of red blood cells causes anemia (see left). A reduction in normal white blood cells leaves the body unable to fight off infection. A lack of platelets prevents blood from clotting at injury sites, leading to excessive bleeding. The cancerous cells often spread in the bloodstream, causing enlargement of the lymph nodes, spleen, and liver. Leukemia may be either acute or chronic. It is usually treated with chemotherapy, and sometimes with radiotherapy followed by a stem-cell transplant. The outlook depends on the type and severity, but treatment is more likely to be successful in children.

BLOOD CELL PRODUCTION

All blood cells are produced in the bone marrow—soft, fatty tissue found in the center of large, flat bones such as the shoulder blades, ribs, breastbone, and pelvis. They develop from a single type called a stem cell. Red blood cells carry oxygen to the tissues. Lymphocytes (a type of white blood cell) fight infection. Platelets help the blood to clot at injury sites, reducing blood loss.

ACUTE LYMPHOBLASTIC LEUKEMIA

In acute lymphoblastic leukemia (ALL), cancerous, immature lymphocytes, called lymphoblasts, multiply uncontrollably and build up in the bone marrow. As a result, the production of normal blood cells is disrupted, so the levels fall too low. In addition, the lymphoblasts invade the bloodstream, where they multiply further and carry the cancer to other organs and tissues in the body.

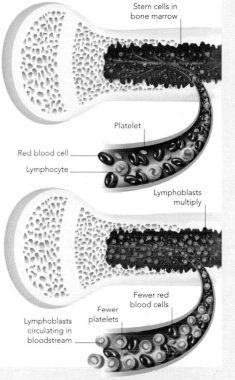

Stem cells in bone marrow

Platelet

Red blood cell

Lymphocyte

Lymphoblasts multiply

Fewer red blood cells

Fewer platelets

Lymphoblasts circulating in bloodstream

LYMPHOMA

ORIGINATING IN THE LYMPHATIC SYSTEM, LYMPHOMA IS A CANCER INVOLVING CELLS CALLED LYMPHOCYTES.

Like the blood, the lymphatic system contains lymphocytes, the white blood cells that help the body fight infection. In lymphoma, these cells become cancerous and multiply in a lymph node. The cancer may spread to tissues such as the spleen and bone marrow, and to other nodes. Lymphomas are grouped as either Hodgkin or non-Hodgkin, and there are nearly 100 different forms. Most cause swelling of lymph nodes in the neck, armpits, or groin; fever; fatigue; and night sweats. Lymphoma in a group of nodes may be treated with radiation therapy; if widespread, chemotherapy may be used.

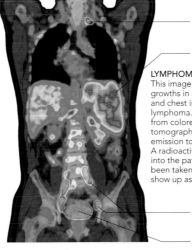

Cancer in lymph node

Cancerous tissue in spleen

LYMPHOMA SCAN

This image shows malignant growths in a patient's abdomen and chest in a type of non-Hodgkin lymphoma. The image is formed from colored composite computed tomography (CT) and positron emission tomography (PET) scans. A radioactive substance, injected into the patient's bloodstream, has been taken up by the tumors, which show up as patches of intense pink.

Cancerous lymph nodes

Bladder

HODGKIN LYMPHOMA

This form of lymphoma involves enlarged, abnormal cells called Reed–Sternberg cells. The cause is not known. Hodgkin lymphoma usually affects people aged from about 15 to 30 and 55 to 70 years old. The most common symptom is enlarged lymph nodes. Others include fatigue, itchy skin, or a rash. Some people have fever, night sweats, weight loss, or pain in the lymph nodes after drinking alcohol. The condition increases vulnerability to infection since the cells of the immune system are unable to function properly. The doctor may order blood tests for anemia and take a biopsy from a swollen lymph node to look for cancerous cells. CT scans and a bone marrow biopsy may be performed to see if the disease has spread. Treatment includes radiation therapy and chemotherapy.

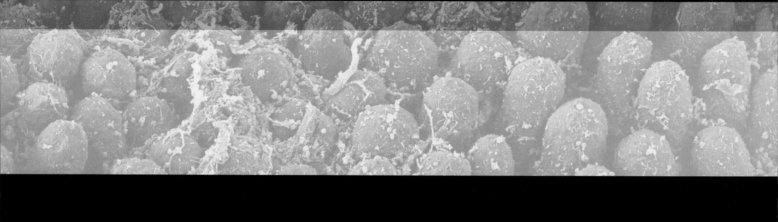

PEOPLE ARE PROBABLY MORE AWARE OF THEIR DIGESTIVE
SYSTEM THAN ANY OF OTHER SYSTEM, NOT LEAST BECAUSE
OF ITS FREQUENT MESSAGES. HUNGER, THIRST, APPETITE,
GAS, AND THE FREQUENCY AND NATURE OF BOWEL
MOVEMENTS ARE ALL ISSUES AFFECTING DAILY LIFE.
EATING WELL, ALONG WITH REGULAR EXERCISE, IS ONE
OF THE BEDROCKS OF GOOD HEALTH. PLENTY OF FRESH

DIGESTIVE SYSTEM

DIGESTIVE ANATOMY

THE DIGESTIVE SYSTEM CONSISTS OF A LONG PASSAGEWAY, KNOWN AS THE ALIMENTARY CANAL OR DIGESTIVE TRACT, AND ASSOCIATED ORGANS, INCLUDING THE LIVER, GALLBLADDER, AND PANCREAS. THE DIGESTIVE TRACT STARTS AT THE MOUTH AND CONTINUES THROUGH THE ESOPHAGUS AND INTESTINES TO THE ANUS. ALONG ITS COURSE FOOD IS BROKEN DOWN AND NUTRIENTS EXTRACTED, WHILE WASTE MATERIALS ARE DISPOSED OF.

After being eaten, or ingested, food embarks on a journey. It can take up to 24 hours to cover a distance of 30 ft (9 m), through various muscular tubes and chambers. The process begins at the mouth, where food is initially crushed and ground down by the teeth during chewing. The resulting ball, or bolus, of food continues down the throat (pharynx), then travels through the food tube (esophagus) to the stomach, small intestine, large intestine, and anus. In the small intestine, chemical digestion breaks down food into molecules small enough to be absorbed into the bloodstream. What cannot be digested is compacted as feces in the large intestine and eliminated as stool through the anus. Food travels through the system by a process of muscular contraction called peristalsis (see p.199). In addition to the digestive tract, the digestive system includes several glands: the spit-making salivary glands; the pancreas, which produces powerful digestive juices; and the body's major nutrient processor, the liver.

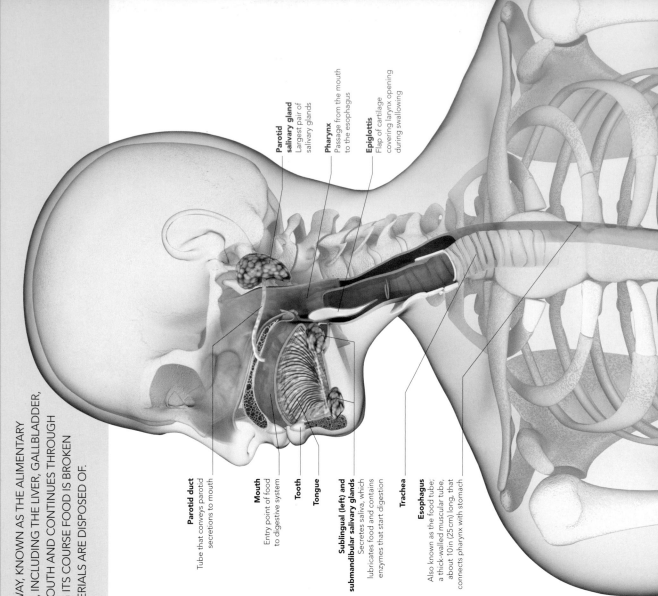

Parotid salivary gland
Largest pair of salivary glands

Pharynx
Passage from the mouth to the esophagus

Epiglottis
Flap of cartilage covering larynx opening during swallowing

Parotid duct
Tube that conveys parotid secretions to mouth

Mouth
Entry point of food to digestive system

Tooth

Tongue

Sublingual (left) and submandibular salivary glands
Secretes saliva, which lubricates food and contains enzymes that start digestion

Trachea

Esophagus
Also known as the food tube; a thick-walled muscular tube, about 10 in (25 cm) long, that connects pharynx with stomach

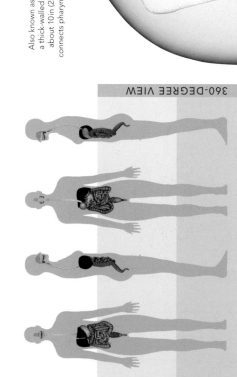

360-DEGREE VIEW

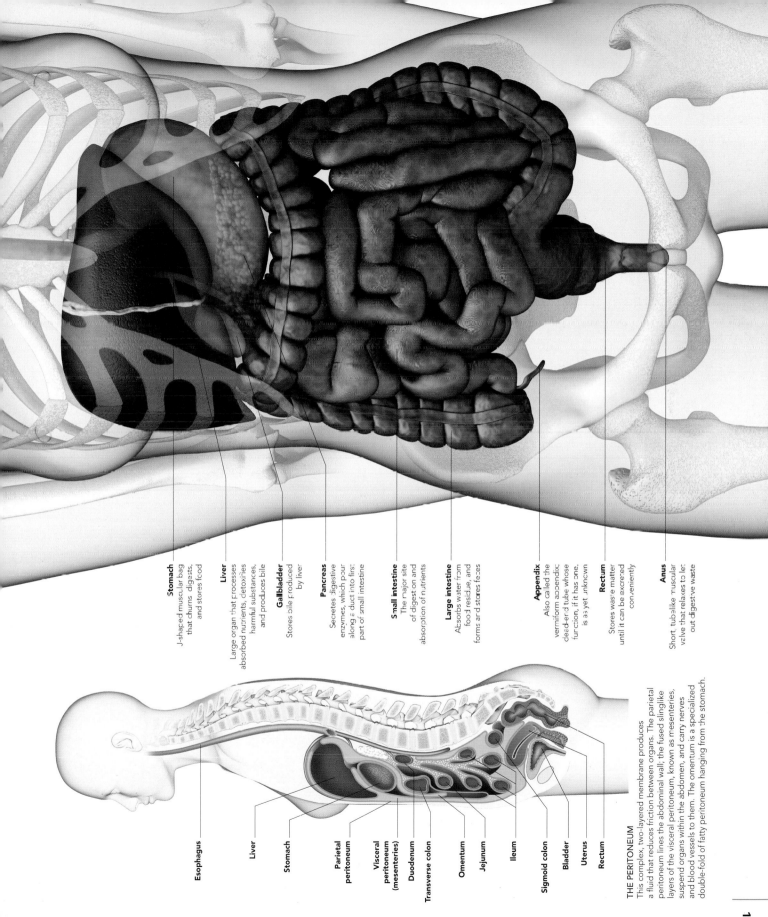

Stomach
J-shaped muscular bag that churns, digests, and stores food

Liver
Large organ that processes absorbed nutrients, detoxifies harmful substances, and produces bile

Gallbladder
Stores bile produced by liver

Pancreas
Secretes digestive enzymes, which pour along a duct into first part of small intestine

Small intestine
The major site of digestion and absorption of nutrients

Large intestine
Absorbs water from food residue, and forms and stores feces

Appendix
Also called the vermiform appendix; dead-end tube whose function, if it has one, is as yet unknown

Rectum
Stores waste matter until it can be excreted conveniently

Anus
Short, tube-like muscular valve that relaxes to let out digestive waste

Esophagus

Liver

Stomach

Parietal peritoneum

Visceral peritoneum (mesenteries)

Duodenum

Transverse colon

Omentum

Jejunum

Ileum

Sigmoid colon

Bladder

Uterus

Rectum

THE PERITONEUM
This complex, two-layered membrane produces a fluid that reduces friction between organs. The parietal peritoneum lines the abdominal wall; the fused slinglike layers of the visceral peritoneum, known as mesenteries, suspend organs within the abdomen, and carry nerves and blood vessels to them. The omentum is a specialized double-fold of fatty peritoneum hanging from the stomach.

MOUTH AND THROAT

THE PROCESS OF DIGESTION STARTS WHEN FOOD ENTERS THE MOUTH. FOOD IS CHEWED, LUBRICATED BY SALIVA, AND MOVED ABOUT BY THE TONGUE. IN ABOUT ONE MINUTE, THE FOOD BECOMES A SOFT, MOIST MASS CALLED A BOLUS. EACH BOLUS IS SWALLOWED THROUGH THE THROAT (PHARYNX) AND PASSES INTO THE ESOPHAGUS.

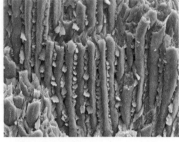

ENAMEL SURFACE
This microscope image shows enamel, which is a hard material made of U-shaped enamel prisms packed with the crystalline mineral substance hydroxyapatite.

TEETH

There are four types of tooth, each of which has a different role. The incisors, at the front, are chisel-shaped with sharp edges for cutting, while the pointed canines or "eye teeth" are designed for tearing. The premolars, with their two ridges, and the flatter molars toward the back of the mouth, which are the largest and strongest teeth, crush and grind food. The portion of the tooth above the gum is the crown; the part embedded in the jawbone is known as the root; and the area where these two meet, at the gum or gingival surface, is called the neck of the tooth. The crown's outside layer is made of a tough bonelike material called enamel, which is the hardest substance in the body. Beneath it is a layer of softer but still strong tissue called dentine, which is shock-absorbing. At the center of the tooth, the soft dental pulp contains blood vessels and nerves. Below the gum, bonelike cementum and periodontal ligament tissues secure the tooth in the jawbone.

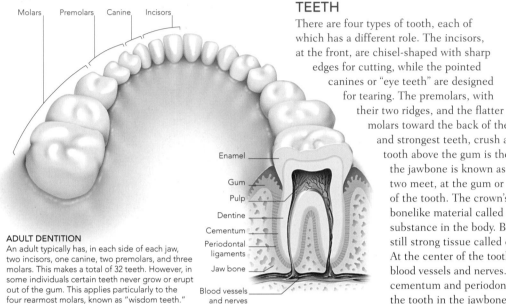

Molars · Premolars · Canine · Incisors

Enamel
Gum
Pulp
Dentine
Cementum
Periodontal ligaments
Jaw bone
Blood vessels and nerves

ADULT DENTITION
An adult typically has, in each side of each jaw, two incisors, one canine, two premolars, and three molars. This makes a total of 32 teeth. However, in some individuals certain teeth never grow or erupt out of the gum. This applies particularly to the four rearmost molars, known as "wisdom teeth."

SWALLOWING

The process of swallowing begins as a voluntary action when a bolus of food is pushed by the rear of the tongue to the back of the mouth. Swallowing usually takes place after a period of chewing; to swallow a solid item such as a tablet without chewing demands concentration. It is easier to swallow a tablet with water, because drinks are usually gulped down right after entering the mouth. Automatic reflexes control subsequent stages of swallowing, as the muscles of the throat contract and move the bolus back and down, and squeeze it into the top of the esophagus. A flap of cartilage known as the epiglottis prevents food from going down "the wrong way" into the larynx and the trachea, where it would cause choking.

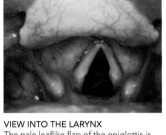

VIEW INTO THE LARYNX
The pale leaflike flap of the epiglottis is visible at the top of this image. Below it is the inverted "V" of the vocal cords.

BREATHE OR SWALLOW

The pharynx is a dual-purpose passageway: for air when breathing, and food, drinks, and saliva when swallowing. Nerve signals from the brain operate the muscles of the mouth, tongue, pharynx, larynx, and upper esophagus to prevent food from entering the trachea. If food is inhaled, irritation of the airway triggers the coughing reflex to expel inhaled particles and prevent choking. The complex muscle movements of swallowing are a voluntary reflex and also occur when solid matter contacts touch sensors at the back of the mouth.

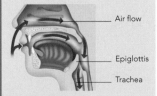

Air flow

Epiglottis

Trachea

DUAL INTAKE
Breathing occurs through the nose or the mouth. Their passageways meet at the throat, and air flows into the trachea.

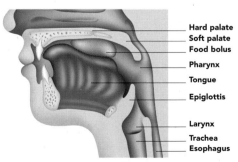

Hard palate
Soft palate
Food bolus
Pharynx
Tongue
Epiglottis
Larynx
Trachea
Esophagus

1 PHARYNGEAL STAGE
Before the food bolus reaches the back of the mouth, the epiglottis is raised in its normal position, allowing free flow of air from the nasal cavity to the trachea. The esophagus is relaxed.

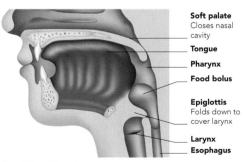

Soft palate
Closes nasal cavity
Tongue
Pharynx
Food bolus
Epiglottis
Folds down to cover larynx
Larynx
Esophagus

2 ESOPHAGEAL STAGE
The larynx rises and meets the tilted epiglottis, closing the trachea, and the soft palate lifts to close the nasal cavity. Food enters the esophagus and is pushed downward.

ANATOMY OF THE MOUTH AND THROAT

The interior of the lips, cheeks, and oral cavity is lined with tough, firmly anchored mucous membrane and a type of tissue called nonkeratinized squamous epithelium. Cells here multiply rapidly to replace those rubbed away when biting, chewing, and swallowing. The front underside of the tongue has a fleshy central ridge, the frenulum, which connects to the floor of the mouth. The tongue is the body's most flexible muscle. Within it are three pairs of intrinsic muscles; and outside, three pairs of extrinsic muscles run from the tongue to other parts of the throat and neck. The root of the tongue anchors to the lower jaw (mandible) and to the curved hyoid bone in the neck. The rear of the mouth leads to the middle part of the throat, the oropharynx. The whole throat or pharynx, from its nasal to laryngeal regions, is about 5 in (13 cm) long in a typical adult.

NOSE, MOUTH, AND THROAT
The roof of the mouth, or oral cavity, is formed by shelves of the maxillary and palatine bones of the skull, together known as the hard palate. This extends rearward as the soft palate, which contains skeletal muscle fibers that allow it to flex when swallowing. The central posterior part of the soft palate extends into a small "finger," the uvula, which can be seen through the open mouth, dangling down from the back, where it helps direct food downward.

SALIVARY GLANDS

Saliva is produced by three pairs of salivary glands: the parotids, in front of and just below each ear; the submandibulars, on the inner sides of the lower jawbone (mandible); and the sublinguals in the floor of the mouth, below the tongue. In addition, numerous small accessory glands are found in the mucous membranes lining the mouth and tongue. Although saliva is composed of 99.5 percent water, it also contains important solutes such as amylase, a digestive enzyme that begins breakdown of starches, and salts. Saliva lubricates food to make chewing and swallowing easier, and it keeps the mouth moist between periods of eating.

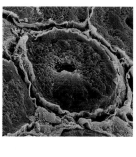

SALIVARY GLAND STRUCTURE
Many small, rounded glandular units called acini (brown), separated by connective tissue (pink), discharge their saliva into tiny central ducts. Acinar ducts converge to become the main saliva-carrying glandular ducts.

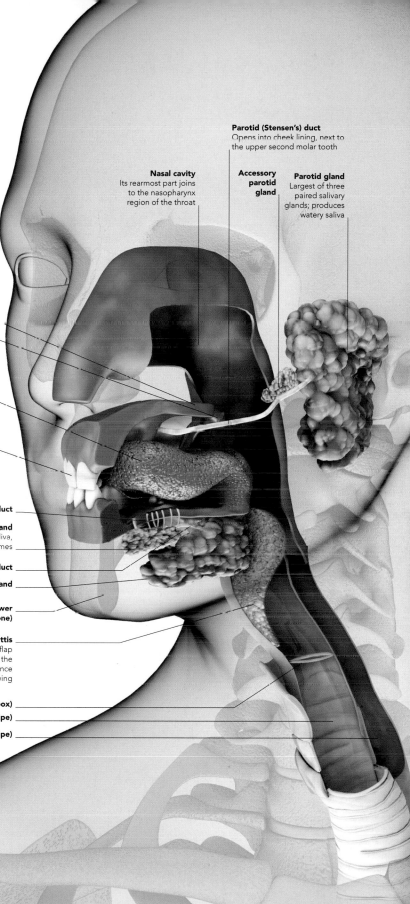

Parotid (Stensen's) duct
Opens into cheek lining, next to the upper second molar tooth

Nasal cavity
Its rearmost part joins to the nasopharynx region of the throat

Accessory parotid gland

Parotid gland
Largest of three paired salivary glands; produces watery saliva

Soft palate

Uvula

Tongue
Moves food around when chewing, contains taste buds, and helps form distinct words in speech

Teeth
Bite off and chew food into a moist, soft pulp, ready to be swallowed

Sublingual duct

Sublingual gland
Produces viscous saliva, which contains enzymes

Submandibular duct

Submandibular gland

Mandible (lower jawbone)

Epiglottis
Cartilaginous flap that blocks off the larynx entrance during swallowing

Larynx (voicebox)

Trachea (windpipe)

Esophagus (food pipe)

STOMACH AND SMALL INTESTINE

AFTER THE MOUTH, THROAT, AND ESOPHAGUS, THE NEXT MAJOR SECTIONS OF THE DIGESTIVE TRACT ARE THE STOMACH AND THE SMALL INTESTINE. THE STOMACH STORES 2½ PINTS (1.5 LITERS) OR MORE OF FOOD FROM A MEAL AND DIGESTS IT BOTH PHYSICALLY AND CHEMICALLY. THE SMALL INTESTINE CONTINUES THE CHEMICAL BREAKDOWN AND IS THE MAIN SITE FOR ABSORBING THE RESULTING NUTRIENTS INTO THE BLOODSTREAM.

STOMACH STRUCTURE

The stomach is the widest part of the digestive tube. It is a muscular-walled, J-shaped sac in which food is stored, churned, and mixed with gastric juices secreted by its lining. This process begins moments after food enters the stomach from the esophagus, through the gastroesophageal junction. Gastric juices include digestive enzymes and hydrochloric acid, which not only breaks down food but also kills potentially harmful microbes. The smooth muscle layers of the stomach wall contract to combine and squeeze the semiliquid mix of food and gastric juices.

LAYERS OF THE STOMACH WALL
The stomach wall has four main layers; the mucosa, submucosa, muscularis, and serosa. The mucosa has deep infolds (gastric pits) that contain the gastric glands. Mucous cells in the upper part of each pit secrete a mucous lining to stop the stomach from digesting itself. Deep in the pits lie cells which produce acid (parietal cells), enzymes (zymogenic or chief cells), and hormones (neuroendocrine cells).

FOLDS AND PITS
With the normal coating of mucus removed in this magnified image (right), the folds (rugae) of the stomach lining are clearly visible.

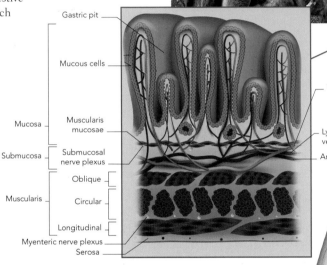

Gastric pit
Mucous cells
Mucosa
Submucosa
Muscularis
Muscularis mucosae
Submucosal nerve plexus
Oblique
Circular
Longitudinal
Myenteric nerve plexus
Serosa
Vein
Lymphatic vessel
Artery

MOVEMENT OF FOOD

Swallowing triggers relaxation of muscles at the gastroesophageal junction, enabling food to enter the stomach from the esophagus. Waves of contractions (peristalsis) of the smooth muscle layers in the stomach wall mix and move food through the stomach. (Similar peristaltic waves propel digestive contents through the whole tract.) The stomach produces up to 5 pints (3 liters) of gastric juices daily. As the food is liquefied, small amounts—just a teaspoonful at a time—are squirted through the stomach's outlet, the pyloric sphincter, into the first part of the small intestine, the duodenum.

PERISTALSIS
Waves of muscle contraction propel food through the tract (see right). The circular muscle contracts and relaxes in sequence producing a "traveling wave" known as peristalsis.

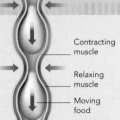

Contracting muscle
Relaxing muscle
Moving food

Duodenum
First and shortest section of the small intestine, about 10 in (25 cm) long

STOMACH FILLING AND EMPTYING
The stomach expands like a balloon as it fills with food and drink from a meal. Gases produced by chemical breakdown of food, and swallowed air, also collect in and expand the stomach. Those in the highest part of the stomach are expelled by belching (burping or eructation).

Chyme

1 AFTER A MEAL
Muscles of the stomach wall mix food with gastric juices and churn it to form chyme.

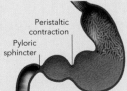

Peristaltic contraction
Pyloric sphincter

2 1–2 HOURS LATER
Peristaltic waves move the liquid stomach contents toward the pyloric sphincter.

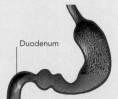

Duodenum

3 3–4 HOURS LATER
The pyloric sphincter opens at intervals to let small quantities of chyme into the duodenum.

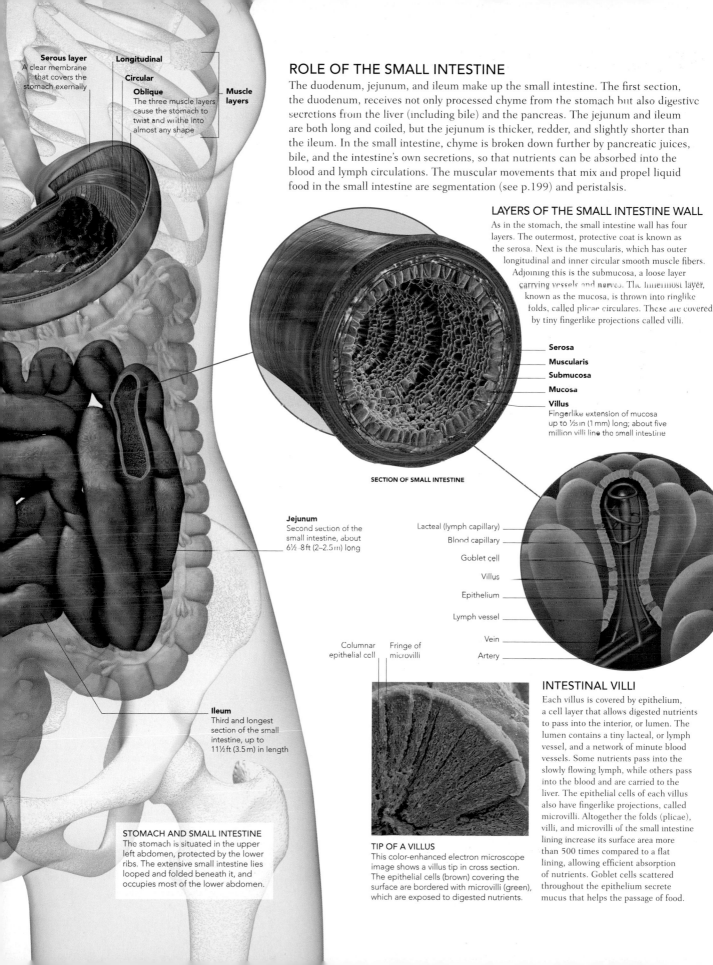

ROLE OF THE SMALL INTESTINE

The duodenum, jejunum, and ileum make up the small intestine. The first section, the duodenum, receives not only processed chyme from the stomach but also digestive secretions from the liver (including bile) and the pancreas. The jejunum and ileum are both long and coiled, but the jejunum is thicker, redder, and slightly shorter than the ileum. In the small intestine, chyme is broken down further by pancreatic juices, bile, and the intestine's own secretions, so that nutrients can be absorbed into the blood and lymph circulations. The muscular movements that mix and propel liquid food in the small intestine are segmentation (see p.199) and peristalsis.

LAYERS OF THE SMALL INTESTINE WALL

As in the stomach, the small intestine wall has four layers. The outermost, protective coat is known as the serosa. Next is the muscularis, which has outer longitudinal and inner circular smooth muscle fibers. Adjoining this is the submucosa, a loose layer carrying vessels and nerves. The innermost layer, known as the mucosa, is thrown into ringlike folds, called plicae circulares. These are covered by tiny fingerlike projections called villi.

Serosa

Muscularis

Submucosa

Mucosa

Villus
Fingerlike extension of mucosa up to 1⁄25 in (1 mm) long; about five million villi line the small intestine

SECTION OF SMALL INTESTINE

Serous layer
A clear membrane that covers the stomach externally

Longitudinal

Circular

Oblique
The three muscle layers cause the stomach to twist and writhe into almost any shape

Muscle layers

Jejunum
Second section of the small intestine, about 6½–8 ft (2–2.5 m) long

Ileum
Third and longest section of the small intestine, up to 11½ ft (3.5 m) in length

Lacteal (lymph capillary)

Blood capillary

Goblet cell

Villus

Epithelium

Lymph vessel

Vein

Artery

Columnar epithelial cell

Fringe of microvilli

INTESTINAL VILLI

Each villus is covered by epithelium, a cell layer that allows digested nutrients to pass into the interior, or lumen. The lumen contains a tiny lacteal, or lymph vessel, and a network of minute blood vessels. Some nutrients pass into the slowly flowing lymph, while others pass into the blood and are carried to the liver. The epithelial cells of each villus also have fingerlike projections, called microvilli. Altogether the folds (plicae), villi, and microvilli of the small intestine lining increase its surface area more than 500 times compared to a flat lining, allowing efficient absorption of nutrients. Goblet cells scattered throughout the epithelium secrete mucus that helps the passage of food.

TIP OF A VILLUS
This color-enhanced electron microscope image shows a villus tip in cross section. The epithelial cells (brown) covering the surface are bordered with microvilli (green), which are exposed to digested nutrients.

STOMACH AND SMALL INTESTINE
The stomach is situated in the upper left abdomen, protected by the lower ribs. The extensive small intestine lies looped and folded beneath it, and occupies most of the lower abdomen.

LIVER, GALLBLADDER, AND PANCREAS

THE LIVER IS THE BODY'S LARGEST INTERNAL ORGAN AND HAS A CRUCIAL ROLE IN THE MANUFACTURE, PROCESSING, AND STORAGE OF MANY CHEMICALS. IT PRODUCES THE DIGESTIVE FLUID BILE, WHICH IS THEN STORED IN THE GALLBLADDER. THE PANCREAS SECRETES VITAL DIGESTIVE ENZYMES.

STRUCTURE AND FUNCTION OF THE LIVER

Weighing about 3⅓ lb (1.5 kg), the dark red, wedge-shaped liver fills the upper right abdomen below the diaphragm. At a microscopic level, the liver's structural units (lobules) are made up of sheets of liver cells (hepatocytes), tiny branches of the hepatic artery and vein, and bile ducts. Nutrient-rich blood arrives from the intestines via the hepatic portal system (see opposite) and filters through the lobules. The liver has over 250 individual functions, the most important of which are storing and releasing blood sugar (glucose) for energy; sorting and processing vitamins and minerals; breaking down toxins into less harmful substances; and recycling old blood cells.

Hepatic vein
Drains all blood from liver into inferior vena cava

Inferior vena cava
Vein that transports blood from liver and lower body to heart just above

Right liver lobe
Forms around two-thirds of the liver's total bulk

Hepatic duct
Drains bile toward the gallbladder

Central vein
Cross section of lobule
Exterior of lobule

Branch of portal vein
Brings nutrient-rich blood from digestion

Central vein
Carries away processed blood for waste disposal

Sinusoid
Receives blood from hepatic portal vein and hepatic artery

Hepatocyte
Filters blood and makes bile

Branch of bile duct
Channels bile fluid away from liver for digestion

Branch of hepatic artery
Brings oxygen-rich blood to liver

Hepatic portal vein
Supplies blood from intestinal tract to liver

Artery **Bile duct** **Vein**

Gallbladder
Storage bag for liver's bile fluid

LIVER LOBULES
The six-sided lobules nestle together and have blood- and bile-collecting vessels around their exteriors.

INSIDE A LOBULE
Hepatocytes filter incoming blood into constituents destined for bile ducts, storage, or waste disposal.

LIVER FUNCTIONS

Most of the liver's tasks are concerned with metabolism. They include: the breakdown of digestive products; the storage of the resulting products; the circulation of substances such as vitamins and minerals; and the construction of complex molecules, such as enzymes.

BILE PRODUCTION	Liver cells secrete bile into small canals called bile canaliculi, which drain into bile ducts running between the lobules. These bile ducts converge to form the common hepatic duct, which conveys bile to the gallbladder for storage.
NUTRIENT PROCESSING	The liver removes nutrients from the blood. It converts simple sugars into glycogen—a process called glycogenesis—and synthesizes amino acids.
GLUCOSE REGULATION	The liver maintains blood glucose levels by converting fat and proteins into glucose. This process is called gluconeogenesis.
DETOXIFICATION	Harmful substances in the blood, such as alcohol and some other poisons, are detoxified. Waste products and unwanted amino acids are converted into urea.
PROTEIN SYNTHESIS	The liver synthesizes blood-clotting proteins and proteins for the fluid part of blood (plasma).
MINERAL AND VITAMIN STORAGE	The liver is a reservoir of minerals such as iron and copper, and vitamins including A, B_{12}, D, E, and K.
BLOOD WASTE DISPOSAL	Bacteria and general foreign particles are eliminated.
RECYCLING BLOOD CELLS	Old red blood cells are broken down and their constituents reused.

Pancreas
Hidden behind lower stomach and transverse colon

LIVER ARCHITECTURE
In this electron micrograph at a magnification of around 300 times, sheets of hepatocytes can be seen radiating from the central canal. This canal contains the central vein.

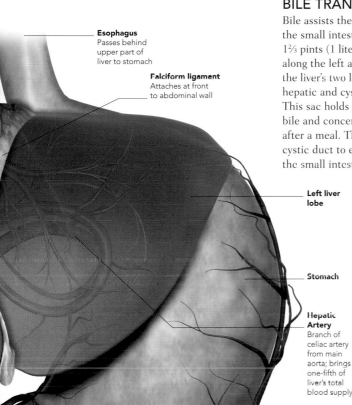

Esophagus
Passes behind upper part of liver to stomach

Falciform ligament
Attaches at front to abdominal wall

Left liver lobe

Stomach

Hepatic Artery
Branch of celiac artery from main aorta; brings one-fifth of liver's total blood supply

Transverse colon

BILE TRANSPORT

Bile assists the breakdown of fats (lipids) in the small intestine. The liver secretes up to 1⅔ pints (1 liter) of bile daily. The bile passes along the left and right hepatic ducts from the liver's two lobes, then along the common hepatic and cystic duct to the gallbladder. This sac holds about 1⅔ fl oz (50 ml) of bile and concentrates it, ready for release after a meal. The bile flows along the cystic duct to enter the first part of the small intestine, the duodenum.

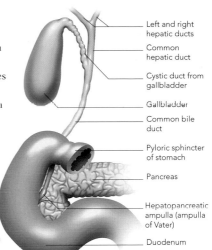

Left and right hepatic ducts

Common hepatic duct

Cystic duct from gallbladder

Gallbladder

Common bile duct

Pyloric sphincter of stomach

Pancreas

Hepatopancreatic ampulla (ampulla of Vater)

Duodenum

DUAL DUCTS
The common bile duct joins the pancreatic duct at the hepatopancreatic ampulla, which empties into the duodenum.

THE PANCREAS

The head end of this gland nestles in a loop of the duodenum, its main body lies behind the stomach, and its tapering tail sits above the left kidney, below the spleen. Each day, the pancreas produces around 2⅔ pints (1.5 liters) of digestive juice containing enzymes that break down lipids, proteins, and carbohydrates. The fluid flows into the main and accessory pancreatic ducts, which empty the juices into the duodenum.

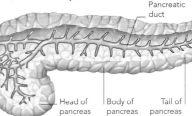

Pancreatic duct

PANCREATIC STRUCTURE
The pancreas is up to 6 in (15 cm) long, soft and flexible, and gray-pink in color.

Head of pancreas

Body of pancreas

Tail of pancreas

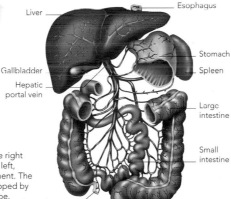

Liver

Esophagus

Stomach

Spleen

Gallbladder

Hepatic portal vein

Large intestine

Small intestine

Appendix

Rectum

EXTERNAL ANATOMY
The liver has two main lobes, the right one being much larger than the left, separated by the falciform ligament. The gallbladder is completely enveloped by the lower portion of the right lobe.

HEPATIC PORTAL SYSTEM
Veins from almost every part of the digestive tract, even the lower esophagus, converge to form the hepatic portal vein that enters the liver. In this view, some organs have been removed to reveal the blood vessels.

THE HEPATIC PORTAL CIRCULATION

The liver is unusual in that it receives two blood supplies. The hepatic artery delivers oxygen-rich blood to the liver. In addition, the hepatic portal vein supplies the liver with oxygen-poor, nutrient-rich blood from the digestive tract, before this blood returns to the heart and is pumped throughout the body. This enables the liver to prevent toxins absorbed in the intestines from reaching the rest of the body, and to regulate the levels of many other substances in the bloodstream. Veins from several organs, including the intestines, pancreas, stomach, and spleen, drain into the hepatic portal vein. It is around 3 in (8 cm) long and supplies up to four-fifths of the blood into the liver. The flow rate increases after a meal, but falls during physical activity as blood is diverted from the abdominal organs to skeletal muscles.

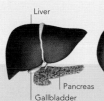

Liver

Pancreas

Gallbladder

FRONT

RIGHT SIDE

REAR

LEFT SIDE

360-DEGREE VIEW

LARGE INTESTINE

THE LARGE INTESTINE IS THE FINAL PART OF THE DIGESTIVE TRACT AND COMPRISES THREE MAIN REGIONS—THE CECUM, COLON, AND RECTUM. THE CECUM IS A SHORT POUCH THAT LINKS THE SMALL INTESTINE TO THE COLON, WHICH IS ABOUT 5 FT (1.5 M) LONG. THE COLON CHANGES LIQUID DIGESTIVE WASTE PRODUCTS FROM THE SMALL INTESTINE INTO A MORE SOLID FORM THAT THE BODY EXCRETES AS FECES VIA THE RECTUM AND ANUS.

ROLE OF THE COLON

When the chemical breakdown of food in the small intestine is complete (see pp.194–95), almost all the nutrients vital for bodily functions have been absorbed. The waste product from this process is partially digested, liquefied food (chyme). This passes from the small intestine, through the ileocecal valve, into the cecum. From there, it reaches the first part of the colon, the ascending colon. The main function of the colon is to convert the liquid chyme into semisolid feces for storage and disposal. Sodium, chloride, and water are absorbed through the lining of the colon into blood and lymph, making the feces less watery. Bicarbonate and potassium are secreted by the colon to replace the sodium and chloride. There are also billions of symbiotic or "friendly" microorganisms within the colon.

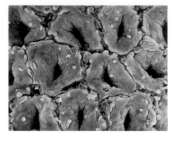

COLONIC GLANDS
This microscope image (magnified 120 times) shows the openings to tubular glands lining the colon. They absorb water from feces.

Ascending colon
Section of colon rising up right side of abdomen

LAYERS OF THE COLON WALL

The colon wall has several layers. The first is an outer coating (serosa). Inside this is the muscularis layer, which comprises two bands of smooth muscle fibers, longitudinal and then circular. These are responsible for colonic movements. The next layer is the submucosa, which has many small lobes of lymphatic tissue called lymphoid nodules. Innermost is the undulating mucosa. This contains goblet cells in intestinal glands, which secrete lubricating mucus to ease the passage of feces.

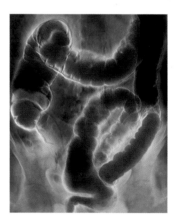

IMAGING THE COLON
This contrast X-ray of the large intestine was made by passing barium—a fluid opaque to X-rays—into the bowel via the rectum.

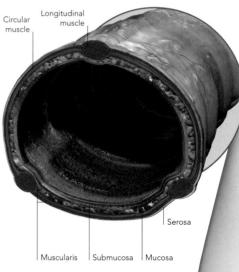

Circular muscle

Longitudinal muscle

Serosa

Muscularis   Submucosa   Mucosa

Cecum
Pouchlike entrance to large intestine

Ileocecal valve
Controls flow of liquefied food from small intestine

GUT FLORA

Trillions of microorganisms, mainly bacteria, live in the intestinal tract—chiefly in the large intestine. They are known as gut flora (or gut microbiota) and have a vital role in human health and disease. They produce enzymes that break down certain food components, especially the plant fiber cellulose, which human enzymes cannot digest. In this way, the bacteria feed on the undigested fiber in fecal material, provide nutrients that can be absorbed into the body, and help reduce the amount of feces. As part of their metabolism, the gut flora also produce vitamins K and B, and the gases hydrogen, carbon dioxide, hydrogen sulfide, and methane. In addition, the flora help control harmful microbes which may enter the digestive system, and assist the immune system in fighting disease by promoting formation of antibodies and the activity of lymphoid tissue in the colonic lining. Overall the gut flora and the body exist in a mutually beneficial partnership (symbiosis). When feces are excreted, at least one-third of their weight is composed of these bacteria.

BACTERIA IN THE COLON
This electron microscope image (magnified over 2,000 times) shows clusters of rodlike bacteria on the lining of the colon.

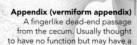

Appendix (vermiform appendix)
A fingerlike dead-end passage from the cecum. Usually thought to have no function but may have a role in maintaining normal gut flora

PARTS OF THE COLON
The three sections of the colon form an almost rectangular "frame," with the loops and coils of the small intestine inside it, the stomach and liver above, and the rectum below.

COLONIC MOVEMENT

Unlike the small intestine, the colon's long muscle does not form a complete tube within its wall. Instead, the muscle is concentrated into three bands called taenia coli. These run the length of the colon and form puckerings or pouches called haustra. The muscular movements in the walls of the colon mix and propel feces along the digestive tract toward the rectum. The movement of feces varies in rate, intensity, and nature, depending mainly on the stage of digestion of the contents. The three main types of motion are known as segmentation, peristaltic contractions, and mass movements. Fecal material passes more slowly through the colon than through the small intestine, enabling the reabsorption into the blood of up to 4¼ pints (2 liters) of water every day.

Transverse colon
Highest section of colon, just below stomach, passing across upper abdomen

Haustra
Pouches that give colon its puckered appearance

Feces

Taeni coli
Bands of longitudinal muscle running length of colon

Descending colon
Section of colon that passes down left side of abdomen

Sigmoid colon
Final colonic section, making an S-shaped bend to meet rectum

Rectum
Final part of large intestine, holds feces waiting to be passed out through the anus

Bladder

Prostate gland
Present only in male

External anal sphincter
Composed of skeletal (striped) muscle; mainly voluntary

Anus
Valvelike exit from end of digestive tract

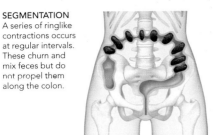

SEGMENTATION
A series of ringlike contractions occurs at regular intervals. These churn and mix feces but do not propel them along the colon.

PERISTALTIC CONTRACTIONS
Small waves of movement called peristaltic contractions (see p.194) propel feces toward the rectum. The muscles behind the contents contract, while those in front relax.

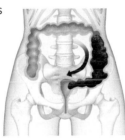

MASS MOVEMENTS
These extra-strong peristaltic waves move from the middle of the transverse colon. They happen two or three times a day and drive feces into the rectum.

RECTUM, ANUS, AND DEFECATION

The rectum is around 5 in (12 cm) long, and it is normally empty except just before and during defecation. Below the rectum lies the anal canal, which is around 1½ in (4 cm) long. In the walls of the anal canal are two strong sets of muscles forming short tubes—the internal and external anal sphincters. During defecation, peristaltic waves in the colon push feces into the rectum, which triggers the defecation reflex. Contractions push the feces along, and the anal sphincters relax to allow them out of the body through the anus.

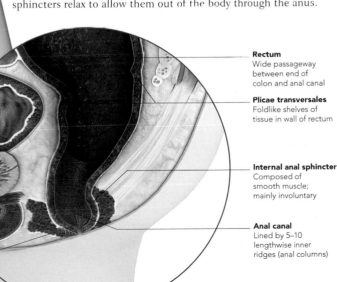

Rectum
Wide passageway between end of colon and anal canal

Plicae transversales
Foldlike shelves of tissue in wall of rectum

Internal anal sphincter
Composed of smooth muscle; mainly involuntary

Anal canal
Lined by 5–10 lengthwise inner ridges (anal columns)

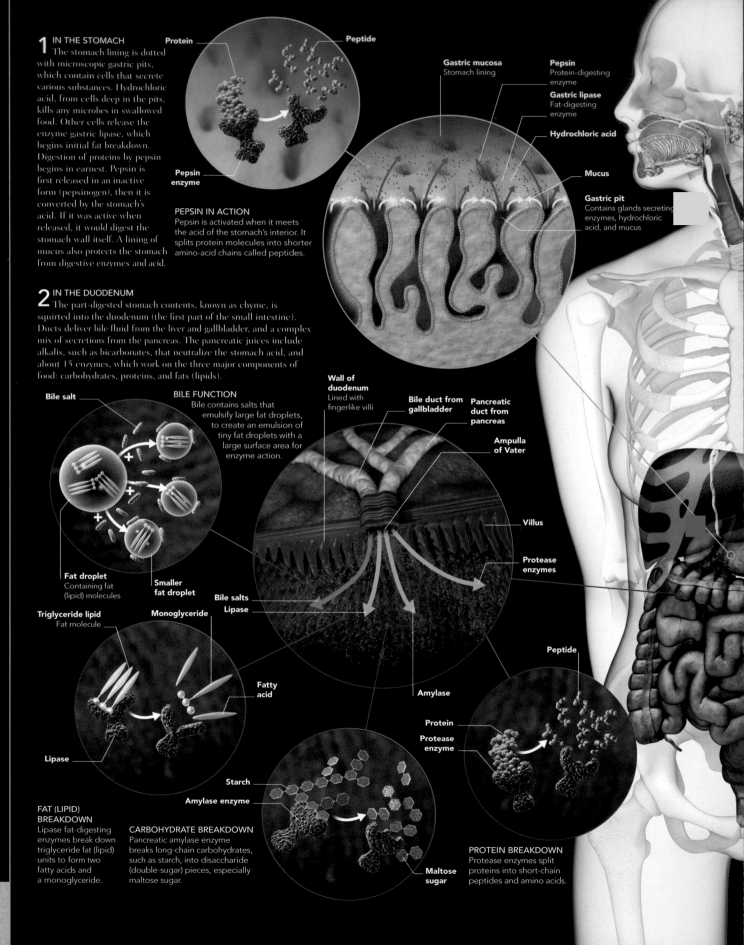

1 IN THE STOMACH

The stomach lining is dotted with microscopic gastric pits, which contain cells that secrete various substances. Hydrochloric acid, from cells deep in the pits, kills any microbes in swallowed food. Other cells release the enzyme gastric lipase, which begins initial fat breakdown. Digestion of proteins by pepsin begins in earnest. Pepsin is first released in an inactive form (pepsinogen), then it is converted by the stomach's acid. If it was active when released, it would digest the stomach wall itself. A lining of mucus also protects the stomach from digestive enzymes and acid.

Protein

Peptide

Pepsin enzyme

PEPSIN IN ACTION
Pepsin is activated when it meets the acid of the stomach's interior. It splits protein molecules into shorter amino-acid chains called peptides.

Gastric mucosa
Stomach lining

Pepsin
Protein-digesting enzyme

Gastric lipase
Fat-digesting enzyme

Hydrochloric acid

Mucus

Gastric pit
Contains glands secreting enzymes, hydrochloric acid, and mucus

2 IN THE DUODENUM

The part-digested stomach contents, known as chyme, is squirted into the duodenum (the first part of the small intestine). Ducts deliver bile fluid from the liver and gallbladder, and a complex mix of secretions from the pancreas. The pancreatic juices include alkalis, such as bicarbonates, that neutralize the stomach acid, and about 15 enzymes, which work on the three major components of food: carbohydrates, proteins, and fats (lipids).

Bile salt

BILE FUNCTION
Bile contains salts that emulsify large fat droplets, to create an emulsion of tiny fat droplets with a large surface area for enzyme action.

Wall of duodenum
Lined with fingerlike villi

Bile duct from gallbladder

Pancreatic duct from pancreas

Ampulla of Vater

Fat droplet
Containing fat (lipid) molecules

Smaller fat droplet

Bile salts

Lipase

Villus

Protease enzymes

Triglyceride lipid
Fat molecule

Monoglyceride

Fatty acid

Amylase

Peptide

Lipase

Protein

Protease enzyme

FAT (LIPID) BREAKDOWN
Lipase fat-digesting enzymes break down triglyceride fat (lipid) units to form two fatty acids and a monoglyceride.

Starch

Amylase enzyme

CARBOHYDRATE BREAKDOWN
Pancreatic amylase enzyme breaks long-chain carbohydrates, such as starch, into disaccharide (double-sugar) pieces, especially maltose sugar.

Maltose sugar

PROTEIN BREAKDOWN
Protease enzymes split proteins into short-chain peptides and amino acids.

DIGESTION

THE DIGESTIVE PROCESS INVOLVES A SERIES OF PHYSICAL AND CHEMICAL ACTIONS THAT BREAK DOWN THE COMPONENTS OF FOOD INTO NUTRIENT PARTICLES SMALL ENOUGH FOR ABSORPTION.

Vigorous physical digestion of food, mashing and churning, occurs in the mouth, but becomes progressively less important in successive sections of the digestive tract. The stomach also breaks food into small particles physically using muscular movement, but like the mouth, secretes digestive chemicals (enzymes), too. By the time the pulverized food and enzymes (chyme) reach the duodenum (the first part of the small intestine), many food particles are already microscopically small, yet not small enough to pass across cell membranes into the body tissues. Chemical digestion then takes over in importance, with large molecules split into even smaller, absorbable particles that can enter the bloodstream.

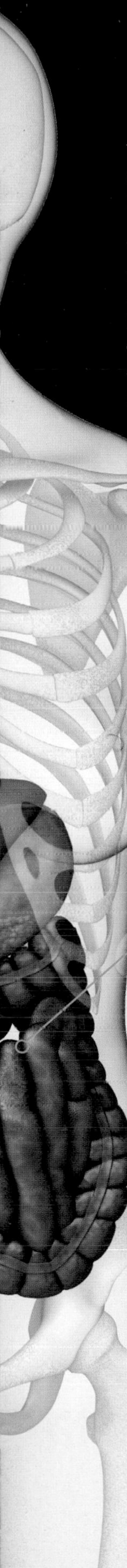

HOW ENZYMES WORK

An enzyme is a biological catalyst—a substance that boosts the rate of a biochemical reaction, but remains unchanged. Most enzymes are proteins. They affect the reactions of digestive breakdown, and also the chemical changes that release energy and build new materials for cells and tissues. Each enzyme has a specific shape due to the way its long chains of subunits (amino acids) fold and loop. The substance to be altered (the substrate) fits into a part of the enzyme known as the active site. For digestion, the enzyme may undergo a slight change in 3-D configuration that encourages the substrate to break apart at specific bonds between its atoms.

Active site

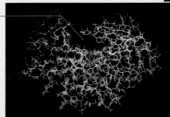

PEPSIN
A computer model of this digestive enzyme shows the active site as the gap at the top. A protein molecule slots in here and breaks apart.

3 IN THE SMALL INTESTINE

After the duodenum, the remainder of the small intestine is the site for the final breakdown of food substances and their absorption into the blood and lymphatic fluids. The pancreatic juices and bile fluids continue to work, but the small intestine releases few further enzymes into its inner passage, the lumen. Instead, its enzymes act within the lining cells, and on their surfaces. These enzymes include lactase and maltase, which break down the double (disaccharide) sugars, lactose and maltose, into single-unit glucose and galactose. Intestinal peptidases convert short peptide chains (originally from proteins) into their subunits, amino acids. The fingerlike villi of the intestine lining have surface cells bearing smaller projections of their own (microvilli) where some of the final changes occur.

Lumen
Fluid-filled space inside small intestine

Villus

Capillary of villus

ABSORPTION ACROSS VILLI
The fingerlike villi of the small intestine lining (left) provide a large area for the absorption of the products of digestion. These substances are shown here accumulating in the bloodstream from left to right.

Lacteal
Lymph capillary of villus

Direction of blood flow

Wall of small intestine

Fatty acid

Small intestine lumen

Epithelial cell membrane
Formed into "brush" of microvilli

CLOSE-UP OF VILLUS SURFACE
Short-chain fatty acids, glucose, and amino acids pass through the intestine's epithelial (lining) cells (above) and then into a blood capillary (red). Larger fatty acids are reassembled into triglyceride lipids, packaged, and passed into a lymph capillary (lacteal, purple).

DIGESTIVE JOURNEY
Each part of the digestive tract has its own conditions to further the dismantling of food substances into their subunits. Simple salts and minerals, such as sodium, potassium, and chloride, do not need digestion. They are mostly dissolved rapidly and absorbed in the small intestine.

Epithelial (lining) cell of small intestine wall

Glucose

Short-chain fatty acid

Amino acid

Lipid package

EXTREME CLOSE UP OF CELL MEMBRANE
The enzymes that complete digestion are embedded in the surface membrane of the intestine's epithelial cells (below). The resulting amino acids and sugars are then absorbed through dedicated protein channels in the membrane, while fatty acids simply pass straight through.

Short-chain fatty acid
Simply diffuses across cell membrane

Small intestine lumen

Epithelial cell membrane

Maltase enzyme
Splits (double) maltose into (single) glucose

Glucose
Passes across membrane through channel protein

Peptidase enzyme
Splits peptides into amino acids

Amino acids
Pass across membrane through channel protein in twos and threes

Epithelial cell interior

FATES OF NUTRIENTS

The digestive process takes, on average, 24–36 hours but can vary hugely between people and depending on what is eaten. Food is in the stomach for 2–4 hours, and in the small intestine for 1–5 hours. The final stages of digestion and waste compaction in the large intestine may take 12–24 hours, or even more. Different breakdown products are available for absorption at different times.

	MOUTH	STOMACH	SMALL INTESTINE	LARGE INTESTINE
PROTEINS		Hydrochloric acid and pepsin break protein into peptide chains	Peptidases snip peptides into amino acids for absorption	
CARBOHYDRATES	Salivary amylase begins starch digestion during chewing	Stomach acid inactivates salivary amylase	Enzymes, such as pancreatic amylase, yield simple sugars	
FAT (LIPIDS)		Gastric lipase splits lipids into fatty acids, and monoglycerides	Pancreatic lipase products enter lacteals	
FIBER				Soluble fiber broken down— not absorbed
SOLUBLE				
INSOLUBLE				
WATER		Small amounts absorbed by stomach lining	Absorbed by small intestine lining	Most water absorbed by large intestine
FAT-SOLUBLE VITAMINS (A,D,K)			Emulsified by bile salts and absorbed	Further absorption, K manufacture by gut bacteria
WATER-SOLUBLE VITAMINS (B, C)			Dissolve and are absorbed relatively easily	Continued absorption
MINERALS				Most minerals dissolve easily as inorganic salts for uptake in the small intestine and the colon
IRON				
SODIUM				
CALCIUM				

NUTRIENTS AND METABOLISM

THE BODY'S INTERNAL BIOCHEMICAL REACTIONS, CHANGES, AND PROCESSES ARE TERMED METABOLISM. DIGESTION PROVIDES THE NUTRIENTS AS RAW MATERIALS, WHICH ENTER METABOLIC PATHWAYS IN ALL CELLS AND TISSUES.

TAKING IN NUTRIENTS

"Nutrients" encompass all substances that are useful to the body. These include complex chemicals broken down to release energy, chiefly carbohydrates and fats; proteins, which are mainly for building the structural parts of cells; and vitamins and minerals, which ensure healthy functioning. The digestive system absorbs the nutrients into the blood and lymph at different stages along the tract. The blood flow from the major absorption sites of the intestines is along the hepatic portal vein (see p.197) to the liver. This large gland is the chief processor of nutrients. According to the body's needs, the liver breaks down some nutrients into even smaller, simpler molecules, stores others, and releases others into the circulation.

FINAL STAGES OF DIGESTION

The colon (large intestine, see p.198) is the last main site for breakdown and uptake of nutrients, including minerals, salts, and some vitamins. A considerable amount of water, mainly from the digestive juices, is also reabsorbed. Fiber, such as pectin and cellulose, gives bulk to the digestive remnants, and allows the walls to grip the residues as they are compressed into feces awaiting expulsion. Fiber also helps delay the absorption of some molecules, including sugars, spreading out their uptake through time rather than in one short "rush." In addition, fiber binds with some fatty substances, such as cholesterol, and helps prevent their overabsorption.

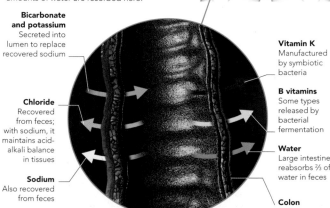

CECUM
Each day about 3½–17 fl oz (100–500 ml) of digestive fluids, undigested leftovers, rubbed-off intestinal linings, and other matter enters the first chamber of the large intestine, the cecum. Considerable amounts of water are resorbed here.

Bicarbonate and potassium
Secreted into lumen to replace recovered sodium

Chloride
Recovered from feces; with sodium, it maintains acid- alkali balance in tissues

Sodium
Also recovered from feces

Vitamin K
Manufactured by symbiotic bacteria

B vitamins
Some types released by bacterial fermentation

Water
Large intestine reabsorbs ⅔ of water in feces

Colon

BREAKDOWN AND BUILDING UP

Catabolism is the breaking apart of more complex molecules into simpler ones. This happens as part of energy production, for example, when glucose or fats are split apart to release energy. The converse is anabolism, which is the construction of complex molecules from simpler ones. For example, amino acids are linked together to make peptide chains, which then combine to form proteins, as part of protein synthesis (production).

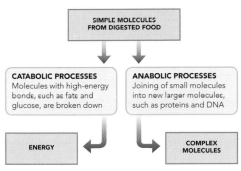

INTERPLAY
Metabolism is a complex interplay of construction and destruction, with many molecules being recycled as they pass between the two processes.

FUNCTIONS OF VITAMINS AND MINERALS

Vitamins are organic substances that are mostly incorporated into coenzymes—molecules that assist and support enzymes in the control of metabolic processes. Regular vitamin intake is required because only a few vitamins can be manufactured in the body. Minerals are simple inorganic substances such as calcium, iron, chloride, and iodine. They are needed both for general metabolism and specialized uses, such as iron for hemoglobin in red blood cells.

BLOOD CLOTTING	BLOOD CELL FORMATION AND FUNCTIONING	HEALTHY TEETH	HEALTHY EYES
Vitamin K	Vitamins B_6 and B_{12}	Vitamins C and D	Vitamin A
Calcium	Vitamin E	Calcium	Zinc
Iron	Folate	Phosphorus	
	Copper	Fluorine	
	Iron	Magnesium	
	Cobalt	Boron	

HEALTHY SKIN AND HAIR	HEART FUNCTIONING	BONE FORMATION	MUSCLE FUNCTIONING
Vitamin A	Vitamin B_1 (Thiamine)	Vitamin A	Vitamin B_1 (Thiamine)
Vitamin B_2 (Riboflavin)	Vitamin D	Vitamin C	Vitamin B_6
Vitamin B_3 (Niacin)	Inositol	Vitamin D	Vitamin B_{12}
Vitamin B_6 (Pyridoxine)	Calcium	Fluorine	Vitamin E
Vitamin B_{12}	Potassium	Calcium	Vitamin B_7 (Biotin)
Biotin	Magnesium	Copper	Calcium
Sulfur	Selenium	Phosphorus	Potassium
Zinc	Sodium	Magnesium	Sodium
	Copper	Boron	Magnesium

HOW THE BODY USES FOOD

The three major food components yield different breakdown products. Carbohydrates (starches and sugars) can be reduced to the simple sugar glucose; proteins are cut into polypeptide chains, peptides, and finally single amino acids; fats (lipids) are reduced to fatty acids and glycerol. The major use of glucose is as the body's most adaptable and readiest source of energy. Uses of fatty acids include forming the bi-lipid membranes around and inside cells (see p.27). Amino acids are reassembled into the body's own proteins, both structural (collagen, keratin, and similar tough substances) and functional (enzymes). However, the body can adapt and divert nutrients to different uses as conditions dictate.

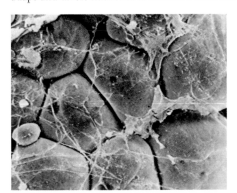

FAT TISSUE
Relatively, fatty substances, or lipids, are the body's most concentrated energy store, producing the most energy when metabolized. Adipose tissue consists of cells replete with fat droplets, stored for times of shortage.

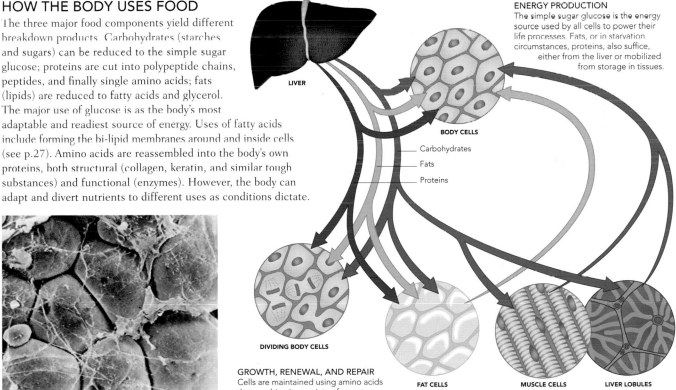

LIVER

BODY CELLS

Carbohydrates
Fats
Proteins

ENERGY PRODUCTION
The simple sugar glucose is the energy source used by all cells to power their life processes. Fats, or in starvation circumstances, proteins, also suffice, either from the liver or mobilized from storage in tissues.

DIVIDING BODY CELLS

GROWTH, RENEWAL, AND REPAIR
Cells are maintained using amino acids that combine in a variety of ways to build up different protein structures, fats to produce membranes, and glucose to provide the energy. Dividing cells for growth or repair require increased supplies of these nutrients.

FAT CELLS

MUSCLE CELLS

LIVER LOBULES

ENERGY STORAGE
Surplus glucose is converted into glycogen, which can be stockpiled in the liver and muscle cells. Fatty acids are a concentrated energy store. They are derived directly from dietary fats, from conversion of excess amino acids, or from conversion of glucose.

UPPER DIGESTIVE TRACT DISORDERS

MANY PROBLEMS OF THE ESOPHAGUS (FOOD TUBE) AND STOMACH RELATE TO THE CORROSIVE PROPERTIES OF THE ACIDIC STOMACH CONTENTS. THE UNDERSTANDING AND TREATMENT OF SEVERAL DIGESTIVE DISORDERS HAS BEEN REVOLUTIONIZED IN THE PAST TWO DECADES BY THE DISCOVERY THAT THEY ARE LINKED TO THE PRESENCE OF THE BACTERIUM *HELICOBACTER PYLORI*.

GINGIVITIS

INFLAMMATION OF THE GUMS, OR GINGIVAE, IS ONE OF THE MOST COMMON OF ALL HEALTH PROBLEMS.

The usual cause of gingivitis is poor oral hygiene. Dental plaque (a film of bacteria) builds up around the base of the teeth, where the crowns meet the gum. The gums become purplish red and swollen, and they bleed easily when brushed. Left untreated, the gum may pull away from the tooth neck, producing a pocket where bacteria can collect and cause infection. The main treatment is dental attention to remove the plaque.

GASTRIC REFLUX

ACIDIC STOMACH CONTENTS CAN FLOW BACK INTO THE ESOPHAGUS, CAUSING DISCOMFORT KNOWN AS HEARTBURN.

Heartburn is a common symptom and often occurs after overeating or drinking too much alcohol, or in pregnant women. Sometimes, however, the problem persists or increases in severity and requires medical attention. If reflux is long term, it can cause inflammation in the esophagus. Obesity and smoking both increase the likelihood of gastric reflux. The symptoms can also develop in people with a hiatus hernia (see opposite).

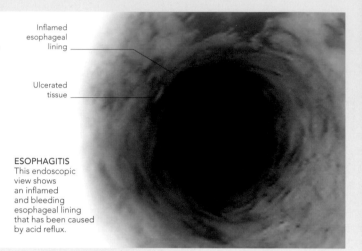

Inflamed esophageal lining

Ulcerated tissue

ESOPHAGITIS
This endoscopic view shows an inflamed and bleeding esophageal lining that has been caused by acid reflux.

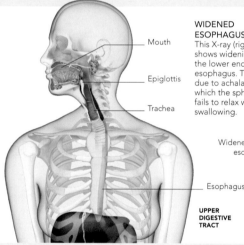

Mouth

Epiglottis

Trachea

Esophagus

UPPER DIGESTIVE TRACT

WIDENED ESOPHAGUS
This X-ray (right) shows widening of the lower end of the esophagus. This is due to achalasia, in which the sphincter fails to relax with swallowing.

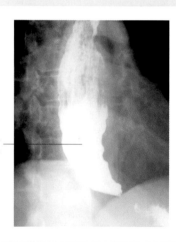

Widened lower esophagus

ACHALASIA

THIS MUSCLE DISORDER OF THE ESOPHAGUS CAUSES DIFFICULTY SWALLOWING AND DELAYS OR PREVENTS FOOD FROM PASSING THROUGH TO THE STOMACH.

Achalasia is caused by failure of the muscular ring (sphincter) at the lower end of the esophagus to relax on swallowing, combined with poor coordination of the contractions in the muscular wall of the esophagus that propel food along to the stomach. Gradually, the lower esophagus distends, causing symptoms such as difficulty swallowing, discomfort or pain behind the breastbone, and regurgitation of undigested food, especially at night when lying down. Treatments include widening the sphincter using an inflatable balloon, botox injections, drugs to relax the muscles, and surgery to cut muscle tissue in the lower esophagus.

CANCER OF THE ESOPHAGUS

A MALIGNANT TUMOR THAT OCCURS IN THE ESOPHAGUS IS OFTEN LINKED TO SMOKING AND EXCESS ALCOHOL.

The symptoms of esophageal cancer may not be apparent at first. Difficulty swallowing solid foods, and then fluids, is a common symptom. Later, food may be regurgitated and spill over into the lungs, causing a cough. Eventually the cancer may spread through the esophagus wall to nearby structures. Treatment involves surgery to remove the tumor or to insert a tube in the narrowed area to help swallowing.

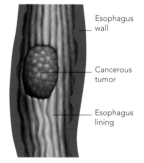

Esophagus wall

Cancerous tumor

Esophagus lining

ESOPHAGEAL TUMOR
An esophageal tumor physically narrows or blocks the passageway for swallowed food. It can be detected by endoscopy or a barium X-ray.

FOOD POISONING

CONSUMING CONTAMINATED FOOD OR DRINKS CAN RESULT IN DIARRHEA, VOMITING, AND ABDOMINAL PAIN.

Most people have had an episode of food poisoning, often when traveling overseas. Contaminated food may taste normal, the symptoms appearing hours or days later. Mostly, episodes are mild and clear up in a few days. However, some more serious infections, such as salmonella, may require treatment with antibiotics and rehydration. Careful preparation, storage, and cooking of food helps avoid these problems.

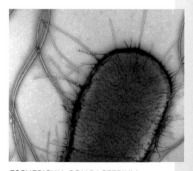

ESCHERICHIA COLI BACTERIUM
If *E. coli* bacteria contaminate food, such as meat or water, they cause an episode of food poisoning. Infection with *E. coli* can be serious, particularly in small children.

GASTRITIS

INFLAMMATION OF THE STOMACH LINING, CALLED GASTRITIS, CAUSES DISCOMFORT OR PAIN, AND NAUSEA AND VOMITING.

Sudden onset (acute) gastritis may be caused by overindulging, especially in alcohol, or by medications known for their effect on the stomach lining, such as aspirin. Chronic gastritis develops over the longer term and may be due to repeated insult to the lining by alcohol, tobacco, or drugs. Another common cause is the bacterium *Helicobacter pylori*. Gastritis usually gets better with medication and by removing the underlying cause.

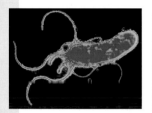

COMMON CULPRIT
More than 50 percent of people have *H. pylori* in their stomach lining. If the bacteria cause symptoms, they can be eradicated with antibiotics.

STOMACH CANCER

A CANCEROUS TUMOR IN THE STOMACH LINING IS MORE LIKELY IN THOSE WHO SMOKE, HAVE A HIGH-SALT DIET, OR ARE INFECTED WITH *H. PYLORI*.

Stomach cancer is more common in people over 50 years of age and in males. This type of cancer spreads (metastasizes) to other parts of the body rapidly and has often done so before the symptoms are noticed. These include upper abdominal discomfort or pain, especially after eating, along with nausea and vomiting, and loss of appetite and weight loss. Anemia may also develop due to bleeding from the stomach lining. If the cancer is caught early enough, surgical treatment can be successful.

PEPTIC ULCERS

PEPTIC ULCERS ARE ERODED, INFLAMED AREAS EITHER IN THE LINING OF THE STOMACH OR THE FIRST PART OF THE SMALL INTESTINE (THE DUODENUM) THAT CAUSE PAIN.

Most peptic ulcers are associated with *Helicobacter pylori* bacteria. These damage the mucous lining that normally protects against the powerful acidic juices in the stomach and first part of the duodenum. Other contributory factors include alcohol, smoking, certain medications, family history, and diet. Upper abdominal pain is a common symptom. With a duodenal ulcer this is often worse before a meal and relieved by eating; in a gastric ulcer, eating aggravates the pain.

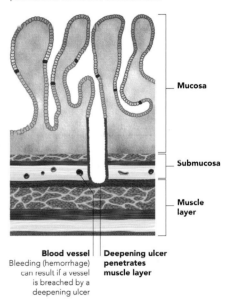

SITES OF PEPTIC ULCERS
A common site for ulcers is in the first part of the duodeum (duodenal bulb). In the stomach, most ulcers develop in the lesser curvature.

EARLY ULCER
If the protective mucous barrier coating the stomach lining breaks down, gastric juices containing strong acid and enzymes come into contact with mucosal cells.

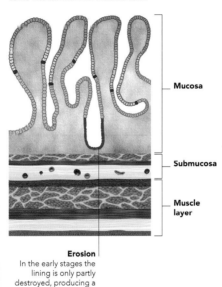

Erosion
In the early stages the lining is only partly destroyed, producing a shallow area of damage

PROGRESSIVE ULCERATION
A true ulcer penetrates the entire lining (mucosal layer) as well as the submucosa and muscle layers. In severe cases, it can perforate the stomach or duodenal wall.

Blood vessel
Bleeding (hemorrhage) can result if a vessel is breached by a deepening ulcer

Deepening ulcer penetrates muscle layer

HIATUS HERNIA

WEAKNESS IN THE GAP IN THE DIAPHRAGM THROUGH WHICH THE ESOPHAGUS PASSES ALLOWS PART OF THE STOMACH TO PROTRUDE INTO THE CHEST CAVITY.

The diaphragm is a muscular sheet that separates the abdomen from the chest cavity. Normally the stomach lies completely beneath the diaphragm, but in people with a hiatus hernia, its upper section protrudes up through the normally taut aperture, or gap (hiatus), through which the lower esophagus passes. The hiatus helps the esophageal sphincter (ring of muscle at lower end of esophagus) prevent acidic stomach contents from passing up into the lower esophagus, so any symptoms of a hernia are those of gastric reflux (see opposite). There are two types of hiatus hernia: sliding and paraesophageal. Sliding hernias usually have no symptoms and it is estimated that they are present in around a third of all people over 50. In rare cases, however, paraesophageal hernias can cause severe pain and require surgery.

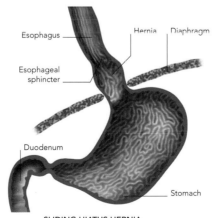

SLIDING HIATUS HERNIA
This is the most common type of hiatus hernia and occurs when the junction between the esophagus and stomach slides up through the diaphragm.

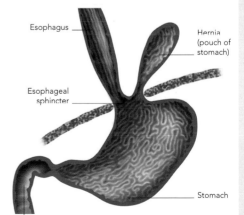

PARAESOPHAGEAL HIATUS HERNIA
In about 1 in 10 hernias, a pouchlike part of the stomach is pushed upward through the diaphragm and lies adjacent to the lower esophagus.

LIVER, GALLBLADDER, AND PANCREAS DISORDERS

THE LIVER, PANCREAS, AND GALLBLADDER ARE ALL VITAL ORGANS IN THE DIGESTION, ABSORPTION, AND METABOLISM OF FOOD, DRINKS, AND MEDICATIONS. AS WITH ALL OTHER ORGANS, THEY ARE VULNERABLE TO INFECTION, TOXIC DAMAGE, AND MALIGNANCY. OFTEN, AS IN THE CASE OF ALCOHOLIC LIVER DISEASE AND HEPATITIS, DISEASES ARE RELATED TO LIFESTYLE BEHAVIOR AND AS SUCH ARE PREVENTABLE.

ALCOHOLIC LIVER DISEASE

Over a period of many years, regular excessive alcohol consumption can lead to serious liver damage. Men generally drink more heavily than women and should therefore statistically be more likely to develop alcohol-related liver disease. However, women do not metabolize alcohol as efficiently as men and are more vulnerable to its side effects. The toxic effects of certain chemicals in alcohol can damage the liver in different ways, and in some people these toxic effects can increase the risk of developing liver cancer.

PROGRESSION OF THE DISEASE

ALCOHOL CAN CAUSE A WIDE RANGE OF LIVER DISEASES, DEPENDING ON THE NUMBER OF YEARS OF HEAVY DRINKING.

Almost all long-term, heavy drinkers have what is known as a "fatty liver." When alcohol is broken down into its various constituents (metabolized), it produces fat. Globules of fat become lodged in the liver cells, causing them to swell. Fatty liver does not cause any symptoms, but blood test results may be abnormal. If a person stops drinking at this stage, the fat disappears and the liver may eventually return to normal. However, continued heavy drinking can lead to alcoholic hepatitis, in which the liver becomes inflamed. Symptoms vary from none at all, to acute illness and jaundice. The final stage of alcoholic liver damage is cirrhosis, which can be fatal. Often the only treatment option at this stage is a liver transplant.

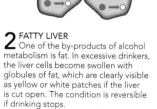

Acetaldehyde

Alcohol

Liver cell

Water

1 HOW DAMAGE OCCURS
When alcohol (ethanol) is broken down, a substance called acetaldehyde is formed. It is thought that this chemical binds with proteins in the liver cell, which may cause damage, inflammation, and fibrosis.

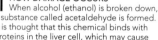

Fat-laden cell   Liver cell

2 FATTY LIVER
One of the by-products of alcohol metabolism is fat. In excessive drinkers, the liver cells become swollen with globules of fat, which are clearly visible as yellow or white patches if the liver is cut open. The condition is reversible if drinking stops.

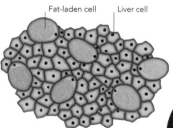

Damaged tissue

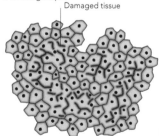

3 ALCOHOLIC HEPATITIS
With continued excessive drinking, fatty liver may develop into hepatitis. The liver becomes inflamed and infiltrated with leucocytes (white blood cells). Liver cells may become severely damaged and die.

4 CIRRHOSIS
In this final stage of alcoholic liver disease, the permanent fibrosis and scarring of the liver tissue is life threatening. As the cells are permanently damaged, the liver is unable to carry out its normal functions.

Scar tissue

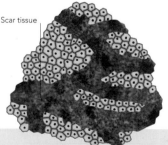

PORTAL HYPERTENSION

RAISED PRESSURE IN THE BLOOD VESSELS SUPPLYING THE LIVER CAUSES DISTENDED VEINS IN THE ESOPHAGUS AND STOMACH.

One of the complications of liver cirrhosis is portal hypertension. As the tissue becomes progressively scarred and fibrosed, it obstructs the flow of blood into the liver from the portal vein, a large vessel carrying blood from the digestive tract. Pressure builds up in the vein and can cause other vessels "upstream" to become distended. Among these are veins in the abdomen, rectum, and those that supply the esophagus with blood. The swollen veins, or varices, protrude into the esophagus and may bleed. In some cases, only slight oozing occurs. In others, a major hemorrhage causes massive vomiting of blood. Not everyone who has liver cirrhosis develops portal hypertension and esophageal varices. In those who do develop the condition, the varices can be treated with drugs to reduce the blood pressure or injected with a sclerosing (hardening) agent, much like that used to treat varicose veins.

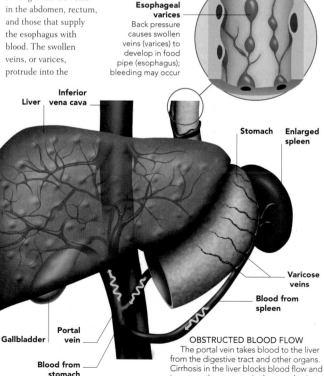

Esophageal varices
Back pressure causes swollen veins (varices) to develop in food pipe (esophagus); bleeding may occur

Liver

Inferior vena cava

Stomach

Enlarged spleen

Varicose veins

Blood from spleen

Gallbladder

Portal vein

Blood from stomach

OBSTRUCTED BLOOD FLOW
The portal vein takes blood to the liver from the digestive tract and other organs. Cirrhosis in the liver blocks blood flow and increases the pressure in the portal vein. Back pressure causes veins "upstream" and in the esophagus to distend.

HEPATITIS

HEPATITIS IS AN INFLAMMATION OF THE LIVER THAT CAN BE CAUSED BY A NUMBER OF DIFFERENT VIRUSES.

Viral hepatitis can be either acute (sudden onset) or chronic (long-term). Although acute hepatitis may clear up in a few weeks, it can progress to the chronic form. The most common type is hepatitis A, which is caused by ingesting contaminated food or water. Hepatitis B is mainly spread via infected blood, but the virus is also found in semen and can be sexually transmitted. Hepatitis C virus is also transmitted through blood and many individuals have been infected by blood transfusions. Most commonly acquired through IV drug use, symptoms vary from feeling mildly unwell to jaundice and liver failure.

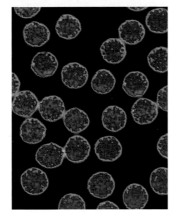

HEPATITIS
This image showing spherical hepatitis B viruses is magnified about 200,000 times. The virus is one of the many causes of the liver disorder, acute hepatitis.

LIVER ABSCESS

A RARE CONDITION IN WHICH PUS-FILLED CAVITIES DEVELOP IN THE LIVER TISSUE, OFTEN DUE TO THE SPREAD OF INFECTION FROM THE ABDOMEN.

A liver abscess can be caused by an amebic or bacterial infection that spreads from elsewhere in the body via the blood. The origin of the infection varies but may be an infected appendix or gallbladder. However, the cause is often unknown. Some people have very few symptoms and the abscess can go undetected for several weeks. In others, the condition can result in severe pain, vomiting, weight loss, and a high fever. Liver abscesses are usually drained of pus using a large needle. Once the bacterium is identified, the infection is treated with antibiotics.

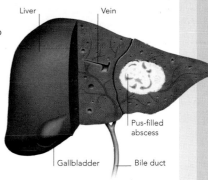

Liver — Vein — Pus-filled abscess — Gallbladder — Bile duct

INFECTED ABSCESS
This rare type of abscess can occur alone or in groups. It usually develops following an infection elsewhere in the body, which spreads via the blood to the liver. It can be successfully drained of pus, using a needle and a syringe.

GALLSTONES

SMALL, HARD MASSES FORMED FROM BILE CAN OCCUR IN THE GALLBLADDER. THEY CAUSE PAIN WHEN THEY MOVE ON AND BECOME LODGED IN ADJACENT DUCTS.

In developed countries, the majority of gallstones consist primarily of cholesterol, a fatty substance that is processed in the liver and stored in the gallbladder as one of the constituents of bile. Gallstones can develop if the normal "mix" of bile is altered and the cholesterol content is high. They are far more common in women and are unusual before the age of 30. Most people with gallstones have no symptoms at all. It is only when a stone becomes lodged in one of the ducts leaving the gallbladder that symptoms occur. The main symptom is pain, which varies in intensity and often develops after a fatty meal, when bile is released from the gallbladder to help digestion. For symptomatic gallstones, the treatment is usually cholecystectomy (often by laparoscopic surgery) to remove the gallbladder.

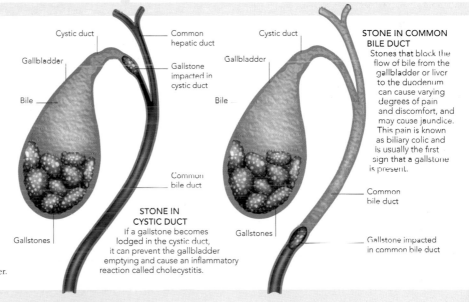

Cystic duct — Gallbladder — Bile — Common hepatic duct — Gallstone impacted in cystic duct — Common bile duct — Gallstones

STONE IN CYSTIC DUCT
If a gallstone becomes lodged in the cystic duct, it can prevent the gallbladder emptying and cause an inflammatory reaction called cholecystitis.

Cystic duct — Gallbladder — Bile — Common bile duct — Gallstones — Gallstone impacted in common bile duct

STONE IN COMMON BILE DUCT
Stones that block the flow of bile from the gallbladder or liver to the duodenum can cause varying degrees of pain and discomfort, and may cause jaundice. This pain is known as biliary colic and is usually the first sign that a gallstone is present.

CANCER OF THE PANCREAS

AN INCREASINGLY COMMON MALIGNANT TUMOR, OFTEN LINKED TO SMOKING, RESULTS IN PANCREATIC CANCER.

Tumors of the pancreas are categorized into those that occur in its body or tail of this organ and those that develop in its head. Cancer in the head of the pancreas blocks the flow of bile and is therefore more likely to cause jaundice, while a malignancy in the body or tail commonly produces pain in the upper abdomen. This type of cancer is more common in people who smoke and is seen more often in men. Cancer of the pancreas has a poor prognosis and usually the treatment aims to relieve symptoms.

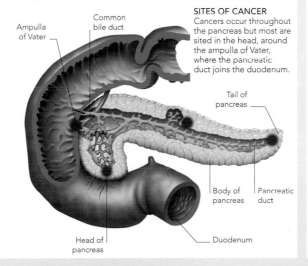

Ampulla of Vater — Common bile duct — Tail of pancreas — Body of pancreas — Pancreatic duct — Head of pancreas — Duodenum

SITES OF CANCER
Cancers occur throughout the pancreas but most are sited in the head, around the ampulla of Vater, where the pancreatic duct joins the duodenum.

PANCREATITIS

A SERIOUS INFLAMMATION OF THE PANCREAS, WHICH CAN BE CAUSED BY GALLSTONES OR BY DRINKING EXCESSIVE AMOUNTS OF ALCOHOL.

Pancreatitis can be either acute or chronic. In both types, the inflammation is triggered by the enzymes that the pancreas itself manufactures to aid digestion of food in the duodenum. These enzymes become activated while still inside the pancreas and begin to digest the tissue. There are many causes of acute pancreatitis, the most common of which are gallstones, alcohol, some drugs, and certain infections, such as mumps. Chronic pancreatitis is usually associated with long-term alcoholism. In both types, the main feature is pain. In acute pancreatitis this is particularly severe and may be accompanied by nausea and vomiting.

LOWER DIGESTIVE TRACT DISORDERS

INFECTIONS OF THE LOWER DIGESTIVE TRACT—THE COLON, RECTUM, AND ANUS—ARE AMONG THE MOST COMMON DIGESTIVE CONDITIONS. THEY ARE A MAJOR CAUSE OF DEATH IN DEVELOPING NATIONS, BUT USUALLY CAUSE ONLY MINOR PROBLEMS IN DEVELOPED COUNTRIES. OTHER DIGESTIVE DISORDERS, SUCH AS CANCER AND INTESTINAL INFLAMMATION, CAUSE PROBLEMS WORLDWIDE.

IRRITABLE BOWEL SYNDROME

THIS COMBINATION OF INTERMITTENT ABDOMINAL PAIN, CONSTIPATION, AND DIARRHEA AFFECTS AS MANY AS ONE IN FIVE PEOPLE DURING THEIR LIVES.

Often called IBS, this is one of the most common of all digestive complaints. It occurs mainly in people aged 20–30 years, and is twice as frequent in females than males. Its precise cause is unclear, but it is thought to involve abnormal muscular movements within the intestine. The factors that trigger IBS may include a bout of gastroenteritis or sensitivity to particular substances, such as caffeine, alcohol, high-fat foods, or artificial sweeteners. There also seems to be a genetic component, because some families have a history of IBS. The symptoms of IBS include diarrhea, constipation, abdominal pain, and, in particular, bloating and large quantities of intestinal gas; these problems can be made worse by anxiety, depression, or stress. The pain is often in the lower left area of the abdomen, and it may be relieved by passing gas or feces. IBS is generally a long-term complaint, but it is usually intermittent and rarely serious.

INFLAMMATORY BOWEL DISEASE

THIS ENCOMPASSES TWO CONDITIONS WITH SIMILAR SYMPTOMS: ULCERATIVE COLITIS AND CROHN'S DISEASE.

Both of these conditions involve serious inflammation of the intestines. There may be an underlying problem with the immune system, which makes it attack the body's own intestinal tissues. There is also a tendency for ulcerative colitis and Crohn's disease to run in families. However, their detailed causes remain unclear. Most cases are long-term and begin between the ages of 15 and 30. Symptoms common to both conditions include abdominal pain, diarrhea, appetite loss, fever, intestinal bleeding, and weight loss. Treatment involves antidiarrheal and anti-inflammatory drugs, and sometimes surgery. The operation, called a colectomy, removes the worst affected portions of the large intestine.

Stricture

Large intestine

Terminal ileum

Area of inflammation

Cecum

Rectum

CROHN'S DISEASE
In Crohn's disease, patches of inflammation with ulceration may occur anywhere in the digestive tract, from the mouth to the anus. It also causes narrowings (strictures) in the intestines. The area where the small and large intestines meet, including the terminal ileum and cecum, is often affected.

Inflamed large intestine

Cecum

Inflamed rectum

ULCERATIVE COLITIS
Inflammation and ulceration can affect the rectum alone—when it may be known as proctitis—or part or all of the colon as well. Development in the lining of open, sorelike ulcers can cause blood, and sometimes pus, to appear in the feces or even on their own with no fecal matter.

DIVERTICULAR DISEASE

DIVERTICULAR DISEASE INCLUDES DIVERTICULOSIS—POUCHES THAT FORM IN THE WALL OF THE COLON.

Most people with diverticular disease are aged over 50 and have eaten a low-fiber diet for many years, with consequent straining to pass hard stools. The problem becomes more frequent with age. The lowest part of the colon, the sigmoid colon, is most commonly affected, but the entire colon can be involved. In diverticulosis, patches of the intestinal wall bulge outward into blind-ending pouches known as diverticula. About 95 percent of people with diverticular disease show no symptoms, but some people have abdominal pain and irregular bowel habits. Diverticulitis occurs when the pouches become inflamed, causing severe pain, fever, and constipation. As in IBD (above), the pain is often in the lower left abdomen and may fade after passing gas or stools.

1 HARD FECES
Soft, bulky feces are able to pass easily along the colon. If feces are hard and dry, usually due to lack of fiber or "roughage" in the diet, the contractions of the smooth muscle layers of the colon must increase in force, putting pressure on the walls of the colon.

Hard, dry feces

Wall of colon

Blood vessel

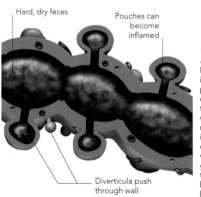

Hard, dry feces

Pouches can become inflamed

2 POUCHES FORM
Eventually, the increased pressure pushes small areas of intestinal lining through points of weakness in the muscle of the intestinal wall, often near a blood vessel. The pea- to grape-sized pouches that form easily trap bacteria and may become inflamed.

Diverticula push through wall

APPENDICITIS

AN INFLAMED APPENDIX CAUSES ACUTE PAIN THAT USUALLY STARTS IN THE UPPER- OR MID-ABDOMEN, AND IS COMMON IN CHILDREN AND ADOLESCENTS.

Other symptoms of appendicitis include mild fever, nausea, vomiting, and perhaps loss of appetite and frequent urination. In many cases the inflammation progresses so quickly that the person with appendicitis needs urgent hospitalization. Surgical removal of the appendix, called an appendectomy, is one of the most commonly performed emergency operations. Left untreated, an inflamed appendix can rupture, causing peritonitis (inflammation of the abdominal lining) and abscesses.

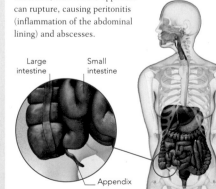

Large intestine

Small intestine

Appendix

COLORECTAL CANCER

CANCER OF THE COLON, RECTUM, OR BOTH, IS ONE OF THE MOST COMMON CANCERS IN INDUSTRIALIZED COUNTRIES. RISK FACTORS INCLUDE FAMILY HISTORY AND AGING.

A malignant tumor in the intestinal wall can often start as a polyp (see right) in the intestinal lining. A high-fat, low-fiber diet, excessive alcohol, lack of exercise, and obesity increase the likelihood of this cancer. Symptoms are a change in bowel habits and stool consistency, abdominal pain, loss of appetite, fecal blood, and a sensation of not fully emptying the bowels. Colorectal cancer can be detected by screening programs, including fecal tests for blood and endoscopic examination (sigmoidoscopy). If detected and treated early, the chances of survival of five years or longer are high.

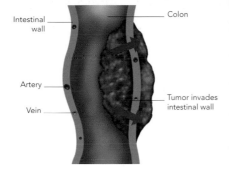

Intestinal wall · Colon · Artery · Vein · Tumor invades intestinal wall

COLONIC TUMOR
Over time, malignant tumors grow and invade the intestinal wall, from where the cancer can spread to other parts of the body via the bloodstream.

INTESTINAL POLYPS

SLOW-GROWING, USUALLY NONCANCEROUS GROWTHS, LOCATED IN THE LARGE INTESTINE, THAT PROJECT FROM THE MUCUS MEMBRANE LINING.

These are common in later years; one person in three over the age of 60 may be affected by intestinal polyps. Most people show no symptoms, but polyps may cause diarrhea, rectal bleeding, and perhaps anemia. Most cases are successfully detected by colonoscopy and removed, but then need regular checks since there is an increased risk of colorectal cancer (see left).

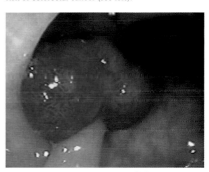

INTESTINAL OBSTRUCTION

OBSTRUCTION OF THE INTESTINE CAUSES ABDOMINAL PAIN AND DISTENSION, ABSENCE OF FECAL EXCRETION OR GAS, VOMITING, AND SOMETIMES DEHYDRATION.

Digestive material may be prevented from moving along the intestine due to physical blockage or perhaps paralysis of the smooth muscles in the intestinal wall. Causes include pressure from a tumor, or severe inflammation, as in Crohn's disease, that may narrow the intestine so that it is effectively blocked. Some hernias, intussusception (see panel below), and volvulus are further possibilities. Sometimes the muscles fail to contract, perhaps due to mesenteric infarction, serious abdominal peritonitis, or major abdominal surgery. To stabilize the condition and confirm the diagnosis, urgent hospitalization is needed. Treatment often involves surgery to relieve the blockage and sometimes to remove a part of the bowel.

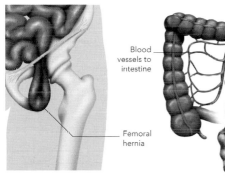

Blood vessels to intestine · Femoral hernia

FEMORAL HERNIA
The intestine slips through the narrow femoral canal and becomes trapped, causing an obstruction and severe pain.

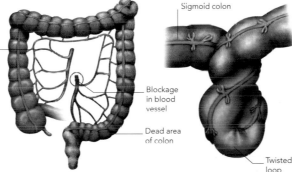

Blockage in blood vessel · Dead area of colon

MESENTERIC INFARCTION
A segment of intestine, deprived of blood due to blockage of a vessel in the mesentery, soon starts to die.

Sigmoid colon · Twisted loop

VOLVULUS
Intermittent intestinal twisting causes severe pain, distension, and vomiting; surgery is needed.

INTUSSUSCEPTION

Intestinal obstruction in young children, especially boys under two years of age, can be due to intussusception. Part of the intestine telescopes in on itself, forming a tube within a tube. Symptoms include vomiting, abdominal pain, paler skin, and the passage of bloodstained mucus. The condition progresses rapidly and needs urgent medical attention. It can be both diagnosed and often unblocked by a barium enema.

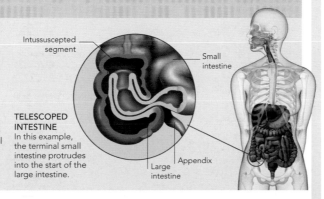

Intussuscepted segment · Small intestine

TELESCOPED INTESTINE
In this example, the terminal small intestine protrudes into the start of the large intestine.

Large intestine · Appendix

HEMORRHOIDS

VARICOSED (SWOLLEN AND ENGORGED) VEINS THAT PROTRUDE FROM THE RECTAL OR ANAL LINING ARE KNOWN AS HEMORRHOIDS.

Rectal or anal bleeding and discomfort are commonly associated with hemorrhoids. Causes include constipation from a low-fiber diet, and straining to pass feces, which may make the blood vessels of the rectum and anus swell. In pregnancy, the growing baby has a similar effect. The symptoms vary greatly in severity, and may include a discharge of mucus from the anus with anal itching. Treatments include ointments, injections, banding, laser therapy, and surgery.

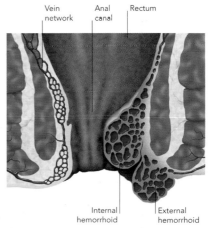

Vein network · Anal canal · Rectum · Internal hemorrhoid · External hemorrhoid

HEMORRHOIDS
The vein network on the left side of this diagram is normal. The blood vessels to the right are swollen and have formed internal and external hemorrhoids.

THOUSANDS OF METABOLIC PROCESSES IN MYRIAD
BODY CELLS PRODUCE HUNDREDS OF WASTE PRODUCTS.
THE URINARY SYSTEM REMOVES THEM BY FILTERING
AND CLEANSING THE BLOOD AS IT PASSES THROUGH THE
KIDNEYS. ANOTHER VITAL FUNCTION IS THE REGULATION
OF THE VOLUME, ACIDITY, SALINITY, CONCENTRATION, AND
CHEMICAL COMPOSITION OF BLOOD, LYMPH, AND OTHER
BODY FLUIDS. UNDER HORMONAL CONTROL, THE KIDNEYS
CONTINUALLY MONITOR WHAT THEY RELEASE INTO THE
URINE TO MAINTAIN A HEALTHY CHEMICAL BALANCE.
DISORDERS OF THE SYSTEM CAN BE SUBTLE, SO URINATION-
RELATED SYMPTOMS SHOULD BE PROMPTLY REPORTED.

URINARY SYSTEM

URINARY ANATOMY

THE URINARY SYSTEM IS COMPOSED OF A PAIR OF KIDNEYS, A PAIR OF URETERS, A BLADDER, AND A URETHRA. THESE COMPONENTS TOGETHER CARRY OUT THE URINARY SYSTEM'S FUNCTION OF REGULATING THE VOLUME AND COMPOSITION OF BODY FLUIDS, REMOVING WASTE PRODUCTS FROM THE BLOOD, AND EXPELLING THE WASTE AND EXCESS WATER FROM THE BODY IN THE FORM OF URINE.

The two kidneys are reddish organs, resembling beans in shape, situated on either side of the abdomen just above the waist and toward the back of the body. The kidneys contain microscopic filtering units that remove waste, unwanted minerals, and excess water from the blood as urine. Each kidney is connected to the bladder by a long tube called a ureter, which transports urine away from the kidney. The bladder is a hollow, muscular organ, situated centrally in the pelvis, that stores urine until it can be released. When empty, the bladder resembles a deflated balloon, gradually becoming spherical, and then pear-shaped, as it fills up. At a certain volume, stretch receptors in its wall transmit nervous impulses that initiate a conscious desire to urinate. The urethra then conducts urine from the bladder to the outside.

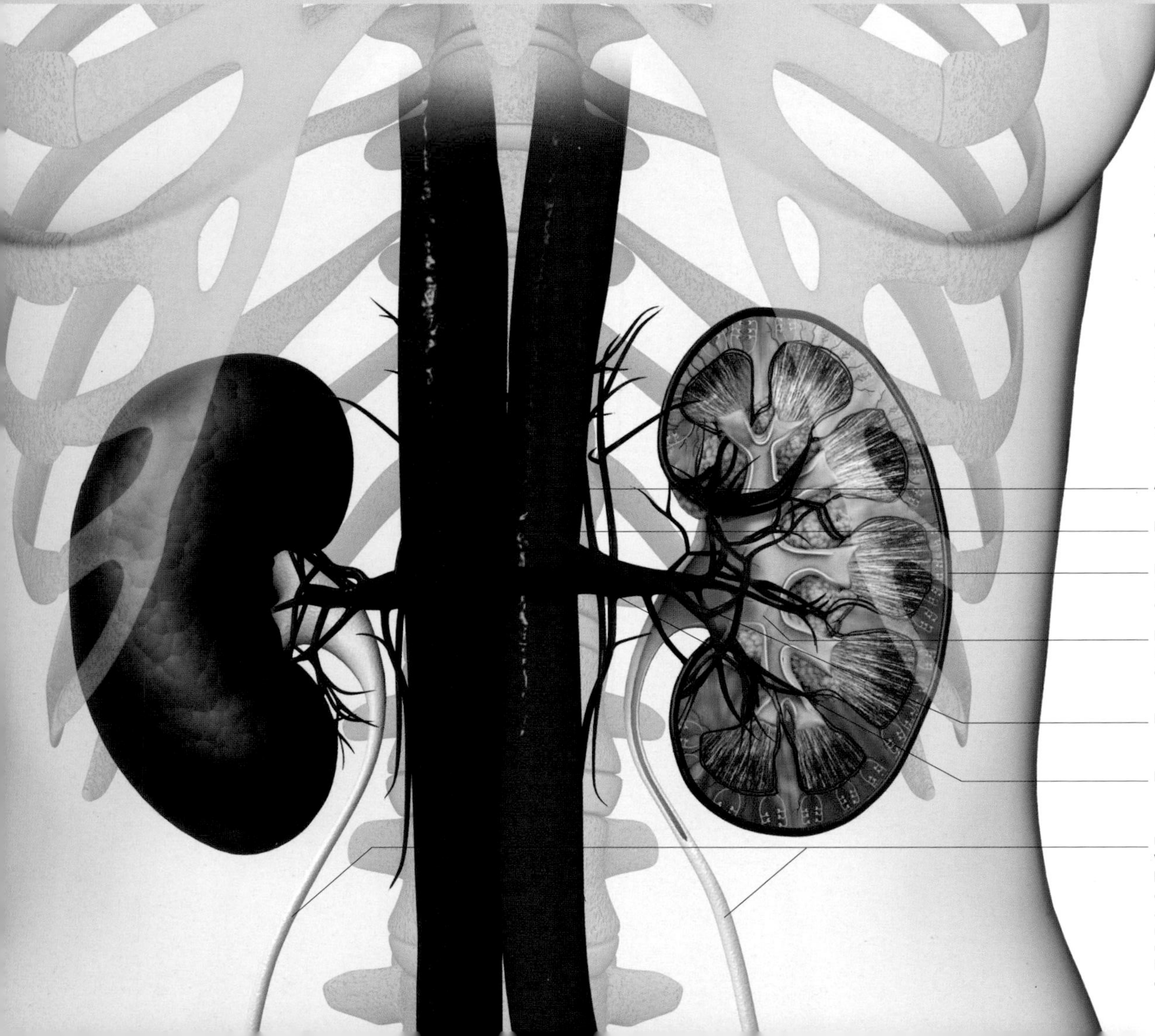

Aorta

Inferior vena cava

Kidney
Each is about 10–12.5 cm (4–5 in) long, and contains about 1 million filtering units

Renal pelvis
Funnel-shaped chamber in which urine collects before passing down the ureter

Renal artery

Renal vein

Ureters
Vessels conveying urine from kidneys to bladder; their walls have three layers: the outer layer, composed of connective and adipose (fat) tissue; the middle layer has muscular fibres, which contract to propel urine to the bladder; the inner, mucosal layer secretes mucus to prevent its cells coming into contact with urine

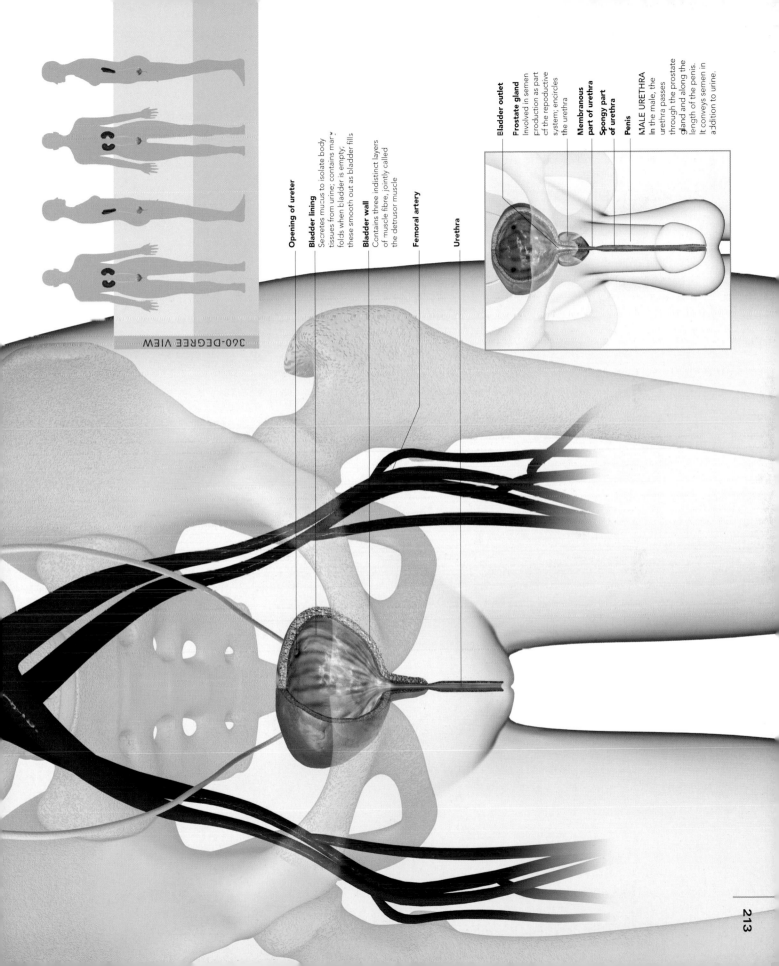

Opening of ureter

Bladder lining
Secretes mucus to isolate body tissues from urine; contains many folds when bladder is empty; these smooth out as bladder fills

Bladder wall
Contains three indistinct layers of muscle fibre, jointly called the detrusor muscle

Femoral artery

Urethra

Bladder outlet

Prostate gland
Involved in semen production as part of the reproductive system; encircles the urethra

Membranous part of urethra

Spongy part of urethra

Penis

MALE URETHRA
In the male, the urethra passes through the prostate gland and along the length of the penis. It conveys semen in addition to urine.

KIDNEY STRUCTURE

THE KIDNEYS ARE PAIRED ORGANS AT THE UPPER REAR OF THE ABDOMINAL CAVITY, ON EITHER SIDE OF THE SPINAL COLUMN. THEIR FUNCTIONS INCLUDE FILTERING WASTE PRODUCTS FROM THE BLOOD. THE WASTE IS EXCRETED—ALONG WITH EXCESS WATER—AS URINE.

INSIDE THE KIDNEY

Each kidney is protected by three outer layers: a tough external coat of fibrous connective tissue, the renal fascia; a layer of fatty tissue, the adipose capsule; and inside this, another fibrous layer, the renal capsule. The main body of the kidney also has three layers: the renal cortex, which is packed full of knots of capillaries known as glomeruli and their capsules; next is the renal medulla, which contains capillaries and urine-forming tubules; and a central space where the urine collects, known as the renal pelvis. The glomeruli, capsules, and tubules are parts of the kidney's million-plus microfiltering units, called nephrons.

GLOMERULUS
In this microscope image the glomerulus is colored pink. This tangled system of capillaries forms the first part of the nephron, and oozes a filtrate fluid, which is collected by the cuplike Bowman's (glomerular) capsule around it.

NEPHRON

Each microfiltering unit, or nephron, spans the cortex and medulla. The glomerulus, capsule, proximal and distal tubules, and the smaller urine collecting ducts are in the former. The latter contains mainly the long tubule loops of Henle and the larger urine collecting ducts.

Glomerulus
Ball-shaped capillary mass is the vascular beginning of a nephron

Renal tubule
A long, much-folded, and looped tube where urine is concentrated

Capillaries
These run from the glomerulus and reabsorb essential nutrients, minerals, salts, and water

Loop of Henle

Urine-collecting duct
Larger collecting vessel fed by renal tubules

Renal cortex

Renal medulla

Renal cortex
Outer region of kidney packed with microscopic spherical structures (the capsule-enveloped glomeruli) lending it a granular appearance

Renal medulla
Composed mainly of capillary networks around long loops of renal tubules

Renal column
Region of cortex tissue separating the renal pyramids

Renal pyramid
Cone-shaped regions of medulla between the renal columns

Major calyx
Several minor calyces (cup-shaped cavities forming the renal pelvis) merge to form major calyces

Minor calyx
Urine from main collecting ducts empties from the renal papilla into a minor calyx

Renal pelvis
Funnel-shaped tube narrowing into upper end of ureter, and into which the major calyces merge

Renal capsule
Thin covering of fibrous tissue around the whole kidney

Renal artery
Supplies blood to kidney; branches from the aorta (main artery of the body)

Renal vein
Removes cleaned blood, which then drains into inferior vena cava (main lower vein in the body)

Renal hilus
Junction at which the renal blood vessels and ureter pass into the kidney

Arcuate arteries and veins
Vessels forming archlike connections between the cortex and medulla

Ureter
Muscular-walled transport tube for urine, leading down to the urinary bladder

Renal papilla
Apex of the renal pyramid

KIDNEY CROSS SECTION
This cutaway shows the kidney's main layers, the cortex and the medulla, which form segments known as renal pyramids. The renal artery and vein circulate huge amounts of blood—about 2½ pints per min (1.2 liters/min) at rest, which is up to one-quarter of the heart's total output.

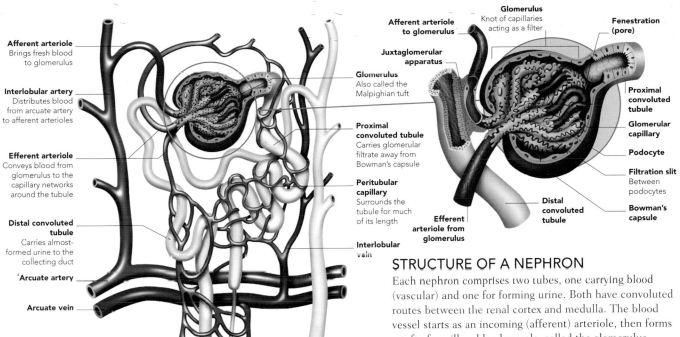

Afferent arteriole
Brings fresh blood to glomerulus

Interlobular artery
Distributes blood from arcuate artery to afferent arterioles

Efferent arteriole
Conveys blood from glomerulus to the capillary networks around the tubule

Distal convoluted tubule
Carries almost-formed urine to the collecting duct

Arcuate artery

Arcuate vein

Capillary network around loop of Henle

Ascending limb

Afferent arteriole to glomerulus

Juxtaglomerular apparatus

Glomerulus
Knot of capillaries acting as a filter

Fenestration (pore)

Glomerulus
Also called the Malpighian tuft

Proximal convoluted tubule
Carries glomerular filtrate away from Bowman's capsule

Peritubular capillary
Surrounds the tubule for much of its length

Interlobular vein

Proximal convoluted tubule

Glomerular capillary

Podocyte

Filtration slit
Between podocytes

Bowman's capsule

Distal convoluted tubule

Efferent arteriole from glomerulus

Urine-collecting duct

Descending limb

Loop of Henle
Positioned in the renal medulla

BLOOD FILTRATION
One end of the renal tubule is a cup-shaped membrane, Bowman's capsule, which envelops the glomerulus. The other end joins a straight urine-collecting tubule. The capsule is about 1/125in (0.2 mm) in diameter. All of a kidney's tubules end to end would stretch 50 miles (80 km). Blood is circulated by the arcuate vessels that run between the renal cortex and medulla.

STRUCTURE OF A NEPHRON

Each nephron comprises two tubes, one carrying blood (vascular) and one for forming urine. Both have convoluted routes between the renal cortex and medulla. The blood vessel starts as an incoming (afferent) arteriole, then forms a tuft of capillary blood vessels, called the glomerulus. This leads to the outgoing (efferent) arteriole, then the peritubular capillaries, and finally a venule that carries the blood away. The renal tube begins with Bowman's capsule. Next, the proximal convoluted tubule dips into the renal medulla and back up to the cortex in a long U-shape—the loop of Henle. Back in the cortex, it winds again as the distal convoluted tubule and joins one of the larger urine-collecting ducts. Between the distal convoluted tubule and afferent arteriole is the juxtaglomerular apparatus (JGA). The JGA helps regulate blood flow into the glomerulus and also produces the hormone renin, which plays a role in regulating blood volume and urine composition.

REGULATION OF URINE PRODUCTION

The amount, composition, and concentration of urine is determined principally by two hormones: ADH (antidiuretic hormone, or vasopressin) and aldosterone. ADH, released by the pituitary gland, acts on the kidneys to reduce urine volume and increase its concentration. Aldosterone, released by the adrenal glands, acts on the kidneys to reduce sodium and water in the urine and increase potassium.

HORMONAL CONTROL OF URINE PRODUCTION

Antidiuretic hormone (ADH) and aldosterone are the two main hormones that affect urine production. The levels of these hormones are altered so that the amounts of water, solutes such as sodium, and wastes in urine are increased or decreased as needed to maintain a constant internal environment.

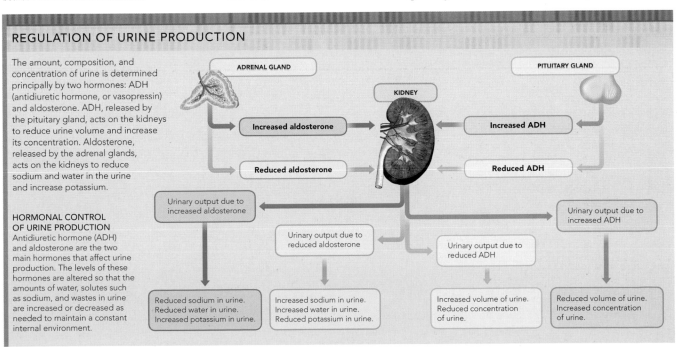

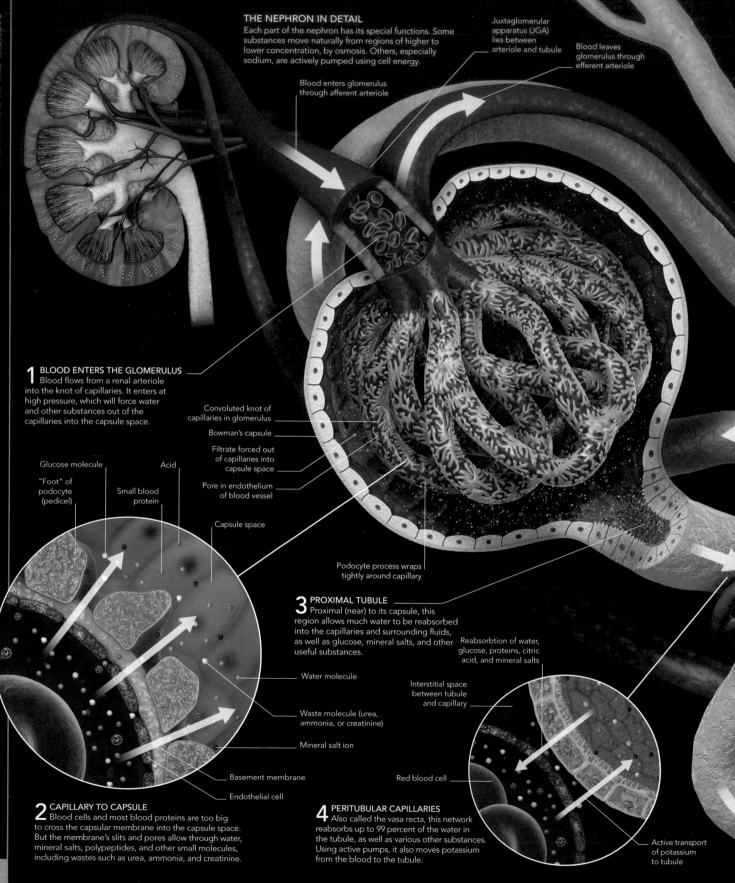

THE NEPHRON IN DETAIL

Each part of the nephron has its special functions. Some substances move naturally from regions of higher to lower concentration, by osmosis. Others, especially sodium, are actively pumped using cell energy.

Blood enters glomerulus through afferent arteriole

Juxtaglomerular apparatus (JGA) lies between arteriole and tubule

Blood leaves glomerulus through efferent arteriole

1 BLOOD ENTERS THE GLOMERULUS

Blood flows from a renal arteriole into the knot of capillaries. It enters at high pressure, which will force water and other substances out of the capillaries into the capsule space.

Convoluted knot of capillaries in glomerulus

Bowman's capsule

Filtrate forced out of capillaries into capsule space

Pore in endothelium of blood vessel

Capsule space

Glucose molecule

Acid

"Foot" of podocyte (pedicel)

Small blood protein

Podocyte process wraps tightly around capillary

3 PROXIMAL TUBULE

Proximal (near) to its capsule, this region allows much water to be reabsorbed into the capillaries and surrounding fluids, as well as glucose, mineral salts, and other useful substances.

Reabsorbtion of water, glucose, proteins, citric acid, and mineral salts

Water molecule

Waste molecule (urea, ammonia, or creatinine)

Mineral salt ion

Basement membrane

Endothelial cell

Interstitial space between tubule and capillary

Red blood cell

Active transport of potassium to tubule

2 CAPILLARY TO CAPSULE

Blood cells and most blood proteins are too big to cross the capsular membrane into the capsule space. But the membrane's slits and pores allow through water, mineral salts, polypeptides, and other small molecules, including wastes such as urea, ammonia, and creatinine.

4 PERITUBULAR CAPILLARIES

Also called the vasa recta, this network reabsorbs up to 99 percent of the water in the tubule, as well as various other substances. Using active pumps, it also moves potassium from the blood to the tubule.

FILTRATION IN THE KIDNEYS

INSIDE A KIDNEY, EACH MICROSCOPIC NEPHRON IS A COMPLEX NETWORK OF COILED CAPILLARIES AND WINDING TUBES. DOZENS OF SUBSTANCES MOVE TO AND FRO TO ELIMINATE WASTES AND FINE-TUNE URINE COMPOSITION.

The arrangement of each kidney's million-plus nephrons, like their filtering process, is highly organized. The arteriole taking blood into the glomerulus then leaves again to form a capillary network around the tubule. The tubule, capillaries, and surrounding fluid exchange many substances in several stages, according to the composition of the tubule's filtrate and the balance of water, minerals, and other components in the blood and fluid. The tubule revisits its original capsule at the JGA (see p.215), as part of the hormone control system, before joining a urine-collecting duct.

INSIDE THE BOWMAN'S CAPSULE

Bowman's capsule—the expanded, bowl-like end of a renal tubule—cups the capillary knot of the glomerulus. Substances passing from blood in the capillaries into the capsule space must pass through layers known together as the capsular membrane. First are the endothelial cells lining the inside of the capillary. Next is the basement membrane forming the outside of the capillary. Footlike processes join onto this from octopus-like cells known as podocytes coating the capillaries. Between these feet, or pedicels, are filtration slits that allow through water, glucose, urea, and other small molecules.

PODOCYTES
Each podocyte's branches end in fine pedicels. These form contacts with pedicels from other podocytes and also the capillary basement membrane.

7 COLLECTING DUCT
Fine adjustment of urine composition continues in the collecting-duct system. About 5 percent of all the water and sodium being reabsorbed into the blood is recovered here.

Transport from blood to tubule

Urine-collecting duct

Distal tubule

Urine accumulates in collecting duct

6 DISTAL TUBULE
Distal to (far from) its capsule, this region may see water go in or out of the tubule, depending on the concentration of water already in the tubule, while hydrogen and potassium ions move to regulate both blood and urine pH. Acids, amines, and ammonia compounds may also be transported into the tubule.

Tubule lined by single layer of epithelial cells

Transport of substances out of tubule

Reabsorption, mostly of water

Some transport from blood to tubule may also occur

Capillary network (vasa recta) around tubules and loop of Henle

5 DESCENDING LOOP
As the loop of Henle dips into the renal medulla, more water moves from the tubule into the blood, as well as small amounts of mineral salts and some urea and creatinine. Some acids and amines may move into the tubule, while ammonia can go in either direction.

Tubule narrows in loop of Henle

8 VENOUS FLOW
Blood flowing away from the nephrons carries 99 percent of its original water, 98 percent of its sodium, calcium, and chloride, and about 40 percent of its urea.

Proximal tubule

Blood leaves nephron to join renal vein

URINARY DISORDERS

SOME PARTS OF THE URINARY TRACT ARE SUSCEPTIBLE TO INFECTIONS, RESULTING IN CONDITIONS
SUCH AS CYSTITIS. SOME CHRONIC KIDNEY DISEASES ARE ALSO CAUSED BY INFECTION. KIDNEY FAILURE
CAN NOW BE TREATED WITH RENAL REPLACEMENT THERAPY, EITHER BY DIALYSIS OR TRANSPLANTATION.
HOWEVER, COMMON SYMPTOMS, SUCH AS INCONTINENCE, MAY STILL BE TROUBLESOME.

URINARY TRACT INFECTIONS

ALL ORGANS IN THE URINARY TRACT CAN BE AFFECTED
BY INFECTION; ALTHOUGH USUALLY CONFINED TO ONE
AREA, IT CAN SPREAD THROUGH THE SYSTEM.

The urine flowing through the urinary tract moves
in one direction, from the kidneys through the ureters
to the bladder and then through the urethra to leave
the body. During urination, the flow from the bladder
is rapid and copious, but for long periods urine is
stagnant in the bladder. Infections can enter the body
through the urethra and spread to the bladder, and
sometimes up the ureters to the kidneys. The adult
female urethra is 1½ in (4 cm) long, compared to the
male's at 8 in (20 cm). This short length and the
proximity of its outlet to the anus (which allows
bacteria from the anal area to enter the urethra)
together account for females' greater susceptibility
to urinary infection. One of the most common
urinary infections is inflammation of the bladder,
known as cystitis. Its main symptoms are burning
pain and a frequent need to urinate but often with
little urine on each occasion.

CYSTITIS
In this color-enhanced
micrograph of a bladder
lining affected by
cystitis, bacteria
(yellow rods) colonize
the lining's inner
surface (blue),
causing inflammation.
The lining secretes
strands of mucus
(orange) and may also
be damaged so that it
leaks blood (red cells),
which stains the urine pink.

SITES OF DISORDERS
Although each of the urinary organs
is affected by its own characteristic
diseases, a disorder of any single
organ can also affect other parts
of the system. For example, kidney
stones may damage the ureters,
and obstruction to the outflow
of urine may damage the kidneys
as a result of back pressure.

Pyelonephritis
An acute infection
of the urine-
collecting system
of the kidney

**Diabetic
nephropathy**
Changes to capillaries
in the kidneys, which
may lead to kidney
failure; caused
by long-term
diabetes mellitus

Glomerulonephritis
Inflammation of the
filtering units of the
kidney (glomeruli);
often related to an
autoimmune process

Reflux
The forcing of urine
up the ureters by
back pressure; can be
caused by blockage
of the urethra; also
occurs in children
when the ureters
are too lax

INCONTINENCE

A TENDENCY TO LEAK URINE, URINARY INCONTINENCE
MOST COMMONLY OCCURS IN WOMEN, ELDERLY PEOPLE,
AND THOSE WITH BRAIN OR SPINAL CORD DAMAGE.

Females may be susceptible to incontinence because
of a weakness in the pelvic floor muscles after childbirth.
There are different types. In stress incontinence, weak
pelvic floor muscles allow small quantities of urine to
escape during exertion, such as running, or activities
that raise intra-abdominal pressure, such as coughing.
In urge incontinence, the urgent desire to urinate is
triggered by irritable bladder muscle that causes the
bladder to contract and expel all its urine. In overflow
incontinence, a blockage in the urethra or a weak
bladder muscle results in buildup of urine that then
leaks out. Total incontinence is the complete loss of
bladder function due to a nervous system disorder,
such as multiple sclerosis.

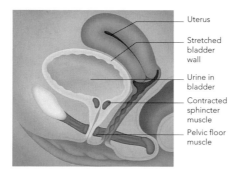

Uterus

Stretched
bladder
wall

Urine in
bladder

Contracted
sphincter
muscle

Pelvic floor
muscle

NORMAL BLADDER
A healthy bladder expands like a balloon
as it fills with urine. The sphincter muscles
and surrounding pelvic floor muscles keep
the exit closed. Nerve signals from stretch
sensors in the bladder wall travel to the
brain, signaling the need for emptying.

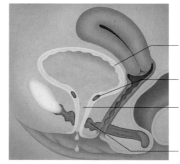

Contracted
bladder
wall

Relaxed
sphincter
muscle

Urethra

Weakened
pelvic floor
muscle

STRESS INCONTINENCE
To empty the bladder, the sphincter and
pelvic floor muscles relax, and the detrusor
muscle in the bladder wall contracts, forcing
urine along the urethra. In incontinence,
weak muscles may allow this to happen
without proper control, so urine leaks out.

KIDNEY STONES

CONCENTRATED SUBSTANCES IN THE URINE MAY FORM CRYSTALLINE DEPOSITS, KNOWN AS KIDNEY STONES OR RENAL CALCULI, WITHIN THE KIDNEY.

Kidney stones are solid, mineral-rich objects that grow due to the process of coming out of solution (precipitation) of chemicals, such as calcium salts, in urine. Kidney stones can take years to form and grow in various shapes and sizes. A stone may stay in the kidney and cause few problems, but it can increase the risk of urinary tract infection.

DETECTING A KIDNEY STONE
After a dye is injected, an X-ray (pyelogram) can reveal kidney stones. Here, the dense material of a stone is clearly visible in the right kidney (orange, left of image).

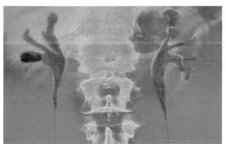

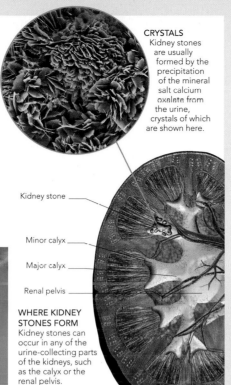

CRYSTALS
Kidney stones are usually formed by the precipitation of the mineral salt calcium oxalate from the urine, crystals of which are shown here.

Kidney stone

Minor calyx

Major calyx

Renal pelvis

WHERE KIDNEY STONES FORM
Kidney stones can occur in any of the urine-collecting parts of the kidneys, such as the calyx or the renal pelvis.

BLADDER TUMORS

MOST TUMORS OF THE BLADDER BEGIN AS SUPERFICIAL WARTLIKE GROWTHS, CALLED PAPILLOMAS; UNTREATED, THEY CAN BECOME CANCEROUS AND SPREAD.

Bladder tumors are more common in people who smoke, and in men. If they enlarge, they can cause difficulty urinating, hematuria (blood in the urine), and increased risk of urinary tract infection. If the tumors become cancerous, they may spread to adjacent organs, such as the rectum, and through the bloodstream to more distant parts of the body.

BLADDER TUMOR
A large bladder tumor (white area, below) may block the outlet from the bladder to the urethra, causing complete retention of urine. This requires urgent medical treatment.

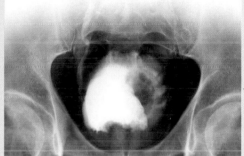

KIDNEY FAILURE

KIDNEY (OR RENAL) FAILURE OCCURS WHEN THE KIDNEY CAN NO LONGER CARRY OUT ITS VITAL FUNCTION OF REMOVING WASTE PRODUCTS FROM THE BLOOD.

There are different types of kidney failure, affecting one kidney or both. Symptoms are caused by buildup of waste products. Acute kidney failure comes on rapidly and may be due to problems such as blood loss, a heart attack, toxins, or a kidney infection. Symptoms include reduced urine output, drowsiness, headache, nausea, and vomiting. Chronic kidney failure develops slowly. It may be due to polycystic kidney disease or long-standing high blood pressure. Symptoms include frequent urination, breathlessness, skin irritation, nausea, vomiting, and muscle twitches and cramps. In end-stage failure, the kidneys have lost all function, and dialysis or kidney transplant is necessary.

POLYCYSTIC KIDNEY
Usually inherited, polycystic kidney disease produces many cysts, or fluid-filled sacs, in the kidney. The kidney enlarges, becomes shaped irregularly, and loses its blood-filtering functions.

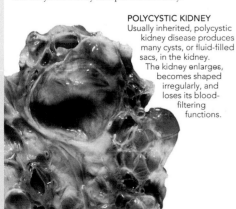

DIALYSIS

Dialysis involves filtering the blood of a person with kidney failure. There are several variations of the process. Hemodialysis and hemofiltration take blood outside the body for treatment. Peritoneal dialysis takes place inside the body (see right). In hemodialysis, blood from the body is pumped by a kidney machine through a filter. The filter contains a semipermeable membrane immersed in a solution known as dialysate. Smaller molecules, such as urea and similar waste products, pass through the membrane into the dialysate while larger, useful molecules such as proteins are retained. The filtered blood returns to the body and the dialysate is discarded. The procedure takes 3 or 4 hours to complete.

PERITONEAL DIALYSIS
The peritoneal membrane in the abdomen acts as a filter. Dialysate is passed into the peritoneal cavity and 4–6 hours later is drained out. Waste products pass from the capillaries of the peritoneal cavity through the peritoneal membrane into the dialysate.

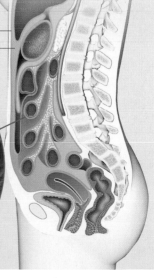

Peritoneal membrane

Dialysate

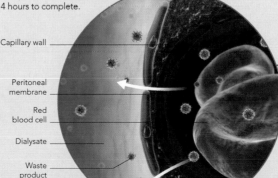

Capillary wall

Peritoneal membrane

Red blood cell

Dialysate

Waste product

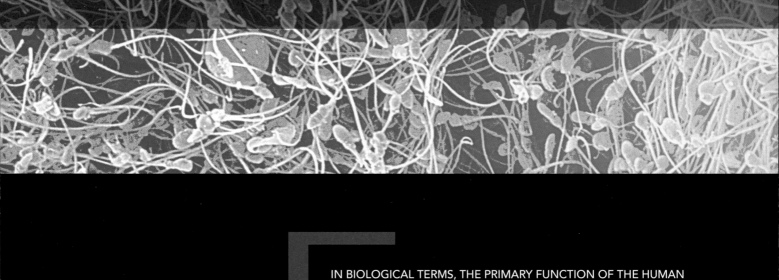

IN BIOLOGICAL TERMS, THE PRIMARY FUNCTION OF THE HUMAN
BODY IS TO REPLICATE ITSELF, AND THE SEXUAL AND PARENTING
INSTINCTS ARE AMONG THE STRONGEST OF OUR BASIC DRIVES.
AS SCIENCE WIDENS THE GAP BETWEEN SEX AND REPRODUCTION,
WE CAN CHOOSE MORE WAYS OF HAVING ONE WITHOUT THE
OTHER. THESE DEVELOPMENTS, ALONG WITH THE CONTRASTS
BETWEEN PROGRESSIVE SOCIETIES AND TRADITIONAL CULTURES,

REPRODUCTION AND BIRTH

MALE REPRODUCTIVE SYSTEM

OF THE BODY'S MAJOR SYSTEMS, THE REPRODUCTIVE SYSTEM IS THE ONE THAT DIFFERS MOST BETWEEN THE SEXES, AND THE ONLY ONE THAT DOES NOT FUNCTION UNTIL PUBERTY. THE MALE SYSTEM PRODUCES SEX CELLS (GAMETES) CALLED SPERM. UNLIKE FEMALE EGG MATURATION, WHICH OCCURS IN CYCLES AND CEASES AT MENOPAUSE, SPERM PRODUCTION IS CONTINUOUS, REDUCING GRADUALLY WITH AGE.

THE REPRODUCTIVE ORGANS

In males, the reproductive organs include the penis, the testes, a number of storage and transport ducts, and some supporting structures. The two oval-shaped testes lie outside the body in a pouch of skin called the scrotum, where they can maintain the optimum temperature for sperm production—approximately 5°F (3°C) lower than body temperature. Testes are glands that are responsible for the manufacture of sperm and the sex hormone testosterone. From each testis, sperm pass into a coiled tube—the epididymis—for the final stages of maturation. Sperm are stored in the epididymides until they are either broken down and reabsorbed, or ejaculated—forced by movement of seminal fluid from the accessory glands (see opposite) down a duct called the vas deferens.

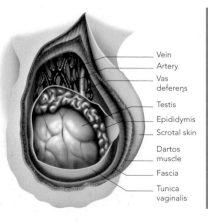

- Blood vessel
- Scrotum
- Vas deferens
- Epididymis
- Seminiferous tubules within testis

INSIDE THE SCROTUM
The scrotum contains two testes where sperm are manufactured within tubes called seminiferous tubules, and the two epididymides where sperm are stored. Each epididymis is a tube about 20 ft (6 m) long, tightly coiled and bunched into a length of just 2 in (4 cm).

SCROTAL LAYERS

Each testis is covered by a thin tissue layer, the tunica vaginalis, around which is a layer of connective tissue called fascia. A muscle layer called the dartos muscle relaxes in hot weather to drop the testes away from the body to keep them cool. In cold weather the muscle contracts to draw up the testes so that they do not become too chilled. The spermatic cord suspends each testis within the scrotum, and contains the testicular artery and vein, lymph vessels, nerves, and the sperm-carrying vas deferens.

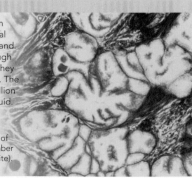

- Vein
- Artery
- Vas deferens
- Testis
- Epididymis
- Scrotal skin
- Dartos muscle
- Fascia
- Tunica vaginalis

PATHWAY FOR SPERM

During ejaculation, waves of muscle contraction squeeze the sperm in their fluid from the epididymis along the loop of the vas deferens. The vas deferens is joined by a duct from the seminal vesicle, one of the male accessory glands, to form the ejaculatory duct. The left and right ejaculatory ducts join the urethra within the prostate, another accessory gland. In the male, the urethra is a dual-purpose tube that carries urine from the bladder during urination and sperm from the testes. During ejaculation, however, the sphincter at the base of the bladder is closed because of high pressure in the urethra.

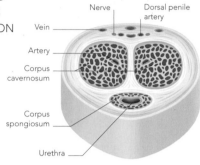

- Nerve
- Dorsal penile artery
- Vein
- Artery
- Corpus cavernosum
- Corpus spongiosum
- Urethra

PENILE ERECTION
During arousal, large quantities of arterial blood enter the corpus spongiosum and corpus cavernosum, compressing the veins. As a result, blood cannot drain from the penis and it becomes hard and erect.

MAKING SPERM

Each testis is a mass of more than 800 tightly looped and folded vessels known as seminiferous tubules. Inside each tubule, sperm begin as bloblike cells called spermatogonia lining the inner wall. These pass through a larger stage, as primary spermatocytes, then become smaller as secondary spermatocytes, and begin to develop tails as spermatids. As all of this happens, they move steadily toward the middle of the tubule. The spermatids finally develop into ripe sperm with long tails. Thousands of sperm are produced every second, each taking about two months to mature.

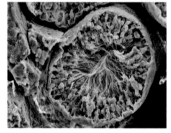

SEMINIFEROUS TUBULE
Sperm take shape as they move toward the center of the tubule. Their long tails can be seen in this cross section.

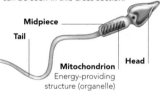

- Midpiece
- Tail
- Head
- Mitochondrion — Energy-providing structure (organelle)

SPERM CELL
A sperm is about 1/500 in (0.05 mm) long but most of this is the tail. The sperm head is only 1/5000 in (0.006 mm), about the same size as a red blood cell.

SEMEN

Seminal fluid, or semen, is sperm mixed with fluid added by several glands including the prostate gland. The prostate secretes fluid through tiny ducts to mix with sperm as they are ejaculated down the urethra. The final mix has around 300–500 million sperm in 1/15–1/6 fl oz (2–5 ml) of fluid.

PROSTATE GLAND
This microscopic view of a section of prostate gland tissue shows a number of secretory ducts (orange and white).

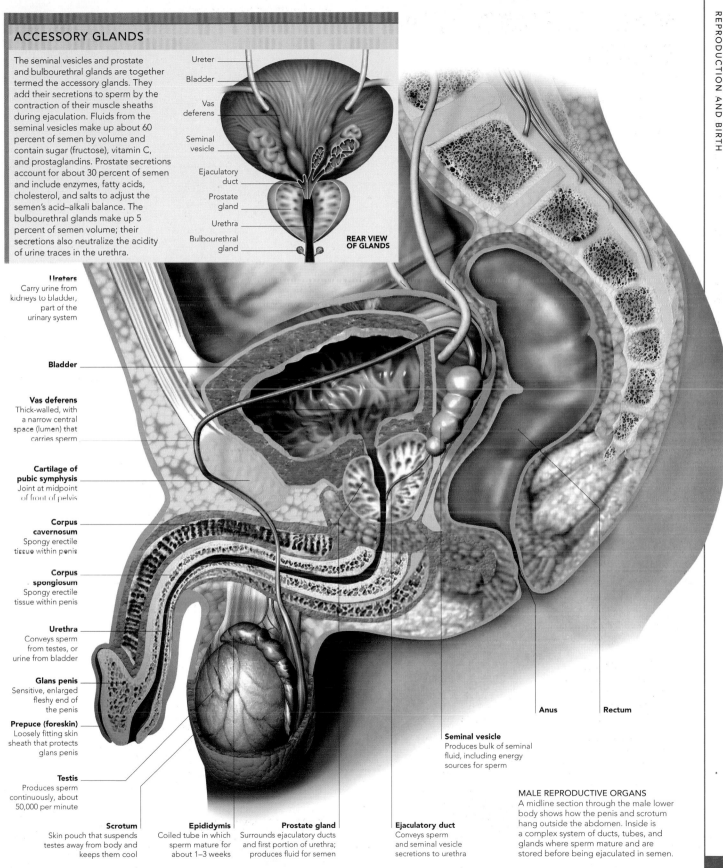

ACCESSORY GLANDS

The seminal vesicles and prostate and bulbourethral glands are together termed the accessory glands. They add their secretions to sperm by the contraction of their muscle sheaths during ejaculation. Fluids from the seminal vesicles make up about 60 percent of semen by volume and contain sugar (fructose), vitamin C, and prostaglandins. Prostate secretions account for about 30 percent of semen and include enzymes, fatty acids, cholesterol, and salts to adjust the semen's acid–alkali balance. The bulbourethral glands make up 5 percent of semen volume; their secretions also neutralize the acidity of urine traces in the urethra.

Ureter

Bladder

Vas deferens

Seminal vesicle

Ejaculatory duct

Prostate gland

Urethra

Bulbourethral gland

REAR VIEW OF GLANDS

Ureters
Carry urine from kidneys to bladder, part of the urinary system

Bladder

Vas deferens
Thick-walled, with a narrow central space (lumen) that carries sperm

Cartilage of pubic symphysis
Joint at midpoint of front of pelvis

Corpus cavernosum
Spongy erectile tissue within penis

Corpus spongiosum
Spongy erectile tissue within penis

Urethra
Conveys sperm from testes, or urine from bladder

Glans penis
Sensitive, enlarged fleshy end of the penis

Prepuce (foreskin)
Loosely fitting skin sheath that protects glans penis

Testis
Produces sperm continuously, about 50,000 per minute

Scrotum
Skin pouch that suspends testes away from body and keeps them cool

Epididymis
Coiled tube in which sperm mature for about 1–3 weeks

Prostate gland
Surrounds ejaculatory ducts and first portion of urethra; produces fluid for semen

Ejaculatory duct
Conveys sperm and seminal vesicle secretions to urethra

Anus

Rectum

Seminal vesicle
Produces bulk of seminal fluid, including energy sources for sperm

MALE REPRODUCTIVE ORGANS

A midline section through the male lower body shows how the penis and scrotum hang outside the abdomen. Inside is a complex system of ducts, tubes, and glands where sperm mature and are stored before being ejaculated in semen.

FEMALE REPRODUCTIVE SYSTEM

UNLIKE THE MALE, THE FEMALE REPRODUCTIVE ORGANS ARE SITED ENTIRELY INSIDE THE LOWER ABDOMEN. THEIR FUNCTION IS TO RIPEN AND RELEASE AN EGG AT REGULAR INTERVALS, AND, IF THE EGG IS FERTILIZED, TO PROTECT AND NOURISH THE EMBRYO AND FETUS. NO EGGS ARE MANUFACTURED AFTER BIRTH—A FEMALE IS BORN WITH A FULL SET.

EGG RELEASE
A color-enhanced electron micrograph showing an egg (red) being released from its follicle into the abdominal cavity. Tendrils (fimbriae) at the end of each fallopian tube guide the egg into the tube.

REPRODUCTIVE TRACT

The female reproductive glands (ovaries) are located within the abdomen. From puberty, they mature and release the female sex cells (gametes), known as egg cells or ova. This release occurs roughly once a month as part of the menstrual cycle (see p.218). The ripe egg travels along the fallopian tube to the uterus, the muscular sac in which, if fertilized, it develops into an embryo and then a fetus. Unfertilized eggs, and the uterine lining, leave via the vagina. The ovaries also make the female sex hormone estrogen.

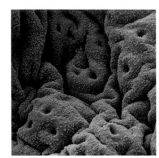

ENDOMETRIUM
An electron micrograph of the thick, folded, glandular endometrium (uterus lining), blood-rich and ready to receive a fertilized egg.

BREASTS

Both females and males have breasts (mammae), which contain modified sweat glands known as mammary glands. In females these are much larger and more developed than in males and produce milk at childbirth. Each breast contains 15–20 lobes of compound areolar glands, each lobe resembling a bunch of grapes on a long stalk. The cells of the glands secrete milk, which flows along merging lactiferous ducts toward the nipple. The breast also contains a widespread drainage system of lymph vessels (see p.174).

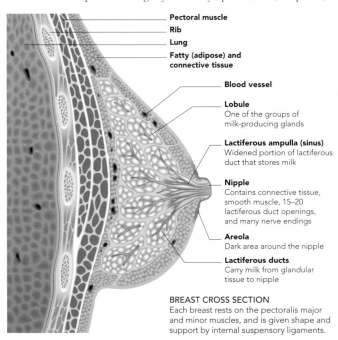

Pectoral muscle
Rib
Lung
Fatty (adipose) and connective tissue

Blood vessel

Lobule
One of the groups of milk-producing glands

Lactiferous ampulla (sinus)
Widened portion of lactiferous duct that stores milk

Nipple
Contains connective tissue, smooth muscle, 15–20 lactiferous duct openings, and many nerve endings

Areola
Dark area around the nipple

Lactiferous ducts
Carry milk from glandular tissue to nipple

BREAST CROSS SECTION
Each breast rests on the pectoralis major and minor muscles, and is given shape and support by internal suspensory ligaments.

OVULATION

An ovary contains thousands of immature egg cells. During each menstrual cycle, follicle-stimulating hormone (FSH) causes one egg to begin development; this takes place inside a primary follicle. The follicle enlarges as its cells proliferate, and begins to fill with fluid, becoming a secondary follicle that moves to the ovary's surface. It also increases its production of the hormone estrogen. A surge of luteinizing hormone (LH) causes the follicle to rupture and release the ripe egg—this is ovulation. The lining of the empty follicle thickens into a corpus luteum—a temporary source of hormones.

Primary follicle
Early stage of development, containing primary oocyte (unripe egg cell)

Secondary follicle
Mature stage of development, containing secondary oocyte (ripened egg)

Ovarian ligament
Stabilizes position of ovary within abdomen

INSIDE AN OVARY
The ovary contains undeveloped eggs, eggs in follicles at various stages of maturation, and empty follicles forming corpora lutea. The bulk of the glandular tissue surrounding these follicles is known as the stroma.

Egg

Corpus luteum
An empty follicle, filled with hormone-producing cells

VULVA

The external genital parts of the female are together known as the vulva. They are sited under the mons pubis, a mound of fatty tissue that covers the junction of the two pubic bones, the pubic symphysis. Outermost in the vulva are the flaplike labia majora, with the smaller, foldlike labia minora within them. Both are called "labia" due to their resemblance to lips. The labia majora contain fatty and connective tissue, sebaceous glands, smooth muscle, and sensory nerve endings. At puberty their exposed surfaces begin to grow hairs. Within the vulva are the openings to the vagina and the urethra. At the front end of the labia minora is the clitoris. Like the male penis, it is sensitive and engorges with blood during sexual arousal.

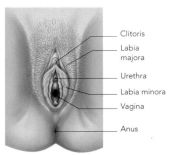

Clitoris
Labia majora
Urethra
Labia minora
Vagina
Anus

EXTERNAL GENITALS
The external genitals have a protective role, preventing infection from reaching the urethra or vagina but allowing urine to exit.

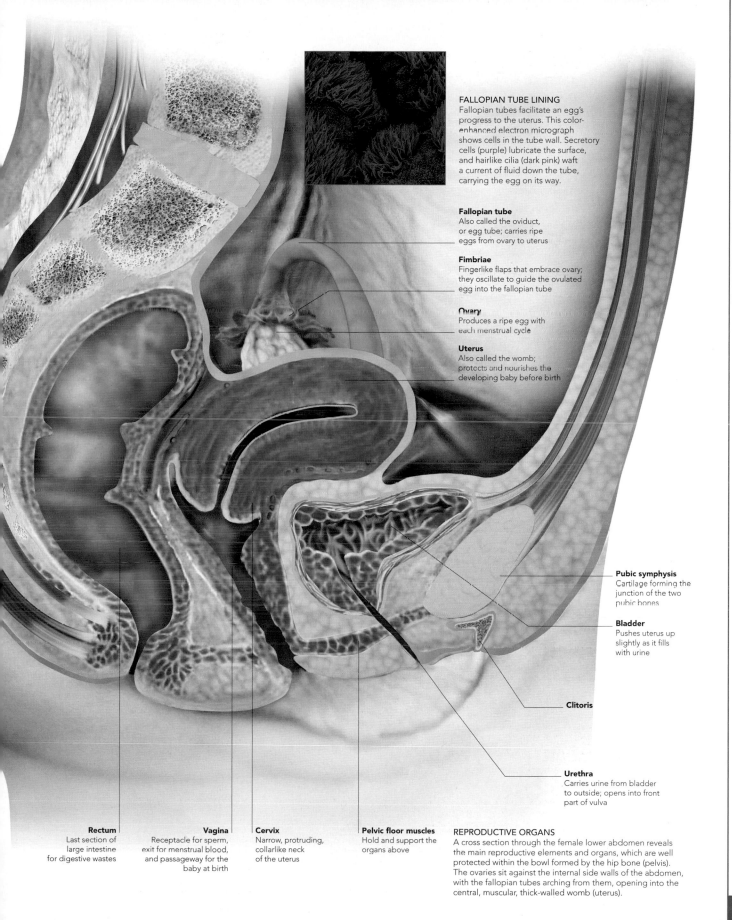

FALLOPIAN TUBE LINING
Fallopian tubes facilitate an egg's progress to the uterus. This color-enhanced electron micrograph shows cells in the tube wall. Secretory cells (purple) lubricate the surface, and hairlike cilia (dark pink) waft a current of fluid down the tube, carrying the egg on its way.

Fallopian tube
Also called the oviduct, or egg tube; carries ripe eggs from ovary to uterus

Fimbriae
Fingerlike flaps that embrace ovary; they oscillate to guide the ovulated egg into the fallopian tube

Ovary
Produces a ripe egg with each menstrual cycle

Uterus
Also called the womb; protects and nourishes the developing baby before birth

Pubic symphysis
Cartilage forming the junction of the two pubic bones

Bladder
Pushes uterus up slightly as it fills with urine

Clitoris

Urethra
Carries urine from bladder to outside; opens into front part of vulva

Rectum
Last section of large intestine for digestive wastes

Vagina
Receptacle for sperm, exit for menstrual blood, and passageway for the baby at birth

Cervix
Narrow, protruding, collarlike neck of the uterus

Pelvic floor muscles
Hold and support the organs above

REPRODUCTIVE ORGANS
A cross section through the female lower abdomen reveals the main reproductive elements and organs, which are well protected within the bowl formed by the hip bone (pelvis). The ovaries sit against the internal side walls of the abdomen, with the fallopian tubes arching from them, opening into the central, muscular, thick-walled womb (uterus).

CONCEPTION TO EMBRYO

AFTER THE UNION OF EGG AND SPERM AT FERTILIZATION, THE EMBRYONIC CELLS REPEATEDLY DIVIDE AND IMPLANT INTO THE UTERUS LINING, WHERE THE EMBRYO DEVELOPS ITS OWN SUPPORT SYSTEM, THE PLACENTA.

The first eight weeks of development within the uterus are known as the embryo stage, in which the fertilized egg becomes a tiny human body, no larger than a thumb. The fertilized egg develops into an enlarging cluster of cells, the blastocyst. Some of these cells will form the baby's body; others become the protective membranes or the placenta, which nourishes the embryo and removes waste products.

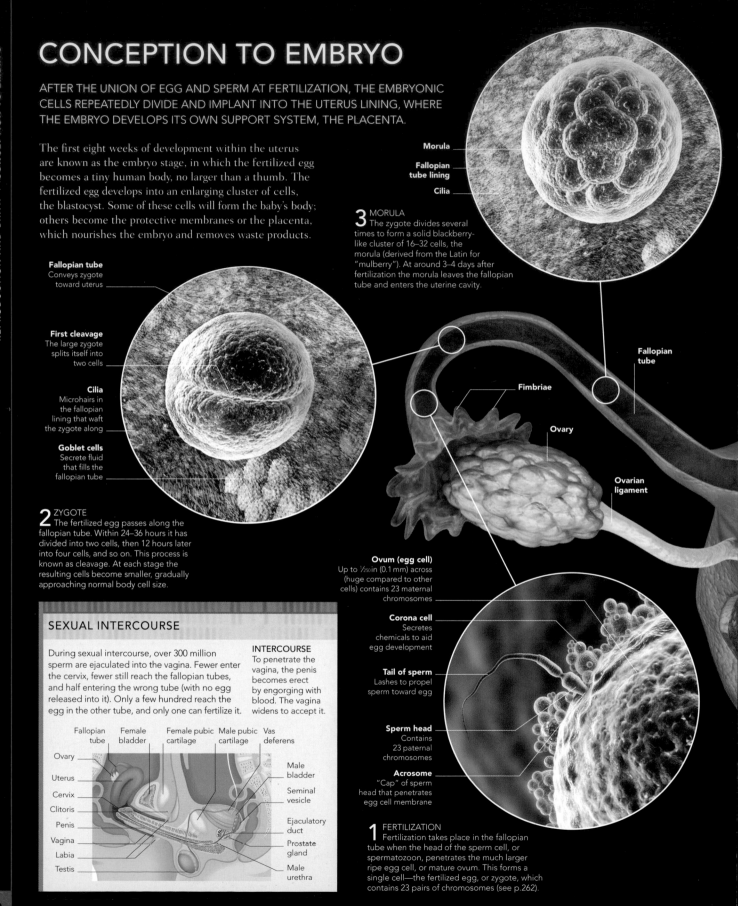

Morula

Fallopian tube lining

Cilia

3 MORULA
The zygote divides several times to form a solid blackberry-like cluster of 16–32 cells, the morula (derived from the Latin for "mulberry"). At around 3–4 days after fertilization the morula leaves the fallopian tube and enters the uterine cavity.

Fallopian tube
Conveys zygote toward uterus

First cleavage
The large zygote splits itself into two cells

Cilia
Microhairs in the fallopian lining that waft the zygote along

Goblet cells
Secrete fluid that fills the fallopian tube

2 ZYGOTE
The fertilized egg passes along the fallopian tube. Within 24–36 hours it has divided into two cells, then 12 hours later into four cells, and so on. This process is known as cleavage. At each stage the resulting cells become smaller, gradually approaching normal body cell size.

Fallopian tube

Fimbriae

Ovary

Ovarian ligament

Ovum (egg cell)
Up to ½₅₀ in (0.1 mm) across (huge compared to other cells) contains 23 maternal chromosomes

Corona cell
Secretes chemicals to aid egg development

Tail of sperm
Lashes to propel sperm toward egg

Sperm head
Contains 23 paternal chromosomes

Acrosome
"Cap" of sperm head that penetrates egg cell membrane

1 FERTILIZATION
Fertilization takes place in the fallopian tube when the head of the sperm cell, or spermatozoon, penetrates the much larger ripe egg cell, or mature ovum. This forms a single cell—the fertilized egg, or zygote, which contains 23 pairs of chromosomes (see p.262).

SEXUAL INTERCOURSE

During sexual intercourse, over 300 million sperm are ejaculated into the vagina. Fewer enter the cervix, fewer still reach the fallopian tubes, and half entering the wrong tube (with no egg released into it). Only a few hundred reach the egg in the other tube, and only one can fertilize it.

INTERCOURSE
To penetrate the vagina, the penis becomes erect by engorging with blood. The vagina widens to accept it.

Fallopian tube
Female bladder
Female pubic cartilage
Male pubic cartilage
Vas deferens

Ovary
Uterus
Cervix
Clitoris
Penis
Vagina
Labia
Testis

Male bladder
Seminal vesicle
Ejaculatory duct
Prostate gland
Male urethra

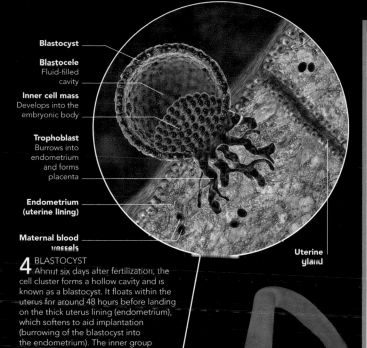

Blastocyst

Blastocele
Fluid-filled cavity

Inner cell mass
Develops into the embryonic body

Trophoblast
Burrows into endometrium and forms placenta

Endometrium (uterine lining)

Maternal blood vessels

Uterine gland

4 BLASTOCYST
About six days after fertilization, the cell cluster forms a hollow cavity and is known as a blastocyst. It floats within the uterus for around 48 hours before landing on the thick uterus lining (endometrium), which softens to aid implantation (burrowing of the blastocyst into the endometrium). The inner group of cells will become the embryo itself.

Myometrium

Endometrium (lining of the uterus)

Cervix

Vagina

5 EMBRYONIC DISK
Within the inner cell mass, an embryonic disk forms. This separates the cell cluster into the amniotic cavity, which develops into a sac that will fill with fluid and fold around to cover the embryo, and the yolk sac, which helps transport nutrients to the embryo during the second and third weeks. The disk develops three circular sheets called the primary germ layers—ectoderm, mesoderm, and endoderm—from which all body structures will derive.

Endoderm
Forms linings of digestive, respiratory, and urogenital tracts, some glands such as thyroid and thymus, also ducts of liver and pancreas, and inner ear lining

Ectoderm
Develops into skin epidermis, hair, nails, tooth enamel, central nervous system, sense organ receptor cells, and parts of eyes, ears, and nasal cavity

Mesoderm
Forms skin dermis, bone, muscle, cartilage, connective tissue, heart, blood cells and vessels, lymph cells and vessels, spleen, and some glands

Maternal blood sinus
Loose, saclike space filled with the mother's blood

Endometrium
Blood-rich uterine lining

Trophoblast
Mass of embryonic cells that burrows and extends into uterine lining to become the placenta

Implantation scar

Embryonic disk

Yolk sac

Amniotic cavity

GROWING EMBRYO

As development proceeds, cells continue to divide. They move to form groups that will become tissues and organs. They also specialize to different cell types, as genes in their chromosomes are switched off or on. In general, development is head-down, with the brain and head taking shape early, then the body, followed by the arms as small buds, and lastly the legs. By the end of the embryonic stage, eight weeks after fertilization, all major organs and body parts are formed. From this time on, the baby is known as a fetus. Each of the top row of diagrams shows the embryo at life size.

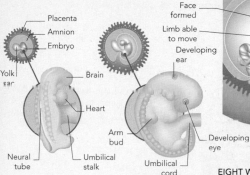

Placenta
Amnion
Embryo
Yolk sac
Brain
Heart
Neural tube
Umbilical stalk
Arm bud
Umbilical cord
Face formed
Limb able to move
Developing ear
Developing eye

THREE WEEKS
The neural tube forms. It will become the spinal cord, enlarged at one end as the brain. The simple tubelike heart pulsates. The embryo is ⁴⁄₅₀–⁵⁄₅₀ in (2–3 mm) long.

FOUR WEEKS
The four-chambered heart beats, sending blood through simple vessels. Intestines, liver, pancreas, lungs, and limb buds can be seen. The embryo is about ⅛ in (4–5 mm) long.

EIGHT WEEKS
At this stage the face and neck take shape, the back straightens, and fingers and toes can be differentiated clearly. The embryo starts to move. It is now around 1–1⅛ in (25–30 mm) in length.

FETAL DEVELOPMENT

FROM THE EIGHTH WEEK OF PREGNANCY UNTIL BIRTH, THE DEVELOPING BABY IS KNOWN AS A FETUS. WITH THE MAJOR ORGANS ALREADY HAVING FORMED DURING EARLIER EMBRYONIC DEVELOPMENT, THE FETAL STAGE IS MAINLY ONE OF PHYSICAL GROWTH, MATURATION, AND ADDING SMALLER DETAILS, SUCH AS HAIR AND NAILS.

FETAL CHANGES: WEEKS 8–24

Pregnancy can be divided into three roughly equal time spans of three months, called trimesters. The first trimester takes in the embryonic stage and the first four weeks of fetal development. In the embryonic stage, the head and brain grow fast and control and coordination nerve pathways form. During early fetal growth, these pathways begin to regulate the functions of major body systems, such as the musculoskeletal and urinary systems.

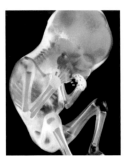

Early in the second trimester, the face takes on a more familiar appearance, and by the end of the second trimester the limbs have lengthened significantly, thereby giving the fetus more babylike proportions.

THE FETAL SKELETON
In this image of a 16-week-old fetus, ossified bone appears yellow; the gaps between the ends of long bones are filled with cartilage, which forms a matrix in which ossified bone develops.

THE PLACENTA

The placenta, or afterbirth, is the unborn baby's life-support system. It transfers oxygen, energy-giving glucose, essential nutrients, and other vital substances from the mother. The maternal and fetal blood do not make direct contact in the placenta but are separated by a barrier of cell layers, which form the outermost chorionic membrane. However, this barrier is thin enough for oxygen, nutrients, and some infection-fighting antibodies to pass through from pools of maternal blood in the placental lacunae (spaces) to fetal blood. This fetal blood has arrived, low in oxygen and nutrients, along two umbilical arteries inside the umbilical cord. It flows back to the baby, refreshed with oxygen and nutrients, along the umbilical vein. The placenta is fully developed by 16–18 weeks. At delivery, on average it weighs 14–21 oz (400–600 g), has a diameter of 8–9 in (20–22 cm), and is about 1 in (2.5 cm) thick.

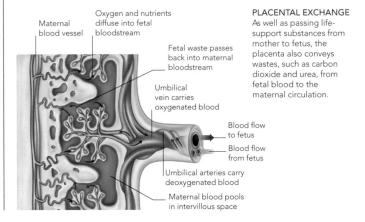

Maternal blood vessel

Oxygen and nutrients diffuse into fetal bloodstream

Fetal waste passes back into maternal bloodstream

Umbilical vein carries oxygenated blood

Blood flow to fetus

Blood flow from fetus

Umbilical arteries carry deoxygenated blood

Maternal blood pools in intervillous space

PLACENTAL EXCHANGE
As well as passing life-support substances from mother to fetus, the placenta also conveys wastes, such as carbon dioxide and urea, from fetal blood to the maternal circulation.

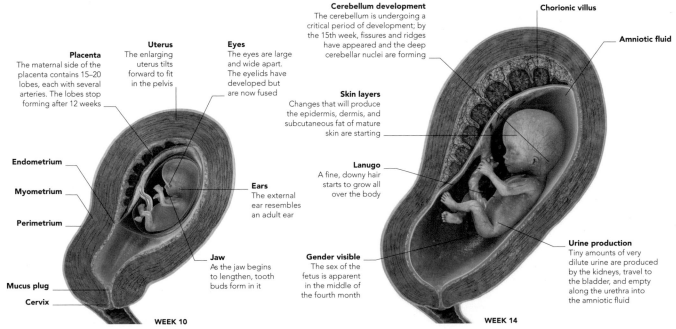

Placenta
The maternal side of the placenta contains 15–20 lobes, each with several arteries. The lobes stop forming after 12 weeks

Uterus
The enlarging uterus tilts forward to fit in the pelvis

Eyes
The eyes are large and wide apart. The eyelids have developed but are now fused

Endometrium

Myometrium

Perimetrium

Ears
The external ear resembles an adult ear

Mucus plug

Cervix

Jaw
As the jaw begins to lengthen, tooth buds form in it

WEEK 10

Cerebellum development
The cerebellum is undergoing a critical period of development; by the 15th week, fissures and ridges have appeared and the deep cerebellar nuclei are forming

Chorionic villus

Amniotic fluid

Skin layers
Changes that will produce the epidermis, dermis, and subcutaneous fat of mature skin are starting

Lanugo
A fine, downy hair starts to grow all over the body

Gender visible
The sex of the fetus is apparent in the middle of the fourth month

Urine production
Tiny amounts of very dilute urine are produced by the kidneys, travel to the bladder, and empty along the urethra into the amniotic fluid

WEEK 14

WEEKS 8–10
The fetus's crown–rump (head and body) length is 2–2⅜ in (5–6 cm), and its heart rate is 170–180 beats per minute. Head growth is slowing. The neck lengthens, the head lifts from the chest, and the kidneys start to function at the end of week 10.

WEEKS 11–14
The fetus has a crown–rump length of 4¾ in (12 cm) and weighs 3½ oz (100 g). The heartbeat strengthens as its rate slows to 150–160 per minute. Up to a quarter of a million new brain cells form every minute.

MATERNAL CHANGES DURING PREGNANCY

From conception, the mother's body starts to change to prepare the uterus for pregnancy and to adapt to the future needs of the developing baby. Early signs of pregnancy are typically a missed period; tender, enlarged breasts; nausea, possibly with vomiting ("morning sickness"); and sometimes cravings for odd foods. A noticeable abdominal bulge usually appears towards the end of the first trimester. As pregnancy progresses, the growing fetus compresses the intestines, bladder, and lungs, leading to abdominal discomfort, frequent urination, and shortness of breath. The breasts continue to enlarge, eventually producing colostrum toward the end of pregnancy.

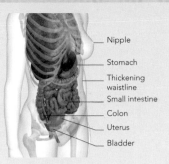

- Nipple
- Stomach
- Thickening waistline
- Small intestine
- Colon
- Uterus
- Bladder

FIRST TRIMESTER (WEEKS 1–10)
Pressure from the enlarging uterus is gentle at first but produces symptoms such as backache and constipation. The mother's blood circulation, breathing rate, and metabolism speed up, so she may feel hot

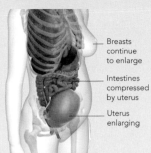

- Breasts continue to enlarge
- Intestines compressed by uterus
- Uterus enlarging

SECOND TRIMESTER (WEEKS 11–24)
Symptoms such as nausea and fatigue tend to diminish from the start of the second trimester. The mother's heartbeat rate and strength and blood volume increase to cope with increased fetal demands.

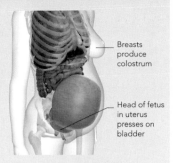

- Breasts produce colostrum
- Head of fetus in uterus presses on bladder

THIRD TRIMESTER (WEEKS 25–38)
Increased weight may cause back pain. Near full term, the fetus's head moves down in the pelvis (engages), relieving pressure on the mother's lungs but increasing pressure on her bladder.

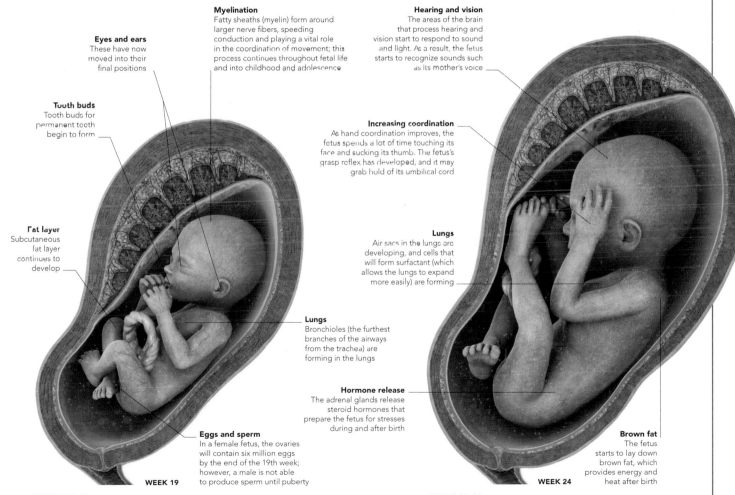

Eyes and ears
These have now moved into their final positions

Tooth buds
Tooth buds for permanent teeth begin to form

Fat layer
Subcutaneous fat layer continues to develop

Myelination
Fatty sheaths (myelin) form around larger nerve fibers, speeding conduction and playing a vital role in the coordination of movement; this process continues throughout fetal life and into childhood and adolescence

Hearing and vision
The areas of the brain that process hearing and vision start to respond to sound and light. As a result, the fetus starts to recognize sounds such as its mother's voice

Increasing coordination
As hand coordination improves, the fetus spends a lot of time touching its face and sucking its thumb. The fetus's grasp reflex has developed, and it may grab hold of its umbilical cord

Lungs
Air sacs in the lungs are developing, and cells that will form surfactant (which allows the lungs to expand more easily) are forming

Lungs
Bronchioles (the furthest branches of the airways from the trachea) are forming in the lungs

Hormone release
The adrenal glands release steroid hormones that prepare the fetus for stresses during and after birth

Eggs and sperm
In a female fetus, the ovaries will contain six million eggs by the end of the 19th week; however, a male is not able to produce sperm until puberty

Brown fat
The fetus starts to lay down brown fat, which provides energy and heat after birth

WEEK 19

WEEKS 15–19
The fetus has a crown–rump length of 10 in (25 cm) and weighs 12⅜ oz (350 g). At this time, the mother usually becomes aware of the fetus's movements, known as "quickening."

WEEK 24

WEEKS 20–24
The fetus has a crown–rump length of 13¾ in (35 cm) and weighs 26½ oz (750 g). Part of the weight gain is subcutaneous brown fat, which starts to accumulate around the shoulder blades.

FETAL CHANGES: WEEKS 25–38

Following the complex development of body systems during the first two trimesters, the third trimester is chiefly a time of maturation and further growth. As early as week 25, fine details such as eyebrows and eyelashes are evident. Nearly all body organs soon function. However, with no air to breathe, the lungs remain collapsed. During the final six weeks, they produce increasing amounts of a substance known as surfactant, which will help them expand and fill with air after birth. In its final 10 weeks, the fetus doubles its weight, mainly due to deposits of subcutaneous fat, reaching an average of $7^{1}/_{2}$ lb (3.4 kg) at full term.

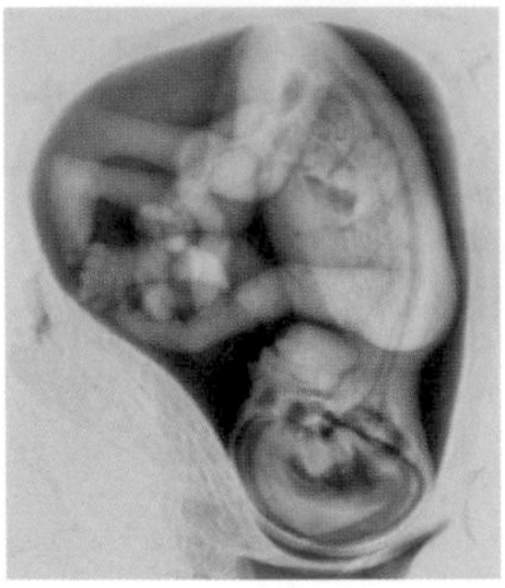

DEVELOPING NERVOUS SYSTEM

The early brain grows rapidly but has a smooth surface. By week 25, it just begins to show its familiar gyri (bulges) and sulci (grooves), as can be seen in this 3-D MRI scan. The striplike spinal cord (colored brown in this color-enhanced image) is, in effect, an extension of the brain; together they form the central nervous system.

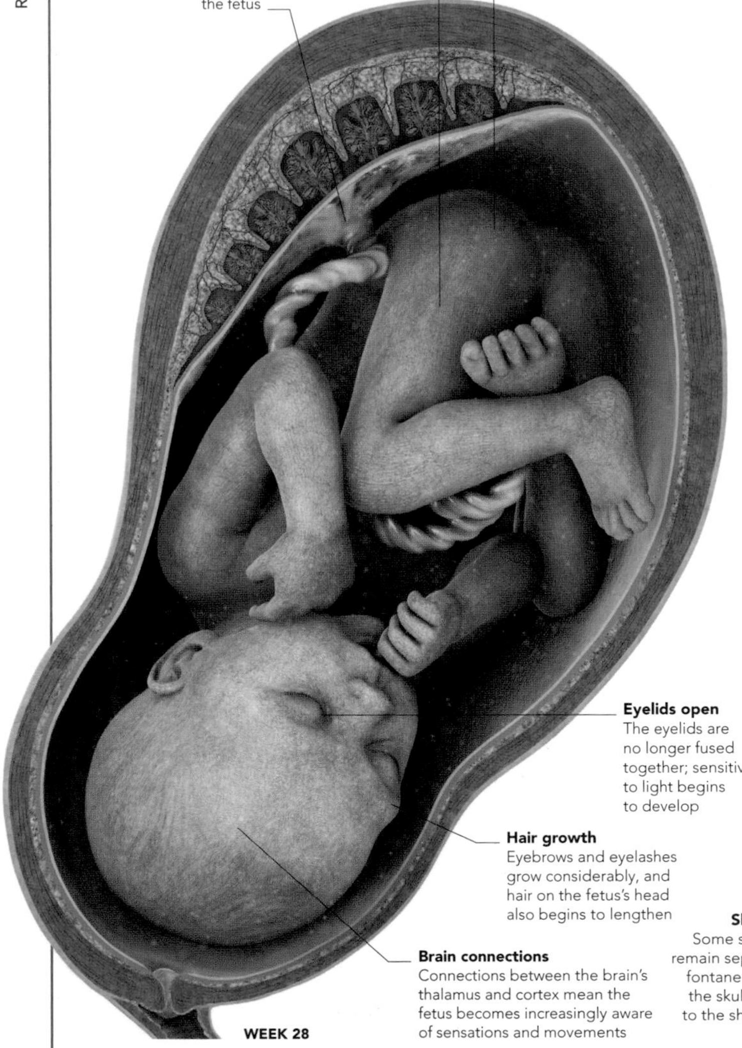

Maternal artery

Maternal vein

Chorionic villus

Umbilical cord

Digestive system
The gut has now developed to the point that milk can be digested

Lungs
Two major changes occur this month: surfactant is produced from the 33rd week onward; and the development of the blood–air barrier means that gas exchange is now possible after birth

Skin
The skin becomes thicker and less translucent; in pale-skinned babies, it also changes color from red to pink

Transfer of antibodies
Immunities are now efficiently transported from the mother to the fetus

Fat accumulation
The fetus continues to gain weight rapidly; the majority of this is fat

Covering of vernix
The fetus's body is now covered in a greasy, protective substance called vernix

Eyelids open
The eyelids are no longer fused together; sensitivity to light begins to develop

Hair growth
Eyebrows and eyelashes grow considerably, and hair on the fetus's head also begins to lengthen

Brain connections
Connections between the brain's thalamus and cortex mean the fetus becomes increasingly aware of sensations and movements

Skull bones
Some skull bones remain separated by fontanelles so that the skull can mold to the shape of the birth canal

Sucking reflex
The sucking reflex develops, in preparation for feeding after birth

WEEK 28

WEEK 33

WEEKS 25–28
The fetus is still small enough to move around quite freely in the uterus. It also has a regular wake–sleep pattern, and rests or sleeps for about half the time.

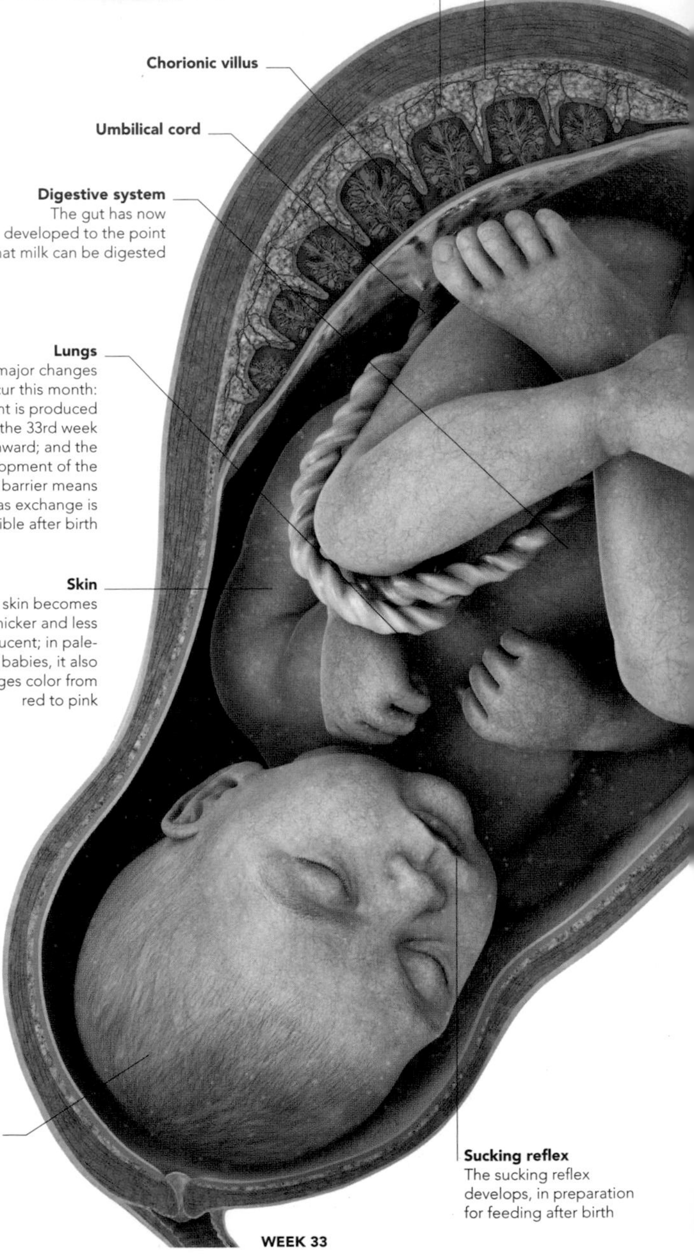

WEEKS 29–33
The fetus's crown–rump length is now $17^{3}/_{4}$ in (45 cm) and it weighs about $5^{1}/_{4}$ lb (2.4 kg). It continually swallows small amounts of amniotic fluid, which is filtered by its kidneys and excreted back into the fluid.

CHANGING PROPORTIONS

During the first trimester, the nervous system undergoes a critical period of development. As a result, the brain and head grow quickly until they reach as much as half of the total length of the fetus's body. In the fifth month of pregnancy, the fetus's trunk and limbs enter a phase of rapid growth, so that the size of the head begins to look more adultlike relative to the rest of the body. From then until birth, the fetus's head grows relatively little in comparison to the huge growth the body undergoes during this time.

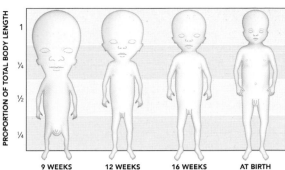

PROPORTION OF TOTAL BODY LENGTH

1 · ¾ · ½ · ¼

9 WEEKS · **12 WEEKS** · **16 WEEKS** · **AT BIRTH**

CHANGING RATES OF GROWTH
During the first trimester, the fetal head grows more quickly than the body. Relative growth of the head then slows and fetal proportions become increasingly adultlike.

Perimetrium

Chorionic villus

Umbilical cord
This will be clamped and cut in the third stage of labor

Amnion

Chorion

Amniotic fluid
The volume of this shock-absorbing liquid has reduced in the weeks leading up to the birth

Weight gain
The fetus continues to gain about 1 oz (28 g) a day during the final month of pregnancy

Myometrium
This powerful muscular outer layer of the uterus is responsible for contractions in labor

Hair
Tiny vellus body hairs replace earlier lanugo hairs in the last weeks. Some babies have profuse head hair, others hardly any

Cervix
This stays tightly closed until birth is near; it will then begin to soften, thin, and dilate. The plug of mucus in the cervix loosens and falls out just before labor starts

WEEK 38 (FULL TERM)

WEEKS 34–38
The fetus spends a lot of time flexing its rib and diaphragm muscles. These movements are thought to be involved in producing growth factors needed during development of the lungs. Reflexes, such as turning to face sounds, are well developed.

ULTRASOUND SCANNING

Scanning the fetus with high-frequency sound waves (ultrasound) is carried out routinely during pregnancy to assess progress and detect any fetal abnormalities. Precise timings vary, but the first scan is usually performed 8–14 weeks after fertilization to determine the due date. The second scan, usually done at 18–21 weeks, is an anomaly scan to check for structural abnormalities. A comparatively recent development is the 3-D scan, which is obtained by digitally adding together successive 2-D scans to produce a detailed, in-depth image of the fetus.

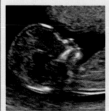 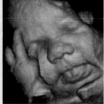

FETAL ULTRASOUND SCANS
Standard 2-D ultrasound scans, such as that of a 12-week-old fetus (left), can reveal details like the fetus's sex, age, and size. 3-D scans, such as that of the head of a full-term fetus (right), can give a much more detailed view of fetal features. A 2-D scan is usually sufficient for routine checks of the development and health of the fetus.

PREPARING FOR BIRTH

CHANGES OCCUR IN THE BODY DURING LATE PREGNANCY, SIGNALING THE APPROACH OF CHILDBIRTH. THE HEAD OF THE FETUS DROPS LOWER INTO THE PELVIS; THE EXPECTANT MOTHER MAY EXPERIENCE WEIGHT LOSS; AND THERE MAY BE SOME EARLY CONTRACTIONS OF THE UTERUS.

MULTIPLE PREGNANCY AND FETAL POSITIONS

The presence of more than one fetus in the uterus is called a multiple pregnancy. Twins occur in about one in 80 pregnancies, and triplets in about one in 8,000. Both events are becoming more common, partly due to improved prenatal care and also to increasing use of fertility methods such as IVF (in vitro fertilization). After about 30 weeks, the most common fetal position is head down, facing the mother's back, with the neck flexed forward. Such a position eases passage through the birth canal. However, about 1 in 30 full-term deliveries is breech, in which the baby's buttocks emerge before the head.

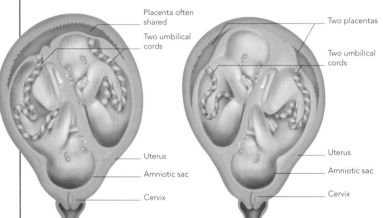

Placenta often shared
Two umbilical cords
Uterus
Amniotic sac
Cervix

Two placentas
Two umbilical cords
Uterus
Amniotic sac
Cervix

MONOZYGOTIC TWINS
A single fertilized egg, or zygote, forms an embryo that splits into two. Each develops into a fetus. The two have the same genes and sex and often share one placenta. They look alike and are known as "identical" twins.

DIZYGOTIC TWINS
Two eggs are fertilized and develop separately, each with its own placenta. They may be different or the same sex. Also called "fraternal twins," they have the same degree of resemblance as any brothers and sisters.

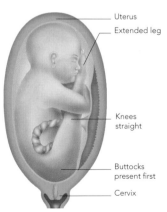

Uterus
Extended leg
Knees straight
Buttocks present first
Cervix

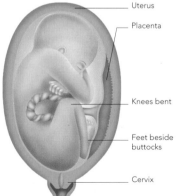

Uterus
Placenta
Knees bent
Feet beside buttocks
Cervix

FRANK BREECH
In frank breech, also called incomplete breech, the baby fails to turn head-down in the uterus. The hips are flexed and the legs are straight, extending alongside the body so that the feet are positioned beside the head.

COMPLETE BREECH
The baby's legs are flexed at the hips and knees, so the feet are next to the buttocks. This occurs less commonly than frank breech. The incidence of breech delivery is much higher among premature babies.

CHANGES IN THE CERVIX

The cervix is the firm band of muscle and connective tissue that forms the necklike structure at the bottom of the uterus. In late pregnancy, the cervix softens in readiness for childbirth. Sporadic uterine tightenings, known as Braxton-Hicks contractions, help thin the cervix so that it merges with the uterus's lower segment. Braxton–Hicks contractions are usually painless, and occur through much of pregnancy, becoming noticeable only after midterm.

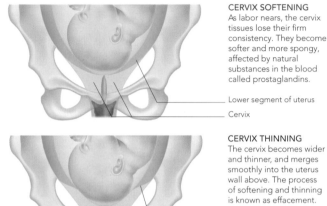

CERVIX SOFTENING
As labor nears, the cervix tissues lose their firm consistency. They become softer and more spongy, affected by natural substances in the blood called prostaglandins.

Lower segment of uterus
Cervix

CERVIX THINNING
The cervix becomes wider and thinner, and merges smoothly into the uterus wall above. The process of softening and thinning is known as effacement.

Cervix merging with uterus
Cervix thinned

CONTRACTIONS

When pregnancy reaches full-term, the uterus is the largest and strongest muscle in the female body. When its muscle fibers shorten, with the eventual aim of expelling the fetus, it is known as a uterine contraction or simply a "contraction." True contractions, as opposed to Braxton-Hicks contractions, are regular and become steadily more frequent, more painful, and longer-lasting. The main area of contraction is in the uterine fundus (upper uterus), which stretches, causing the lower uterus and cervix to thin. Judging when true labor has started can be difficult due to "false alarms."

Evenly spaced
Occasional, irregular Braxton–Hicks contraction

INTENSITY OF CONTRACTION
0   10   20   **MINUTES**
PREGNANCY 20TH WEEK

Contractions increase
Regular, mild contractions occur

INTENSITY OF CONTRACTION
0   10   20   **MINUTES**
PRELABOR 36TH WEEK

Approaching labor
Interval between contractions decreases

INTENSITY OF CONTRACTION
0   10   20   **MINUTES**
EARLY LABOR 40TH WEEK

PROGRESS OF CONTRACTIONS
Gentle, partial contractions occur through much of pregnancy, but true contractions begin only late in pregnancy. At first, they are well spaced out and relatively low in strength. As labor intensifies, the rate and duration of contractions increase and they put more downward pressure on the baby.

LABOR

IN MEDICAL TERMS, LABOR USUALLY MEANS THE FULL PROCESS OF GIVING BIRTH. IT CAN BE DESCRIBED IN THREE PHASES OR STAGES: ONSET OF CONTRACTIONS TO FULL DILATION OF THE CERVIX; DELIVERY OF THE BABY; AND DELIVERY OF THE PLACENTA (AFTERBIRTH).

ENGAGEMENT

During the last weeks of pregnancy, the baby's presenting part (the part that will emerge first, usually the head) descends into the bowl-like cavity of the pelvis. This process is called engagement. When it happens, many women feel a sensation of dropping and "lightening." This is because the movement of the baby lowers the upper uterus, resulting in less pressure on the diaphragm, which makes it easier for the mother to breathe. Engagement usually takes place at around 36 weeks during a first pregnancy, however, it may not happen until the onset of labor during subsequent pregnancies.

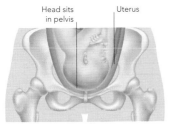

BEFORE THE HEAD ENGAGES
Before the head of the baby engages, the top of the uterus reaches up to the sternum, or breastbone. The widest section of the baby's head has not yet passed through the inlet of the pelvis into the cavity.

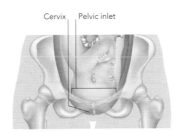

AFTER ENGAGEMENT
The baby's head descends through the inlet of the pelvis and becomes slotted into, or engaged, within the pelvic cavity. The overall position of the uterus drops, and the baby's head rests against the uterine cervix.

INDUCTION

If a pregnancy continues too long past the due date—usually 10–14 days—then labor may need to be medically induced. This may also be advised if the health of the mother or baby, or both, is at risk. There are several induction methods, depending partly on the stage of labor. They include inserting a vaginal suppository, artificially breaking the water, or giving an injection of a hormone that stimulates contractions of the uterus.

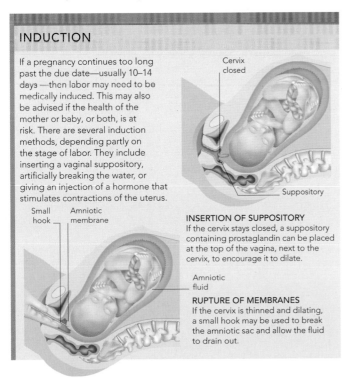

INSERTION OF SUPPOSITORY
If the cervix stays closed, a suppository containing prostaglandin can be placed at the top of the vagina, next to the cervix, to encourage it to dilate.

RUPTURE OF MEMBRANES
If the cervix is thinned and dilating, a small hook may be used to break the amniotic sac and allow the fluid to drain out.

CERVICAL DILATION

The first stage of labor begins with the onset of regular, painful contractions of the uterus, which cause the cervix to dilate. These occur mainly in the upper uterus so that it shortens and tightens, and this pulls and stretches the lower uterine segment and cervix. On average, for a first baby, the cervix dilates at the rate of about ½in (1cm) per hour; progress is usually more rapid for subsequent babies. In most women, the cervix is fully dilated when it opens to about 4in (10cm).

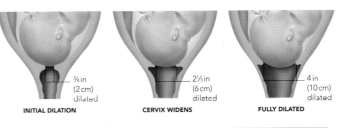

INITIAL DILATION ¾in (2cm) dilated · **CERVIX WIDENS** 2⅓in (6cm) dilated · **FULLY DILATED** 4in (10cm) dilated

SIGNS OF EARLY LABOR

Every woman's experience is different, but generally there are three signs that labor is starting: a "show," contractions, and the water breaking. Before labor begins (usually less than 3 days), the mucous plug in the cervix, which acts as a seal during pregnancy, is passed as a blood-stained or brownish discharge (the "show"). As contractions become stronger and more regular, the membranes retaining the amniotic fluid rupture (break), allowing the fluid (water) to leak out via the birth canal.

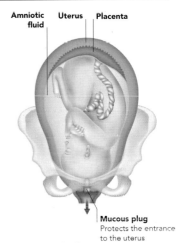

1 THE "SHOW"
For most of a pregnancy, the mucous plug in the cervix prevents microbes from entering the uterus. As the cervix widens slightly, the plug loosens and falls out.

Mucous plug Protects the entrance to the uterus

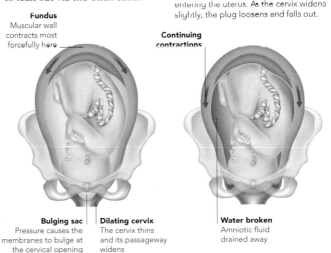

2 CONTRACTIONS
Coordinated muscular contractions are generated in the upper part of the uterus, the fundus. This helps to gradually open, or dilate, the cervix.

Fundus Muscular wall contracts most forcefully here · **Bulging sac** Pressure causes the membranes to bulge at the cervical opening · **Dilating cervix** The cervix thins and its passageway widens

3 WATER BREAKS
The amniotic sac (membrane) around the baby ruptures, or breaks, allowing colorless amniotic fluid to pass out through the birth canal.

Water broken Amniotic fluid drained away

DELIVERY

THE CULMINATION OF PREGNANCY AND LABOR, DELIVERY OF THE BABY AND THE PLACENTA INVOLVES A COMPLICATED SEQUENCE OF EVENTS THAT ULTIMATELY SEPARATE CHILD FROM MOTHER, COMPLETING THE TRANSITION TO AN INDEPENDENT EXISTENCE.

THREE STAGES OF CHILDBIRTH

The first contractions of labor start in response to the secretion of hormones and, during that first stage of labor, uterine contractions pull the cervix until it is merely a thin sheet of tissue and is fully dilated to approximately 4 in (10 cm). The membranes of the amniotic sac that protected the fetus in the uterus rupture, a process referred to as the water breaking. The second stage, delivery, involves synchronized efforts of the mother's contractions and the baby's shifts in position in order to fit its large head into the birth canal and then travel along it to the outside world. After the baby is born and the umbilical cord clamped and cut, is the third stage of labor, in which the placenta, or "afterbirth," is delivered, often with the help of a midwife or obstetrician gently pulling on the cord.

NORMAL DELIVERY

Newborn babies are usually covered with a combination of blood, mucus, and vernix (the greasy covering that protected the fetus in the uterus). This baby's umbilical cord has not yet been clamped and cut.

PELVIC SHAPES

The female pelvis tends to be better adapted to the process of child-bearing and delivery than a man's. Nevertheless, there is a wide variation in shape, with some shapes making childbirth easier than others. A round, shallow shape (gynecoid), the classic "female pelvis," has a generous capacity, and usually results in fewer problems. At the other extreme is a triangular, or android, pelvis, which is less spacious and can cause difficulties with childbirth.

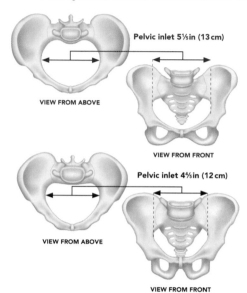

Pelvic inlet 5⅛ in (13 cm)

VIEW FROM ABOVE

VIEW FROM FRONT

Pelvic inlet 4⅘ in (12 cm)

VIEW FROM ABOVE

VIEW FROM FRONT

GYNECOID PELVIS

A gynecoid pelvis is shallow, allowing the uterus to expand as the fetus grows. The round shape of the wider pelvic inlet provides more room for the head of a fetus to pass through at delivery.

ANDROID PELVIS

An android (triangular) pelvis is most like a man's in shape. It can be difficult for a woman with this shape of pelvis to have a vaginal delivery unless her baby is small.

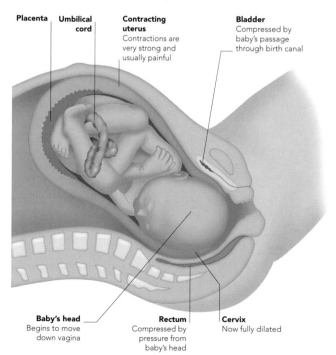

Placenta

Umbilical cord

Contracting uterus
Contractions are very strong and usually painful

Bladder
Compressed by baby's passage through birth canal

Baby's head
Begins to move down vagina

Rectum
Compressed by pressure from baby's head

Cervix
Now fully dilated

1 DILATION OF THE CERVIX

Delivery begins when the cervix is completely dilated. The baby turns toward the mother's spine so that the widest part of the baby's skull is aligned with the widest part of the mother's pelvis. As the baby tucks in its chin, it starts moving out of the uterus and into the vagina, which stretches to accommodate the baby's head.

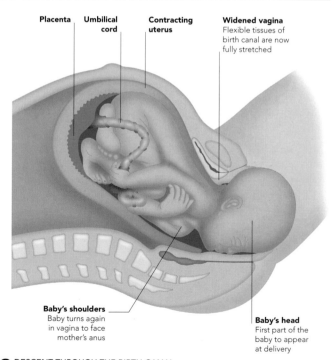

Placenta

Umbilical cord

Contracting uterus

Widened vagina
Flexible tissues of birth canal are now fully stretched

Baby's shoulders
Baby turns again in vagina to face mother's anus

Baby's head
First part of the baby to appear at delivery

2 DESCENT THROUGH THE BIRTH CANAL

As the baby descends through the birth canal, the top of the head appears for the first time. This stage is called "crowning," and the baby has usually turned again, toward the mother's anus this time, so that the emerging head can negotiate the bend in the fully stretched vagina. Birth is usually imminent at this point.

FETAL MONITORING

If delivery is not proceeding as expected, the obstetrician (a physician who specializes in pregnancy and childbirth) can monitor the heart rate of the fetus to determine whether it is in distress. Fetal monitoring can be done with a stethoscope or, occasionally, Doppler ultrasound scans. To achieve more conclusive results, electronic fetal monitoring (EFM), also known as cardiotocography, can be done, either externally with two devices strapped to the mother's abdomen, or internally with an electrode clipped to the baby's head and hooked up to the electronic fetal monitor. It is now possible to monitor the heart rate remotely so the mother can remain mobile during labor.

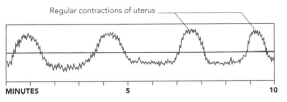

MONITORING THE BABY
External electronic monitors record the baby's heart rate and mother's contractions.

FETAL HEART RATE
The response of the baby's heart rate to the mother's contractions is followed with electronic fetal monitoring; the baby's heart rate should increase during uterine contractions.

Baseline rate 120 bpm

Increase in rate with each contraction

CONTRACTIONS
This tracing showing the intensity of the mother's contractions, and the one above, for the fetal heart rate, shows that the baby responds to the mother's contractions.

Regular contractions of uterus

EPIDURAL ANALGESIA

One of the most commonly used methods of pain relief during labor and delivery, epidural analgesia, is delivered via a needle into the space between the vertebrae and the spinal column in the lower (lumbar) region of the back. It affects the nerve fibers that detect contraction pains. A lower-dose variant known as a mobile or walking epidural reduces pain without removing sensation, allowing women to move around during labor and participate actively in the delivery.

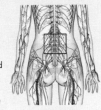

LOCATOR

Spinal cord

Tip of catheter

Vertebra

Fluid

Epidural space

INSERTION OF THE CATHETER
A catheter is passed through a fine needle and inserted into the epidural space. It is left in place, allowing drugs to be administered throughout labor.

Placenta

Umbilical cord

Contracting uterus
Strong contractions continue to push the baby out

Baby's shoulders
First one then the other shoulder emerges

Emerging head
May be guided and supported by an obstetrician as it emerges from the mother's vagina

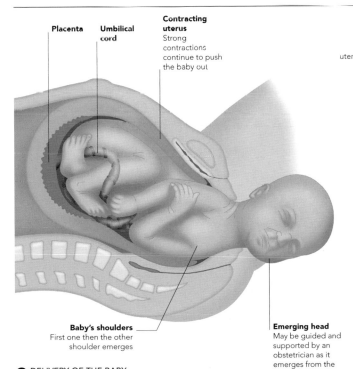

3 DELIVERY OF THE BABY
As the baby's head emerges from the mother's body, the obstetrician checks that the umbilical cord is not wrapped around the baby's neck. Mucus is cleared from its nose and mouth so that it can breathe. The baby rotates again so that its shoulders are in position to slip out easily, one shoulder quickly followed by the other one.

Detaching placenta
Separates from uterus 5–15 minutes after delivery

Cord traction
The placenta is eased out by gently pulling on the cord while pressing the lower abdomen to provide traction

Rectum
No longer compressed

Birth canal
Starts to return to normal size

4 DELIVERY OF THE PLACENTA
The uterus resumes mild contractions soon after the baby is born, sealing shut any blood vessels that are still bleeding. The placenta separates from the lining of the uterus and is eased out by gently pulling on the umbilical cord while pressing on the lower abdomen. The mother may be given a drug to accelerate this process.

AFTER THE BIRTH

OVER THE SPACE OF 40 WEEKS, THE FERTILIZED OVUM DEVELOPS INTO A COMPLEX, MULTICELLULAR HUMAN BEING, AND HAS PROGRESSED FROM FETUS TO NEWBORN BABY. ALL ORGAN SYSTEMS ARE IN PLACE, BUT SOME WILL QUICKLY ADAPT TO LIFE WITHOUT AN UMBILICAL CORD, WHILE OTHERS WILL NOT COMPLETE DEVELOPMENT UNTIL ADOLESCENCE.

NEWBORN ANATOMY

The anatomy of a newborn baby is characterized by various special features that will facilitate its further growth and development outside the mother's uterus. A baby grows faster in the first year after birth than at any other time in its life. Fibrous fontanelles, which separate the skull bones, allow the skull to increase in volume as the brain grows; these fontanelles will start to harden to bone (ossify) at about 18 months, but the process will not be finished until the child is about six years old. Cartilage in the joints and at the end of long bones adapt the skeletal system for rapid growth. At birth, the thymus gland is at its largest relative size since it has been the center for development of the immune system in the fetus. Likewise, the liver is enlarged as in the fetus it was the sole producer of red blood cells. This task will now be taken over by the bone marrow.

APGAR SCORE

To assess if a newborn needs emergency care, the Apgar score tests five criteria both one and five minutes after birth. In dark-skinned babies, "color" refers to the mouth, palms, and soles of the feet.

SIGN	SCORE: 0	SCORE: 1	SCORE: 2
HEART RATE	None	Below 100	Over 100
BREATHING RATE	None	Slow or irregular; weak cry	Regular; strong cry
MUSCLE TONE	Limp	Some bending of limbs	Active movements
REFLEX RESPONSES	None	Grimace or whimpering	Cry, sneeze, or cough
COLOR	Pale or blue	Blue extremities	Pink

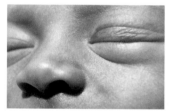

PUFFY EYES
A newborn's eyelids are often puffy. Some babies develop pink eye soon after birth, caused either by a blocked tear duct or by infection with bacteria in the birth canal.

VERNIX
This greasy white substance over the fetus' body stops the skin from wrinkling due to exposure to amniotic fluid while in the uterus. It is washed or wiped off after birth.

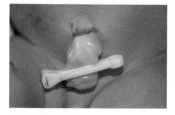

UMBILICAL CORD
The link between fetus and maternal placenta, the umbilical cord, has two arteries and a vein in a jellylike covering. It is clamped and cut shortly after birth.

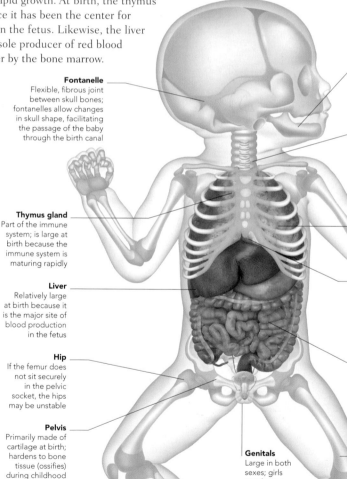

Fontanelle
Flexible, fibrous joint between skull bones; fontanelles allow changes in skull shape, facilitating the passage of the baby through the birth canal

Thymus gland
Part of the immune system; is large at birth because the immune system is maturing rapidly

Liver
Relatively large at birth because it is the major site of blood production in the fetus

Hip
If the femur does not sit securely in the pelvic socket, the hips may be unstable

Pelvis
Primarily made of cartilage at birth; hardens to bone tissue (ossifies) during childhood

BABY SKELETON
At birth, a baby has some 300 bones, but some of these fuse during childhood and adolescence, leaving adults with only 206. Cartilaginous areas of the baby's skeleton are shown in blue and those composed of bone are shown in off-white. A few areas remain cartilage throughout the person's life.

Jaw
Contains fully formed primary (milk) teeth within jawbone; in most cases, teeth do not start to erupt until the baby is six months old

Neck
Undeveloped muscles cannot support the large, heavy head in the first few weeks

Lung
With the first breath, the baby's lungs fill with air, expand, and regular breathing (respiration) begins

Heart
Changes in structure at birth to enable blood to circulate through the lungs rather than through the placenta

Intestines
Excretes the first fecal material as a thick, sticky, greenish black mixture of bile and mucus, called meconium

Genitals
Large in both sexes; girls may have a slight vaginal discharge

Femur
Long bone of the thigh; only the shaft has hardened into bone at birth; the ends are still cartilage to allow for growth

Foot
At birth, most of the bones in the foot are cartilage, and the foot may be turned in or out depending on the baby's position in the uterus

CIRCULATION IN THE UTERUS

As the placenta provides oxygen and nutrients, fetal circulation has anatomical variations ("shunts") to allow blood to bypass not-yet-functioning liver and lungs. The ductus venosus shunts incoming blood through the liver to the right atrium, which shunts it through a gap, the foramen ovale, to the left atrium (mostly bypassing the right ventricle) and onward to the body. Blood that makes it into the right ventricle passes into the pulmonary artery but is shunted into the aorta by the ductus arteriosus, thus bypassing the lungs.

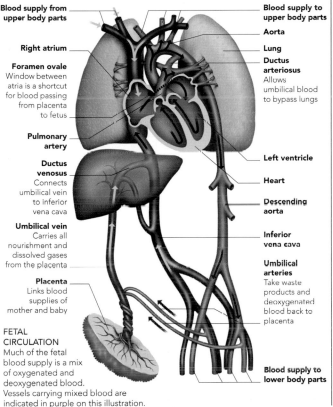

Blood supply from upper body parts

Right atrium

Foramen ovale
Window between atria is a shortcut for blood passing from placenta to fetus

Pulmonary artery

Ductus venosus
Connects umbilical vein to inferior vena cava

Umbilical vein
Carries all nourishment and dissolved gases from the placenta

Placenta
Links blood supplies of mother and baby

Blood supply to upper body parts

Aorta

Lung

Ductus arteriosus
Allows umbilical blood to bypass lungs

Left ventricle

Heart

Descending aorta

Inferior vena cava

Umbilical arteries
Take waste products and deoxygenated blood back to placenta

Blood supply to lower body parts

FETAL CIRCULATION
Much of the fetal blood supply is a mix of oxygenated and deoxygenated blood. Vessels carrying mixed blood are indicated in purple on this illustration.

CIRCULATION AT BIRTH

At birth, the baby takes its first breaths and the umbilical cord is clamped. This forces the circulatory system into a monumental response to convert itself immediately to obtain its oxygen supply via the lungs. Blood is sent to the lungs to retrieve oxygen, and the pressure of this blood returning from the lungs into the left atrium forces shut the foramen ovale between the two atria, thus establishing normal circulation. The ductus arteriosus, the ductus venosus, and the umbilical vein and arteries, close up and become ligaments.

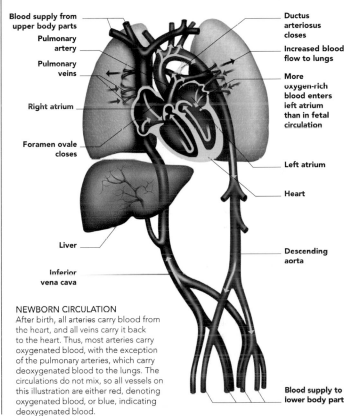

Blood supply from upper body parts

Pulmonary artery

Pulmonary veins

Right atrium

Foramen ovale closes

Liver

Inferior vena cava

Ductus arteriosus closes

Increased blood flow to lungs

More oxygen-rich blood enters left atrium than in fetal circulation

Left atrium

Heart

Descending aorta

Blood supply to lower body parts

NEWBORN CIRCULATION
After birth, all arteries carry blood from the heart, and all veins carry it back to the heart. Thus, most arteries carry oxygenated blood, with the exception of the pulmonary arteries, which carry deoxygenated blood to the lungs. The circulations do not mix, so all vessels on this illustration are either red, denoting oxygenated blood, or blue, indicating deoxygenated blood.

CHANGES IN THE MOTHER

As with a newborn baby, physiological changes take place in the mother after birth that her body has prepared for during the pregnancy. The process of enhancing breast tissue in anticipation of breast-feeding begins early in the pregnancy: the breasts enlarge visibly and the alveoli in each of the milk-producing glands (lobules) swell and multiply. From three months into the pregnancy, the breasts are able to produce colostrum, a fluid rich in antibodies (which help protect a newborn from allergies and respiratory and gastrointestinal infections), water, protein, and minerals. After the birth, colostrum supplies a breast-fed baby with nutrition until the mother's milk begins to flow several days later. Soon after birth, the uterus begins to shrink to its prepregnancy size—a process that is helped along by breast-feeding.

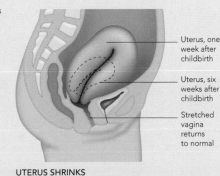

Uterus, one week after childbirth

Uterus, six weeks after childbirth

Stretched vagina returns to normal

UTERUS SHRINKS
After delivery of the baby and the placenta in the second and third stages of labor, hormones in the mother's body cause her uterus and vagina to shrink back to their normal size and position in her body.

BEFORE PREGNANCY

Lobule

DURING PREGNANCY AND LACTATION

New and enlarged lobule

LACTATION
During pregnancy, lobules (milk-producing glands) increase in size and number in preparation for breast-feeding the baby. By the end of the first trimester, they can produce colostrum, the yellow fluid that provides antibodies to protect against allergy and respiratory and gastrointestinal infections in the newborn.

FEMALE REPRODUCTIVE DISORDERS

PROBLEMS CAN ARISE IN ANY PART OF THE FEMALE REPRODUCTIVE TRACT AS WELL AS IN ONE OR BOTH BREASTS. MANY OF THE DISORDERS ARE HARMLESS, AND SOME ARE EVEN SYMPTOMLESS. HOWEVER, THE COMPLEX STRUCTURES OF THE FEMALE REPRODUCTIVE SYSTEM ARE SUBJECTED TO WIDE HORMONAL FLUCTUATIONS AND THE PHYSIOLOGICAL STRESSES OF PREGNANCY AND CHILDBIRTH. THEY ARE ALSO SUSCEPTIBLE TO MORE SERIOUS DISORDERS, INCLUDING VARIOUS TYPES OF CANCER.

BREAST LUMPS

A BREAST LUMP IS ANY SOLID OR SWOLLEN AREA THAT CAN BE FELT OR SEEN IN THE BREAST TISSUE; ONLY ABOUT 1 IN 10 BREAST LUMPS IS DUE TO CANCER.

Breast lumps are an extremely common problem and very few women can claim not to have found one at some time. Generalized breast lumpiness is especially common when breasts change shape during puberty, pregnancy, and in the days before menstruation. Nonspecific lumpiness may be associated with tenderness and is usually related to the hormonal fluctuations of the menstrual cycle; this is often known as fibrocystic disease. A single breast lump may be a fibroadenoma, which is an overgrowth of one or more milk-producing lobules; this condition is noncancerous. A more defined lump may be a breast cyst, which is a fluid-filled sac in the breast tissue. A painful lump in the breast may be an abscess, a collection of pus, caused by an infection. Only a small percentage of lumps are a symptom of breast cancer. It is very important for all women to become aware of their breast shape and how it changes through the menstrual cycle. Ideally, breast familiarity should start around the age of 20 and continue throughout life. There is no evidence that formalized self-examination has greater benefits in detecting cancer than a more relaxed approach of awareness. The emphasis is on knowing what is normal for the individual, what to be aware of, looking and feeling for changes, and reporting changes immediately. From the age of around 40, women should have a mammogram every 1–2 years.

Fibroadenoma
A common noncancerous breast lump

Cyst
One or more fluid-filled sacs within breast

Fatty tissue

Nonspecific lumpiness
Usually related to menstruation; often called fibrocystic disease

TYPES OF BREAST LUMP
Different types of breast lump cause varying amounts of pain and tenderness; often they are symptom-free. They may occur individually or in groups, and more than one type of lump may be present at the same time. Many types of noncancerous breast lumps do not require treatment.

BREAST CANCER

Cancer of the breast is the most common female cancer. The risk increases with age, doubling every 10 years. The causes are unclear, but a number of risk factors have been identified. The female hormone, estrogen, plays a role and women with higher exposure, for example through having an early puberty, late menopause, or no children, have a higher risk. Age is significant, with many more cases occurring over the age of 50. Faulty genes are also a known cause. A breast lump, usually painless, is often the first sign of breast cancer.

Cancerous tumor
Ragged, uneven borders are typical of a cancerous growth

BREAST CANCER
Mammogram of a female breast showing a tumor (white mass). Mammography is a special X-ray technique used to visualize breast tissue as a means of cancer screening.

ENDOMETRIOSIS

ENDOMETRIAL TISSUE FROM THE UTERUS CAN BECOME ATTACHED TO OTHER ORGANS IN THE PELVIC CAVITY.

Endometriosis affects up to 1 in 10 women of childbearing age. It can cause debilitating pain and very heavy menstrual periods; in severe cases, the condition can lead to fertility problems. The endometrium, the lining of the uterus, is shed approximately once every month as part of the menstrual cycle. Endometriosis is when small areas of endometrial tissue grow outside the uterus, most commonly on the ovaries and in the pelvis. These pieces of tissue respond to hormonal changes and bleed during menstruation. However, since the blood cannot leave the body through the vagina, its normal exit, it irritates nearby tissues, causing pain and eventually forming scars. The cause of the disorder is unknown.

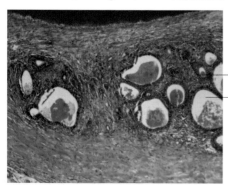

Bleeding

Enlarged gland

ENDOMETRIOSIS
A light micrograph view shows a section through the vaginal lining. The abnormal tissue responds to hormones and bleeds during menstruation.

CERVICAL CANCER

CANCER OF THE CERVIX IS A CANCEROUS GROWTH THAT OCCURS IN THE LOWER END OF THE CERVIX (NECK OF UTERUS).

Cancer of the cervix is one of the most common cancers diagnosed in women worldwide, although it is much less common in developed countries. Most cases of cervical cancer are now thought to be due to infection with the human papilloma virus (HPV), with those most at risk being smokers and women with a weakened immune system. Cell changes detectable at an early, precancerous stage by a cervical smear (Pap) screening test can be treated to prevent cancer developing. Regular screening is therefore a vitally important preventive measure. During a cervical smear test, a scraping of cells is taken from the surface of the cervix and these cells are examined under a microscope in the cytology laboratory. They may also be tested for the presence of HPV. If precancerous cells are found, the affected part of the cervix may be removed. There are a variety of surgical techniques to do this. In some cases a hysterectomy may be performed to remove the uterus.

CERVICAL CANCER
This microscope image of a cervical smear shows cells with abnormal growth (dysplasia). If not treated, these changes may potentially lead to cervical cancer.

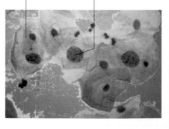

Abnormal cells

Enlarged nucleus
A sign of rapid cell division

OVARIAN CYSTS

FLUID-FILLED SWELLINGS THAT GROW ON OR IN ONE OR BOTH OF THE OVARIES ARE TERMED OVARIAN CYSTS.

Most ovarian cysts are noncancerous swellings filled with fluid that occur on or within an ovary. They are a common problem and are most likely to occur in women of childbearing age. Small cysts are often symptom-free, but if they grow they can press on nearby structures, causing problems such as abdominal pain and a frequent need to urinate. Many cysts disappear of their own accord, but larger ones, particularly if they are causing symptoms, may require surgical removal. Different types of cysts occur, but the most common is a follicular cyst. This occurs in one of the egg-producing follicles within the ovary and may grow to be 2 in (5 cm) across. Another type occurs in multiple form and is called polycystic ovary syndrome. Other less common forms are also found. Occasionally there are complications that require urgent medical attention. For example, the cyst may rupture or become twisted; it may grow extremely large, causing massive abdominal distension; or, very rarely, a cyst may undergo cell changes and develop into ovarian cancer.

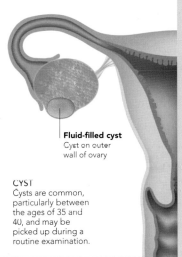

Fluid-filled cyst
Cyst on outer wall of ovary

CYST
Cysts are common, particularly between the ages of 35 and 40, and may be picked up during a routine examination.

OVARIAN CANCER

CANCER OF THE OVARY IS A MALIGNANT GROWTH THAT MAY DEVELOP IN EITHER ONE OR BOTH OVARIES.

Although cancer of the ovary is not the most common cancer of the female reproductive system, it causes more deaths per year than many of the other types because symptoms tend to occur only when the disease is advanced and has spread to other parts of the body. This makes treatment more complex and less likely to succeed. When symptoms do occur, they may include pain and swelling in the abdomen and the need to pass urine frequently. The cancer is more common between the ages of 50 and 70, and is extremely rare in women under the age of 40. Women who have never had children and those with a close relative with the disease are most at risk. Rarely, an ovarian cyst may develop into cancer. Presently, there is no effective screening for ovarian cancer. However, women in the high risk groups will be carefully monitored so, should the disease develop, it is caught at an early and treatable stage.

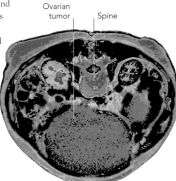

Ovarian tumor   Spine

CANCER OF THE OVARY
A CT scan of a section through a woman's body showing a large ovarian tumor (green, lower center). Also seen are the kidneys (yellow), spine (pink, center), ribs (pink at edges), and body fat (blue). A cancer of this size would cause symptoms due to pressure on surrounding structures and suggests that the disease has spread.

FIBROIDS

NONCANCEROUS TUMORS THAT OCCUR WITHIN THE WALL OF THE UTERUS ARE CALLED FIBROIDS.

Fibroids are very common, occurring in about one-third of women of childbearing age. They can occur singly or in groups and range in size from pea-sized to as large as a grapefruit. Small fibroids are unlikely to cause any problems, but larger ones may result in prolonged and heavy menstrual bleeding, and increasingly severe period pain. Large fibroids can distort the uterus, which may cause infertility, or put pressure on other organs, such as the bladder or rectum.

SITES OF FIBROIDS
Fibroids can occur in any part of the uterus wall and are named according to their site, for example in the cervix, or in the tissues they occur in.

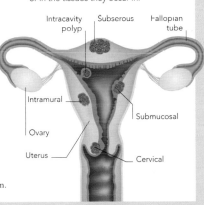

Intracavity polyp   Subserous   Fallopian tube

Intramural

Ovary

Uterus

Submucosal

Cervical

UTERINE CANCER

CANCER OF THE UTERUS OCCURS WHEN A TUMOR ORIGINATES IN THE LINING OF THE UTERUS (THE ENDOMETRIUM).

Uterine cancer is most likely to affect women between the ages of 55 and 65. Although the causes are unclear, there are definite risk factors, which include being overweight, a late menopause (after age 52), and not having children. In premenopausal women, symptoms are heavier-than-normal menstrual bleeding, or bleeding between periods or after intercourse; postmenopausal women may have renewed bleeding. In most cases, the treatment for cancer of the uterus is a hysterectomy.

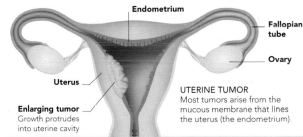

Endometrium

Fallopian tube

Ovary

Uterus

Enlarging tumor
Growth protrudes into uterine cavity

UTERINE TUMOR
Most tumors arise from the mucous membrane that lines the uterus (the endometrium).

PROLAPSED UTERUS

PROLAPSE OF THE UTERUS OCCURS WHEN THE LIGAMENTS AND MUSCLES HOLDING IT IN PLACE ARE WEAKENED, ALLOWING IT TO DISPLACE DOWNWARD.

Prolapse of the uterus is more likely to occur after menopause, when low estrogen levels affect the ability of the ligaments to retain the uterus. Childbirth, obesity, and straining while coughing or opening the bowels are contributing factors. The uterus protrudes down into the vagina, and in severe cases may travel as far as the vulva. Symptoms may include a feeling of fullness in the vagina, pain in the lower back, and difficulty urinating or opening the bowels.

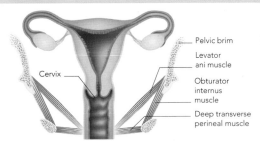

Cervix   Pelvic brim   Levator ani muscle   Obturator internus muscle   Deep transverse perineal muscle

Prolapsed uterus

Weakened muscle
Weakened muscle

NORMAL UTERUS
The uterus is kept in place by muscles and ligaments. Regular Kegel exercises are important to maintain their strength and avoid prolapse.

PROLAPSED UTERUS
In this case of uterine prolapse, the uterus has slipped down into the vagina. The wall of the vagina may also prolapse.

MALE REPRODUCTIVE DISORDERS

THE MALE REPRODUCTIVE TRACT IS SUBJECT TO A RANGE OF DISORDERS. THOSE AFFECTING THE EXTERNAL, VISIBLE PARTS ARE USUALLY APPARENT AT AN EARLY STAGE. HOWEVER, PROBLEMS AFFECTING INTERNAL PARTS OF THE SYSTEM, SUCH AS THE PROSTATE GLAND, MAY NOT BE NOTICED UNTIL A LATER STAGE, WHEN A SUCCESSFUL OUTCOME MAY BE MORE DIFFICULT TO ACHIEVE.

HYDROCELE

THE MEMBRANE SURROUNDING THE TESTIS CAN BECOME FILLED WITH FLUID CAUSING SWELLING, OR A HYDROCELE.

Each testis is surrounded by a double-layered membrane, which under normal conditions contains a small amount of fluid. In a hydrocele, an excessive amount of fluid forms, causing the testis to appear swollen. The condition occurs most frequently in infants and elderly people. The cause of hydrocele is not usually known, although infection, inflammation, or injury to the testis are possible triggers. A hydrocele does not usually cause any pain, but a dragging sensation due to the increased size and weight of the scrotum may be apparent. In younger people with hydrocele, the condition often gets better without the need for treatment. However, if the condition is causing discomfort, the hydrocele may be surgically removed or, for those who are not fit enough for surgery, the fluid may be drained from the area using a needle and syringe.

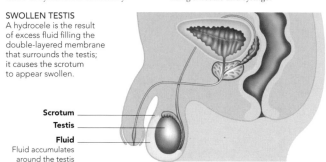

SWOLLEN TESTIS
A hydrocele is the result of excess fluid filling the double-layered membrane that surrounds the testis; it causes the scrotum to appear swollen.

Scrotum

Testis

Fluid
Fluid accumulates around the testis

TESTICULAR CANCER

CANCEROUS TUMORS GROWING WITHIN ONE OF THE TESTES COMMONLY AFFECT YOUNG MEN.

Cancer of the testis is one of the most commonly occurring cancers in men aged between 20 and 40. Although it is curable if discovered at an early stage, the cancer can spread to the lymph nodes and to other parts of the body if not treated. Symptoms of testicular cancer include a hard, painless lump in the testis; a change in the size and appearance of the testis; or a dull ache in the scrotum. There are two main types of testicular cancer, seminoma and non-seminoma. They develop from the sperm-producing cells of the testis. Since early treatment of the cancer is vital and has a very high cure rate, all men should regularly examine their testes; any swellings or changes in the scrotal skin should be reported urgently. Soft lumps or painful swellings are likely to be caused by a cyst or infection.

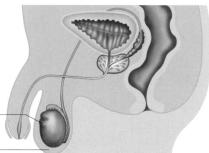

TUMOR ON TESTIS
A tumor of this size on the outer wall of the testis would be clearly felt through the thin outer skin and layers of the scrotum.

Tumor
A tiny growth on the testis

Scrotum

PROSTATE DISORDERS

CONDITIONS THAT AFFECT THE PROSTATE GLAND RANGE FROM INFLAMMATION AND BENIGN ENLARGEMENT TO SERIOUS DISORDERS, SUCH AS CANCER.

The prostate gland lies just beneath the bladder and surrounds the upper part of the urethra (the tube that connects the bladder with the penis). This chestnut-sized organ produces secretions that are added to the sperm-containing fluid, semen. Disorders affecting the prostate are very common and tend to occur in the middle and later years of a man's life. The most serious condition, prostate cancer, is a tumorous growth within the gland. Although potentially life threatening, it tends to occur most commonly in elderly men, in whom it often grows slowly and may never cause symptoms. However, new diagnostic techniques allow the condition to be detected in much younger men who need to receive treatment. Enlargement of the prostate gland is extremely common, and most men over the age of 50 have some degree of such growth. Although considered part of the aging process, the condition can cause distressing urinary symptoms if the gland constricts the urethra. Symptoms may include frequent urination, delay in starting to urinate, weak flow, dribbling, and a feeling of incomplete emptying. Prostatitis (see right) is a common condition, often caused by infection.

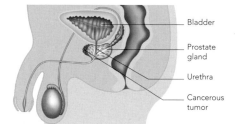

Bladder

Prostate gland

Urethra

Cancerous tumor

PROSTATE CANCER
A cancerous tumor of this size growing on the prostate gland is unlikely to cause immediate problems; but as it grows it may put pressure on the urethra, causing urinary symptoms, and may spread to other parts of the body.

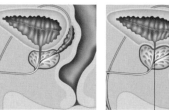

NORMAL PROSTATE

ENLARGED PROSTATE

ENLARGED PROSTATE
A normal prostate gland fits snugly around the urethra and abuts the bladder; enlargement can squash the urethra.

Enlarged prostate presses on urethra

PROSTATITIS

Inflammation of the prostate gland, or prostatitis, can be acute or chronic. The acute type is less common; severe symptoms come on suddenly, but these usually clear up quickly. Symptoms may include fever, chills, and pain around the base of the penis, in the lower back, and during defecation. Chronic prostatitis features longstanding but often mild symptoms that are difficult to treat, such as groin and penis pain, pain on ejaculation, blood in semen, and painful urination. Possibly caused by a bacterial infection from the urinary tract, both types are most common in men between 30 and 50 years.

CAUSATIVE BACTERIUM
This electron micrograph shows the bacterium *Enterococcus faecalis*, implicated in prostatits. It is a normal, harmless inhabitant of the human gut.

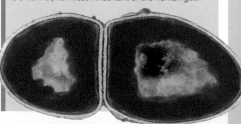

SEXUALLY TRANSMITTED INFECTIONS

SEXUALLY TRANSMITTED INFECTIONS (STI), ALSO KNOWN AS SEXUALLY TRANSMITTED DISEASES (STDS), ARE INFECTIONS THAT ARE PASSED FROM PERSON TO PERSON BY SEXUAL ACTIVITY. GENITAL, ANAL, AND ORAL SEX CAN ALL PASS ON AN INFECTION TO ANOTHER PERSON. STIS CAN USUALLY BE SUCCESSFULLY TREATED WITH DRUGS, AND "SAFE SEX" IS AN IMPORTANT PREVENTIVE MEASURE.

GONORRHEA

GENITAL INFLAMMATION CAUSED BY THE BACTERIUM *NEISSERIA GONORRHOEAE* IS KNOWN AS GONORRHEA.

Although gonorrhea tends to be more prevalent among males, it can also affect women. The main sites of infection are the urethra and, in women, the cervix. Symptoms often do not appear, but if they do commonly include a discharge of pus from the penis or vagina and pain on urination. Women may also experience lower abdominal pain and irregular vaginal bleeding. Occasionally the infection spreads to other parts of the body, such as the joints (via the bloodstream). If the disease is left untreated, it can cause infertility in women.

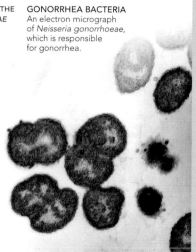

GONORRHEA BACTERIA
An electron micrograph of *Neisseria gonorrhoeae*, which is responsible for gonorrhea.

PELVIC INFLAMMATORY DISEASE (PID)

IN PID, THE FEMALE REPRODUCTIVE TRACT BECOMES INFLAMED, USUALLY AS A RESULT OF AN STI.

PID is a common cause of pelvic pain in young women; other possible symptoms are fever, heavy or prolonged menstrual periods, and pain during intercourse. Sometimes, there are no symptoms. Usually it is the result of an STI such as chlamydia or gonorrhea.

Infection after childbirth or a pregnancy termination are also possible causes. The inflammation starts in the vagina and spreads to the uterus and fallopian tubes. In severe cases, the ovaries are also infected. Left untreated, there may be damage to the fallopian tubes, causing infertility and an increased risk of ectopic pregnancy (see p.244).

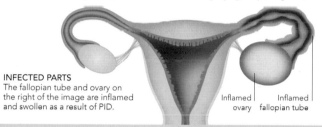

INFECTED PARTS
The fallopian tube and ovary on the right of the image are inflamed and swollen as a result of PID.

Inflamed ovary | Inflamed fallopian tube

NONGONOCOCCAL URETHRITIS

ALSO CALLED NONSPECIFIC URETHRITIS, THIS MALE STI IS CAUSED BY AN INFECTION OTHER THAN GONORRHEA.

Nongonococcal urethritis (NGU) is one of the most common STIs affecting men throughout the world. Typically, features include inflammation of the urethra (the tube leading from the bladder to the tip of the penis), with or without a discharge of pus; inflammation and soreness at the end of the penis; and pain when urinating, particularly when the urine is concentrated first thing in the morning. In about half of all

cases, the agent that is responsible for causing NGU is the bacterium *Chlamydia trachomatis*; this bacterium can also infect women, leading to chlamydial infection. Other possible causes of NGU include the bacterium *Ureaplasma urealyticum*; the protozoan *Trichomonas vaginalis*; the fungus *Candida albicans*; the genital warts virus (human papillomavirus, HPV); and the genital herpes viruses (herpes simplex viruses HSV1 and HSV2). It is important for both partners to seek treatment to prevent reinfecting one another. To effectively prevent STDs, sexually active people should limit their number of sexual partners, and use a condom for penetrative sex.

SYMPTOMS OF NGU
The main problem is inflammation of the urethra, which causes pain and soreness at the external opening on the penis and painful urination. If the infection spreads, the testis and epididymis may also become swollen.

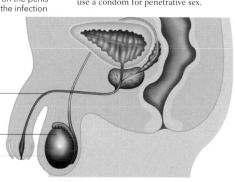

Urethra
Inflammation of urethra causes pain when urinating

Testis
May become swollen if infection spreads

Epididymis
Sometimes also becomes inflamed

SYPHILIS

A BACTERIAL INFECTION OF THE GENITAL ORGANS, SYPHILIS CAN AFFECT BOTH MALES AND FEMALES.

Infamous in history, syphilis has declined dramatically in the years since antibiotics became available. The causative bacterium, *Treponema pallidum*, enters the body by the genital routes and affects the reproductive organs; it spreads to other parts of the body and, if left

untreated, can cause death. The first sign is a highly infectious sore (chancre) on the penis or vagina, along with swollen lymph nodes. The next stage involves a rash and wart-like patches on the skin, with flu-like symptoms. With no treatment, it can proceed to a final, possibly fatal, stage characterized by personality changes, mental illness, and nervous system disorders. Today, the disease rarely progresses to this stage.

CHLAMYDIA INFECTION

INFECTION BY THE BACTERIUM *CHLAMYDIA TRACHOMATIS* CAUSES CHLAMYDIAL INFECTION IN WOMEN.

Chlamydial infection is a very common STI. It occurs only in women, although this bacterium causes nongonococcal urethritis in men. Invasion by bacteria causes inflammation of the reproductive organs, and symptoms include vaginal discharge, a frequent urge to urinate, lower abdominal pain, and pain during intercourse. Chlamydial infection can lead to pelvic inflammatory disease, if left untreated, and may then cause infertility. A swab taken from the cervix can reveal the presence of the bacteria.

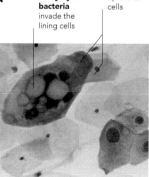

Chlaymydia bacteria invade the lining cells

Epithelial cells

BACTERIA IN CERVICAL SMEAR
This micrograph (x400) of a cervical smear shows *Chlamydia trachomatis* bacteria (pink cells within large blue cell).

INFERTILITY DISORDERS

IF A COUPLE IS UNABLE TO CONCEIVE AFTER A YEAR OF HAVING UNPROTECTED SEX, ONE OR BOTH
PARTNERS MAY HAVE A FERTILITY PROBLEM. THE LIKELIHOOD OF FERTILITY DISORDERS INCREASES
WHEN COUPLES WAIT UNTIL THEY REACH THEIR 30s OR 40s TO START A FAMILY, BY WHICH TIME
NATURAL FERTILITY IS IN DECLINE. FOR COUPLES UNWILLING TO ACCEPT CHILDLESSNESS,
DIFFERENT TYPES OF FERTILITY ASSISTANCE AND TREATMENT ARE AVAILABLE.

CAUSES OF FEMALE INFERTILITY

In about one in three infertility cases, the problem lies with the
woman's reproductive system. There may be a physical problem,
such as damage to a fallopian tube that prevents the egg from
reaching its destination; ovulation problems, in which an egg is not
released on a monthly basis; implantation problems, in which the
uterus has abnormalities that are incompatible with supporting a
fetus; or the cervix may be inhospitable to sperm. However, the
cause may be unknown or it may be a combination of problems.

DAMAGED FALLOPIAN TUBE

FALLOPIAN TUBE DAMAGE, WHICH
RESULTS IN SCARRING AND DISTORTION,
CAN PREVENT MOVEMENT OF EGGS.

The fallopian tube may become
blocked as a result of endometriosis, in
which fragments of the uterine lining
(endometrium) become embedded in
the tube tissue. Pelvic inflammatory
disease (PID), often caused by a sexually
transmitted infection, such as chlamydia
(see p.229), may go unnoticed at the
time of infection but scarring due to
the inflammation can cause problems
with fertility later. An intrauterine
contraceptive device can increase
the risk of PID developing. Usually,
only one tube is affected, and therefore
every other month the woman has a
chance to conceive.

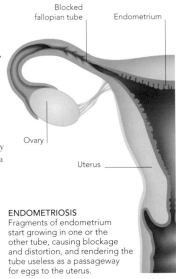

ENDOMETRIOSIS
Fragments of endometrium
start growing in one or the
other tube, causing blockage
and distortion, and rendering the
tube useless as a passageway
for eggs to the uterus.

ABNORMALITIES OF THE UTERUS

IMPLANTATION OF THE FERTILIZED
EGG MAY BE PREVENTED IF THERE IS A
PHYSICAL PROBLEM WITH THE UTERUS.

Although structural abnormalities
of the uterus are rare, they have the
potential to cause problems with fertility.
The uterus may not have developed
correctly during the fetal stage and
may be misshapen or malformed as
a result. Large and numerous
noncancerous tumors (fibroids) in
the muscular wall of the uterus can
encroach on the space within the
organ and cause the uterus to become
distorted. Surgery on the uterus or
pelvic inflammatory disease can also
affect the structure of the uterus and
possibly lead to problems with
conception later.

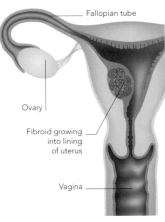

FIBROID
A very large, benign (noncancerous)
growth is seen pushing the uterine wall
inward with the result of reducing the
size of the cavity within the uterus and
distorting the shape of the organ.

CERVICAL PROBLEMS

THE CERVIX CAN CAUSE FERTILITY
PROBLEMS BY BEING HOSTILE TO SPERM
OR BECAUSE IT HAS A PHYSICAL DEFECT.

The cervix, or neck of the womb,
produces mucus that is usually
thick; just before ovulation, when
the level of estrogen increases, the
mucus turns less viscous to allow
sperm to penetrate. If estrogen levels
are low or if there is infection within
the reproductive tract, the mucus may
remain thick and impregnable to sperm.
Another problem that may make the
cervix inhospitable is that sometimes
a woman's immune system forms
antibodies to her partner's sperm,
which will then damage or kill the
sperm in the cervix. Polyps, fibroids,
narrowing (stenosis), and distortion
are other problems of the cervix that
may be related to infertility.

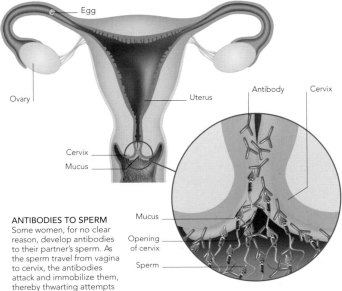

ANTIBODIES TO SPERM
Some women, for no clear
reason, develop antibodies
to their partner's sperm. As
the sperm travel from vagina
to cervix, the antibodies
attack and immobilize them,
thereby thwarting attempts
at fertilization.

OVULATION PROBLEMS

EGGS MAY NOT BE RELEASED AT ALL
OR ONLY INTERMITTENTLY, CAUSING
DIFFICULTY WITH CONCEPTION.

Ovulation is the release of a mature
egg ready for fertilization. Normally,
an egg is released from alternate ovaries
approximately every month. Any deviation
from this pattern has the potential
to cause problems with fertility. The
precise problem can range from complete
absence of egg release to infrequent
release. The problem is common because
it depends on a complex interaction
of hormones. Factors that influence
the hormones include pituitary and
thyroid gland disorders, polycystic
ovary syndrome, long-term use of oral
contraceptives, being very overweight
or underweight, excessive exercise,
and stress. Premature menopause
is another possible cause.

CAUSES OF MALE INFERTILITY

As with female infertility, male causes account for about one in three cases (with the remainder having no known cause). The problem may be with the quality of the sperm or with the transportation of sperm from the testes via the epididymides and the vas deferens before ejaculation. Problems with ejaculation, which may be a result of illness or psychological problems, prevent sperm from reaching the vagina either because erection cannot be achieved or maintained, or because of retrograde ejaculation.

PROBLEMS WITH SPERM PRODUCTION

SPERM MAY BE PRODUCED IN LOW QUANTITIES OR MAY BE DEFORMED OR UNABLE TO SWIM PROPERLY; ALL REDUCE THE LIKELIHOOD OF CONCEPTION.

Huge numbers of sperm must be produced in order for fertilization to occur; men in whom this does not happen are said to have a low sperm count. Microscopic examination can reveal this problem and can also look at the size, shape, and movement (motility) of individual sperm. Problems in any of these areas can cause reduced fertility. If only a small volume of semen is produced per ejaculation, fertility may also be reduced.

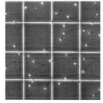

NORMAL SPERM COUNT

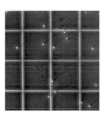

LOW SPERM COUNT

DIFFICULT PASSAGE OF SPERM

DISTORTION OR BLOCKAGE OF ANY OF THE TUBES THAT CARRY SPERM FROM TESTIS TO PENIS CAN REDUCE FERTILITY.

Sperm has a long and tortuous journey from its source in the testis until it is ejaculated. Narrowing, blockage, or other distortion of any of the tubes, including the epididymis and vas deferens, that make up this network can slow or completely block the passage of sperm. There are various causes of this problem, but infection of the male reproductive system is most likely. Some sexually transmitted infections (STIs, see p.241), most notably gonorrhea, can cause inflammation of the tubes, which leaves scar tissue that can distort their structure and affect their sperm-carrying ability.

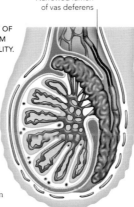

Narrowed lumen of vas deferens

INFLAMED VAS DEFERENS
Damage to the vas deferens, one of the tubes that transports sperm, can prevent or slow down its passage. Infection, usually by a sexually transmitted infection, can be responsible for such damage.

EJACULATION PROBLEMS

ERECTILE DYSFUNCTION AND RETROGRADE EJACULATION CAN BOTH AFFECT FERTILITY.

A number of ejaculation problems prevent sperm from arriving in the vagina by the normal means, making fertilization impossible. The most common of these is erectile dysfunction (difficulty in achieving or maintaining an erection), which may be a result of diabetes mellitus, a spinal cord disease, impaired blood flow, certain drugs, or psychological problems. Another problem, retrograde ejaculation, causes semen to flow back into the bladder because of faulty valves; this can be a complication of surgery for partial or complete removal of the prostate gland. Various treatments are available that can help reduce erectile dysfunction, depending on the nature of the problem.

IN-VITRO FERTILIZATION

A METHOD OF ASSISTED CONCEPTION, IN-VITRO FERTILIZATION BRINGS SPERM AND EGG TOGETHER OUTSIDE THE BODY.

Since the first "test-tube baby" was born through in vitro fertilization (IVF) in 1978, this method of assisted conception has become commonplace. IVF is performed if the fallopian tubes are blocked or if the cause of infertility cannot be found or treated. The method involves extracting eggs from an ovary. Eggs are artificially ripened by the use of hormones so that more than one is available for fertilization, increasing the chance of success. The eggs are then mixed with a sample of sperm from the woman's partner, or a donor, and incubated at normal body temperature for up to six days. One or two fertilized eggs are then injected directly into the woman's uterus through a thin tube that passes through the vagina and cervix. Treatment is successful if one or both of the eggs implants in the uterine wall. IVF results in pregnancy in up to 29 percent of attempts.

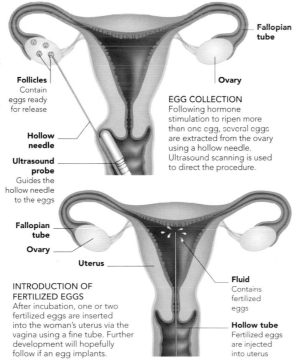

Fallopian tube

Ovary

Follicles
Contain eggs ready for release

Hollow needle

Ultrasound probe
Guides the hollow needle to the eggs

EGG COLLECTION
Following hormone stimulation to ripen more than one egg, several eggs are extracted from the ovary using a hollow needle. Ultrasound scanning is used to direct the procedure.

Fallopian tube

Ovary

Uterus

Fluid
Contains fertilized eggs

Hollow tube
Fertilized eggs are injected into uterus

INTRODUCTION OF FERTILIZED EGGS
After incubation, one or two fertilized eggs are inserted into the woman's uterus via the vagina using a fine tube. Further development will hopefully follow if an egg implants.

INTRACYTOPLASMIC SPERM INJECTION

A refined version of IVF, intracytoplasmic sperm injection (ICSI) can be used in male infertility when conventional assisted techniques have failed. A sperm cell is injected directly into a single mature egg in a laboratory. The procedure is very delicate and involves the use of microinstruments under a microscope. Success rates are up to 50 percent in one menstrual cycle, and only one embryo develops.

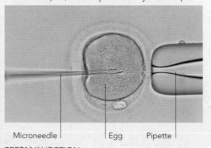

Microneedle   Egg   Pipette

SPERM INJECTION
This micrograph shows a sperm being injected into an egg. The round egg cell is being injected with a microneedle. The tip of a pipette holds the egg securely in place.

PREGNANCY AND LABOR DISORDERS

THE MAJORITY OF PREGNANCIES AND BIRTHS PROCEED WITHOUT ANY MAJOR PROBLEMS AND RESULT IN HEALTHY, FULL-TERM BABIES. HOWEVER, PROBLEMS THAT CAN ARISE IN NORMALLY HEALTHY WOMEN DURING THIS TIME MAY ENDANGER THE HEALTH OF BOTH MOTHER AND BABY. FEW DISORDERS OF PREGNANCY AND LABOR HAVE ANY PERMANENT PHYSICAL EFFECT ON EITHER MOTHER OR BABY.

ECTOPIC PREGNANCY

AN ECTOPIC PREGNANCY IS ONE THAT BEGINS IN A SITE OUTSIDE THE UTERUS, USUALLY IN A FALLOPIAN TUBE.

About 2 in 100 pregnancies is ectopic; this problem is more common in women under 30. The fertilized egg does not implant in the uterine lining but develops in one of the fallopian tubes, or more rarely in another area. Normal embryonic development is not possible and the pregnancy usually fails. The embryo must be surgically removed to avoid rupture of the fallopian tube and internal bleeding.

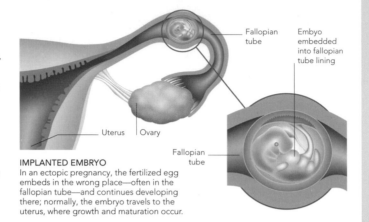

Fallopian tube

Embyo embedded into fallopian tube lining

Uterus | Ovary

Fallopian tube

IMPLANTED EMBRYO
In an ectopic pregnancy, the fertilized egg embeds in the wrong place—often in the fallopian tube—and continues developing there; normally, the embryo travels to the uterus, where growth and maturation occur.

PREECLAMPSIA

HIGH BLOOD PRESSURE AND FLUID RETENTION ARE CHARACTERISTIC OF THIS CONDITION OF PREGNANCY.

A condition that occurs in 5–10 percent of pregnancies, preeclampsia is particularly common in the weeks leading up to the birth. Features include high blood pressure, fluid retention, and protein in the urine. Preeclampsia is usually easy to treat but if left unchecked can proceed to a life-threatening problem called eclampsia, which can cause headaches, visual disturbances, seizures, and eventually coma.

PLACENTAL PROBLEMS

FUNCTIONAL OR POSITIONAL PROBLEMS OF THE PLACENTA PRIOR TO DELIVERY ARE TERMED PLACENTAL PROBLEMS.

Two main problems can affect the placenta: placenta previa, in which the placenta covers the opening of the cervix out of the uterus; and placental abruption, in which the placenta separates from the uterine wall. The degree of severity in placenta previa

depends on how much of the cervix is covered; it ranges from marginal, which may cause few problems, to complete, which is a serious conditon. Placental abruption usually comes on suddenly and can be life-threatening for the fetus because essential supplies are compromised. Both conditions can cause vaginal bleeding, but in less severe cases, symptoms may go unnoticed.

PLACENTA PREVIA
Complete placenta previa, as shown here, is a serious condition in which the cervix is covered entirely by the placenta. A less severe form involves a low-lying placenta that only partially obstructs the exit from the uterus.

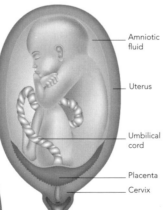

Amniotic fluid

Uterus

Umbilical cord

Placenta

Cervix

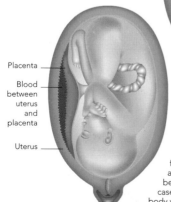

Placenta

Blood between uterus and placenta

Uterus

PLACENTAL ABRUPTION
Premature separation of the placenta from the uterine wall may be concealed, as shown here, in which case blood collects between the uterus and placenta. In other cases, the blood escapes and leaves the body via the vagina, revealing the problem.

MISCARRIAGE

ALSO CALLED SPONTANEOUS ABORTION, MISCARRIAGE IS THE UNINTENDED END OF A PREGNANCY BEFORE WEEK 24.

Miscarriage is very common, occurring in 1 in 4 pregnancies. Most miscarriages occur in the first 14 weeks of pregnancy and over half of these are due to a genetic or fetal abnormality. Later miscarriages have a variety of causes, ranging from physical problems with the cervix or uterus to severe infection. Smoking, alcohol, or drug abuse may also be factors. If three or more occur consecutively, it is known as recurrent miscarriage.

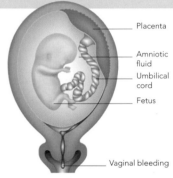

Placenta

Amniotic fluid

Umbilical cord

Fetus

Vaginal bleeding

THREATENED MISCARRIAGE
The fetus remains alive and the cervix is closed although there is some blood loss. It may proceed to full miscarriage, when the fetus dies, or a successful birth.

POLYHYDRAMNIOS

THIS CONDITION CAN OCCUR WHEN AN EXCESSIVE AMOUNT OF AMNIOTIC FLUID BATHES THE FETUS WITHIN THE UTERUS.

In polyhydramnios, excess amniotic fluid builds up in the uterus, causing abdominal pain or discomfort. Polyhydramnios may be chronic, with fluid accumulating slowly over several weeks or, much more rarely, acute, where the problem develops over a few days. Causes include maternal diabetes mellitus, multiple pregnancy, and fetal malformations. The excess fluid allows the baby to move more, increasing the likelihood of an abnormal presentation, and the chances of a premature labor.

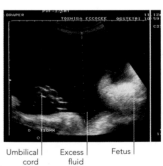

Umbilical cord

Excess fluid

Fetus

EXCESS AMNIOTIC FLUID
This ultrasound scan of a fetus indicates an excess of amniotic fluid, which can cause problems for both mother and baby.

ABNORMAL PRESENTATION

ANY DEVIATION FROM THE HEAD-DOWN, BACKWARD-FACING DELIVERY POSITION OF THE BABY IS CONSIDERED ABNORMAL.

Eighty percent of babies adopt the normal position for birth with the head down and facing toward the mother's back. The baby usually achieves this by week 36. Other babies are in a position that may cause problems during labor. Breech (see p.232) and occipitoposterior positions (see right) are the most common. In a breech birth, the baby's buttocks present first. Some presentations may allow the umbilical cord to drop through the birth canal and cause fetal distress. The cervix and vagina are more vulnerable to tears if the presentation is abnormal.

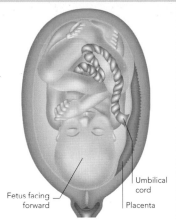

Fetus facing forward

Umbilical cord

Placenta

OCCIPITOPOSTERIOR POSITION
Although the baby's head is facing down, as is normal, the baby is turned 180° toward the front. The majority move to face the mother's back during the course of labor.

PRETERM LABOR

LABOR THAT BEGINS BEFORE THE 37TH WEEK OF PREGNANCY IS CALLED PRETERM, OR PREMATURE, LABOR.

Most pregnancies last for about 40 weeks, but delivery during the final three weeks is considered full term. Labor occurring before 37 weeks is preterm and results in a premature baby. Premature labor rarely causes maternal problems, but the earlier the birth, the greater the problems encountered by the baby. The cause is not always known, but multiple births, polyhydramnios (see opposite), and fetal abnormalities are known trigger factors. Sometimes premature labor can be halted or delayed, giving the baby more time in the womb.

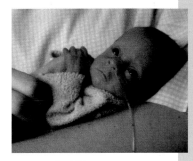

PREMATURE BABY
This premature baby is being fed through a nasogastric tube because his sucking reflex has not yet developed and swallowing ability is poor. Other features are his tiny size, wrinkled and yellow skin, and disproportionately large eyes.

PROBLEMS DURING DELIVERY

SEVERAL PROBLEMS CAN LENGTHEN THE SECOND STAGE OF LABOR OR PREVENT THE PROGRESSION OF NORMAL DELIVERY.

The second stage of labor, or delivery, starts when the cervix has reached full dilation at 4 in (10 cm) and ends with the birth of the baby. Problems at this stage may happen, particularly in a first pregnancy. Some originate in the first stage of labor. These include weak uterine contractions and abnormal presentation, so that the fetus cannot put pressure on the cervix to help dilation. Also, a lengthy first stage can exhaust the mother to such an extent that she has little strength to push in the second stage. Other problems arise in the second stage itself. The baby's passage through the birth canal may be delayed because it is not in the optimal position for delivery. There may be a problem with the baby actually passing through the pelvis; this could be because the baby is especially large or because the mother has a small or irregular pelvis. Once the baby has reached the vaginal opening, problems with delivery may occur if the tissues cannot stretch sufficiently to let the head out. In spite of these potential problems, a normal or assisted vaginal delivery is often possible; however, under certain circumstances, a cesarean section (see right) may be the only option.

CESAREAN SECTION

If delivery is proving difficult, in some cases with multiple births, or if the mother has a good medical reason for avoiding vaginal delivery, a cesarean section is performed. This involves removing the baby and the placenta from the uterus through an incision in the lower abdomen. The procedure is often carried out under epidural anesthesia so that the mother remains conscious and can interact with her baby soon after birth.

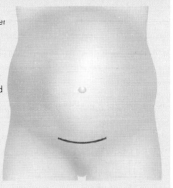

HORIZONTAL INCISION
An incision is made just below the pubic hairline, through which the surgeon can remove baby and placenta.

ASSISTED DELIVERY

IF DELIVERY IS NOT PROCEEDING SMOOTHLY OR QUICKLY ENOUGH, ONE OF TWO TYPES OF ASSISTED DELIVERY, EITHER VACUUM SUCTION OR FORCEPS, MAY BE USED.

Assisted delivery means physically helping the baby out of the womb through the birth canal. It may be necessary if the mother is too exhausted to push the baby out or if the baby becomes stuck or distressed. The possible methods involve vacuum suction or forceps. In each case, the instrument is used to pull the head clear of the vaginal opening, after which the delivery will proceed as normal. To enlarge the birth opening to allow entry of the instrument and make extracting the head easier, an episiotomy is usually performed. Under local anesthetic, an incision is made in the perineum (the tissue between the vagina and anus). Cutting the tissue avoids the ragged tear that may result in the absence of intervention. The cut is made at an angle in order to avoid the anus.

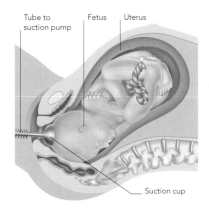

Tube to suction pump

Fetus

Uterus

Suction cup

VACUUM SUCTION DELIVERY
A suction cup is placed on the baby's head and its connecting tube attached to a pump that is switched on. With each contraction, the obstetrician gently pulls the baby toward the outside world.

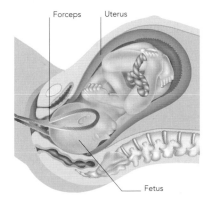

Forceps

Uterus

Fetus

FORCEPS DELIVERY
Spoon-shaped obstetric forceps are carefully placed around the baby's head. As the mother pushes, the obstetrician pulls gently on the forceps until the baby's head reaches the vagina.

FROM AN ENERGETICALLY KICKING, HIGHLY VOCAL NEWBORN
BABY TO THE DIMINISHING PHYSICALITY AND FADING FACULTIES
OF OLD AGE, THE TYPICAL HUMAN BODY FOLLOWS A FAMILIAR
PATH THROUGH LIFE. INHERITANCE AND EARLY CIRCUMSTANCES

GROWTH AND DEVELOPMENT

DEVELOPMENT AND AGING

THE BODY PASSES THROUGH SEVERAL STAGES DURING ITS LIFETIME. THE INITIAL PERIOD IS ONE OF GROWTH AND MATURATION; THIS IS FOLLOWED BY A PERIOD OF RELATIVE STABILITY, THEN INCREASING SENESCENCE, OR DETERIORATION, LEADING FINALLY TO DEATH.

CHANGES THROUGH LIFE

Every body develops and functions according to the set of instructions in its genetic material. This blueprint is unique to each individual (except between identical twins) and is the foundation on which other factors impinge. Chief among these are the environment and lifestyle. Theories to explain why we age include mistakes when copying DNA as cells reproduce, mounting misrepair of cellular machinery, and accumulation of debris due to the effects of substances such as free radicals. The way each body ages may also be programmed: so-called "aging genes" have been identified that can be switched on or off by environmental and lifestyle factors such as diet.

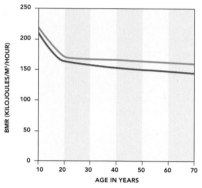

METABOLIC RATE AND AGE
As the body ages, it requires less energy to function. This can be seen in the basal metabolic rate (BMR)—an expression of the amount of energy (measured in kilojoules) needed to fuel the totality of its chemical processes, adjusted for body size (measured in m³).

DISORDERS AND AGING

It may be difficult to untangle "normal" aging processes from the "pathological" changes of recognized disorders. For example, in the brain, blockages in small blood vessels occur more frequently with age, and the blockages may cause death of brain tissue in areas deprived of blood. As these areas of dead tissue accumulate, signs of dementia may occur, but such signs may also be combined with the appearance of amyloid plaques as part of Alzheimer's disease.

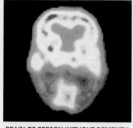

BRAIN OF PERSON WITHOUT DEMENTIA   BRAIN OF PERSON WITH DEMENTIA

BRAIN ACTIVITY IN DEMENTIA
These PET scans show metabolic activity in the brain of a person without dementia (left) and the brain of a person with dementia (right). Red and yellow indicate high activity; blue and purple indicate low activity; black indicates minimal or no activity.

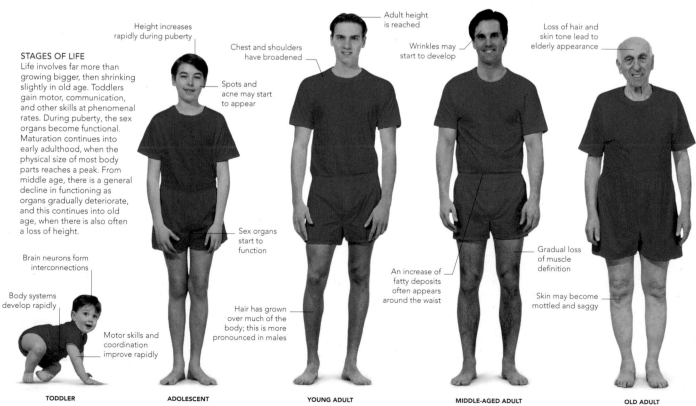

STAGES OF LIFE
Life involves far more than growing bigger, then shrinking slightly in old age. Toddlers gain motor, communication, and other skills at phenomenal rates. During puberty, the sex organs become functional. Maturation continues into early adulthood, when the physical size of most body parts reaches a peak. From middle age, there is a general decline in functioning as organs gradually deteriorate, and this continues into old age, when there is also often a loss of height.

Brain neurons form interconnections

Body systems develop rapidly

Motor skills and coordination improve rapidly

Height increases rapidly during puberty

Spots and acne may start to appear

Sex organs start to function

Hair has grown over much of the body; this is more pronounced in males

Adult height is reached

Chest and shoulders have broadened

Wrinkles may start to develop

An increase of fatty deposits often appears around the waist

Loss of hair and skin tone lead to elderly appearance

Gradual loss of muscle definition

Skin may become mottled and saggy

TODDLER   ADOLESCENT   YOUNG ADULT   MIDDLE-AGED ADULT   OLD ADULT

GENETIC INFLUENCES

There is a vast range of genetic conditions that affect growth and development, and some may also shorten lifespan. It has long been observed that general longevity, and also good health into old age, tends to run in families. This could indicate genetic influences but could also be due to external factors such as a healthy lifestyle, because these also tend to run in families. Research studies on the families of long-lived people, and especially on identical and non-identical twins, have shown that there is a strong genetic component to longevity. Further research has uncovered "life-lengthening" variants of already-known genes. Some of these appear to work by suppressing other "life-shortening" genes that predispose to disorders such as heart problems.

GENETIC GROWTH DISORDER
The 11-year-old girl in the middle has dyschondrosteosis, a growth disorder due to an altered gene. She is short compared to her sisters aged four (left) and 13 (right).

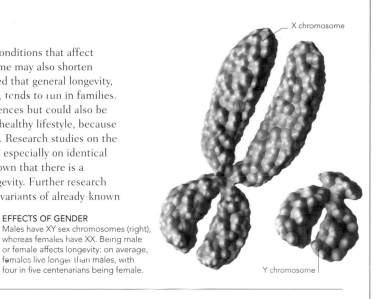

X chromosome

Y chromosome

EFFECTS OF GENDER
Males have XY sex chromosomes (right), whereas females have XX. Being male or female affects longevity: on average, females live longer than males, with four in five centenarians being female.

ENVIRONMENTAL INFLUENCES

Numerous environmental conditions affect development and aging, from drinking water quality to radioactive exposure. Air polluted by traffic exhaust fumes may contain a cocktail of chemicals that affect the lungs, heart, and brain and may even trigger certain forms of cancer. The sun's ultraviolet rays (UVA or UVB) are estimated to cause 90 percent of the symptoms of prematurely aged skin, especially wrinkles—and much of this damage happens in childhood, well before it becomes apparent. Genetics is also a factor because, compared to light-colored skin, dark skin is partly protected by its pigmentation and so is at lower risk of aging signs, skin growths, and cancers such as melanoma.

EFFECTS OF SUN EXPOSURE
Although a suntan may be regarded as healthy, overexposure to sunlight makes skin age early, with furrows, deeper wrinkles, and blemishes. It also increases the chance of developing skin cancer.

EFFECTS OF AIR POLLUTION
Among the harmful agents in polluted air are tiny floating specks known as particulates, mainly from vehicle exhausts and fossil fuels. Lodging deep in the lungs, they may contribute to asthma and lung cancer.

LIFESTYLE INFLUENCES

As a person matures, especially from adolescence to early adulthood, he or she also becomes socially and intellectually more independent. Depending on culture and tradition, he or she assumes more control over lifestyle choices, such as stress levels, time for relaxation, and risk taking. Diet has a major impact on most aspects of health and aging. It affects the way the digestive system processes food, and influences blood cholesterol levels. The most important lifestyle factors to promote wellness and delay the onset of aging include having a healthy body weight, a balanced diet, avoiding alcohol or drinking in moderation, doing regular physical exercise, and not smoking.

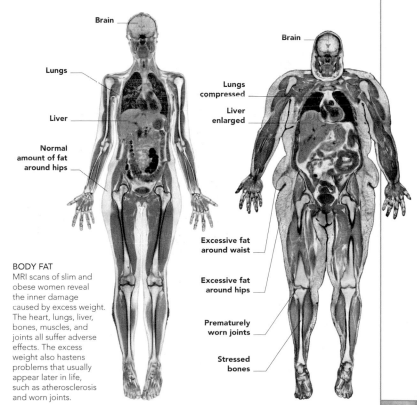

Brain

Lungs

Liver

Normal amount of fat around hips

Brain

Lungs compressed

Liver enlarged

Excessive fat around waist

Excessive fat around hips

Prematurely worn joints

Stressed bones

BODY FAT
MRI scans of slim and obese women reveal the inner damage caused by excess weight. The heart, lungs, liver, bones, muscles, and joints all suffer adverse effects. The excess weight also hastens problems that usually appear later in life, such as atherosclerosis and worn joints.

THE BODY THROUGH LIFE

FROM BIRTH TO DEATH, THE BODY UNDERGOES NUMEROUS CHANGES. MANY OF THESE OCCUR AT ROUGHLY THE SAME TIME DURING LIFE, ALTHOUGH THERE ARE SOME DIFFERENCES BETWEEN INDIVIDUALS AND BETWEEN DIFFERENT PARTS OF THE WORLD.

The effects of development and aging can be seen in all the systems of the body (see chart below). While some changes are common to both men and women, others, particularly those affecting the reproductive system, are gender-specific. Little is known with certainty about why we age. However, it is clear that cells, the body's building blocks, change over time. After they have divided a fixed number of times, their ability to function properly starts to decline. This means that connective tissue becomes less flexible, making organs, blood vessels, and airways less efficient. Cell membranes also change, so tissues are less effective at receiving oxygen and nutrients and at getting rid of wastes.

KEY

- Skeletal system
- Muscular system
- Nervous system
- Endocrine system
- Cardiovascular system
- Respiratory system
- Skin, hair, and nails
- Lymphatic and immune systems
- Digestive system
- Urinary system
- Reproductive system

♂ Male only   ♀ Female only

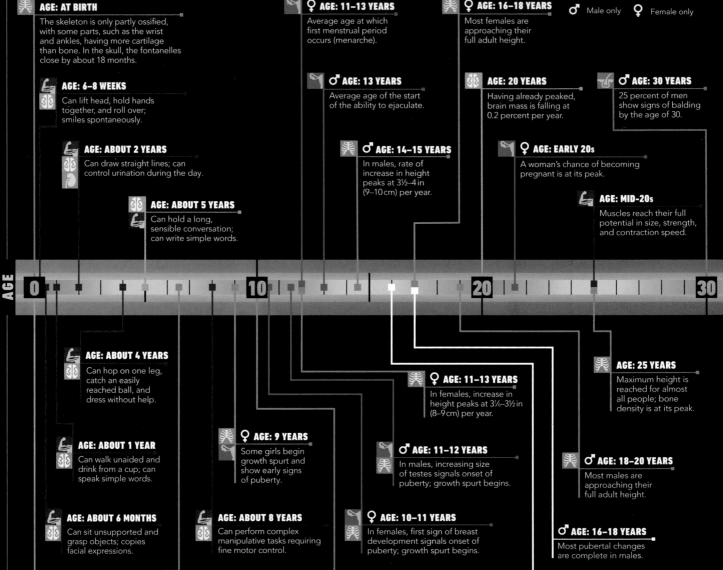

AGE: AT BIRTH
The skeleton is only partly ossified, with some parts, such as the wrist and ankles, having more cartilage than bone. In the skull, the fontanelles close by about 18 months.

AGE: 6–8 WEEKS
Can lift head, hold hands together, and roll over; smiles spontaneously.

AGE: ABOUT 2 YEARS
Can draw straight lines; can control urination during the day.

AGE: ABOUT 5 YEARS
Can hold a long, sensible conversation; can write simple words.

♀ AGE: 11–13 YEARS
Average age at which first menstrual period occurs (menarche).

♂ AGE: 13 YEARS
Average age of the start of the ability to ejaculate.

♂ AGE: 14–15 YEARS
In males, rate of increase in height peaks at 3½–4 in (9–10 cm) per year.

♀ AGE: 16–18 YEARS
Most females are approaching their full adult height.

AGE: 20 YEARS
Having already peaked, brain mass is falling at 0.2 percent per year.

♀ AGE: EARLY 20s
A woman's chance of becoming pregnant is at its peak.

♂ AGE: 30 YEARS
25 percent of men show signs of balding by the age of 30.

AGE: MID-20s
Muscles reach their full potential in size, strength, and contraction speed.

AGE
0   10   20   30

AGE: ABOUT 4 YEARS
Can hop on one leg, catch an easily reached ball, and dress without help.

AGE: ABOUT 1 YEAR
Can walk unaided and drink from a cup; can speak simple words.

AGE: ABOUT 6 MONTHS
Can sit unsupported and grasp objects; copies facial expressions.

AGE: AT BIRTH
A baby is born able to see, hear, and perform reflex actions such as sucking, grasping, urinating, and defecating.

♀ AGE: 9 YEARS
Some girls begin growth spurt and show early signs of puberty.

AGE: ABOUT 8 YEARS
Can perform complex manipulative tasks requiring fine motor control.

AGE: 6–7 YEARS
Height reaches about two-thirds of adult stature; first adult teeth appear.

♀ AGE: 11–13 YEARS
In females, increase in height peaks at 3¼–3½ in (8–9 cm) per year.

♂ AGE: 11–12 YEARS
In males, increasing size of testes signals onset of puberty; growth spurt begins.

♀ AGE: 10–11 YEARS
In females, first sign of breast development signals onset of puberty; growth spurt begins.

♂ AGE: 10 YEARS
Some boys begin growth spurt and show early signs of puberty.

AGE: 25 YEARS
Maximum height is reached for almost all people; bone density is at its peak.

♂ AGE: 18–20 YEARS
Most males are approaching their full adult height.

♂ AGE: 16–18 YEARS
Most pubertal changes are complete in males.

♀ AGE: 15–17 YEARS
Most pubertal changes are complete in females.

LIFE EXPECTANCY

There are many factors that affect life expectancy. Women usually survive longer than men, probably due to the protective effects of hormones released before menopause. Around the world, average life expectancy varies, from less than 50 years in parts of Africa to more than 80 in Japan, Canada, Australia, and parts of Europe. This is due to genetic tendencies, lifestyle factors, resources, and the prevalence of infectious diseases. Historically, lifespans have been increased by improvements in sanitation, healthcare, and nutrition, among other factors.

LIFE EXPECTANCY AROUND THE WORLD
This chart shows the life expectancy of people living in the world's 25 most highly populated countries. Life expectancy is lowest in poor countries and those affected by war, and highest in the developed world.

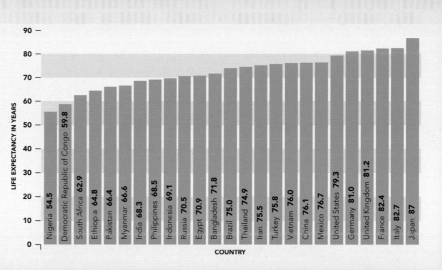

LIFE EXPECTANCY IN YEARS / COUNTRY

Nigeria 54.5 · Democratic Republic of Congo 59.8 · South Africa 62.9 · Ethiopia 64.8 · Pakistan 66.4 · Myanmar 66.6 · India 68.3 · Philippines 68.5 · Indonesia 69.1 · Russia 70.5 · Egypt 70.9 · Bangladesh 71.8 · Brazil 75.0 · Thailand 74.9 · Iran 75.5 · Turkey 75.8 · Vietnam 76.0 · China 76.1 · Mexico 76.7 · United States 79.3 · Germany 81.0 · United Kingdom 81.2 · France 82.4 · Italy 82.7 · Japan 87

 AGE: EARLY 40s
Bones start to lose mineral content; gradual loss of height begins; in males, testosterone level starts to decline.

♀ **AGE: 50–52 YEARS**
Most women pass through the menopause.

AGE: 60–65 YEARS
Loss of muscle mass accelerates; total blood volume is 20–25 percent lower than that at age 25; sense of smell starts to diminish noticeably.

AGE: EARLY 80s
The heart's pumping capacity is about 40–50 percent of that at age 25; four in five people have noticeable hearing loss.

AGE: 30 YEARS
The reserve capacity of most organs starts to decline by about 1 percent per year.

♂ **AGE: MID-50s**
Number of taste buds starts to decline significantly in males.

AGE: EARLY 70s
The heart's pumping capacity is about 70–75 percent of that at age 30.

AGE: 30 YEARS
The prefrontal cortex of the brain is now fully developed, enabling improved decision-making.

♂ **AGE: EARLY 50s**
About half of men show signs of male-pattern baldness; red-blood-cell level starts to decline in males.

♂ **AGE: 79.0 YEARS**
Highest average life expectancy at birth for males born 2000–2010 (Japan).

♀ **AGE: 40 YEARS**
A woman's chance of becoming pregnant has more than halved since the age of 20.

♀ **AGE: LATE 40s**
Menstrual cycles start to get shorter.

♀ **AGE: 69.5 YEARS**
Average worldwide life expectancy at birth for females born 2000–2010.

AGE: MID-40s
Hearing loss affects about one in five people; bile production starts to decline.

♂ **AGE: 65.0 YEARS**
Average worldwide life expectancy at birth for males born 2000–2010.

♀ **AGE: 86.1 YEARS**
Highest average life expectancy at birth for females born 2000–2010 (Japan).

AGE: 42–44 YEARS
Vision problems due to presbyopia become noticeable; females start to lose significant numbers of taste buds.

AGE: 60 YEARS
About one in five people lost all their natural teeth; red-blood-cell level begins to decline in females.

♂ **AGE: 90 YEARS**
In males, total muscle mass is almost halved from that at age 20.

CHILDHOOD

THE FIRST YEARS OF LIFE ARE A PERIOD OF RAPID GROWTH AND DEVELOPMENT AS PHYSICAL SIZE INCREASES, ORGANS AND BODY SYSTEMS MATURE, AND BASIC PHYSICAL SKILLS, SUCH AS MANIPULATION, WALKING, AND TALKING, AND INTELLECTUAL ABILITIES DEVELOP.

BONES AND MUSCLES

Growth is fastest during infancy—usually defined as the period from birth to 12 months. As an approximate guide, birth height doubles during the first two years; and birth weight doubles by four months and triples by one year, although weight gain tends to vary more between individuals. Most height gain is in the long bones of the legs. These elongate at specialized regions known as epiphyseal, or growth, plates. Before birth, most bones form as cartilage, which gradually hardens, or ossifies, into true bone at primary ossification centers. After birth, secondary ossification occurs near the bones' ends. Muscles also grow rapidly, and their contours gradually become more defined as the amount of body fat diminishes.

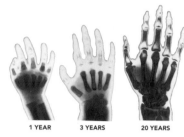

1 YEAR **3 YEARS** **20 YEARS**

CARTILAGE TO BONE
These X-rays of the hand and wrist show bony tissue as darker pink, blue, or purple. At one year, the bones in the wrist and fingers are largely made of cartilage. At three years, ossification of the cartilage is well under way, and by 20 years the bones are fully formed.

ORGAN MATURATION
Many of the organs and body systems are relatively undeveloped at birth and need to mature before they can function fully. For example, connections between neurons in the nervous system have to form for motor coordination, and enzyme production by the digestive system has to increase to enable solid food to be digested.

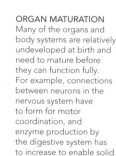

Brain
Connections between neurons form rapidly

Teeth
Primary teeth gradually replaced by adult teeth

Thymus
An important part of the immune system in childhood where T cells (a type of white blood cell) mature

Muscles
Grow longer and bulkier. Muscular coordination improves as the nervous system develops

Digestive system
Matures rapidly to enable solid foods to be digested from about 6 months old

Bones
Grow and ossify

Articular cartilage
Smooth tissue protecting the bone end

Secondary ossification center

Epiphysis (head)

Epiphyseal (growth) plate

Periosteum

Epiphyseal (growth) plate
Produces new cartilage

Epiphyseal line
Ossifies during late adolescence, marking the end of bone lengthening

Blood vessel

Marrow cavity

Diaphysis (shaft)

LONG BONE OF A NEWBORN
The shaft, or diaphysis, is turning to hard bone from the primary ossification center in the middle of the developing bone, and has a marrow cavity. The bulbous end or head, the epiphysis, is entirely cartilage, and relatively soft.

LONG BONE OF A CHILD
A secondary ossification center inside the epiphysis begins to change the surrounding cartilage to hardened, mineralized bone tissue. The epiphyseal plate between the shaft and the head of the bone produces new cartilage.

LONG BONE OF AN ADULT
By about 18–20 years, all zones have hardened into true bone, with the epiphyseal plate represented by a line of dense, bony tissue. The only remaining cartilage is smooth, slippery articular cartilage, which covers the articular surface of the bone.

AGE **0** **5**

AGE: 6–8 WEEKS
Can lift head, hold hands together, and roll over; smiles spontaneously.

AGE: ABOUT 6 MONTHS
Can sit unsupported and grasp objects; copies facial expressions.

AGE: ABOUT 1 YEAR
Can walk unaided and drink from a cup; can speak simple words.

AGE: ABOUT 2 YEARS
Can control urination during the day; can draw straight lines.

AGE: ABOUT 4 YEARS
Can hop on one leg, catch an easily reached ball, and dress without help.

THE NERVOUS SYSTEM

The size of a fetus's head is limited by having to fit through the mother's pelvis during birth. At this stage, the brain is one-quarter of its adult size; it has almost its full complement of nerve cells, or neurons, but these have not yet developed large numbers of interconnections. After birth, the brain's growth speeds up. Just two years later, it is already four-fifths of its adult size, and millions of interconnections have been made to establish nerve pathways—for example, to control muscles for motor skills. The basic structure of the brain is complete by three years but some parts, such as regions of the prefrontal and parietal cortex, continue to mature and develop. Maturation of the reticular formation in the brainstem leads to less distractibility and a lengthening attention span.

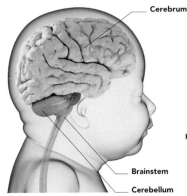

BRAIN AT BIRTH
The pattern of bulging ridges (gyri) and grooves (sulci) on the large upper cerebrum is increasing in complexity. The cerebellum, involved in coordination, is less developed.

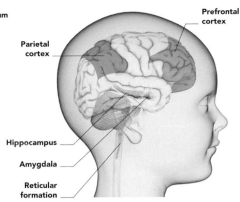

BRAIN AT THREE YEARS
Due to development of the hippocampus and amygdala, the ability to retain memories is increasing swiftly. Attention span also lengthens as the reticular formation matures.

THE DIGESTIVE AND IMMUNE SYSTEMS

For the first few months after birth, a baby produces a relatively low level of digestive enzymes, as a result of which it cannot properly digest solid food and must be fed on milk. The digestive system is usually mature enough to accept selected solids from about four to six months. The immune system is also immature at birth, although certain types of disease-fighting antibodies have already been transferred from the mother through the placenta. After birth, antibodies in breast milk help the infant's gut lining resist harmful microbes. These antibodies provide passive immunity until, from about three months, the infant's immune system makes more of its own antibodies. During childhood, the thymus also plays an important role in immunity, helping to produce mature white blood cells called T cells (see p.177).

VACCINATION
In most regions, infants and children are vaccinated (usually by injection) against a range of infections. Vaccines help protect against infection by stimulating the body's immune system.

DENTAL DEVELOPMENT

The first set of teeth, known as the primary or deciduous dentition, erupts through the gums in a set order, from about six months into the third year. In general, apart from the canines, the teeth appear from front to back. However, the exact times and order vary, and occasionally a baby is born with one or more teeth. These primary teeth loosen and fall out as the adult, or permanent, teeth erupt through the gums. This starts at about six years of age. The set of 32 permanent teeth is complete when the third molars (wisdom teeth) appear in the late teens or early twenties. In some people, however, the third molars never erupt.

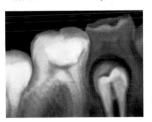

TOOTH ERUPTION
This colored X-ray shows a permanent, tooth (green) erupting under a child's primary teeth.

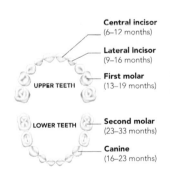

PRIMARY DENTITION
The baby teeth consist of two incisors, one canine, and two molars, in each half of each jaw, totaling 20 teeth. The age at which they erupt is given above. The teeth are shed from front to back.

Central incisor (6–12 months)
Lateral Incisor (9–16 months)
First molar (13–19 months)
Second molar (23–33 months)
Canine (16–23 months)
UPPER TEETH
LOWER TEETH

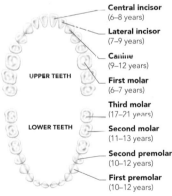

ADULT DENTITION
The first to erupt are the first molars and the central incisors, from about 6 years. There are 32 teeth in the complete set.

Central incisor (6–8 years)
Lateral incisor (7–9 years)
Canine (9–12 years)
First molar (6–7 years)
Third molar (17–21 years)
Second molar (11–13 years)
Second premolar (10–12 years)
First premolar (10–12 years)
UPPER TEETH
LOWER TEETH

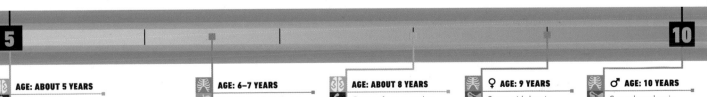

5 | 10

AGE: ABOUT 5 YEARS
Can hold a long, sensible conversation; can write simple words.

AGE: 6–7 YEARS
Height reaches about two-thirds of adult stature; first adult teeth appear.

AGE: ABOUT 8 YEARS
Can perform complex manipulative tasks requiring fine motor control.

♀ **AGE: 9 YEARS**
Some girls begin growth spurt and show early signs of puberty.

♂ **AGE: 10 YEARS**
Some boys begin growth spurt and show early signs of puberty.

ADOLESCENCE

ADOLESCENCE IS THE TRANSITIONAL PERIOD BETWEEN CHILDHOOD AND ADULTHOOD. ITS KEY EVENT IS PUBERTY, THE PERIOD OF SEXUAL AND PHYSICAL MATURATION THAT TYPICALLY BEGINS AT AROUND 10 OR 11 IN GIRLS, AND 12 OR 13 IN BOYS.

PUBERTY

Puberty is the period in which the reproductive system matures and begins to function. It is also accompanied by other changes, including a growth spurt and the appearance of secondary sexual features, such as pubic hair and, especially in boys, deepening of the voice. Puberty is triggered by gonadotropin-releasing hormone (GnRH) from the hypothalamus. This stimulates the pituitary gland to release luteinizing hormone (LH) and follicle-stimulating hormone (FSH). These, in turn, stimulate the sex organs to produce sex hormones, which are responsible for the changes of puberty in both sexes.

GROWTH SPURT
Puberty usually begins earlier in girls, who start to grow rapidly in height before boys begin their pubertal growth spurt. However, growth is more marked and lasts longer in boys, as a result of which men end up taller, on average, than women.

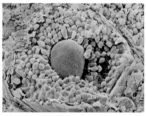

OVARIAN FOLLICLES

SEMINIFEROUS TUBULES

EGG AND SPERM PRODUCTION
At puberty, the ovaries start to form mature follicles (the collection of blue cells in the top micrograph), each containing a ripe egg (red). In the seminiferous tubules of the testes, sperm (green cells in the lower micrograph) develop and move away from their supporting cells (orange).

Brain
Gray matter decreases in volume due to loss of rarely-used neural pathways

Larynx
Lengthening of cartilage in the larynx and thickening of tissue in the vocal cords causes deepening of the voice, most noticeable in males

Skin
Skin and hair become oily, due to hormonal changes

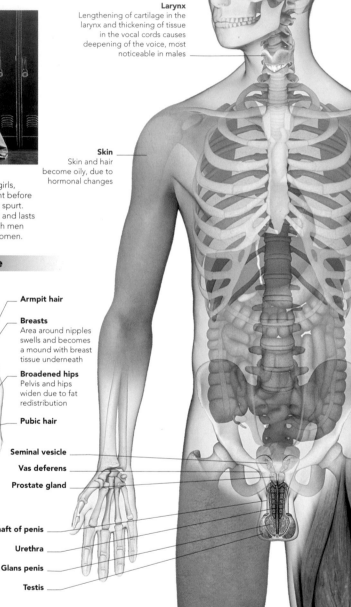

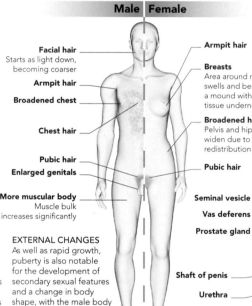

Male | Female

Facial hair
Starts as light down, becoming coarser

Armpit hair

Broadened chest

Chest hair

Pubic hair

Enlarged genitals

More muscular body
Muscle bulk increases significantly

Armpit hair

Breasts
Area around nipples swells and becomes a mound with breast tissue underneath

Broadened hips
Pelvis and hips widen due to fat redistribution

Pubic hair

Seminal vesicle

Vas deferens

Prostate gland

Shaft of penis

Urethra

Glans penis

Testis

EXTERNAL CHANGES
As well as rapid growth, puberty is also notable for the development of secondary sexual features and a change in body shape, with the male body becoming more muscular and the female body becoming more rounded.

AGE

10

15

♀ **AGE: 10–11**
First signs of breast development signals onset of puberty in females; growth spurt begins.

♂ **AGE: 11–12**
Increasing size of testes signals the onset of puberty in males; growth spurt begins.

♀ **AGE: 11–13**
Increase in height peaks at 3⅛–3½ in (8–9 cm) per year.

♀ **AGE: 11–13**
Average age at which first menstrual period occurs (menarche).

♂ **AGE: 13**
Average age of onset of the ability to ejaculate.

♂ **AGE: 14–15**
Increase in height peaks at 3½–4 in (9–10 cm) per year.

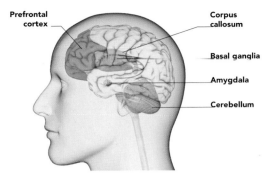

Prefrontal cortex

Corpus callosum

Basal ganglia

Amygdala

Cerebellum

TEENAGE BRAIN
The teenage brain's prefrontal cortex, which is not yet mature, is closely connected to the basal ganglia and cerebellum. These are important in motor skills and refined movements, which may explain why teenagers often seem clumsy.

THE NERVOUS SYSTEM

Through the teenage years, the brain's sensory and language centers become fully mature, so the individual is well equipped to deal with a range of social and intellectual challenges. However, the prefrontal cortex, which is involved in planning and in assessing risks and outcomes, is still developing. Until the prefrontal cortex takes more control as adolescence comes to an end, the amygdala, which processes emotions, has a relatively dominant influence. This may be one reason why adolescents seem to lack judgement and tend toward impulsiveness. As adolescence progresses, the corpus callosum connecting the two hemispheres thickens, allowing increased information-processing skills.

13 YEARS   **18 YEARS**

"PRUNING" GRAY MATTER
Gray matter—the brain's immense network of interconnected nerve cells—peaks during childhood, then decreases during adolescence as rarely-used neural pathways are "pruned." These MRI scans show plentiful gray matter in green and lower amounts in blue and purple. The scans indicate that brain areas performing more advanced functions, such as the frontal lobes, seem to mature later.

THE MENSTRUAL CYCLE

The principal sign of female sexual maturity is the onset of menstruation. For a few days each month, termed the period or menstruation, bleeding occurs from the vagina as the uterine lining (endometrium) is shed at the start of each cycle. Afterward, the endometrium thickens again. At the start of the cycle, follicle-stimulating hormone (FSH) stimulates egg follicles in the ovary. They secrete estrogen, which triggers the production of luteinizing hormone (LH); this matures the egg and triggers its release (ovulation). The now empty follicle (corpus luteum) secretes progesterone. If the egg is not fertilized, the corpus luteum dies, progesterone levels fall, menstrual bleeding occurs, and the cycle begins again.

Bones
Ossification of the bones is completed

PHYSICAL MATURATION
Most parts of the body are affected by the hormones of puberty. In males, the chief hormone is testosterone from the testes. In females, estrogen from the ovaries, along with FSH and LH from the pituitary gland, have the greatest influence.

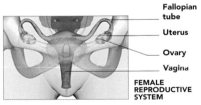

Fallopian tube

Uterus

Ovary

Vagina

FEMALE REPRODUCTIVE SYSTEM

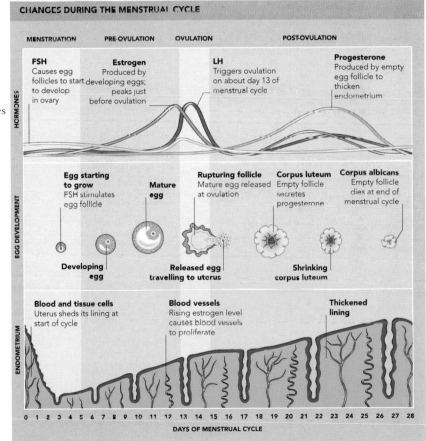

CHANGES DURING THE MENSTRUAL CYCLE

MENSTRUATION   PRE-OVULATION   OVULATION   POST-OVULATION

HORMONES

FSH
Causes egg follicles to start to develop in ovary

Estrogen
Produced by developing eggs; peaks just before ovulation

LH
Triggers ovulation on about day 13 of menstrual cycle

Progesterone
Produced by empty egg follicle to thicken endometrium

EGG DEVELOPMENT

Egg starting to grow
FSH stimulates egg follicle

Mature egg

Rupturing follicle
Mature egg released at ovulation

Corpus luteum
Empty follicle secretes progesterone

Corpus albicans
Empty follicle dies at end of menstrual cycle

Developing egg

Released egg travelling to uterus

Shrinking corpus luteum

ENDOMETRIUM

Blood and tissue cells
Uterus sheds its lining at start of cycle

Blood vessels
Rising estrogen level causes blood vessels to proliferate

Thickened lining

0 1 2 3 4 5 6 7 8 9 10 11 12 13 14 15 16 17 18 19 20 21 22 23 24 25 26 27 28

DAYS OF MENSTRUAL CYCLE

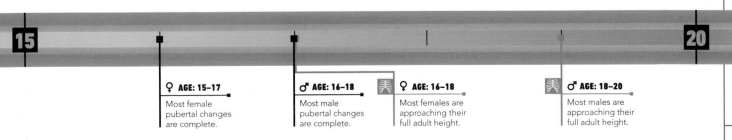

15   **20**

♀ **AGE: 15–17**
Most female pubertal changes are complete.

♂ **AGE: 16–18**
Most male pubertal changes are complete.

♀ **AGE: 16–18**
Most females are approaching their full adult height.

♂ **AGE: 18–20**
Most males are approaching their full adult height.

EARLY ADULTHOOD

BY THE AGE OF 20–23, MOST PEOPLE HAVE REACHED THEIR GREATEST PHYSICAL HEIGHT. THERE FOLLOWS A PERIOD OF CONSOLIDATION AS THE BODY'S OTHER PARTS AND ORGANS REACH THEIR FULL MATURITY. DURING THE 20S AND EARLY 30S, PHYSICAL STRENGTH AND FERTILITY IN WOMEN ARE AT THEIR PEAK.

BONES AND MUSCLES

On average, due to differences in the timing of puberty, the skeleton reaches its natural potential earlier in females than in males. After the teenage growth spurt, gain in height slows at an increasing rate, reaching zero in most people by the age of 25. However, bones and muscles may continue to change for years to come, mainly in response to stresses and strains, such as a very physical occupation or vigorous sport. Long bones increase in diameter, and flatter bones thicken between their widest surfaces. Bones may also stiffen due to changes in their composition, with the more crystalline structure of calcium phosphate-hydroxyapatite laid down at the expense of collagen and other protein components. Muscles reach their full potential in size, strength, and contraction speed by the mid-20s.

OPTIMUM RESPONSE
The body's all-round physical potential is most attainable during its third decade. As its owner makes increasing demands, the body responds more quickly than it could do later, laying foundations for future physical health.

Brain
Begins to diminish from maximum volume

Mandible
Becomes hardest bone in body

Muscles
Respond quickly to changing patterns of use

Tendons
Reach greatest strength and elasticity

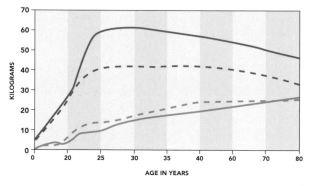

	Lean body mass (male)
– – –	Lean body mass (female)
	Fat (male)
– – –	Fat (female)

MUSCLE AND FAT
The amount of lean body mass—mainly muscle—peaks in the mid-20s then plateaus or slowly declines. In contrast, body fat increases throughout life.

(Graph: KILOGRAMS vs AGE IN YEARS; vertical axis 0–70, horizontal axis 0, 20, 25, 30, 35, 40, 60, 70, 80)

PEAK PERFORMANCE
In most physical sports, competitors are at their peak between 20 and 30 years of age. The Jamaican sprinter Usain Bolt turned 22 as he was shattering world and Olympic records in Beijing in 2008.

AGE 20 — **25** — **30**

AGE: 20
Having already peaked, brain mass is falling at 0.2 percent per year.

♀ AGE: EARLY 20s
A woman's chance of becoming pregnant is at its peak.

AGE: 25
Maximum height is reached for almost all people. Bone density is at its peak.

AGE: MID-20s
Muscles reach full potential in size, strength, and contraction speed.

✸ THE NERVOUS SYSTEM

The brain reaches its maximum weight between the ages of 15 and 18 years. By 20-plus, this organ has already started to shrink, but initially at a very slow rate, less than 0.2 percent per year. However, it retains its ability to make or reinforce thousands of new neural connections daily, both for mental tasks and also for skilled muscle control. Coupled with increasing experience and better abilities in "learning to learn," overall mental functioning continues to improve. Activity patterns in the adult brain also differ from those seen in the teens. The frontal lobes are more active when processing emotional information, evident in greater maturity and perceptiveness. The major nerves change little in the 20s, although they may show signs of damage if the body is exposed to toxins, such as heavy metals from pollution.

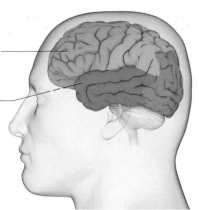

Frontal lobe
Although one of the first brain areas to shrink, this also becomes more active

Temporal lobe
This region of cortex also begins to shrink at a relatively early stage

BRAIN SHRINKAGE
The first areas of the brain to shrink are the frontal lobes, involved in planning, memory, and decisions, and the temporal lobes, which participate in smell, hearing, language, and memory.

✸ SKIN

As long as it has not been overexposed to sunlight or other damaging agents, such as chemicals, the skin changes only slightly before the mid-20s. Production of the major skin proteins collagen and elastin slows, and the latter also becomes less "springy." Turnover of skin cells starts to decrease, and the surface layers of dead cells are shed more slowly. All these changes are slight and often go unnoticed until the 30s or 40s.

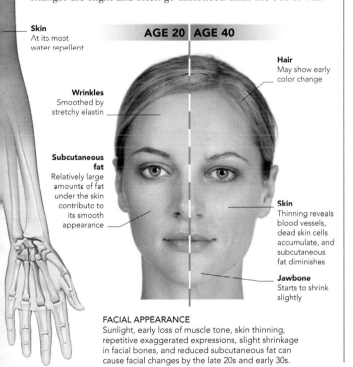

Skin
At its most water repellent

AGE 20 | **AGE 40**

Hair
May show early color change

Wrinkles
Smoothed by stretchy elastin

Subcutaneous fat
Relatively large amounts of fat under the skin contribute to its smooth appearance

Skin
Thinning reveals blood vessels, dead skin cells accumulate, and subcutaneous fat diminishes

Jawbone
Starts to shrink slightly

FACIAL APPEARANCE
Sunlight, early loss of muscle tone, skin thinning, repetitive exaggerated expressions, slight shrinkage in facial bones, and reduced subcutaneous fat can cause facial changes by the late 20s and early 30s.

✸ FERTILITY

Statistics on fertility are skewed by the use of contraceptive methods and, in some areas, by poor nourishment. However, the 20s is usually regarded as the most fertile decade. The ovaries are very responsive to the hormonal changes that bring about ovulation, the eggs released are more likely to be viable, the oviducts or fallopian tubes have not yet accumulated natural scarlike changes with age, and the lining of the uterus builds up rapidly with greatest blood flow to receive a fertilized egg. From the early 20s, the chance of conceiving declines while the probability of infertility simultaneously rises.

—— Likelihood of pregnancy
—— Likelihood of infertility

DECREASING CHANCES
For well-nourished, healthy women, the chance of becoming pregnant falls from around 85 percent aged 20 to about 36 percent by 40. Simultaneously, likelihood of infertility rises tenfold, from 3 to 32 percent.

THE FERTILE WINDOW

The prospects for many younger women to have a baby have been transformed by techniques such as in vitro fertilization (IVF). Here, an ovum is at the point of fertilization during IVF. The fertilized ovum will later be transferred to the uterus, to implant and develop into a fetus. IVF has also extended the age range for becoming a mother—the "fertile window"—into the 40s, 50s, and in a few rare cases, beyond.

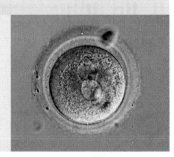

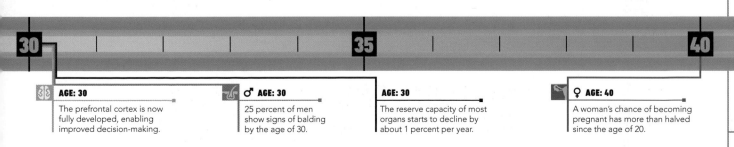

30 | **35** | **40**

✸ **AGE: 30**
The prefrontal cortex is now fully developed, enabling improved decision-making.

♂ **AGE: 30**
25 percent of men show signs of balding by the age of 30.

AGE: 30
The reserve capacity of most organs starts to decline by about 1 percent per year.

♀ **AGE: 40**
A woman's chance of becoming pregnant has more than halved since the age of 20.

40–60 MIDDLE AGE

AFTER THE RIGORS OF ADOLESCENCE AND YOUNG ADULTHOOD, AND BEFORE SIGNIFICANT AGING SETS IN, THE TWO DECADES BETWEEN 40 AND 60 ARE GENERALLY A TIME OF STASIS. THIS PHASE SHOWS THE LEAST OF THE MAJOR CHANGES DURING A BODY'S LIFETIME.

SKIN AND HAIR

Most people develop skin creases and furrows by the age of 40, and these are especially noticeable when making facial expressions. In about half of people, fully formed wrinkles are apparent by 50. These appear where the skin is regularly stretched and compressed, especially at the corners of the eyes and mouth (see p.76), due partly to the skin's reduced elasticity and "springback" as its content of the stretchy protein elastin declines. Exposure to sunlight (and other sources of ultraviolet light, such as sunlamps) encourages wrinkle formation. Other factors that increase wrinkling include smoking, an inherited predisposition, and having a paler skin color. The skin also produces less of the oily-waxy sebum from its sebaceous glands, which may make the skin feel dry and rough. Hair loss in men is common, but thinning of the hair also occurs in women: about half of all women have noticeable thinning by the age of 50.

MALE-PATTERN BALDNESS

The typical form of hair loss in men is known as male-pattern baldness. The front hairline recedes and thinning occurs on the crown. The bald areas then grow and link, to leave hair only around the sides and rear of the scalp, as has happened to the man shown on the right. On average, male-pattern baldness affects about two out of three men by the age of 60. It is associated with the male sex hormone testosterone and is also greatly influenced by genetic factors.

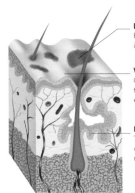

EYE WRINKLES
The skin around the eyes is relatively thin, with less cushioning fat; as a result, creases from habitual movements appear more readily.

Liver spots
Pigmentation patches in areas exposed to sun

Wrinkles
Creases resulting from thinning and loss of elasticity of the skin

Dermis
Thinner; contains fewer collagen fibers, causing reduced elastic recoil

THE STRUCTURE OF OLDER SKIN
The outer layer or epidermis thins, with lowered cell turnover, and collagen and elastin are lost from the dermis beneath.

MIDLIFE CHANGES
Aging in most body systems is still slow, but gradually accelerates after age 50. The cumulative effects of lifestyle factors, such as smoking, often also start to become apparent at this time.

Eyes
The lens begins to lose elasticity, causing difficulty in focusing

Ovaries
At menopause, the ovaries stop responding to FSH and produce less progesterone and estrogen

Joints
Ligaments around joints begin to lose elasticity

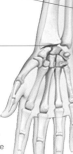

AGE | 40 | 45 | 50

AGE: EARLY 40s
Bones start to lose mineral content; gradual loss of height begins; in males, testosterone level starts to decline.

AGE: 42–44
Vision problems due to presbyopia become noticeable; females start to lose significant numbers of taste buds.

AGE: MID-40s
Hearing loss affects about one in five people; bile production starts to decline.

♀ **AGE: LATE-40s**
Menstrual cycles start to get shorter.

THE NERVOUS SYSTEM

Loss of nerve cells in the brain and peripheral nerves continues at a gradually increasing rate between the ages of 40 and 60 years. There is loss or degeneration of synapses between neurons, and levels of neurotransmitters are reduced. The speed of nerve signals also decreases—one reason why reflexes and reaction speeds slow down. In the sense organs, the sensory nerve cells deteriorate and, when they die, are not replaced. Sight, hearing, smell, and taste are all affected. The sense of touch, including sensitivity to heat, cold, and pain also begins to deteriorate.

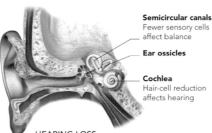

Semicircular canals
Fewer sensory cells affect balance

Ear ossicles

Cochlea
Hair-cell reduction affects hearing

HEARING LOSS
Loss of the delicate sensory hair cells in the cochlea can result from excessive exposure to prolonged loud noise (especially mid- to high frequencies), side effects from medications, and genetic predisposition.

Lens degeneration
Stiffening of the crystalline structure contributes to presbyopia; clouding of the lens (cataract) can also impair vision

CHANGES IN THE EYE
Presbyopia, in which the lens cannot focus on nearby objects, is common from the age of 40–50 years. The ciliary muscles around the lens weaken, also impairing focusing.

Muscles
Muscle mass decreases

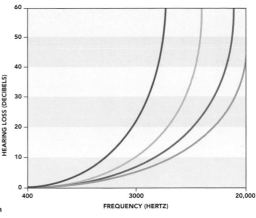

HIGH FREQUENCY HEARING LOSS
As the eardrum becomes more rigid, the joints between the tiny ear ossicles stiffen, and the cochlea becomes less efficient at transforming sounds into nerve signals, hearing becomes duller and more distorted. This particularly affects high-frequency sounds.

— Age 20
— Age 30
— Age 50
— Age 70

BONES AND MUSCLES

Several age-related changes to bones and muscles commonly begin in middle age, including osteoporosis (see p.64), when decline in sex hormones disrupts the balance between bone breakdown and reconstruction. This most commonly affects women following menopause but also occurs in men. Bone loses its mineral content, becoming less dense and more fragile. The muscles also become weaker and stiffer, which affects both voluntary body movements and automatic ones, such as inhalation and exhalation.

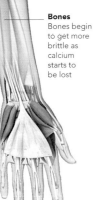

Bones
Bones begin to get more brittle as calcium starts to be lost

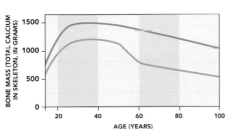

LOSS OF BONE MASS
From a peak at around 30 years old, minerals (especially calcium) are slowly lost from the bones. The dip in bone mass in women aged 50–60 is linked to menopause (see right).

— Men
— Women

MENOPAUSE

Menopause signals the end of a woman's ability to conceive by natural means. It results from decreasing production of sex hormones, especially estradiol (the main form of estrogen) and progesterone from the ovaries. In developed countries, the average age of menopause is 50–52 years, while in less developed countries it may be less than 45 years. Menopausal symptoms may occur sporadically over several years (a period known as perimenopause) and then occur continually for several months. They include hot flashes, night sweats, insomnia, headaches, dry vagina, and irregular bleeding during periods. Menopause is not usually considered complete or definite until a woman has not had a period for one year.

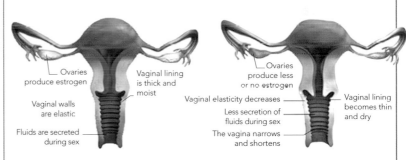

Ovaries
produce estrogen

Vaginal lining
is thick and moist

Vaginal walls
are elastic

Fluids are secreted
during sex

PREMENOPAUSAL VAGINA
Before menopause, the vaginal lining is thick and well lubricated; the walls stretch easily and mucous fluids are secreted.

Ovaries
produce less or no estrogen

Vaginal elasticity decreases

Less secretion of
fluids during sex

The vagina narrows
and shortens

Vaginal lining
becomes thin and dry

POST MENOPAUSAL VAGINA
Declining estrogen levels cause a reduction in vaginal mucus production; the vagina walls lose some elasticity and become thinner.

---

50   **55**   **60**

♀ **AGE: 50–52**
Most women pass through menopause.

♂ **AGE: EARLY 50s**
About half of men show signs of male-pattern baldness; red-blood-cell level begins to decline in males.

♂ **AGE: MID-50s**
Number of taste buds start to decline significantly in males.

AGE: 60
About one in five people have lost all their natural teeth; red-blood-cell level begins to decline in females.

60+ OLD AGE

FROM THE AGE OF ABOUT 60, SIGNS OF AGING BECOME INCREASINGLY APPARENT AS ORGANS DETERIORATE. THE RATE AND PATTERN OF DETERIORATION VARY FROM ONE PERSON TO ANOTHER, BUT EVENTUALLY ONE OR MORE VITAL ORGANS FAILS COMPLETELY, RESULTING IN DEATH.

SKIN AND HAIR

The skin continues to lose tautness and elasticity, resulting in more extensive and prominent wrinkling, especially in areas exposed to the sun. Pigmented age spots also become more common with age, and the skin becomes less sensitive, particularly to gentle touch, heat, cold, and pain. On the scalp, the amount of hair loss varies considerably from person to person. Even if relatively little hair is lost, individual hairs become narrower, which contributes to the thinned appearance of the scalp. At the microscopic scale, the surface of each hair also becomes more irregular, rougher, and pitted, causing the hair to appear dull.

Hair
Typically loses all pigmentation

Forehead
Extensive wrinkling is common, especially in people who have spent a lot of time outdoors

Cheek fat pad
This descends, creating a deeper fold from the nose to the mouth's outer corner

FACIAL APPEARANCE
In old age, the cumulative effects of environmental factors, such as sunlight, and natural aging leave their mark on the facial skin as extensive, often deep wrinkles.

THE NERVOUS SYSTEM

Overall, the brain loses neurons and shrinks in old age, although not uniformly in all areas—each year some brain regions may lose 1–2 percent of their volume whereas others lose almost nothing. Mentally, the typical slow decline in so-called "fluid intelligence," involving speed of thought and rapid memory recall, can be compensated for by using experience and knowledge to make decisions and solve problems, known as "crystallized intelligence." As in the brain, neurons in the peripheral nerves lose connections, reduce their production of neurotransmitters, and function less efficiently.

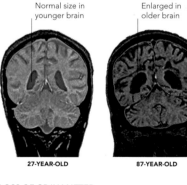

Ventricles
Normal size in younger brain

Ventricles
Enlarged in older brain

27-YEAR-OLD   **87-YEAR-OLD**

LOSS OF GRAY MATTER
There is a general loss of gray matter in old age. In addition, the ventricles (spaces filled with cerebrospinal fluid) become larger, as can be seen in the two MRI scans above.

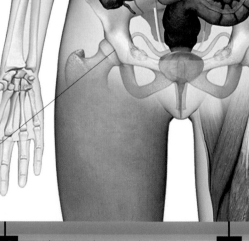

Brain
Loss of gray matter

Eyes
Loss of elasticity of the lens continues; cataracts may also develop

Ears
Less elastic eardrum and fewer sensory nerve endings affect hearing

Heart
The heart rate slows; valves thicken and stiffen

Joints
Joint linings deteriorate due to wear and tear

AGE 60 | | | | 65 | | | | 70 | | | | 75

AGE: 60–65
Loss of muscle mass accelerates; total blood volume is 20–30 percent lower than at age 25; sense of smell starts to diminish noticeably.

♂ AGE: 69.8 YEARS
Average worldwide life expectancy at birth in 2015 for males.

AGE: EARLY 70s
The heart's pumping capacity is about 70–75 percent of that at age 30.

♀ AGE: 74.2 YEARS
Average worldwide life expectancy at birth in 2015 for females.

BONES AND MUSCLES

Through life, the process of bone remodeling rejuvenates overstressed or damaged bone tissue (see pp.54–55). From the early 60s, the balance in this process shifts significantly as deconstruction increases compared to regeneration. In addition, thinning of the bone due to calcium loss accelerates, especially if bones are subjected to less physical stress. The spine, in particular, shows this, as decrease in bone density and deterioration of the intervertebral disks contribute to height loss and a more stooped posture. Other changes include stiffening of ligaments around joints and a general reduction in muscle mass. An average decline in muscle strength of one third between 45 and 70 is due in part to deterioration of muscle fibers, which are replaced by collageneous connective tissue and fat. As a result, the muscles become stiffer and slower to react.

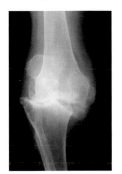

X-RAY OF OSTEOARTHRITIC KNEE
Osteoarthritis of the knee is common in old age as the cartilage over the ends of the leg bones suffers wear and water loss, causing the ends of the bones to rub together painfully.

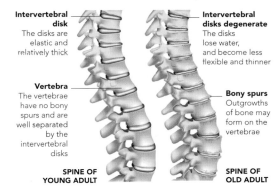

Intervertebral disk
The disks are elastic and relatively thick

Vertebra
The vertebrae have no bony spurs and are well separated by the intervertebral disks

Intervertebral disks degenerate
The disks lose water, and become less flexible and thinner

Bony spurs
Outgrowths of bone may form on the vertebrae

SPINE OF YOUNG ADULT

SPINE OF OLD ADULT

SPINAL CHANGES
Disks of cartilage between the vertebrae (individual spinal bones) lose water with age and become thinner, causing loss of height. The disks are also more brittle and fragile, increasing the likelihood of prolapse (a "slipped disk") or even complete collapse.

ORGAN FUNCTION

Age affects all body organs, but initially most have a reserve—an ability to function beyond their usual needs. With age, this reserve is taken up. In a single individual, the body systems do not all age at the same rate. Futhermore, the systems and organs affected most vary from one person to another, depending on factors such as long-term habits and lifestyle. For example, excessive alcohol intake affects the liver and brain, and lack of physical exercise means bones and muscles deteriorate more rapidly.

Muscles
Collagen fibers and fat accumulate

Bones
Loss of calcium accelerates

AGING BODY SYSTEMS
From the age of 60, most body systems show signs of aging, although they are often complex. For example, loss of taste buds means that sensitivity to salty and sweet flavors fades before sensitivity to sour and bitter.

Liver
Enzyme production and ability to metabolize certain drugs decline

Gallbladder
Bile production decreases; gallstones more likely to form

LIVER FUNCTION
Aging reduces the liver's enzyme production and its ability to metabolize certain drugs, therefore dosages of these drugs may need to be reduced for older people.

KIDNEY EFFICIENCY
An indicator of kidney efficiency is creatinine clearance—the rate at which creatinine (a by-product from muscle tissue) is filtered from blood into the urine. Indicating how well the kidneys process breakdown products, it falls steadily with age.

[Graph: CREATININE CLEARANCE (ML/MIN/1.73 M²) vs AGE IN YEARS, from 30 to 80, declining from about 140 to below 100]

DYING AND DEATH

The fact that different systems and organs deteriorate faster in some people than in others suggests that the rate of aging is affected by a mixture of genes, lifestyle, and disease. Often, one organ in an individual deteriorates fastest, becomes the "weakest link," and leads to the final decline. The traditional definition of death as cessation of heartbeat and breathing has been blurred by modern technology, especially critical care. Death is now defined as the "irreversible cessation of all functions of the entire brain."

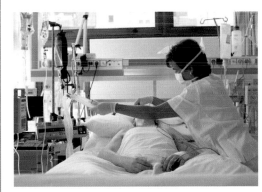

MAINTAINING VITAL FUNCTIONS
Critical care provides organ support, such as artificial ventilation to maintain air flow into and out of the lungs and dialysis for failing kidneys, to maintain body functions when organs no longer function.

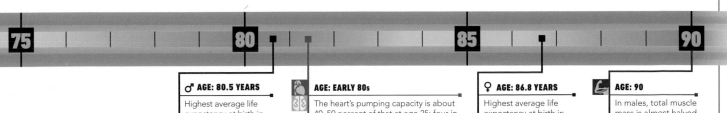

75 **80** **85** **90**

♂ **AGE: 80.5 YEARS**
Highest average life expectancy at birth in 2015 for males (Japan).

AGE: EARLY 80s
The heart's pumping capacity is about 40–50 percent of that at age 25; four in five people have noticeable hearing loss.

♀ **AGE: 86.8 YEARS**
Highest average life expectancy at birth in 2015 for females (Japan).

AGE: 90
In males, total muscle mass is almost halved from that at age 20.

INHERITANCE

THE PASSING OF GENETIC INFORMATION FROM PARENT TO CHILD IS KNOWN AS INHERITANCE. THE INFORMATION IS IN CHEMICAL CODES CARRIED BY DEOXYRIBONUCLEIC ACID (DNA) IN THE SEX CELLS (EGGS AND SPERM).

INHERITANCE OF GENES

Everything needed to specify a person is passed on in our genes. Every gene carries a "blueprint" to make a particular product. Some gene products have a distinctive effect on a person's appearance or biology, for instance, skin pigmentation or eye color. However, gene products can combine (for example, one can control or regulate the production of another) to produce a complex trait, such as athletic ability. Simple features controlled by single genes are inherited in predictable patterns (see pp.264–65). However, complex traits, such as height, are controlled by many genes. Their inheritance is not always predictable: tall parents tend to have tall children but not always. The way information is copied in physical form and passed from parent to offspring is much more clearly understood.

SEQUENCING THE GENOME

By 2003 the Human Genome Project identified all 3 billion base pairs in the full set of human DNA. In 2012 it was realized that a large portion of the DNA instructs for building RNA rather than proteins. A major technique used in DNA sequencing is gel electrophoresis. DNA is extracted from cells, purified, and broken into smaller fragments of known length by chemicals known as restriction enzymes. The DNA fragments are then placed in a gel substance through which an electric current passes. The fragments move at different speeds and separate out through the gel according to their size and electrical charge. They are then stained by chemical dyes and show up as dark stripes, like the bar codes seen below. Computers can read these bar codes to reveal the base pair sequences.

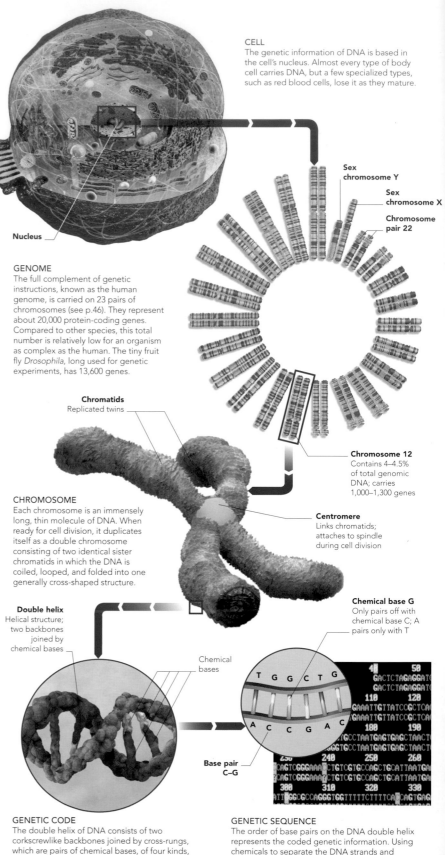

CELL
The genetic information of DNA is based in the cell's nucleus. Almost every type of body cell carries DNA, but a few specialized types, such as red blood cells, lose it as they mature.

Nucleus

Sex chromosome Y

Sex chromosome X

Chromosome pair 22

GENOME
The full complement of genetic instructions, known as the human genome, is carried on 23 pairs of chromosomes (see p.46). They represent about 20,000 protein-coding genes. Compared to other species, this total number is relatively low for an organism as complex as the human. The tiny fruit fly *Drosophila*, long used for genetic experiments, has 13,600 genes.

Chromatids
Replicated twins

Chromosome 12
Contains 4–4.5% of total genomic DNA; carries 1,000–1,300 genes

CHROMOSOME
Each chromosome is an immensely long, thin molecule of DNA. When ready for cell division, it duplicates itself as a double chromosome consisting of two identical sister chromatids in which the DNA is coiled, looped, and folded into one generally cross-shaped structure.

Centromere
Links chromatids; attaches to spindle during cell division

Double helix
Helical structure; two backbones joined by chemical bases

Chemical bases

Chemical base G
Only pairs off with chemical base C; A pairs only with T

Base pair C–G

GENETIC CODE
The double helix of DNA consists of two corkscrewlike backbones joined by cross-rungs, which are pairs of chemical bases, of four kinds, adenine (A), thymine (T), guanine (G), and cytosine (C). The bases always pair in a specific way.

GENETIC SEQUENCE
The order of base pairs on the DNA double helix represents the coded genetic information. Using chemicals to separate the DNA strands and identify the bases, DNA sequencing machines can show the data on screen as lists of letters.

DNA REPLICATION

Apart from carrying genetic information in chemically coded form, as its sequences of base pairs, DNA has another key feature. It can make exact copies of itself in a process known as replication. It does this by separation of the two backbone strands and the bases attached to them, at the bonds between the base pairs. Then each strand acts as a template to build a complementary partner strand. DNA replication takes place before cell division (see right).

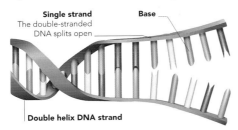

1 SEPARATION
The two strands of the double helix separate at the base pair links. This exposes each base, ready to latch onto its partner in the newly constructed strand.

Single strand
The double-stranded DNA splits open

Base

Double helix DNA strand

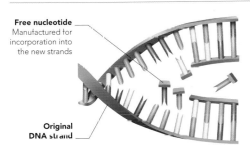

Free nucleotide
Manufactured for incorporation into the new strands

Original DNA strand

2 BASES JOIN
Free nucleotides, each one a base combined with a portion of DNA backbone, join to the two sets of exposed bases. This can only happen in the correct order since A always pairs with T, and C with G.

3 TWO STRANDS FORM
More nucleotides join, linked by a new backbone. Each strand now has a new "mirror-image" partner, giving two double-helices, which are identical to each other and to the original.

Original DNA strand

New DNA strand

New DNA strand forming

MUTATIONS

DNA replication usually works well. However, factors such as radiation, or certain chemicals, may cause a fault, where one or more base pairs do not copy exactly. This change is a mutation. The new base sequence may result in a different protein being built from it, which could cause a problem in the body.

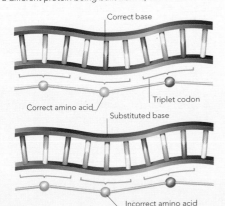

Correct base

Correct amino acid

Triplet codon

Substituted base

Incorrect amino acid

NORMAL GENE
Each set of three base pairs (a triplet codon) specifies which amino acid should be added to the series of amino acids that make normal protein for that gene.

MUTATED GENE
In a point mutation, one base pair is altered. A different amino acid may be specified, which will disrupt the protein's eventual shape and function.

MAKING NEW BODY CELLS

The process by which a cell divides into two identical daughter cells is known as mitosis. First, all the genetic material is duplicated by DNA replication, and each chromosome becomes two identical units or chromatids (see opposite). These double-chromosomes line up across the cell's middle, then separate and move apart, one member migrating to each end of the cell as it splits into two. Mitosis occurs constantly to produce new cells for growth, maintenance, and repair.

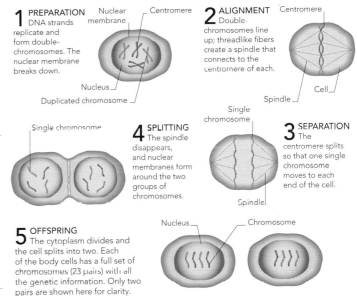

1 PREPARATION
DNA strands replicate and form double-chromosomes. The nuclear membrane breaks down.

Nuclear membrane

Centromere

Nucleus

Duplicated chromosome

2 ALIGNMENT
Double-chromosomes line up; threadlike fibers create a spindle that connects to the centromere of each.

Centromere

Cell

Spindle

Single chromosome

3 SEPARATION
The centromere splits so that one single chromosome moves to each end of the cell.

Spindle

Single chromosome

4 SPLITTING
The spindle disappears, and nuclear membranes form around the two groups of chromosomes.

5 OFFSPRING
The cytoplasm divides and the cell splits into two. Each of the body cells has a full set of chromosomes (23 pairs) with all the genetic information. Only two pairs are shown here for clarity.

Nucleus

Chromosome

MAKING SEX CELLS

Cell division by meiosis produces sex cells: the eggs and sperm. It is similar to mitosis, above, but has an extended series of stages that separate each pair of chromosomes, ending with four daughter cells that have half the normal number of chromosomes, with only one member of a chromosome pair in each. At fertilization, when egg and sperm unite, the chromosome number (23 pairs) is restored. All subsequent cell divisions of the fertilized egg are by mitosis.

1 PREPARATION
DNA strands replicate and coil up in the nucleus, forming X-shaped double-chromosomes.

Duplicated chromosme

2 PAIRING
The matching (homologous) pairs align, make contact, and exchange genetic material.

Matching pair of chromosomes

Duplicated chromosomes

4 TWO OFFSPRING
Each cell has one double-chromosome of each pair, as a random choice during separation.

Spindle

3 FIRST SEPARATION
The threadlike spindle pulls one of each pair to each end as the cell splits.

Chromosome pair separates

5 SECOND SEPARATION
The double-chromosome splits, each half moving to one end of the dividing cell.

Spindle

6 FOUR OFFSPRING
The four sex cells differ from each other and the parent cell in their genetic composition.

Chromosome

Nucleus

Single chromosome

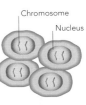

PATTERNS OF INHERITANCE

GENES ARE PASSED FROM ONE GENERATION TO THE NEXT, IN A VAST SEQUENCE OF INHERITANCE. THEY ARE RESHUFFLED AT EACH STAGE SO THAT OFFSPRING ARE UNIQUE, BUT THERE ARE PATTERNS IN THE MODE OF INHERITANCE.

VERSIONS OF GENES

Each cell in a body has a double set of genetic material, as 23 pairs of chromosomes. One chromosome of each pair, and the genes on it, come from the mother. The other chromosome is from the father. So there are, in effect, two versions of every gene in the set: one maternal and one paternal. These versions of genes are called alleles. Inheritance patterns vary depending on how these two versions interact, because they may be identical or slightly different.

TWO BY TWO
Chromosomes have the same sets of genes. But the individual allele on one chromosome may differ slightly from its equivalent allele on the other chromosome.

GENERATIONAL SEQUENCE

In the double set of genes of each person, one single set was inherited from the mother, and one set from the father. In turn, each of the parents had inherited one single set from one grandparent and one from the other grandparent, and so on. The versions of the genes (alleles) are mixed, or reshuffled, at each generation. So, in effect, a child has one-quarter of its genes from each grandparent. This is why a child's inherited features strongly resemble a mixture of the features of the parents, but those of the grandparents less markedly, and so on.

MIXED, BUT NOT BLENDED
Genes are "units" of inheritance that are shuffled at each generation into different combinations. Individual genes do not blend with each other to create new versions.

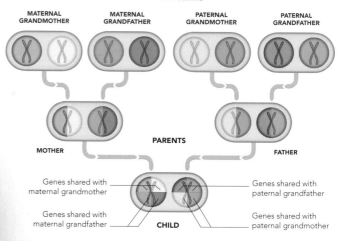

GRANDPARENTS

MATERNAL GRANDMOTHER   MATERNAL GRANDFATHER   PATERNAL GRANDMOTHER   PATERNAL GRANDFATHER

MOTHER   PARENTS   FATHER

Genes shared with maternal grandmother — Genes shared with paternal grandfather

Genes shared with maternal grandfather — Genes shared with paternal grandmother

CHILD

INHERITANCE OF GENDER

The gender of offspring is determined by the two sex chromosomes, numbered as pair 23. In a female, both of these chromosomes are X, and have the same sets of genes. A male has one X chromosome, and the other is the much smaller Y, with male genes. When egg cells form, they each contain an X. When sperm cells form, half their number receive an X, and the other half, a Y. An egg fertilized by an X-carrying sperm has an XX sex chromosome pairing, so the baby is female. If the sperm contains a Y, the pairing is XY, and the baby is male. The gender of offspring is always determined by the father.

BOY OR GIRL?
Gender is determined by the inheritance of sex chromosomes, X and Y (the other 22 chromosome pairs are not shown here).

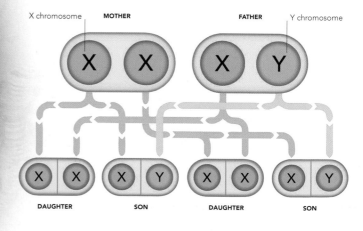

X chromosome   MOTHER   FATHER   Y chromosome

X   X   X   Y

DAUGHTER   SON   DAUGHTER   SON

RECESSIVE AND DOMINANT GENES

Because chromosomes exist as pairs, the genes they carry are also in pairs (one inherited from each parent). Each gene of a pair is known as an allele. In some cases, these two alleles are the same. In others, they are different. One allele may be dominant and "overpower" the other, which is recessive. This complicates the process of inheritance.

RECESSIVE AND RECESSIVE
For a particular body feature, or trait, both parents have two identical alleles—known as "homozygous." Here they are recessive; without a dominant allele, all offspring have the recessive version of the trait.

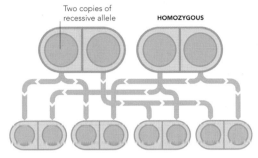
Two copies of recessive allele HOMOZYGOUS

ALL OFFSPRING HAVE RECESSIVE VERSION OF TRAIT

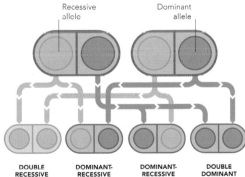
Recessive allele HOMOZYGOUS HETEROZYGOUS Dominant allele

DOUBLE RECESSIVE DOMINANT-RECESSIVE DOUBLE RECESSIVE DOMINANT-RECESSIVE

RECESSIVE AND MIXED
One parent has two recessive alleles, as above; the other is "heterozygous", having two different alleles, one recessive and one dominant. The chance is 1 in 2 that an offspring inherits the recessive trait.

MIXED AND MIXED
In this case, both parents are mixed, or heterozygous, each with one recessive allele and one dominant allele. The chance of an offspring having the recessive trait (no dominant allele) is reduced to 1 in 4.

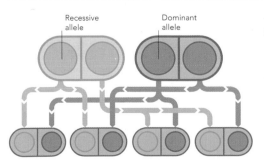
Recessive allele Dominant allele

DOUBLE RECESSIVE DOMINANT-RECESSIVE DOMINANT-RECESSIVE DOUBLE DOMINANT

DOMINANT AND RECESSIVE
This time, both parents are homozygous, but one for the recessive trait, and one for the dominant. All four offspring are heterozygous and show the dominant version of the trait.

Recessive allele Dominant allele

ALL OFFSPRING HAVE DOMINANT VERSION OF TRAIT BUT CARRY RECESSIVE ALLELE

SEX-LINKED INHERITANCE

The pattern of inheritance changes when the alleles for a body feature are carried on the pair of sex chromosomes instead of the ordinary chromosomes. In the case of a female, XX, any dominant and recessive alleles can interact, as shown to the left. But in the male, an allele on the X chromosome may not have its equivalent on the Y chromosome, and vice versa. This means a single allele can be the only one to determine the feature. An example is color-impaired vision, where the problem allele is on the X chromosome.

COLOR-IMPAIRED FATHER AND UNAFFECTED MOTHER
Sex chromosomes combine in four possible ways, governed by chance. Here, any daughter will inherit the color-impairment allele, and will be a carrier, but also has the normal allele on her other X chromosome, to give normal vision. No sons can be affected, nor can they be carriers.

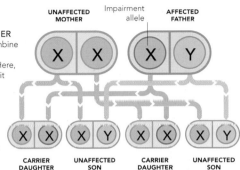
UNAFFECTED MOTHER Impairment allele AFFECTED FATHER

X X X Y

X X X Y X X X Y

CARRIER DAUGHTER UNAFFECTED SON CARRIER DAUGHTER UNAFFECTED SON

CARRIER MOTHER AND UNAFFECTED FATHER
The four possible combinations give one-quarter each for unaffected sons and daughters. There is a one-in-four chance a daughter is a carrier, or that a son inherits the color-impairment allele. He has no second X chromosome and therefore no normal allele, so the result is impaired vision.

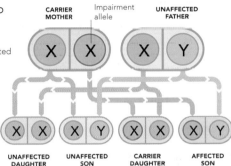
CARRIER MOTHER Impairment allele UNAFFECTED FATHER

X X X Y

X X X Y X X X Y

UNAFFECTED DAUGHTER UNAFFECTED SON CARRIER DAUGHTER AFFECTED SON

MULTIPLE-GENE INHERITANCE

Some body traits follow clear single-gene inheritance patterns. However, the situation becomes more complex in two ways. First, there may not be only two alleles of a gene with a simple dominant-recessive interaction between them. There may be three alleles or more in existence in the general population, although each person can have only two of them. An example is the blood type system with alleles for A, B, and O. Second, a trait may be influenced by more than one gene. These two situations mean a trait can be governed by multiple genes, and for each of these genes, by multiple alleles of the gene—added to which, the genes may interact in different ways, according to which alleles are present in each of them. In such cases, the numbers of possible combinations multiply, making multigene inheritance patterns exceptionally difficult to unravel.

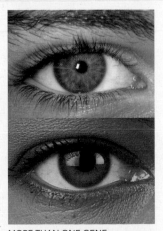

MORE THAN ONE GENE
Eye color was originally viewed as a simple case of single-gene inheritance. But research shows there are probably more than a dozen genes that affect iris color (see p.43).

INHERITED DISORDERS

INHERITED DISORDERS ARE CAUSED BY DEFECTIVE GENES OR ABNORMAL CHROMOSOMES THAT ARE PASSED FROM PARENTS TO CHILDREN. IN CHROMOSOME DISORDERS, THERE IS A PROBLEM IN THE NUMBER OR STRUCTURE OF CHROMOSOMES, WHEREAS IN GENE DISORDERS, THERE IS A FAULT IN ONE OR MORE OF THE GENES THAT ARE CARRIED ON INDIVIDUAL CHROMOSOMES.

CHROMOSOME DISORDERS

Chromosome disorders are inherited disorders that result from either an incorrect number of chromosomes (numerical disorders) being passed on from parent to child or from an alteration in the structure of some of the chromosomes (structural disorders).

Mosaicism is a type of numerical disorder but not every single cell is affected, so the problem may not manifest itself. Errors occur before fertilization when genetic material is swapped during the cell division involved in egg and sperm production.

NUMERICAL

A MISTAKE DURING CELL DIVISION IN EGG OR SPERM CELLS (MEIOSIS) CAN RESULT IN ONE CELL HAVING TOO MANY CHROMOSOMES AND THE OTHER TOO FEW.

This is the most common type of chromosome anomaly, with two-thirds of disorders falling into this group. In many cases, extra or missing chromosomes result in miscarriage. However, there are a few exceptions in which the fetus survives. The most common is Down syndrome, also known as trisomy 21 because the disorder is caused by an extra chromosome 21. Abnormalities in the sex chromosomes have a less severe effect on the embryo, and there may not be any obvious signs of a problem. A girl with an extra X chromosome or a boy with an extra Y chromosome will probably go unnoticed. However, a boy with an extra X chromosome (XXY) will have Klinefelter syndrome, which becomes apparent at puberty with the failure of secondary sexual characteristics to develop. A girl born with only one X chromosome has Turner syndrome.

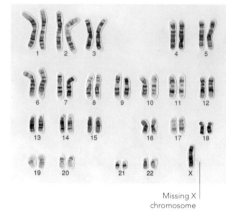

Missing X chromosome

TURNER SYNDROME
This set of chromosomes (karyotype) from a female with Turner's syndrome shows only one X chromosome rather than the two that are normal for a female. Although they are of normal intelligence and have normal life expectancy, girls with this condition have short stature and are usually infertile.

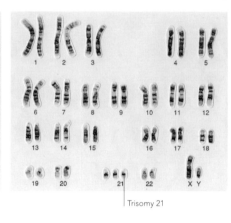

Trisomy 21

DOWN SYNDROME
This karyotype is from a male with Down syndrome, indicated by an extra chromosome 21, known as trisomy 21. This is the most common chromosomal abnormality and causes a characteristic physical appearance, learning difficulties, and often abnormalities of the heart.

STRUCTURAL

DURING CELL DIVISION IN EGG OR SPERM CELLS (MEIOSIS), A SMALL SECTION OF CHROMOSOME MAY BECOME MISPLACED.

During the natural exchange of genetic material between chromosomes, a small section may be deleted, duplicated, or inserted the wrong way round, or inverted. Such structural abnormalities often result in miscarriage, but if the pregnancy proceeds to full term, there may be birth defects depending on the amount and type of altered genetic material. Another problem, translocation, happens when material is swapped between two different chromosomes. If there is no net loss or gain of material, it is known as a balanced translocation, and outward problems are unlikely. In unbalanced translocation, the extra and missing information can lead to defects or, if severe, miscarriage.

NORMAL 7   **NORMAL 21**

PAIRED CHROMOSOMES
In normal chromosomes, the pairs are equally matched; arms are of equal length and the positions of the genes are exactly the same for each of the 22 non-sex chromosome pairs. This occurs whether the chromosome is long (as in chromosome 7) or if it is much shorter (as in chromosome 21).

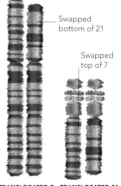

Swapped bottom of 21

Swapped top of 7

TRANSLOCATED 7   **TRANSLOCATED 21**

BALANCED TRANSLOCATION
In a typical translocation, a large part of one chromosome is joined to another; here most of chromosomes 7 and 21 are joined. In a balanced translocation, such as this, no genetic material is lost or gained and no outward abnormality is seen.

MOSAICISM

MOSAICISM IS A MIXTURE OF BODY CELLS, SOME CONTAINING A NORMAL NUMBER OF CHROMOSOMES AND OTHERS CONTAINING AN ABNORMAL NUMBER.

The presence of more than one type of cell in an individual is known as mosaicism. For example, some cells may have the normal number of 46 chromosomes, while others have an extra one, giving a total of 47 chromosomes. If only a few cells have an abnormal number of chromosomes, there are unlikely to be any outward signs of disease and the abnormality would be detectable only by an analysis of a blood sample. With a greater proportion of incorrect cells, disorders may occur. These disorders are the same as with straightforward numerical abnormalities. For example, Down syndrome can be a result of mosaicism if a large number of cells have an extra chromosome 21 (making a total of 47 chromosomes). However, this is a rare cause of numerical disorders, being responsible for only 1 or 2 in 100 cases. Other syndromes, such as Turner and Klinefelter, can also develop as a result of mosaicism.

GENE DISORDERS

Many disorders result from inheriting faulty genes. Defective genes may have mild, moderate, or potentially fatal consequences, or they may have no effect at all. Some genetic disorders are apparent soon after birth or in early life whereas others, such as Huntington's disease, are not discovered until adult life. Types of inheritance (see p.225) include dominant, where only one parent has to carry the defective gene; recessive, where both parents are carriers; and X-linked, where the faulty gene is on the X chromosome.

HUNTINGTON'S DISEASE

A DOMINANT GENE DISORDER, HUNTINGTON'S DISEASE, CAUSES DEGENERATION OF PART OF THE BRAIN.

Huntington's disease is caused by an abnormal dominant gene. Also called Huntington's chorea, the disease causes involuntary movements, personality changes, and progressive dementia. Deterioration occurs over 15–20 years. Symptoms do not usually develop until over the age of 30, by which time the gene may have been passed to the next generation. For this reason, genetic testing and counseling is offered if a relative suffers from the disease.

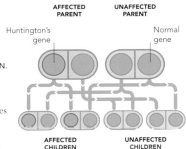

AFFECTED PARENT · Huntington's gene
UNAFFECTED PARENT · Normal gene
AFFECTED CHILDREN
UNAFFECTED CHILDREN

DOMINANT INHERITANCE
In this example, one of the parents has the abnormal gene and the other parent is unaffected. Each child has a 1 in 2 chance of inheriting the faulty gene and therefore of developing the disorder in adulthood.

ALBINISM

A DISORDER CHARACTERIZED BY A LACK OF THE BROWN PIGMENT MELANIN, ALBINISM CAUSES VERY PALE FEATURES.

Albinism's several forms are due to mutation in one or more of several genes, such as oculocutaneous albinism type 2 (OCA2). Such mutations may lead to a fault in an enzyme essential for melanin production. The condition is rare, with an incidence of about 1 in 20,000. Affected individuals have little or no pigment in the skin, hair, and eyes; skin is pale, hair is white, and eyes range from pink to very pale blue. The eyes are sensitive to bright light, and visual impairment is common.

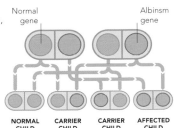

CARRIER PARENTS · Normal gene · Albinsm gene
NORMAL CHILD · CARRIER CHILD · CARRIER CHILD · AFFECTED CHILD

RECESSIVE INHERITANCE
In this example, both parents carry an abnormal gene but do not have the disorder. Their children may be unaffected (1 in 4 chance), may be carriers of the faulty gene (1 in 2), or may have the condition (1 in 4).

COLOR-IMPAIRED VISION

COLOR-IMPAIRED VISION AFFECTS THE ABILITY TO DISTINGUISH BETWEEN TWO COLORS, NOTABLY RED AND GREEN OR, LESS COMMONLY, BLUE AND YELLOW.

A person with color-impaired vision has difficulty distinguishing between two colors because of a defect in the cones in the retina. The different forms are due to faulty genes, mostly on the X chromosome. The more common form, red-green, affects males since they have only one X chromosome, and any defect carried is likely to be expressed. In a female, who has two X chromosomes, a fault on one will be over-ridden by the other, and she becomes a carrier. This type of inheritance is called X-linked recessive. Blue–yellow color impairment is also inherited but not linked to the X chromosome.

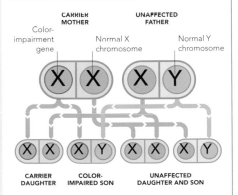

CARRIER MOTHER · Color-impairment gene · Normal X chromosome
UNAFFECTED FATHER · Normal Y chromosome
CARRIER DAUGHTER · COLOR-IMPAIRED SON · UNAFFECTED DAUGHTER AND SON

X-LINKED RECESSIVE INHERITANCE
In this example, a mother carries the abnormal gene on the X chromosome but is unaffected. The sons have a 1 in 2 chance of inheriting the disorder, and the daughters have a 1 in 2 chance of being carriers but will not have the disorder.

CYSTIC FIBROSIS

IN CYSTIC FIBROSIS THE MUCUS-SECRETING GLANDS PRODUCE ABNORMALLY THICK SECRETIONS THAT CAUSE PROBLEMS IN MANY PARTS OF THE BODY.

Cystic fibrosis is a common, severe, inherited disease that causes various health problems and a reduced life expectancy. Symptoms are due to excess mucus in the body, notably in the lungs and pancreas. This causes repeated lung infections and problems digesting food, causing failure to put on weight or grow at the normal rate. Cystic fibrosis is caused by an abnormal gene carried on chromosome 7. It is a recessive disorder, which means a defective gene has to be received from each parent. Prenatal genetic testing and genetic counseling will be offered to parents of one affected child if they are considering having more.

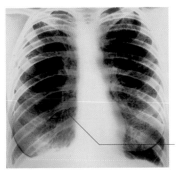

DAMAGE TO LUNGS
In this color-enhanced chest X-ray of the lungs of a patient with cystic fibrosis, the bronchial walls (orange) on either side of the spine (white, centre) are thickened as a result of repeated infection caused by excess production of mucus in the body.

Thickened bronchial walls

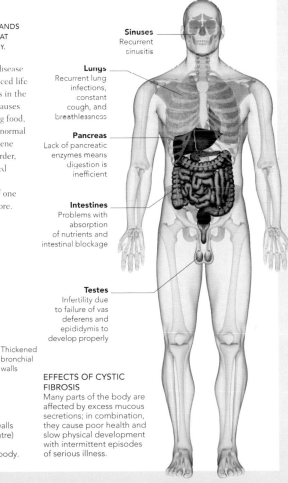

Sinuses
Recurrent sinusitis

Lungs
Recurrent lung infections, constant cough, and breathlessness

Pancreas
Lack of pancreatic enzymes means digestion is inefficient

Intestines
Problems with absorption of nutrients and intestinal blockage

Testes
Infertility due to failure of vas deferens and epididymis to develop properly

EFFECTS OF CYSTIC FIBROSIS
Many parts of the body are affected by excess mucous secretions; in combination, they cause poor health and slow physical development with intermittent episodes of serious illness.

CANCER

CANCER IS NOT A SINGLE DISEASE, BUT A LARGE GROUP OF DISORDERS WITH DIFFERENT SYMPTOMS. NEARLY ALL CANCERS HAVE THE SAME BASIC CAUSE: CELLS MULTIPLY UNCONTROLLABLY BECAUSE THE NORMAL REGULATION OF THEIR DIVISION HAS BEEN DAMAGED. THE FAULTY GENES AT THE ROOT OF THE PROBLEM—AND THERE MUST BE MORE THAN ONE OF THESE GENES—MAY BE INHERITED, OR THEY MAY BE CAUSED BY KNOWN CARCINOGENS (CANCER-CAUSING AGENTS) OR THE AGING PROCESS.

CANCEROUS (MALIGNANT) TUMORS

A CANCER IS A GROWTH OR LUMP THAT DAMAGES SURROUNDING TISSUES AND ORGANS, AND THAT MAY SPREAD TO OTHER PARTS OF THE BODY.

Normally, cells divide and replace themselves at a controlled rate. A malignant, or cancerous, tumor is a mass of abnormal cells that divide excessively quickly and do not carry out the normal functions of their tissue.

These cells are often irregular in size and shape and bear little resemblance to the normal cells from which they arose. This irregular appearance is often used to diagnose cancer during microscopic examination of a small sample of tissue taken from a tumor. A tumor gradually enlarges, crowding out normal cells, pressing on nerves, and infiltrating blood and lymph vessels. It is important to distinguish a malignant tumor from a nonmalignant one, because cancerous cells can spread to other parts of the body.

CANCER CELLS DIVIDING
In this magnified image, a cancerous cell is dividing to form two cells that contain damaged genetic material. If left untreated cancer cells multiply uncontrollably.

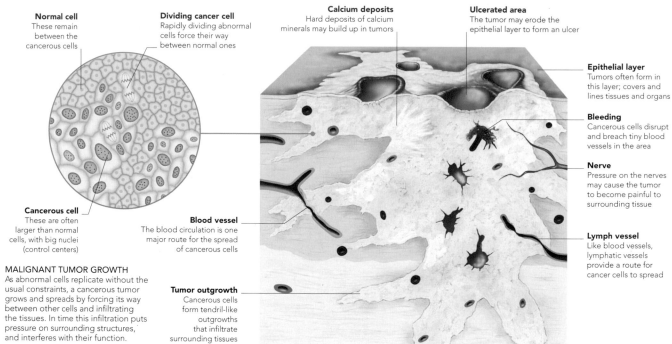

Normal cell
These remain between the cancerous cells

Dividing cancer cell
Rapidly dividing abnormal cells force their way between normal ones

Calcium deposits
Hard deposits of calcium minerals may build up in tumors

Ulcerated area
The tumor may erode the epithelial layer to form an ulcer

Epithelial layer
Tumors often form in this layer; covers and lines tissues and organs

Bleeding
Cancerous cells disrupt and breach tiny blood vessels in the area

Nerve
Pressure on the nerves may cause the tumor to become painful to surrounding tissue

Lymph vessel
Like blood vessels, lymphatic vessels provide a route for cancer cells to spread

Cancerous cell
These are often larger than normal cells, with big nuclei (control centers)

Blood vessel
The blood circulation is one major route for the spread of cancerous cells

Tumor outgrowth
Cancerous cells form tendril-like outgrowths that infiltrate surrounding tissues

MALIGNANT TUMOR GROWTH
As abnormal cells replicate without the usual constraints, a cancerous tumor grows and spreads by forcing its way between other cells and infiltrating the tissues. In time this infiltration puts pressure on surrounding structures, and interferes with their function.

NONCANCEROUS (BENIGN) TUMORS

Benign tumors are caused by changed cells that multiply abnormally and do not carry out their usual functions. However, these tumors are usually self-contained and do not spread into surrounding tissues or elsewhere in the body. If they do not cause problems, they may be left untreated, since treatment itself may carry small risks. But some benign tumors grow so that they become unsightly or compress surrounding areas, in which case treatment may be advised.

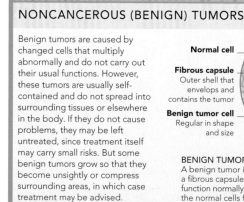

Normal cell

Fibrous capsule
Outer shell that envelops and contains the tumor

Benign tumor cell
Regular in shape and size

BENIGN TUMOR STRUCTURE
A benign tumor is usually contained within a fibrous capsule or "shell." Its cells do not function normally but they resemble more closely the normal cells from which they originated.

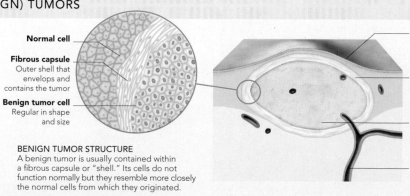

Surrounding tissues
May be distorted but are neither breached nor infiltrated

Fibrous capsule
Forms a boundary that prevents tumor cells from spreading

Body of tumor
May enlarge slowly or rapidly, depending on genetic changes in the cells

Blood vessel
Oxygen and nutrients reach the tumor through a system of blood vessels

HOW CANCER STARTS

CANCERS ARE OFTEN TRIGGERED BY CARCINOGENS (SUCH AS TOBACCO SMOKE AND CERTAIN VIRUSES). HOWEVER, INHERITANCE OF FAULTY GENES ALSO PLAYS A PART.

Cancer-causing agents—carcinogens—damage specific genes (sections of DNA) known as oncogenes that regulate vital processes such as cell division and growth, repair of damaged genes, and the ability of faulty cells to self-destruct. Most damaged genes are repaired as part of normal cell metabolism, but some can gradually be altered, or mutated, by regular exposure to a carcinogenic agent so that they fail to carry out their functions. Damage to oncogenes may cause them to make altered versions of their chemicals within the cell. These work like molecular locks or keys that "trick" the cell into functioning abnormally; eventually the cell may become cancerous and divide to form a tumor.

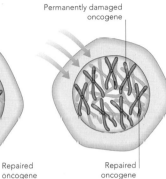

Carcinogen | Normal gene | Newly damaged oncogene | Permanently damaged oncogene | Permanently damaged oncogene

Outer cell membrane | Chromosome | Nucleus | Newly damaged oncogene | Repaired oncogene | Repaired oncogene

1 DAMAGE FROM CARCINOGENS
Carcinogens continually bombard cells and may eventually affect genes on the chromosomes. Usually new damage to oncogenes is limited and soon repaired.

2 PERMANENT DAMAGE
Damage to oncogenes and repair continue, but with time or higher than normal exposure to carcinogens, some of the oncogenes suffer permanent harm.

3 CELL BECOMES CANCEROUS
Eventually, a number of oncogenes are permanently altered. Key cell functions are irreparably affected and the cell "tips over" into cancerous mode.

FORMATION OF A TUMOR

It takes just one cell to undergo cancerous changes for a tumor to form. This unchecked cell divides into two cells, which each do the same, and so on. All resulting cells inherit the cancerous changes. The total cell number doubles on each division. A solid tumor is detectable after 25–30 doublings, when it contains about one billion cells.

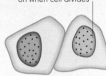

Nucleus with damaged ongenes | Damaged ongenes passed on when cell divides

CANCEROUS CELL | **FIRST DOUBLING**

Abnormal cells multiply to form a solid tumor

TUMOR GROWTH
After just four doublings, there are 16 cells, and after ten doublings, more than 1,000. The doubling time varies according to the tumor type but ranges from days to many years.

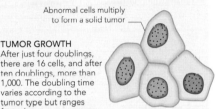

SECOND DOUBLING

HOW CANCER SPREADS

THE DEFINING FEATURE OF A MALIGNANT TUMOR IS ITS ABILITY TO SPREAD, NOT ONLY LOCALLY INTO NEIGHBORING TISSUES, BUT ALSO TO DISTANT SITES.

The spread of cancerous cells to distant body locations is known as metastasis. The initial tumor is called the primary tumor, and those that develop in remote sites are known as secondary tumors or metastases. Secondary tumors do not arise randomly; for example, breast cancer tends to spread to the bones and lungs. To metastasize, cancerous cells must overcome many obstacles, such as scavenging white blood cells and other weapons of the body's immune system. However, once they penetrate healthy tissue, the malignant cells set up their own blood system by invading existing blood vessels and by producing chemicals that stimulate blood vessels to infiltrate the tumor (angiogenesis). The main routes of spread are the body's two "highways" for distributing nutrients and collecting wastes: the blood and lymph systems.

Spread by lymph

The lymphatic system is a network of vessels that contain lymph fluid, and nodes (glands) containing white blood cells. Cancerous cells enter a lymph vessel and travel to a lymph node, where they may develop into a tumor; some cells may be destroyed by the immune system, temporarily halting the spread.

Spread by blood

Primary cancer often spreads to sites in the body that have a good blood supply, such as the liver, lungs, and brain. The liver is a particularly common site since it receives a plentiful supply from the heart and also from the intestines via the portal vascular system (see p.197). When the cancerous cells reach very small blood vessels, they can push between the cells of the vessel wall and invade the tissues beyond.

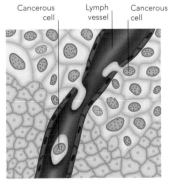

Cancerous cell | Lymph vessel | Cancerous cell

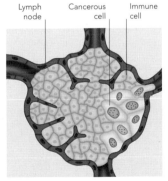

Lymph node | Cancerous cell | Immune cell

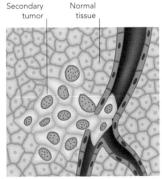

Blood vessel | Cancerous cell | Cancerous cell | Secondary tumor | Normal tissue

1 LYMPH VESSEL BREACHED
As the primary tumor grows, its cells invade adjacent tissues—and the small vessels of the lymphatic system are rarely far away. Cancerous cells enter the lymph fluid and pass along it to the nearest lymph node.

2 TUMOR IN LYMPH NODE
Just one cancerous cell entering a local lymph node can start to divide and grow into a secondary tumor (metastasis). Immune cells here may destroy some of the cancer cells and stop the spread temporarily.

1 BLOOD VESSEL WALL RUPTURED
As a primary tumor expands and infiltrates, some of its cells rupture the walls of blood vessels. The cancerous cells can now detach, to be swept away by the blood and flow around the circulatory system.

2 SECONDARY TUMOR FORMED
Cancerous cells are often bigger than red blood cells and become lodged in narrow vessels distant from the primary site. Here the cells divide, push into surrounding tissues, and establish a secondary tumor.

GLOSSARY

Text terms in *bold italics* refer to other items that appear in the glossary.

A

Abscess
A walled cavity containing *pus*, surrounded by inflamed or dying *tissue*.

Accommodation
The process by which the eyes adjust to focus on nearby or distant objects.

Acoustic neuroma
A *tumor* on the *nerve* that connects the ear and brain.

Acquired immune deficiency syndrome (AIDS)
A condition resulting from infection with the *human immunodeficiency virus* (*HIV*), which is spread by sexual intercourse or infected blood. AIDS results in the loss of resistance to infections and some *cancers*.

Acute
A condition that begins abruptly and may last for a short time. Contrasts with *chronic* conditions.

Adenoids
Clusters of *lymphoid tissue* on each side of the back of the upper part of the throat.

Adipose tissue
Tissue made of specialized cells that store fatty (*lipid*) substances, for energy, physical "padding," and insulation.

Allele
Form or version of a *gene*. For example, the *gene* for eye color has several eye-color versions.

Allergen
Any substance causing an allergic reaction in a person previously exposed to it.

Alveolus (pl. alveoli)
One of many tiny air sacs in the lungs. Gases diffuse in and out of blood through alveolar walls.

Alzheimer's disease
A progressive *dementia* due to loss of *nerve* cells in the brain that affects more than 1 in 10 people over 65.

Amino acid
One of about 20 kinds of building-block subunits of *protein*.

Amniocentesis
The process of withdrawing a sample of fluid from the *uterus* to obtain information on the health and genetic makeup of the *fetus*.

Anemia
A group of conditions in which the amount of *hemoglobin* in the blood is reduced.

Aneurysm
A swelling of an *artery* caused by damage to or weakness in the vessel wall.

Angina
Pain or tightness in the center of the chest brought on by exertion; caused by an inadequate blood supply to the heart muscle.

Angiography
A method of imaging blood vessels in which *X-rays* are taken after a *contrast medium* has been injected.

Angioplasty
Any process used to widen the bore of an *artery* that is narrowed by disease. See also *Balloon angioplasty*.

Antibiotic
A medical drug that acts chiefly against *bacteria*; it has little or no effect against *viruses*.

Antibody
A soluble *protein* that attaches to body invaders, such as *bacteria*, and helps destroy them.

Anticoagulant
A drug used to limit any tendency for blood to clot within *arteries* or *veins*.

Aorta
The central and largest *artery* of the body. It arises from the heart's left *ventricle* and supplies oxygenated blood to all other *arteries* except the pulmonary *artery*.

Aortic valve
A triple-cusped valve at the origin of the *aorta* that allows blood to leave the left *ventricle* of the heart but prevents backward flow.

Appendix
The wormlike structure attached to the large intestine. Its function, if it has one, is as yet unknown.

Aqueous humor
The fluid filling the front chamber of the eye between the back of the *cornea* and front of the iris and *lens*.

Arrhythmia
An irregular heartbeat, due to a defect in the electrical impulses or pathways that control contractions.

Arteriole
A small terminal branch of an *artery* leading to even smaller *capillaries*, which link to the *veins*.

Artery
An elastic, muscular-walled tube that transports blood away from the heart to other body parts.

Arthritis
Inflammation in a joint, causing varying degrees of pain, swelling, redness, and restriction of movement.

Articulation
A joint, or the way in which jointed parts are connected.

Asthma
A disease in which the airways narrow so that breathing becomes intermittently difficult.

Atherosclerosis
A degenerative disease of arteries in which raised plaques of fatty material limit blood flow and cause local blood clotting.

Atrial fibrillation
A disorder in which the *atria* beat very rapidly.

Atrial septal defect
A hole in the wall (the septum) between the upper two chambers of the heart.

Atrium (pl. atria)
One of two thin-walled, upper chambers of the heart.

Autoimmune disease
A disease caused by a defect in the immune system, which attacks the body's own tissues.

Autonomic nervous system (ANS)
The portion of the nervous system controlling unconscious functions such as heartbeat and breathing.

Axon
The long, fiberlike process of a *nerve* cell that conducts impulses away from the cell body; bundles of axons form *nerves*.

B

Bacterium (bacteria pl.)
A type of microorganism with one cell. Only a few of the many species of bacteria cause disease.

Balloon angioplasty
The use of a catheter with an inflatable tip to widen an *artery*.

Basal ganglia
Paired masses, or nuclei, of *nerve* cell bodies lying deep in the brain; concerned with control of movement.

Base
In *nucleic acids* (*DNA*, *RNA*), nitrogen-containing chemical units or nitrogenous bases (adenine, thymine, guanine, cytosine, uracil) the order of which carries genetic information.

Benign
Mild and with no tendency to spread; contrasts with a *malignant* condition.

Beta-blocker
A drug that blocks the action of epinephrine (adrenaline). This slows the *pulse* and reduces blood pressure.

Bile
A greenish brown fluid from the *liver* that is concentrated and stored in the *gallbladder*; helps in digestion of fats.

Biliary system
The network of *bile* vessels formed by the ducts from the *liver* and the *gallbladder*, and the *gallbladder* itself.

Biopsy
A sample of *tissue* from any part of the body that is suspected of disease, taken for microscopic examination.

Blood clot
A mesh of *fibrin*, *platelets*, and blood cells that forms when a blood vessel is damaged.

Boil
An inflamed, *pus*-filled area of skin, which is usually an infected *hair follicle*.

Bolus
A chewed-up quantity of food ready to be swallowed; also, a drug rapidly injected into the bloodstream.

Bone marrow
The fatty *tissue* within bone cavities, which may be red or yellow. Red bone marrow produces *red blood cells*, *platelets*, and most *white blood cells*.

Bradycardia
A slow heart rate. This is normal in athletes but may signal disorders in others.

Brain stem
The lower part of the brain; houses the centers that control vital functions, such as breathing and the heartbeat.

Breech delivery
A buttock-first birth; carries a slightly higher risk to the *fetus* than a head-first birth.

Bronchial tree
The trachea and the branching system of air tubes in the lungs; consists of progressively smaller *bronchi* and bronchioles.

Bronchitis
Inflammation of the lining of the breathing tubes, resulting in a cough that produces large amounts of sputum (phlegm).

Bronchus (pl. bronchi)
One of the larger air tubes in the lungs. Each lung has a main bronchus that divides into smaller branches.

C

Calcium-channel blocker
A drug that limits movement of dissolved calcium across cell membranes; it is used to treat high blood pressure and heart *arrhythmias*.

Cancer
A group of diseases caused by uncontrolled growth of abnormal cells, which may form a *tumor*. It may invade surrounding tissue or spread to other parts of the body (see *metastasis*).

Capillary
One of the tiny blood vessels that link the smallest *arteries* and smallest *veins*.

Carcinoma
A *cancer* of either the inner or the outer surface layer (*epithelium*). Carcinomas commonly occur in the skin, linings of the air tubes, large intestine, breast, *prostate gland*, and *uterus*.

Cardiac
Relating to the heart.

Carpal tunnel syndrome
Numbness and pain in the thumb and middle fingers. It results from pressure on the median *nerve* where it passes through the gap under a *ligament* in front of the wrist.

Cartilage
Common type of connective *tissue*, usually tough and resilient, forming some structural parts, such as the ear and nose, and lining bones inside joints.

Central nervous system (CNS)
The brain and spinal cord; receives and analyzes sensory data, and initiates a response.

Cerebellum
The region of the brain located behind the *brain stem*. It is concerned with balance and the control of fine movement.

Cerebrospinal fluid
A watery fluid that bathes the brain and spinal cord.

Cerebrum
The largest part of the brain; made up of two cerebral hemispheres. It contains the *nerve* centers for thought, personality, the senses, and voluntary movement.

Chemotherapy
Treatment involving powerful chemical drugs, often used to kill cancerous or *malignant* cells.

Chlamydia
Small *bacterium* that causes the eye disease trachoma and *pelvic inflammatory disease*.

Cholecystitis
Inflammation of the *gallbladder*; commonly the result of obstructed outflow of *bile* by a *gallstone*.

Cholecystography
X-ray of the *gallbladder* after a *contrast medium* has been introduced into it.

Cholestasis
A slowing or cessation of the flow of *bile* in the *liver*.

Chorionic villus sampling
Removal of a small piece of *tissue* from the *placenta* for *chromosome* or *gene* analysis; allows early detection of fetal abnormalities.

Chromosome
A threadlike structure, present in all nucleated body cells, that carries the genetic code for the formation of the body. Chromosomes coil into "X" shapes. A normal human body has 23 pairs of chromosomes.

Chronic
A persistent medical condition that usually lasts more than six months and may result in a long-term change in the body; contrasts with *acute*.

Cirrhosis
Replacement of *liver tissue* by fine fibrous *tissue*, which results in hardening and impaired function; may be caused by excessive alcohol consumption or infection.

Cochlea
The coiled structure in the inner ear that contains the organ of Corti, which converts sound vibrations into *nerve* impulses for transmission to the brain.

Collagen

The body's most important structural *protein*, present in bones, *tendons*, *ligaments*, and other connective *tissues*. Collagen fibrils are twisted into bundles called fibers.

Colon

The part of the large intestine that extends from the cecum to the rectum. Its main function is to conserve water by absorbing it from the bowel contents.

Congenital

Present at birth. Congenital disorders may be hereditary or may result from diseases or injuries that occur during fetal life or the birth itself.

Contrast medium

A substance through which *X-rays* are unable to pass.

Cornea

The transparent dome at the front of the eyeball that is the eye's main focusing *lens*.

Coronary

A term meaning "crown." Refers to the arteries that encircle and supply the heart with blood.

Corpus callosum

The wide, curved band of about 20 million *nerve* fibers that connects the two hemispheres of the *cerebrum*.

Corticosteroid

A drug that simulates the natural steroid *hormones* of the outer zone (*cortex*) of the adrenal glands or the *sex hormone*.

Cortex

Outer layer in various *organs*, such as the cerebral cortex (brain), *renal* cortex (*kidney*), and adrenal cortex.

Cranial nerves

The 12 pairs of *nerves* emerging from the brain and *brain stem*. They include the *nerves* for smell, sight, eye movement, facial movement and sensation, hearing, taste, and head movement.

Crohn's disease

An inflammatory disease that affects the *gastrointestinal tract*. Symptoms may include pain, *fever*, and diarrhea.

Cyst

A walled cavity, which is usually spherical, filled with fluid or semisolid matter; usually benign.

Cystadenoma

A harmless, *cyst*like growth of glandular *tissue*.

Cystitis

Inflammation of the urinary bladder, usually caused by infection. Produces frequent, painful urination and, in some cases, incontinence.

Cytoplasm

Watery or jellylike substance that fills the bulk of a cell; it contains many *organelles*.

D

Defibrillation

A strong pulse of electric current applied to the heart to restore its normal rhythm.

Dementia

The loss of mental powers and *memory* as well as the ability to look after oneself; dementia is often a result of degenerative brain disease.

Dermis

The thick inner layer of skin made of connective *tissue*; contains structures such as sweat glands.

Dialysis

The basis of artificial *kidney* machines that separate dissolved substances. A system of filtration across a semipermeable membrane. Permits waste excretion and preservation of essential nutrients.

Diaphragm

The dome-shaped muscular sheet that separates the chest from the abdomen. When the muscle contracts, the dome flattens, increasing chest volume and drawing air into the lungs.

Diastole

The period in the heart cycle when all four chambers are relaxed and the heart is filling with blood.

Diffusion

The natural tendency of fluid substances to spread out, especially when in solution, to give an even concentration.

Digestive system

The mouth, *pharynx*, esophagus, stomach, and intestines. Associated *organs* are the *pancreas*, *liver*, and *gallbladder*, and their ducts.

Diverticular disease

The presence of diverticula, the small sacs that are created by protruson of the intestine's inner lining through the wall.

DNA (Deoxyribonucleic acid)

A chemical with a double-helix structure that carries genetic information in the form of the sequence of its subunits (*bases*).

Dominant

In genetics, when one form (allele) of a *gene* is "stronger" and determines what trait is expressed even in the presence of a *recessive allele* of the same gene.

Dopamine

A chemical messenger (neurotransmitter) in the brain that is involved in the control of body movement.

Down syndrome

A genetic disorder in which a person's cells contain an extra *chromosome* 21 (three instead of the usual two). For this reason the condition is also known as trisomy 21.

Duodenum

The C-shaped first part of the small intestine, into which the stomach empties. Ducts from the *gallbladder*, the *liver*, and the *pancreas* all enter the duodenum.

Dura mater

A tough membrane, the outer layer of the *meninges*, which covers the brain and the spinal cord. It lies over the arachnoid and pia mater and adheres closely to the inside of the skull.

E

Eardrum

The membrane separating the outer ear from the *middle ear* that vibrates in response to sound.

Ectopic pregnancy

Implantation of a fertilized egg in a site other than the uterine lining.

Electrocardiography

Recording and study of the heart's electrical activity.

Electroencephalography

Recording and study of the brain's electrical signals.

Embolus (pl. emboli)

Any material, such as *blood clots*, air bubbles, *bone marrow*, fat, or *tumor* cells, carried away in the bloodstream and capable of causing a blockage in a blood vessel.

Embryo

The developing baby from conception until the eighth week of pregnancy.

Endocarditis

An inflammation that affects either the inner lining of the heart wall or a *heart valve*.

Endocrine gland

A gland that produces *hormones* (chemical messenger substances), which are released directly into the blood rather than along tubes or ducts.

Endorphin
A morphinelike substance produced by the body in times of pain and stress, and also during exercise.

Endoscopy
Insertion of a viewing device into the body, through a natural orifice or an incision, to study the interior, take samples, or carry out treatment.

Enzyme
A *protein* that accelerates a chemical reaction.

Epidermis
The outer layer of the skin; its cells become flatter and scalier toward the surface.

Epiglottis
A leaflike flap of *cartilage* at the entrance of the *larynx*, covering the opening of the airways during swallowing and helps prevent food or liquid from entering the windpipe (*trachea*).

Epilepsy
A disorder featuring episodes of unregulated electrical discharge throughout the brain or in a specific area.

Epithelium
Covering or lining *tissue* that forms sheets and layers around and within many *organs* and other *tissues*.

Esophagitis
An inflammation that affects the esophagus, often caused by the acid reflux into the esophagus.

Estrogen
A sex *hormone* that prepares the uterine lining for an implanted fertilized egg and stimulates the development of a female's secondary sexual characteristics.

Eustachian tube
The tube that connects the back of the nose to the *middle ear* cavity and equalizes air pressure.

F

Fallopian tube
One of the two tubes along which an *ovum* travels to the *uterus*, after release from an *ovary*: the most common site of an *ectopic pregnancy*.

Fertilization
The union of a sperm and an egg, after sexual intercourse or artificial insemination, or in a laboratory test tube.

Fetus
The developing baby from the eighth week after *fertilization* until birth. See *embryo*.

Fever
A body temperature that registers above 98.6°F (37°C), measured in the mouth, or 99.8°F (37.7°C) in the rectum.

Fiberoptics
The transmission of images through bundles of flexible, glass or plastic threads. Some types of endoscope use fiberoptic transmission to view and treat structures that are located far within the human body.

Fibrin
An insoluble *protein* that is converted from the blood *protein* fibrinogen to form a fibrous network, which is a stage in the creation of a *blood clot*.

Fibroid
A *benign tumor* of fibrous and muscular *tissue* growing in the wall of the *uterus*, usually in women over 30. Fibroids are often multiple and may cause symptoms.

Fibrosis
An overgrown scar or connective *tissue* that is formed as the body's natural healing response to any wound or burn. Fibrous *tissue* may modify an *organ's* structure, and thereby impair its effectiveness.

Fistula
An abnormal channel that lies between any part of the interior of the body and the surface of the skin, or between two internal *organs*.

E

Gallbladder
The small, fig-shaped bag lying under the *liver*, into which *bile* secreted by the *liver* passes to be stored.

Gallstone
An oval or faceted mass of cholesterol, calcium, and *bile* pigment, that forms in the *gallbladder*. Gallstones vary in size and are more common in women than in men.

Ganglion
Lumplike group of cell bodies of *nerve* cells (*neurons*) with many interconnections; alternatively, a localized, *cyst*like, fluid-filled lump that is near a *tendon*, joint, or bone.

Gastric juice
A mixture produced by the cells of the stomach that contains hydrochloric acid and digestive *enzymes*.

Gastritis
An inflammation of the stomach lining from any cause, including infection or alcohol.

Gastrointestinal tract
The muscular tube that extends from the mouth, through the *pharynx*, esophagus, stomach, and small and large intestines to the rectum.

Gene
A distinct section of a chromosome that is the basic unit of inheritance. Each gene consists of a segment of deoxyribonucleic acid (*DNA*) containing the code that governs the production of a specific *protein*.

Genome
The full set of *genes*, or hereditary information, for a living organism; the human genome has 20,000–25,000 *genes*.

Glaucoma
An abnormal rise in the pressure of the fluids within the eye that, if left untreated, causes internal damage to the eye that may result in blindness.

Glial tissue
Cells of the nervous system that provide support for *neurons*.

Glucose
A simple sugar obtained by breakdown of long-chain carbohydrates, such as starch, in the diet. Glucose is the main form of energy within the body.

Gonorrhea
A sexually transmitted disease that may cause pelvic inflammation in women and narrowing of the urine outlet tube in men. If untreated, the disease may spread to other parts of the body.

Gout
A metabolic disorder causing attacks of *arthritis*, usually in a single joint.

Gray matter
The darker colored regions of the brain and spinal cord that are composed mainly of *neuron* cell bodies as opposed to their projecting fibers, which form *white matter*.

H

Hair follicle
A pit on the surface of the skin from which hair grows.

Heart-lung machine
A pump and oxygenator that performs the functions of the heart and the lungs during *cardiac* operations.

Heart valve
One of four structures of the heart that allow passage of blood in one direction only.

Hematoma
An accumulation of blood within any part of the body, caused by a torn blood vessel.

Hemoglobin
The *protein* in *red blood cells* that combines with oxygen, carrying it from the lungs throughout the body.

Hemiplegia
Paralysis of one half of the body, from damage to the motor areas in the brain, or to the *nerve* tracts that connect these motor areas to the spinal cord.

Hemophilia
An inherited bleeding disorder caused by deficiency of a clotting *protein*.

Hemorrhage
The escape of blood from a blood vessel, usually as a result of an injury.

Hemorrhoids
Ballooning of *veins* in the anal lining (external hemorrhoids) or in the lower part of the rectum (internal hemorrhoids).

Hepatic
Concerning the liver.

Hepatitis
Inflammation of the *liver*, usually as a result of a viral infection, excess alcohol, or toxic substances. Symptoms include *fever* and *jaundice*.

Hepatocyte
A type of *liver* cell with many functions.

Hernia
Displacement of an organ or tissue out of the cavity in which it usually lies. The most common type is a *hiatus hernia*.

Hiatus hernia
The sliding upward of part of the stomach through the opening in the *diaphragm*.

Hippocampus
A structure in the brain concerned with learning and long-term *memory*.

Homeostasis
Active processes by which an organism maintains a constant internal temperature.

Hormone
A chemical released by the *endocrine glands* and some *tissues*. Acts on specific receptor sites in other parts of the body.

Human immunodeficiency virus (HIV)
The virus that causes AIDS and destroys cells of the immune system, thereby undermining its effectiveness.

Hypothalamus
A small structure that is located at the base of the brain, where the nervous and hormonal systems of the body interact. The hypothalamus is linked to the *thalamus* above and the *pituitary gland* below.

I–K

Ileum
The final segment of the small intestine, where most absorption of nutrients takes place.

Immune deficiency
Any failure of the function of the immune system from causes such as AIDS, *cancer* treatment, or aging.

Immunity
Resistance or protection against disease, especially infection.

Immunosuppressant
A drug that interferes with the production and activity of certain lymphocytes.

Interferon
A *protein* produced by cells to defend against viral infections and some *cancers*.

In vitro fertilization
Fertilization of *ova* in a laboratory container by the addition of sperm or sperm *nuclei*; the resulting *embryos* are introduced into the woman's *uterus*.

Irritable bowel syndrome
Recurrent gas, abdominal discomfort, and alternating constipation and diarrhea. It is often associated with stress.

Jaundice
A yellowing of the skin and whites of the eyes that is due to deposition of *bile* pigment. Jaundice results from altered *liver* function.

Kaposi's sarcoma
A slow-growing *tumor* of blood vessels that affects some people with AIDS. Scattered bluish brown nodules occur on the skin and internally.

Kidney
One of two bean-shaped organs in the back of the abdominal cavity that filter blood and remove wastes, particularly urea.

Killer T cells
White blood cells that can destroy damaged, infected, or *malignant* body cells.

L

Laparoscopy
The visual inspection of the interior of the abdomen, through a narrow optical and illuminating device (endoscope), and often using a video camera.

Larynx
The structure in the neck at the top of the *trachea*, known as the voice box, that contains the *vocal cords*.

Lens
The internal lens of the eye, also called the crystalline lens; it fine-focuses vision by adjusting its curvature. The outer lens is called the *cornea*.

Leukemia
A group of blood disorders in which *malignant white blood cells* grow in *bone marrow* and invade *organs* elsewhere in the body.

Ligament
A band of *tissue* consisting of *collagen* (a tough, fibrous, elastic *protein*). Support bones, mainly in and around joints.

Limbic system
A collection of structures inthe brain that plays a role in the automatic (involuntary) body functions, emotions, and the sense of smell.

Lipid
Fatty or oily substance, insoluble in water, with varied roles in the body, including formation of *adipose tissue*, cell membranes (phospholipid), and steroid *hormones*.

Liver
The large *organ* in the upper right abdomen that performs vital chemical functions, including processing of nutrients from the intestines, manufacture of sugars, *proteins*, and fats; detoxification of poisons; and conversion of waste to *urea*.

Lobe
A rounded projection or subdivision forming part of a larger structure such as the brain, lung, or *liver*.

Lymphatic system
An extensive network of transparent lymph vessels and *lymph nodes*. It returns excess *tissue* fluid to the circulation and combats infections and *cancer* cells.

Lymph node
A small, oval gland packed with *white blood cells* that acts as a barrier to the spread of infection. Nodes occur in series along lymph vessels.

Lymphocyte
White cell that is part of the immune system; it protects against *virus* infections and *cancer*.

Lymphoid tissue
A *tissue* rich in *lymphocytes* found in *lymph nodes*, the *spleen*, intestines, and *tonsils*.

M

Macula
Any small, flat, colored spot on the skin; also the central region of the *retina*; also a structure in the ear vestibule.

Malignant
Refers to a *cancer* that may invade surrounding tissues and spread to other parts of the body. Contrasts with *benign*.

Mammography
The *X-ray* screening of the breasts, using low-radiation *X-rays*; used to detect breast *cancer* at an early stage.

Mastectomy
Surgical removal of part or all of the breast. It is usually performed to treat breast *cancer* and is often followed by *radiation therapy*.

Mastitis
Inflammation of the breast, usually resulting from an infection acquired during breast-feeding. *Bacteria* enter through cracks in the nipples. Symptoms include *fever*, and hardening or tenderness of the breast.

Medulla
The inner part of an *organ*, such as the *kidneys* or adrenal glands. Also refers to the part of the *brain* stem lying immediately above the start of the spinal cord, just in front of the *cerebellum*.

Meiosis
The stage in the formation of the sperm and eggs when chromosomal material is randomly redistributed and the number of *chromosomes* is reduced to 23 instead of the usual 46 found in other body cells.

Memory
The data store for recent and remote experience. Short-term memory stores are small and the contents are soon lost unless repeatedly refreshed. Long-term memory stores are very large but are not always readily accessible.

Meninges
The three membrane layers around the brain and spinal cord, the pia mater on the inside, arachnoid in the middle, and *dura mater* next to the skull.

Meningitis
An inflammation of the *meninges*, usually as a result of an infection with *viruses* or *bacteria*.

Meniscectomy
Surgical removal of a torn or displaced *cartilage* (*meniscus*) from the knee joint; usually carried out with the use of a fiberoptic viewing tube, which is inserted into the joint, and a TV monitor.

Meniscus
A crescent-shaped pad of *cartilage* found in the knee and some other joints.

Menopause
The end of the reproductive period in women, when the *ovaries* have ceased their production of eggs and menstruation has stopped.

Metabolism
The sum of all the physical and chemical processes that take place in the body.

Metastasis
The spread or transfer of any disease, but especially *cancer*, from its original site to another site where the disease process continues.

Microscopy
Examination by a microscope, often to make a diagnosis. Simple techniques use focused light rays and magnifying lenses; in order to achieve higher magnifications, beams of electrons are used.

Middle ear
The air-filled cleft within the temporal bone between the *eardrum* and the outer wall of the inner ear; contains *ossicles*. Also called the tympanic cavity.

Migraine
A neurological disorder causing recurrent severe headaches often accompanied by nausea and visual disturbances.

Miscarriage
A spontaneous ending of a pregnancy before the *fetus* is mature enough to survive outside the *uterus*.

Mitochondrion (pl. mitochondria)
A cell *organelle* involved in the production of energy for cell functions. It contains genetic material (mitochondrial DNA) derived solely from the mother.

Mitosis
The process by which a cell *nucleus* divides to produce two daughter cells, each of which has the identical genetic makeup of the parent cell.

Mitral valve
The valve that lies between the left *atrium* and the left *ventricle* of the heart.

Mole
Any birthmark, pigmented spot, growth, or *congenital* blemish, whether flat, raised, and or hairy, on the skin.

Molecule
A group of atoms joined, or bonded, together. Water (H_2O) has three atoms, two hydrogen (H) and one oxygen (O); large molecules such as *proteins* and *DNA* have millions.

Motor cortex
The part of the surface layer of each hemisphere of the *cerebrum* in which voluntary movement is initiated. The motor *cortex* can be mapped into areas that are linked to particular parts of the body.

Motor neuron
A *nerve* cell that carries the impulses to muscles that cause its movement.

Motor neuron disease
A rare disorder in which *motor neurons* suffer a progressive destruction, resulting in a corresponding loss of movement.

Mucocele
A *cyst*like abnormal sac filled with mucus that arises from a *mucous membrane*.

Mucous membrane
The soft, mucus-secreting epithelial layer lining the tubes and cavities of the body.

Muscular dystrophy
One of several herditary muscle disorders featuring gradual, progressive muscle degeneration and weakening.

Mutation
Change in the genetic material, usually by alteration of one or more *nucleic acid* bases.

Myocardium
The special muscle of the heart, The fibers form a network that can contract spontaneously.

Myofibril
Cylindrical elements within muscle cells (fibers) that consist of thinner filaments, which move to produce muscle contraction.

Myofilament
Long, threadlike *proteins* within *myofibrils* of muscle cells.

N

Nephron
The *kidney's* filtering and tubular system, consisting of a filtration capsule, the glomerulus, and a series of tubules that reabsorbs or excretes water and wastes to control fluid balance.

Nerve
The threadlike projections of individual *neurons* (nerve cells) held together by a fibrous sheath. Nerves carry electrical impulses to and from the brain and spinal cord and other body parts.

Neuron
A single *nerve* cell, the function of which is to transmit electrical impulses.

Nociceptor
A *nerve* ending responding to painful stimuli.

Noninvasive
Any medical procedure that does not involve penetration of the skin or an entry into the body through any of the natural openings.

Nucleic acid
Deoxyribonucleic acid (*DNA*) or ribonucleic acid (*RNA*); chains of *nucleotides*, with genetic information in the order of the bases of the *nucleotides*.

Nucleotide
Building-block subunit of a *nucleic acid* (*DNA*, *RNA*) consisting of a sugar, a phosphate, and a nitrogen-containing base.

Nucleus (pl. nuclei)
The control center of a cell, containing the genetic material, *DNA*. It is bounded by a nuclear membrane.

O

Olfactory nerve
One of two *nerves* of smell that run from the olfactory bulb in the roof of the nose directly into the underside of the brain.

Optic nerve
One of the two *nerves* of vision. Each one has about one million *nerve* fibers running from the *retina* to the brain, carrying visual stimuli.

Organ
Discrete body part or structure with a vital function, for example, the heart, *liver*, brain, or *spleen*.

Organelle
A tiny part inside a cell that has a specific role. The *nucleus*, *mitochondria*, and ribosomes are examples.

Ossicle
One of three tiny bones (the incus, malleus, and stapes) of the *middle ear* that convey vibrations from the *eardrum* to the inner ear.

Ossification
The process of formation, renewal, and repair of bone. Most bones in the body develop from *cartilage*.

Osteoarthritis
A degenerative joint disease that features damage to the *cartilage*-covered, weight-bearing surfaces of the joint.

Osteomalacia
Bone softening caused by defective mineralization, which usually results from poor calcium absorption due to a deficiency of vitamin D.

Osteon
The rod-shaped unit, also called a haversian system, that is the building block of cortical bone.

Osteoporosis
Loss of bone substance, due to bone being reabsorbed faster than it is being formed. The bones become brittle and are fractured easily.

Osteosarcoma
A highly *malignant* form of bone *cancer* that mainly affects adolescents. It often develops near the knee.

Osteosclerosis
Increased bone density that may result from a severe injury, osteoarthritis, or osteomyelitis. It is detected on an X-ray film as an area of extreme weakness.

Otitis media
Inflammation in the middle-ear cavity, often caused by infection that has spread from the nose or throat.

Otosclerosis
A hereditary bone disease affecting the inner ear, in which the foot of the inner *ossicle* becomes fused to the surrounding bone.

Ovary
One of two structures lying at the end of the fallopian tubes on each side of the *uterus*. They store ovarian follicles, release the mature *ova*, and produce the female *sex hormones* (*estrogen* and *progesterone*).

Ovulation
The release of an *ovum* from a mature follicle in the *ovary* about midway through the menstrual cycle; if not fertilized, the egg is shed during menstruation.

Ovum (pl. ova)
The egg cell; if *fertilization* occurs, the ovum develops into an *embryo*.

P

Pacemaker
An electronic device implanted in the chest that delivers short electric pulses via electrodes to regulate the heartbeat.

Paget's disease
A disease that causes bone to become weaker, thicker, and distorted.

Pancreas
A gland behind the stomach that secretes digestive *enzymes* and *hormones* that regulate *glucose* levels.

Paralysis
Loss of the power of movement of part of the body due to a *nerve* or muscle disorder.

Paraplegia
Paralysis of the lower limbs, usually from injury or disease to the spinal cord or brain.

Parasite
An organism that lives in or on another organism (the host), and benefits at the host's expense.

Parasympathetic nervous system
One of the two divisions of the *autonomic nervous system*; it maintains and restores energy, for example by slowing the heart rate.

Parathyroid glands
Two pairs of *endocrine glands*, located behind the thyroid gland, that help control the level of calcium in the blood.

Parietal
A term referring to the wall of a body cavity, rather than its contents.

Parkinson's disease
A neurological disorder featuring involuntary tremor, muscle rigidity, slow movements, small handwriting, and tottering steps.

Parotid glands

The large pair of salivary glands situated, one on each side, above the angles of the jaw just below and in front of the ears.

Pelvic inflammatory disease

Infection of the reproductive *organs* of the female. The cause may be unknown, but it often occurs following a sexually transmitted disease.

Pelvis

The basinlike ring of bones to which the lower end of the *spine* is attached and with which the thigh bones articulate. The term is also used to refer to the soft *tissue* contents.

Peptic ulcer

The local destruction of the lining of the esophagus, stomach, or *duodenum* from the effects of the *bacterium* *Helicobacter pylori*, stomach acid, and digestive *enzymes*.

Pericarditis

An inflammation of the membranous *pericardium* that surrounds the heart. It may cause pain and the accumulation of fluid, called pericardial effusion.

Pericardium

The two layers of membrane that surround the heart. The outer fibrous sac encloses the heart, and roots of the major blood vessels emerging from it. The inner layer attaches to the heart wall.

Periosteum

The tough *tissue* that coats all bone surfaces except joints, from which new bone can be formed; it contains blood vessels and *nerves*.

Peripheral nervous system

All the *nerves* with their coverings that fan out from the brain and spinal cord, linking them with the rest of the body. The system consists of *cranial nerves* and *spinal nerves*.

Peristalsis

A coordinated succession of contractions and relaxations of the muscular wall of a tubular structure, such as the intestines, that moves the contents along.

Peritoneum

The double-layered membrane that lines the inner wall of the abdomen. The peritoneum covers and partly supports the abdominal *organs*. It also secretes a fluid that lubricates the movement of the intestines.

Peritonitis

An inflammation of the *peritoneum* due to *bacteria*, *bile*, pancreatic *enzymes*, or chemicals; sometimes the cause may be unknown.

Phagocyte

A *white blood cell* or similar cell that surrounds and engulfs unwanted matter, such as invading microbes and cellular debris.

Pharynx

The passage leading down from the back of the nose and the mouth to the esophagus; it consists of the nasopharynx, the oropharynx, and the laryngopharynx.

Pituitary gland

A gland hanging from the underside of the brain. It secretes *hormones* that control many other glands in the body and is itself regulated by the *hypothalamus*.

Placenta

The disk-shaped *organ* that forms in the *uterus* during pregnancy. It links the blood supplies of the mother and baby via the *umbilical cord* and nourishes the growing embryo.

Plasma

The fluid part of the blood from which all cells have been removed; contains *proteins*, salts, and various nutrients.

Platelet

A fragment of large cells called megakaryocytes that is present in large numbers in the blood and necessary for blood clotting.

Pleura

A double-layered membrane, the inner layer of which covers the lung and the outer layer lines the chest cavity. A layer of fluid lubricates and enables movement between the two.

Pleural effusion

Accumulation of excessive fluid between the layers of the *pleura*, which separates them and compresses the underlying lung.

Pleurisy

Inflammation of the *pleura*, usually from a lung infection such as *pneumonia*; may lead to adhesion between the pleural membranes, causing pain on inhalation.

Plexus

A network of interwoven *nerves* or blood vessels.

Pneumoconiosis

Any lung-scarring disorder due to inhalation of mineral dust; scarring causes the lungs to be less efficient in supplying oxygen to the blood.

Pneumocystis pneumonia

A lung infection with the opportunistic microorganism *Pneumocystis carinii*; occurs mainly in *immune deficiency* disorders.

Pneumonia

Inflammation of the smaller air passages and *alveoli* of the lungs due to infection or contact with inhaled irritants or toxic material.

Pneumothorax

The presence of air in the space between the *pleura*, which causes the lung to collapse.

Primary

A term describing a disorder that has originated in the affected structure.

Progesterone

A female sex *hormone* secreted by the *ovaries* and *placenta* that allows the *uterus* to receive and retain a fertilized egg.

Prostaglandin

One of a group of fatty acids made naturally in the body that act like *hormones*.

Prostate gland

A male accessory sex gland situated at the base of the bladder and opening into the *urethra*. It secretes some of the fluid in semen.

Prosthesis

Any artificial replacement for a part of the body, internal or external, whether its purpose is functional or cosmetic.

Protein

Huge *molecule* composed of chains of *amino acids*; the basis of many structural materials (keratin, *collagen*), *enzymes*, and antibodies.

Pulmonary artery

The *artery* that conveys deoxygenated blood from the right *ventricle* of the heart to the lungs to be reoxygenated.

Pulse

Rhythmic expansion and contraction of an *artery* as blood is forced through it.

Pus

A yellowish green fluid that forms at the site of a bacterial infection; contains *bacteria*, dead *white blood cells*, and damaged *tissue*.

Q–R

Quadriplegia
Paralysis of both arms, both legs, and the trunk usually caused by severe spinal cord damage in the neck region.

Radiation therapy
Treatment with radiation, such as *X-rays* or radioactive implants, usually against cancerous or *malignant* cells.

Recessive
In genetics, when one form (*allele*) of a *gene* is "weaker" and only determines what trait is expressed when a *dominant allele* of the same gene is not present.

Red blood cells
Biconcave, disk-shaped cells, without nuclei, that contain *hemoglobin*. There are 4–5 million red cells in $\frac{1}{500}$ pint (1 milliliter) of blood.

Renal
Relating to the *kidneys*.

Respiration
1. Bodily movements of breathing.
2. Gas exchange of oxygen for carbon dioxide in the lungs.
3. Similar gas exchange in the *tissues* (cellular respiration).
4. Breakdown of *molecules* such as *glucose* to release their energy for cellular functions.

Reticular formation
Nerve cells scattered throughout the brain stem that are concerned with alertness and direction of attention to external events.

Retina
A light-sensitive layer lining the inside of the back of the eye that converts optical images to *nerve* impulses, which travel into the brain via the *optic nerve*.

Rheumatoid arthritis
A disorder that causes joint deformity and destruction. It typically affects smaller joints first and often in a symmetrical pattern.

Ribosome
Ball-shaped *organelle* within cells involved in building *proteins* from *amino acids*.

RNA
Ribonucleic acid, different forms of which carry out various functions, including transfer of genetic information and manufacture of *proteins*.

Rubella
A mild viral infection, also known as German measles; if it affects a woman in early pregnancy, it can cause serious harm to the *fetus*.

S

Saccharide
The basic unit that makes up carbohydrates.

Saliva
A watery fluid secreted into the mouth by the salivary glands to aid in chewing, tasting, and digestion.

Sarcoma
A *cancer* arising from connective *tissue* (such as bone), muscle, fibrous *tissue*, or blood vessels.

Sciatica
Pain caused by pressure on the sciatic *nerve*; felt in the buttock and back of thigh.

Secondary
A term describing a disorder that follows or results from another disorder (know as the primary disorder).

Septal defect
An abnormal opening in the central heart wall that allows blood to flow from the right side to the left or vice versa.

Serous otitis media
A disorder in which sticky fluid accumulates in the *middle ear* and impedes the movement of the *ossicles*.

Sex hormones
Steroid substances that bring about the development of sexual characteristics. Sex hormones also regulate the menstrual cycle and the production of sperm and eggs.

Sinoatrial node
A cluster of specialized muscle cells in the right *atrium* that acts as the heart's natural *pacemaker*.

Sinus bradycardia
An abnormally slow, but regular, heart rate resulting from a low rate of pacing by the *sinoatrial node*.

Sphincter
A muscle ring, or local thickening of the muscle coat, surrounding an opening in the body.

Spinal fusion
A surgical operation to fuse two or more adjacent *vertebrae* in order to stabilize the *spine*.

Spinal nerves
The 31 pairs of combined motor and sensory *nerves* that emerge from and enter the spinal cord.

Spine
The column of 33 ringlike bones, called *vertebrae*, that divides into seven cervical *vertebrae*, 12 thoracic *vertebrae*, five lumbar *vertebrae*, and the fused *vertebrae* of the sacrum and coccyx.

Spleen
A lymphatic *organ* situated on the upper left of the abdomen that removes and destroys worn-out *red blood cells* and helps fight infection.

Stapedectomy
An operation to relieve deafness caused by *otosclerosis*.

Stem cell
Generalized type of cell, usually fast-dividing, with the potential to become many different kinds of specialized cells.

Stroke
Damage to the brain by deprivation of its full blood supply or leakage of blood from a ruptured vessel; may impair movement, sensation, vision, speech, or intellect.

Subarachnoid hemorrhage
Bleeding from a ruptured *artery* or *aneurysm* lying under the arachnoid layer of the *meninges*.

Subdural hemorrhage
Bleeding between the *dura mater* and arachnoid layers of the *meninges*.

Sublingual glands
The pair of salivary glands that is located on the floor of the mouth.

Submandibular glands
The pair of salivary glands that lie immediately under the jawbone near its angle.

Suture
A surgical stitch that is used to close a wound or incision.

Sympathetic nervous system
One of the two divisions of the *autonomic nervous system*. It prepares the body for "fight or flight" action through the secretion of epineprine by the adrenal gland, for example by constricting the intestinal and skin blood vessels, widening the pupils of the eyes, and increasing the heart rate.

Synapse
The junction between two *nerve* cells, or between a *nerve* cell and a muscle fiber or a gland. Chemical messengers are passed across a synapse to produce a response in a target cell.

Synovial fluid
Thin, slippery, lubricating fluid within a joint.

Synovial joint
A mobile joint with a membrane that produces a lubricating fluid.

Syphilis
A sexually transmitted or *congenital* infection that, if untreated, passes through three stages and can involve serious damage to the nervous system. *Congenital* syphilis is very rare.

T

Taste bud
A spherical nest of receptor cells that is found mainly on the tongue; each responds most strongly to a sweet, salty, sour, or bitter flavor.

Tendinitis
Inflammation of a *tendon*, causing pain and tenderness, usually from injury.

Tendon
A strong band of *collagen* fibers that joins muscle to bone and transmits the pull caused by muscle contraction.

Tenosynovitis
Inflammation of the inner lining of a *tendon* sheath, usually from excessive friction due to overuse.

Testis (pl. testes)
One of a pair of the sperm- and *hormone*-producing sex glands, in the scrotum.

Testosterone
The principal male sex *hormone* produced in the *testis* and in small amounts in the adrenal *cortex* and *ovary*.

Thalamus
A mass of *gray matter* that lies deep within the brain. It receives and coordinates sensory information.

Thorax
The part of the trunk between the neck and the abdomen that contains the heart and the lungs.

Thrombolytic drug
A drug that dissolves *blood clots* and restores blood flow in blocked *arteries*.

Thrombus (pl. thrombi)
A *blood clot* that usually results from damage to a vessel lining.

Tissue
Structure of similar cells with one main function.

Tonsils
Oval masses of *lymphoid tissue* on the back of the throat on either side of the soft palate; help protect against childhood infections.

Trachea
The windpipe. A muscular tube lined with *mucous membrane* and reinforced by about 20 rings of *cartilage*.

Transient ischemic attack (TIA)
A "mini-stroke" that passes completely in 24 hours. An attack can imply danger of a full *stroke*.

Tumor
A *benign* or *malignant* swelling, especially a mass of cells resulting from uncontrolled multiplication.

U

Umbilical cord
The structure that connects the *placenta* to the *fetus*. It provides the immunological, nutritional, and hormonal link with the mother.

Urea
A waste product of the break-down of *proteins*; the nitrogen-containing component of urine.

Urethra
The tube that carries urine from the bladder to the exterior; much longer in the male than in the female.

Urethritis
Inflammation of the lining of the *urethra*, which is usually caused by a sexually transmitted disease.

Urinary tract
The system that forms and excretes urine; made up of the *kidneys*, ureters, bladder, and *urethra*.

Uterus
A hollow muscular structure in which the *fetus* grows and is nourished until birth.

V

Vagina
The passage from the *uterus* to the external genitals that stretches during sexual intercourse and childbirth.

Vagus nerves
The tenth pair of *cranial nerves*; helps control automatic functions such as heartbeat and digestion.

Vas deferens
One of a pair of tubes that lead from the *testis* carrying sperm, which mix with seminal fluid before entering the *urethra*.

Vasectomy
A surgical procedure for male sterilization in which each *vas deferens* is cut and tied.

Vein
A thin-walled blood vessel that returns blood at low pressure to the heart.

Vena cava
One of the two large *veins*, the superior and inferior, that empty into the right *atrium*.

Ventricle
A chamber or compartment, usually fluid-filled. For example, the two *cardiac ventricles* of the heart and four cerebral *ventricles* in the brain.

Vertebra (pl. vertebrae)
One of the 33 bones of the vertebral column (*spine*), also known as disks.

Virus
The tiniest form of infecting particle (germ). It takes over a cell to produce copies of itself.

Vocal cords
One of two sheets of *mucous membrane* stretched across the inside of the *larynx* that vibrate to produce voice sounds when air passes between them.

W–Z

Wart
A contagious, harmless skin growth that is caused by the human papilloma *virus*.

White blood cell
Any of the colorless blood cells that play various roles in the immune system.

White matter
Nerve tissue formed mainly of the projecting fibers, or *axons*, of *neurons* (*nerve* cells).

X chromosome
A sex *chromosome*. Body cells of females have two X chromosomes.

X-ray
Very short-wavelength, invisible electromagnetic energy which, if not carefully controlled, can penetrate and damage body *tissues*; used for imaging and treatment (*radiation therapy*).

Y chromosome
A sex *chromosome*. Its presence is necessary for the development of male characteristics. Body cells of males have one Y and one X *chromosome*.

Zygote
The cell produced when an egg is fertilized by a sperm; contains genetic material for a new person.

INDEX

Page numbers in **bold** indicate extended treatments of a topic.

ACKNOWLEDGMENTS

Dorling Kindersley would like to thank several people for their help in the preparation of this book. Anna Barlow contributed valuable comments on the cardiovascular system. Peter Laws assisted with visualization and additional design work was done by Mark Lloyd. 3-D illustrations were created from a model supplied by Zygote Media Group, Inc. Ben Hoare, Peter Frances, and Ed Wilson all provided editorial assistance. Marianne Markham and Andrea Bagg contributed to the initial development work. For the revised edition, Dorling Kindersley would like to thank: Peter Laws, Alison Gardner, Clare Joyce, and Anna Reinbold for design assistance; Arpita Dasgupta and David Summers for editorial assistance; Liz Moore for picture research; and Jolyon Goddard for development work.

The publisher would like to thank the following for their kind permission to reproduce their photographs:

(Key: a–above; b–below/bottom; c–center; f–far; l–left; r–right; t–top)

Sidebar Images: **8–47 Integrated Body – Corbis:** Digital Art; **48–69 Skeletal System – Wellcome Images:** Prof. Alan Boyde; **70–81 Muscular System – Science Photo Library:** Eye of Science; **82–119 Nervous System – Science Photo Library:** Nancy Kedersha; **120–129 Endocrine System – Wellcome Images:** University of Edinburgh; **130–145 Cardiovascular System – Wellcome Images:** EM Unit / Royal Free Med. School; **146–161 Respiratory System – Science Photo Library:** GJLP; **162–171 Science Photo Library:** Steve Gschmeissner; **172–187 Lymph & Immunity – Science Photo Library:** Francis Leroy, Biocosmos; **188–209 Digestive System – Science Photo Library:** Eye of Science;). **220–245 Reproduction & Life Cycle – Science Photo Library:** Susumu Nishinaga; **246–269 Growth And Development – Science Photo Library:** Science Photo Library. **270–288 Corbis:** Digital Art

Feature Boxes: Corbis: Digital Art

6 Science Photo Library: Sovereign, ISM. **8–9 Corbis:** Digital Art (c). **10–11 Science Photo Library:** François Paquet-Durand. **12 Alamy Images:** Phototake Inc. (c); **Science Photo Library:** CNRI (bl); Dr. P. Marazzi (br). **Wellcome Images:** Prof. R. Bellairs (cr); K. Hodivala-Dilke & M. Stone (tr). **13 Alamy Images:** Chad Ehlers (bl). **Getty Images:** Science Faction / L. Steinmark – CMSP (br). **Science Photo Library: :** Alfred Pasieka (tc); CNRI (bc); Zephyr (cra); K H Fung (tr);); Rvi Medical Physics, Newcastle / Simon Frase (cl) Wellcome Dept. of Cognitive Neurology (cr); **14 Robert Steiner MRI Unit, Imperial College London:** (clb, bl, cl). **14–15 Robert Steiner MRI Unit, Imperial College London. 15 Science Photo Library:** CNRI (br); Sovereign, ISM (tc); Antoine Rosset (cr). **16 Robert Steiner MRI Unit, Imperial College London:** (b). **17 Robert Steiner MRI Unit, Imperial College London:** (bl, bc, br). **Science Photo Library:** (c); Zephyr (cla); BSIP S&I (tr). **18 Robert Steiner MRI Unit, Imperial College London:** (cl, c, b). **19 Robert Steiner MRI Unit, Imperial College London:** (tc, tr). **Science Photo Library:** (c); Dr. Najeeb Layyous (bl); Medimage (br). **20 Robert Steiner MRI Unit, Imperial College London:** (tr, cb, clb). **Science Photo Library:** Simon Fraser (c). **20–21 Robert Steiner MRI Unit, Imperial College London:** (b). **21 Dreamstime.com:** Robert Semnic (ca); **Getty Images:** Abrams, Lacagnina (cra). **Science Photo Library:** K H Fung (cb); Mauro Fermariello (br). **24 Science Photo Library. 25 Dorling Kindersley:** Andy Crawford (cl). **Science Photo Library:** Steve Gschmeissner (bc). **26 Specialist Stock:** PHONE Labat J.M. / F. Rouquette (cr); Volker Steger (tc). **27 Science Photo Library:** François Paquet-Durand. **28 Science Photo Library:** CNRI. **30 Science Photo Library:** Adam Hart-Davis (cr); Adam Hart-Smith (crb). **31 Science Photo Library:** Richard Wehr / Custom Medical Stock Photo. **32 Corbis:** Minden Pictures / Eric Phillips / Hedgehog House (bl). **Getty Images:** Tyler Stableford (br). **Science Photo Library:** Edward Kinsman (tr). **33 Corbis:** Science Faction / Hank Morgan – Rainbow (bl); Franck Seguin (cl). **NASA:** (cr). **35 Corbis:** Alessandro Della Bella (cr); Steve Lipofsky (tl). **37 Dreamstime.com:** Kateryna Kon (tl). **39 Science Photo Library:** Professors P. Motta & T. Naguro (bl). **40 Alamy Images:** Phototake Inc. (br). **Science Photo Library:** Nancy Kedersha / UCLA (cl). **Specialist Stock:** Ed Reschke (bl). **41 Corbis:** Visuals Unlimited (br). **Science Photo Library:** Innerspace Imaging (tr); Claude Nuridsany & Marie Perennou (cl). **Specialist Stock:** Ed Reschke (tl, cr). **Wellcome Images:** David Gregory & Debbie Marshall (br). **42 Science Photo Library:** Lawrence Livermore Laboratory (cra). **43 Alamy Images:** Bjanka Kadic (tr). **46 Science Photo Library:** Biophoto Associates (ca); CNRI (tr). **47 Science Photo Library:** Alain Pol, ISM (cr). **Wellcome Images:** Annie Cavanagh (bl). **52 Science Photo Library:** Steve Gschmeissner (bl). **Wellcome Images:** Prof. Alan Boyde (c). **53 Science Photo Library:** Biophoto Associates (c); Prof. P. Motta / Dept. of Anatomy / University "La Sapienza", Rome (bl). **Wellcome Images:** M.I. Walker (cra). **55 Science Photo Library:** Steve Gschmeissner (cra). **56 Dorling Kindersley:** Philip Dowell / Courtesy of the Natural History Museum, London (tr). **Wellcome Images:** (ca). **57 Science Photo Library:** Eye of Science (tl); GJLP (bc). **58 Science Photo Library:** Simon Brown (cl). **59 Science Photo Library:** Anatomical Travelogue (cl). **62 Science Photo Library:** Sovereign, ISM (cr). **Wellcome Images:** (c). **63 Wellcome Images:** (tl). **64 Wellcome Images:** (tr). **65 Science Photo Library:** CNRI (br); GCa (cr). **Wellcome Images:** (tc, tr). **66 Science Photo Library:** CNRI (tc); Zephyr (bc). **Wellcome Images:** (cr). **67 Science Photo Library:** Biophoto Associates (tl); St Bartholomew's Hospital, London (tr). **68 Science Photo Library:** Princess Margaret Rose Orthopaedic Hospital (ca); Antonia Reeve (br). **69 Mediscan:** (bc). **Science Photo Library:** CNRI (tl). **75 Science Photo Library:** (bl). **Wellcome Images:** M.I. Walker (bl). **76 Getty Images:** Stone / Catherine Ledner (bc). **Specialist Stock:** Ed Reschke (cl). **77 Science Photo Library:** Neil Borden (cla). **78 Science Photo Library:** Steve Gschmeissner (cl). **80 Science Photo Library:** Biophoto Associates (br). **81 Mediscan:** (bl). **Wellcome Images:** (ca). **86 Wellcome Images:** Dr. Jonathan Clarke (cla). **87 Science Photo Library:** Dr. John Zajicek (cra); Nancy Kedersha (tr). **91 Alamy Images:** allOver Photography (bl). **Science Photo Library:** Zephyr (bl). **Specialist Stock:** Alfred Pasieka (cla). **92 Science Photo Library:** CNRI (c). **93 Science Photo Library:** Bo Veisland, MI&I (bl); Zephyr (tc). **94 Science Photo Library:** Sovereign, ISM (bl). **96 Alamy Images:** Phototake Inc, (cl). **100 Science Photo Library:** Steve Gschmeissner (br). **101 Wellcome Images:** (cl, c).**103 Corbis:** Visuals Unlimited (c). **104 Science Photo Library:** Eye of Science (cl). **105 Alamy Images:** Phototake Inc. (bc). **Science Photo Library:** Pascal Goetgheluck (tl). **107 Science Photo Library:** Susumu Nishinaga (tc). **108 Science Photo Library:** Prof. P. Motta / Dept. of Anatomy / University "La Sapienza", Rome (tc). **109 Dorling Kindersley:** (bc). **Louise Thomas:** (tr). **110 Corbis:** Glenn Bartley (cl). **111 Science Photo Library:** Eye of Science (br). **113 Science Photo Library:** Dept. Of Nuclear Medicine, Charing Cross Hospital (bl). **114 Science Photo Library:** Simon Fraser / Royal Victoria Infirmary, Newcastle Upon Tyne (br); Alfred Pasieka (bl). **115 Science Photo Library:** Alfred Pasieka (cr). **116 Alamy Images:** Medical-on-Line (br); Phototake Inc. (fbl, bl). **Dorling Kindersley:** Steve Gorton (cl). **117 Science Photo Library:** Ctesibius, ISM (c). **118 Science Photo Library:** Bo Veisland (tc); Prof. Tony Wright, Institute of Laryngology & Otology (clb). **Wellcome Images:** (cr). **119 Science Photo Library:** Sue Ford (br). **125 Science Photo Library:** Manfred Kage (bc); **Dreamstime.com:** Jlcalvo (br). **127 Alamy Images:** Medical-on-Line (cr). **Wellcome Images:** (br). **128 Dorling Kindersley:** (br). **Specialist Stock:** Ed Reschke (cl). **129 Alamy Images:** Scott Camazine (crb). **Wellcome Images:** (cr). **136 Science Photo Library:** (c). **137 Science Photo Library:** CNRI (bl); Manfred Kage (tr).**138 Alamy Images:** Phototake Inc. (tr). **140 Science Photo Library:** BSIP VEM (cra); Alain Pol, ISM (br). **141 Science Photo Library:** CNRI (cb); Prof. P. Motta / G. Macchiarelli / University "La Sapienza", Rome (tc). **142 Alamy Images:** Medical-on-Line (bl). **Science Photo Library:** (br). **143 Science Photo Library:** Professors P.M. Motta & G. Macchiarelli (bl). **144 Science Photo Library:** James King-Holmes (br). **150 Alamy Images:** Phototake Inc. (bc). **Dorling Kindersley:** Dave King (cl). **Mediscan:** (cl). **Science Photo Library:** BSIP, Cavallini James (tr). **154 Science Photo Library:** Zephyr (cl). **155 Science Photo Library:** CNRI (cra, fcra); **Alamy Stock Photo:** Custom Medical Stock Photo (br). **156 Science Photo Library:** Dr. Gopal Murti (br); Dr. Gary Settles (cl). **Wellcome Images:** R. Dourmashkin (bc). **157 Alamy Images:** Scott Camazine (c). **Science Photo Library:** CNRI (cr). **158 Alamy Images:** Phototake Inc. (cl). **Science Photo Library:** CNRI (bl). **Wellcome Images:** (cr). **159 iStockphoto.com:** (bc); Daniel Fascia (cl). **160 Science Photo Library:** Biophoto Associates (c); CNRI (bc); (br). **161 Science Photo Library:** ISM (tr); James Stevenson (crb). **164 Science Photo Library:** Sheila Terry (cl). **166 Alamy Images:** Phototake Inc. (crb). **Science Photo Library:** J.C. Revy (ca). **167 Science Photo Library:** Steve Gschmeissner (clb); Prof. P. Motta / Dept. of Anatomy / University, "La Sapienza", Rome (cr); Prof. P. Motta / Dept. of Anatomy / University "La Sapienza", Rome (br). **168 Dorling Kindersley:** Steve Gorton (tr); Jules Selmes and Debi Treloar (ca). **Science Photo Library:** Alfred Pasieka (c). **169 Alamy Images:** Medical-on-Line (bl). **Mediscan:** (cl). **Wellcome Images:** (c, bc). **171 Alamy Images:** Ian Leonard (cra); WoodyStock (cla); Shout (clb); Medical-on-Line (bl). **Science Photo Library:** BSIP, Laurent (crb); Dr. P. Marazzi (br). **177 Science Photo Library:** CNRI (bl). **178 Alamy Images:** Phototake Inc. (cra). **180 Alamy Images:** Phototake Inc. (cla). **181 Science Photo Library:** (clb); NIBSC (cra); Eye of Science (br). **Wellcome Images:** (bl). **182 Alamy Images:** Scott Camazine (bc). **183 Science Photo Library:** Eye of Science (cla); David Scharf (tr, br); NIBSC (bc). **184 Alamy Images:** Phototake Inc. (cl). **Science Photo Library:** (br); Dr. P. Marazzi (bc). **Wellcome Images:** Annie Cavanagh (c). **185 Science Photo Library:** Sue Ford (cra). **186 Science Photo Library:** CNRI (br). **Wellcome Images:** (bl). **187 Science Photo Library:** ISM (bc); Manfred Kage, Peter Arnold Inc. (cl). **192 Science Photo Library:** CNRI (c); Eye of Science (tr). **193 Science Photo Library:** Steve Gschmeissner (bl). **194 Science Photo Library:** Eye of Science (ca); **Dreamstime.com:** Guniita (c). **195 Alamy Images:** Phototake Inc. (bc). **196 Science Photo Library:** Prof. P. Motta / Dept. of Anatomy / University "La Sapienza", Rome (br). **198 Science Photo Library:** Prof. P. Motta / Dept. of Anatomy / University "La Sapienza", Rome (ca); Alain Pol, ISM (cl); Professors P. Motta & F. Carpino / University "La Sapienza", Rome (bc). **201 Science Photo Library:** Dr. T. Blundell, Dept. of Crystallography, Birkbeck College (cra). **203 Corbis:** Frank Lane Picture Agency (bl). **204 Mediscan:** (c). **Science Photo Library:** David M. Martin, MD (cra). **Wellcome Images:** David Gregory & Debbie Marshall (br). **205 Science Photo Library:** P. Hawtin, University of Southampton (cla). **207 Science Photo Library:** Alfred Pasieka (tc). **209 Science Photo Library:** David M. Martin, MD (crssa). **214 Wellcome Images:** David Gregory & Debbie Marshall (cl). **217 Science Photo Library:** Manfred Kage (tr). **218 Science Photo Library:** Prof. P.M. Motta et al (cl). **219 Science Photo Library:** CNRI (bl); Steve Gschmeissner (tc); Zephyr (cra) **222 Science Photo Library:** Steve Gschmeissner (cr); Parviz M. Pour (br). **224 Science Photo Library:** Professors P.M. Motta & J. Van Blerkom (tr). **Wellcome Images:** Yorgos Nikas (ca). **225 Alamy Images:** Phototake Inc. (tc). **228 Science Photo Library:** Tissuepix (cl). **230 Science Photo Library:** Simon Fraser (tr). **231 Dept of Fetal Medicine, Royal Victoria Infirmary:** (br, fbr). **232 Science Photo Library:** Alfred Pasieka (c). **234 Science Photo Library:** Keith / Custom Medical Stock Photo (cl). **235 Science Photo Library:** BSIP, Laurent (tl). **236 Alamy Images:** Pavel Filatov (cl); Ross Marks Photography (clb); Shout (bl). **238 Science Photo Library:** Custom Medical Stock Photo (br); Sovereign, ISN (cr); CNRI (bl). **239 Science Photo Library:** GJLP (cra). **240 Science Photo Library:** CNRI (br). **241 Mediscan:** CDC (cla). **Science Photo Library:** (br). **243 Science Photo Library:** James King-Holmes (fcl, cl). **Specialist Stock:** Jochen Tack (br). **244 Mediscan:** Chineze Otigbah (br). **245 Alamy Images:** Janine Wiedel Photolibrary (tr). **248 Getty Images:** Michel Tcherevkoff (b). **Science Photo Library:** Science Source (cr). **249 Alamy Images:** Chad Ehlers (cl). **Corbis:** Karen Kasmauski (bl). **Science Photo Library:** Maria Platt-Evans (tr). **252 Science Photo Library:** (c). **253 Dorling Kindersley:** (cr). **Science Photo Library:** BSIP VEM (bl). **254 Getty Images:** Image Source (cla). **Science Photo Library:** Prof. P.M. Motta, G. Macchiarelli, S.A, Nottola (cl); Susumu Nishinaga (clb). **255 PNAS:** 101(21):8174–8179, May 25 2004, Nitin Gogtay et al, Dynamic mapping of human cortical development during childhood through early adulthood © 2004 National Academy of Sciences, USA. Image courtesy Paul Thompson, UCLA School of Medicine (tr). **256 Corbis:** Sampics (bl). **257 Corbis:** Image Source (clb). **Science Photo Library:** CC Studio (br). **258 Getty Images:** Philip Haynes (bl); Dougal Waters (tc). **259 Science Photo Library:** Gunilla Elam (cb, crb). **260 Corbis:** Heide Benser (ca). **Oregon Brain Aging Study, Portland VAMC and Oregon Health & Science University:** (cb). **261 Getty Images:** Siqui Sanchez (br). **Wellcome Images:** (tl). **262 Corbis:** Andrew Brookes (bl). **Science Photo Library:** Philippe Plailly (br). **265 Alamy Images:** Albert Biest (crb). **Dorling Kindersley:** Jules Selmes and Debi Treloar (br). **266 Wellcome Images:** Wessex Reg. Genetics Centre (c, cr). **267 Science Photo Library:** Simon Fraser (bc). **268 Science Photo Library:** Steve Gschmeissner (tr).

Endpaper: **Wellcome Images:** K. Hardy

All other images © Dorling Kindersley
For further information see:
www.dkimages.com

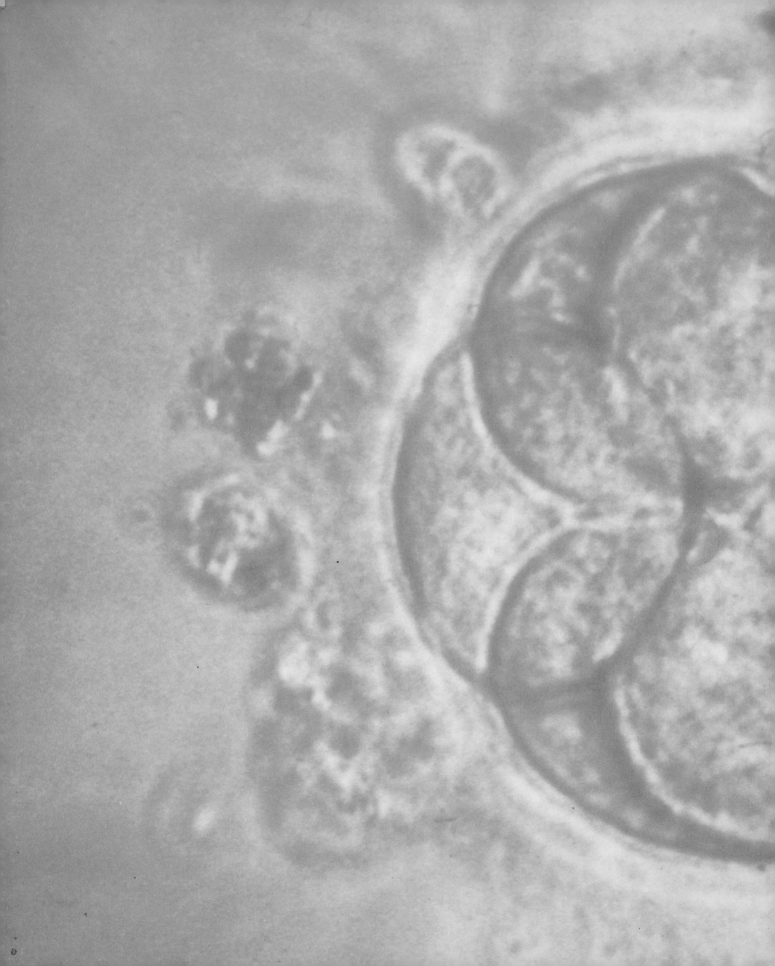